Studies in Logic

Mathematical Logic and Foundations

Volume 117

Computational Paths

The Calculus of Equality

Volume 107
Non-commutative Algebras. Pseudo-BCK Algebreas versus m-pseudo-BCK Algebras
Afrodita Iorgulescu

Volume 108
Semitopology: decentralised collaborative action via topology, algebra, and logic
Murdoch J. Gabbay

Volume 109
The Cognitive Dimension of Social Argumentation. Proceedings of the 4th European Conference on Argumentation, Volume I. Fabio Paglieri, Alessandro Ansani and Marco Marini, eds.

Volume 110
The Cognitive Dimension of Social Argumentation. Proceedings of the 4th European Conference on Argumentation, Volume II. Fabio Paglieri, Alessandro Ansani and Marco Marini, eds.

Volume 111
The Cognitive Dimension of Social Argumentation. Proceedings of the 4th European Conference on Argumentation, Volume III. Fabio Paglieri, Alessandro Ansani and Marco Marini, eds.

Volume 112
Implicative-groups vs. Groups and Generalizations. Second Edition
Afrodita Iorgulescu

Volume 113
Meaning as a Set-theoretic Object. A Gentle Introduction to the Ideas Behind Formal Semantics
Jaroslav Peregrin

Volume 114
Logic and Language: New Frontiers in Analysis and Interpretation
José David García-Cruz, ed.

Volume 115
Four Functors. The 'ancestor' of Mundici's equivalence between unital commutative *l*-groups and MV algebras
Afrodita Iorgulescu

Volume 116
Logic: A Primer
Neil Tennant

Volume 117
Computational Paths. The Calculus of Equality
Arthur Freitas Ramos, Ruy J.G.B. de Quieroz, Anjolina Grisi de Oliveira, Dov M. Gabbay

Studies in Logic Series Editor
Dov Gabbay dov.gabbay@kcl.ac.uk

Computational Paths

The Calculus of Equality

Arthur Freitas Ramos
Ruy J.G.B. de Queiroz
Anjolina Grisi de Oliveira
Dov M. Gabbay

ISBN 978-1-84890-515-3

College Publications
Scientific Director: Dov Gabbay
Managing Director: Jane Spurr

http://www.collegepublications.co.uk

Cover prepared by Laraine Welch

Dedication

To Adam and Luke,
and to all who seek the reasons behind equality.

To Lucas and Pedro.
for being who they are.

To Lydia and Jane,
for their continued support.

Preface

This book arose as a possible continuation of the volume published in 2011 with World Scientific, namely, *The Functional Interpretation of Logical Deduction*, R. de Queiroz, A. de Oliveira & D. Gabbay. At that stage, the idea was to pursue a formal account of logical deduction capable of providing enough tools to deal with the formalisation of deduction in a variety of logics, via the mechanism of labelling formulas in a deduction so as to account for meta-level information which, in some cases, may constitute a crucial device for formal proofs. There was already, of course, a framework for accounting for proof theory and logical deduction in a bidimensional fashion, namely, the formal account of intuitionistic type theory which arose from the so-called Curry–Howard proposition-as-types paradigm, admittedly restricted to the formalisation of intuitionistic logic.

In our view then and now, the functional interpretation of logical connectives is concerned with a certain harmony between, on the one hand, a functional calculus on the expressions built up from the recording of the deduction steps (the *labels*), and, on the other hand, a logical calculus on the formulas. It has been associated with Curry's early discovery of the correspondence between the axioms of intuitionistic implicational logic and the type schemes of the so-called 'combinators' of Combinatory Logic (Curry 1934), and has been referred to as the *formulae-as-types* interpretation. Howard's (1969) extension of the formulae-as-types paradigm to full intuitionistic first order predicate logic (published in 1980) meant that the interpretation has since been referred to as the 'Curry–Howard' functional interpretation. Although Heyting's intuitionistic logic (1930) did fit well into the formulae-as-types paradigm, it seems fair to say that, since Tait's (1965, 1967) intensional interpretations of Gödel's (1958) *Dialectica* system of functionals of finite type, there has been enough indication that the bidimensional framework as in $t : \varphi$, would also be applicable to logics beyond the realm of intuitionism. Ultimately, the foundations of a *functional* approach to formal logic are to be found in Frege's (1893, 1903) system of 'concept writing', not in Curry or Howard, or indeed Heyting.

In an attempt to account for some of the less declarative aspects of certain non-classical logics, in such a way that those aspects could be handled directly in the object language, D. Gabbay set up a novel research programme in his 1996 book on *Labelled Deductive Systems*. The idea, which may be seen as the seeds of a more general alternative to the type-theoretic interpretation of two-dimensional logical systems (i.e. 'terms alongside formulas'), is that the declarative unit of a logical system is to be seen as a *labelled* formula '$t : A$' (read 't labels A'). From this perspective, a logical system is taken to be not simply a calculus of logical deductions on formulas, but a suitably harmonious combination of a functional calculus on the labels and a logical calculus on the formulas. A logic will then be defined according to the meta-level features of the conceptual norm that the logic is supposed to formalise: the allowable logical moves will then be 'controlled' by appropriate constraints on 'what has been done so far' (has the assumption been used at all; have the assumptions been used in a certain order; has the assumption been used more than once; etc.).

One could also regard it as an attempt at a reinterpretation of Frege's formal apparatus of a logical calculus where abstractors and functional operators work harmoniously alongside logical connectives and quantifiers. In other words, the functional interpretation (sometimes referred to as the Curry–Howard–Tait interpretation) can be viewed in a wider perspective of a labelled deductive system which can be used to study a whole range of logics, including some which may not abide by the tenets of the intuitionistic interpretation (e.g. classical implicational logic, many-valued logics, etc.). The result is a labelled natural deduction system which we would like to see as a reinterpretation of Frege's 'functional' account of logic: it is as if the theory of functions of *Grundgesetze* is put together with the theory of predicates of *Begriffsschrift*, in such a way that a formula is true (valid) if and only if a deduction of it can be constructed where the label contains no free variable (i.e. its proof-construction is a 'complete' object, which means that the truth of the formula relies on no assumptions). The weaker the logic, the stricter are the ways in which assumptions may be withdrawn. Classical implicational logic, for example, will have a procedure for withdrawing implicational assumptions depending on the history of the deduction, which its intuitionistic counterpart will not have. So, we need to look for a paradigm for two-dimensional logical systems (terms alongside formulas) which can account for a general perspective on the harmony between withdrawal of assumptions in the logic and abstraction of variables in the functional calculus. We are beginning to find what seems to be a reasonable architecture for such a methodology underlying logical systems based on term calculi: *Grundgesetze* alongside *Begriffsschrift* (1992).

Formal counterpart to purpose, use, consequences. To the extent that one is willing to look for a counterpart to meaning as use, purpose, consequences in formal logic, one needs to ask where the 'explanation of consequences' comes in. Furthermore, since the language of formal logic is not used in a direct 'interaction' between an 'Utterer' and an 'Interpreter' (to borrow Peirce's terms), or an 'Attacker' and a 'Defender' (Lorenzen), or even 'Nature' and 'Myself' (Hintikka), or 'Spoiler' and 'Duplicator' (Ehrenfecht-Fraïssé), then, in order to pinpoint where 'purpose' comes into being, one needs to find where this dichotomy is to be found in the language of predicate logic and in the language for formalising calculable function (i.e., Church's lambda calculus). Coming to the rescue are, on the one hand, Church's notions of 'normal form', abstraction-application, and term-reduction in λ-calculus and, on the other hand, Gentzen's 'introduction vs elimination' rules and proof reduction in Natural Deduction proof systems.[1]

In the context of the analysis of logical constants by rules of inference in Natural Deduction, in the tradition of the verification principle, one finds an attempt to approach the formal counterpart of 'purpose' in P. Martin-Löf's discussion of 'Correctness of assertion and validity of inference'[2]:

> As you can see from its title, my talk falls into two parts, and the order between these two parts is essential: one has to begin with correctness of assertion and then deal with validity of inference only after that. In addition to these two parts, I will add a third part in the middle about the relation between correct assertion and knowledge. Let us begin, then, with the very notion of assertion, usually written like this,
>
> $$\vdash C$$
>
> Throughout, I use C for the content of an assertion, and I use the Frege sign $\vdash$, or the assertion sign in Russell's terminology, for the assertoric force. An assertion in general is just what you obtain by taking a content C and prefixing the assertion sign to it–a purely formal addition that you make.
>
> (...)
>
> The other part here, namely the content C, I have to say something about, because it is given a semantic definition, simply as something to do. (...) As a term for something to do I will generally use task, Aufgabe in German, which was introduced, albeit for propositions (Aussagen) rather than assertoric contents, by Kolmogorov in 1932.

[1]Just as a historical reminder, in both cases (predicate logic and lambda calculus) the actual languages come from the father of formal logic, namely, Gottlob Frege. Indeed, the language of predicate logic comes from his *Begriffsschrift* and the language of the rules for calculating functions from his *Grundgesetze* (with his 'course-of-values' notation).

[2]"This is a slightly edited transcript of a lecture given by Per Martin-Löf on 26 October 2022 at the Rolf Schock Symposium in Stockholm."

> As far as the assertion sign is concerned, this explanation was a purely formal one. From the contentual point of view, we also have to explain what is the purpose of uttering an assertion. I take it that to explain the meaning of a complete sentence is the same as explaining the purpose of the act of uttering it, that is, saying something by means of it. So the question is, What is the purpose of an utterance of the assertion $\vdash C$? To answer this question we have to introduce, not only the speaker, who produces the assertion, but also the hearer, who receives the assertion from the speaker, as indicated in the figure,
>
> $$\text{speaker} \quad \longrightarrow \quad \vdash C \quad \longrightarrow \quad \text{hearer}$$
>
> This is necessary, because we cannot explain the meaning, which is to say the purpose, of an assertion speaking about the speaker alone: the meaning has to do with the interaction between the speaker and the hearer. (Martin-Löf 2022, p.1)[3]

[3]It is appropriate to point out here that this represents a major change in the position held by the proponents of the tradition (Heyting, Kolmogorov, Gentzen, Dummett, Prawitz, Martin-Löf) of the so-called 'verificationism' on which it is based a so-called language-based version of intuitionism:

> So recall the explanations of the meanings of the logical constants, the connectives and the quantifiers, given by Brouwer, Heyting and Kolmogorov: they all follow the common pattern that, whatever the logical constant may be, an explanation is given of what a proof of a proposition formed by means of that logical constant looks like, that is, what is the form, and, more precisely, canonical or direct form, of a proof of a proposition which has that specific logical constant as its outermost sign. It is clear from this what ought to be the general explanation of what a proposition is, namely, that a proposition is defined by stipulating how its proofs, more precisely, canonical or direct proofs, are formed. And, if we take the rules by means of which the canonical proofs are formed to be the introduction rules, I mean, if we call those rules introduction rules as Gentzen did, then his suggestion that the logical constants are defined by their introduction rules is entirely correct, so we may rightly say that a proposition is defined by its introduction rules.

And, in the sequel:

> Now what I would like to point out is that this is an explanation which could just as well be identified with the verification principle, provided that it is suitably interpreted. Remember first of all what the verification principle says, namely, that the meaning of a proposition is the method of its verification. The trouble with that principle, considered as a formula, or as a slogan, is that it admits of several different interpretations, so that there arises the question: how is it to be interpreted? Actually, there are at least three natural interpretations of it. On the first of these, the means of verifying a proposition are simply identified with the introduction rules for it, and there is then nothing objectionable about Wittgenstein's formula, provided that we either, as I just did, replace method by means, which is already plural in form, or else make a change in it from the singular to the plural number: the

As we have emphasised on earlier occasions since the late 1980s, the explanation of the (immediate) consequences of a proposition in the context of Natural Deduction rules is made explicit by the so-called reduction rules, which spell out the effect of the elimination rules on the result of the application of introduction rules. This opens the way to see a parallel with the pragmatist/dialogical/game-theoretic accounts of the semantics of predicate logic: Utterer/Defender/Myself come in via the introduction rules, whereas Interpreter/Attacker/Nature find their counterpart in the elimination rules. The Inversion Principle, in the form of reduction rules, brings them together for logical harmony.

As a matter of fact, the search for a justification for the 'language-based' account of intuitionism often refers to Dummett's defence of using 'Wittgensteinian ideas about language', precisely because this rests upon taking with full seriousness the view of language as an instrument of social communication, and therefore looks very unlike traditional intuitionist accounts, which, notoriously, accord a minimum of importance to language or to symbolism as a means of transmitting thought, and are constantly disposed to slide in the direction of solipsism. Under the professed understanding that Wittgenstein's ideas on 'meaning is use' are connected with the claim that the 'meaning of a proposition is determined by its method of verification', Dummett says that a grasp of the meaning of a statement consists in a capacity to recognise a proof of it when one is presented to us. By pointing to moments of self-correction in which Wittgenstein rejects his brief remark on proof and meaning, and asks 'is the meaning, the point, of a math. proposition really clear if one only understands its proof?', we have identified a prevailing idea (in the *Nachlass*) that the meaning of a statement is rather determined by an explanation of the (immediate) consequences one can draw from it. Here one may wish to start wondering whether saying that the act of recognising a proof of a proposition, when one is presented to us, would not make us slide in the direction of solipsism. But what is one to do with such a presentation or description?

According to Wittgenstein, the solipsist (but also the idealist and the realist) wants to say something that makes no difference. 'Solipsistic = egocentric'. 'Our view of language is a one-sided (*einseitige*) one.'

> meaning of a proposition is the methods of its verification. Interpreted in this way, it simply coincides with the intuitionistic explanation of what a proposition is, or, if you prefer, the Gentzen version of it in terms of introduction rules. For instance, using this manner of speaking, there are two methods of verifying a disjunctive proposition, namely, the two rules of disjunction introduction, and absurdity is defined by stipulating that it admits of no method of verification. (Martin-Löf 2013, p.6)

By the way, as we have remarked in previous opportunities, it is unwarranted to refer to 'verification principle' as represented by conditions of proof as 'Wittgenstein's formula', since he manifested a self-correction a couple of times to a previous remark he had made associating the meaning of a proposition to its method of proof.

Computational paths: a formal calculus of equality. Here we wish to present a framework for studying, on the one hand, the mathematical foundations of *Labelled Deductive Systems* and, on the other hand, to develop a unified approach to the study of equality in mathematics and computer science through the lens of *computational paths*. The central insight is that equality proofs–traditionally treated as mere witnesses of truth–carry computational content that can be manipulated, composed, and reasoned about systematically. As a matter of fact, *computational paths* offer a distinctive perspective on the calculus of equality: by recording how an equality was derived (not just that it holds), one can recover higher-dimensional structure in a proof-irrelevant setting, build a complete homotopy theory with explicit combinatorial foundations, and provide a framework where all coherence conditions are either computationally verified or automatically satisfied. Starting from the definition of a computational path as a pair (s, π) of a step list and a propositional equality [54], we construct the full path calculus: reflexivity, symmetry, transitivity, congruence, transport, dependent application, and operations for products, sums, dependent pairs, and function types. The key insight is that even in a proof-irrelevant setting where the principle of Uniqueness of Identity Proofs (UIP) holds for the type of computational paths, the computational traces remain distinct combinatorial objects which support a rich algebraic structure.

As for the basis for the formulation of the logical system which we call labelled natural deduction, we have developed a series of justified connections between a wide spectrum of concepts and techniques, ranging from the philosophical works of Wittgenstein to a fundamental collection of techniques for formalising proofs in mathematics. This has been a sustained effort to fulfil a task going back at least forty years, and the expectation is to continue with the same determination as always. A recent paper belonging to a series of articles which began in 1987–1988, and was followed by others in the 1990s, has just been published. This view of proofs and meaning first emerged in several papers and led to a call for a rethinking of intuitionistic type theory which gave rise to technical results such as those documented in the book published by World Scientific, as well as in several articles from then up until now (2026), including the present one.

Labelling: a new paradigm for the functional interpretation. The functional interpretation of logical connectives, the so-called Curry–Howard interpretation, provides an adequate framework for the establishment of various calculi of logical inference. Being an 'enriched' system of natural deduction, in the sense that terms representing proof-constructions (the *labels*) are carried alongside formulas, it constitutes an apparatus of great usefulness in the formulation of logical calculi in an operational manner. By uncovering a certain harmony between a functional calculus on the labels and a logical calculus on the formulas, it proves to be instrumental in giving mathematical foundations for systems of logic presentation designed to handle meta-level features at the object-level via a labelling mechanism, such as, e.g. D. Gabbay's (1996) *Labelled Deductive Systems*.

Here we demonstrate that the introduction of 'labels' is of great usefulness, not only for the understanding of the proof-calculus itself, but also for the clarification of its connections with model-theoretic interpretations. For example, the Skolem-type procedures of introducing 'new' names (in order to eliminate the existential quantifier, for example) and 'discharging' them at the end of a deduction (i.e. concluding that the deduction could have been made without those new names previously introduced) are directly reflected in the *elimination* rules for connectives like $\vee$ and $\exists$. From this perspective one realises that for plain systems it may be true that "natural deduction is only satisfactory for $\rightarrow$, $\forall$, $\wedge$; the connectives $\vee$ and $\exists$ receive a very *ad hoc* treatment: the elimination rule for "$\vee$" [and "$\exists$"] is with the presence of an extraneous formula C," (Girard 1989) (page 34) although the situation becomes very different in the presence of labels alongside formulas. In a natural deduction system with labels alongside formulas, we realise that there is nothing 'extraneous' about the arbitrary formula 'C'. For example, the rule of $\exists$-*elimination* (shown below) states that:

1. if an arbitrary formula C can be obtained from a deduction which uses both the individual's name newly introduced ('let it be t') and the assumption that it does have the required property ('let $g(t)$ be the evidence for $P(t)$', g being a newly introduced name for the way in which the evidence for $P(t)$ depends on t),

   then

2. this formula C can be obtained *without* using the new names at all.

Condition (1) is reflected in the label expression alongside the formula C in the premise, which is required to show a free occurrence of the new names (i.e. t and g), and requirement (2) is shown to be met when the new names become bound (by the ´-abstractor) in the label expression alongside the same formula C in the conclusion.

$\vee$-*elimination*

$$\frac{p : A_1 \vee A_2 \qquad \dfrac{[s_1 : A_1]}{d(s_1) : C} \qquad \dfrac{[s_2 : A_2]}{e(s_2) : C}}{\texttt{CASE}(p, \acute{s_1}d(s_1), \acute{s_2}e(s_2)) : C}$$

$\exists$-*elimination*

$$\frac{e : \exists x^D.P(x) \qquad \dfrac{[t : D, g(t) : P(t)]}{d(g,t) : C}}{\texttt{INST}(e, \acute{g}\acute{t}d(g,t)) : C}$$

In our labelled natural deduction, the λ-abstractor is reserved for the constructor of label expressions associated with the connective of implication, i.e. '$\rightarrow$'.

Although all abstractors (λ, $\acute{}$, ε, etc.) originate in Frege's value-range technique, there is an important difference: whereas the λ-abstractor accompanies an introduction of an implication in the logical calculus, here our $\acute{}$-abstractor reflects the discharge of assumptions by binding the corresponding variables, but this discharge is not accompanied by the introduction of a conditional connective on the logical side. In other words, we shall need a more general view of the device of abstraction, which may not necessarily be connected with the implication connective, but rather with the discharge of assumptions in the logical calculus.

In passing, note that the abstractors are all reminiscent of Frege's device of transforming functions into objects by forming value-range terms with the help of the notation '$\acute{\varepsilon} f(\varepsilon)$', where the sign ('$\acute{}$', which is the smooth breathing for Greek vowels, according to Dummett (1991)) plays the rôle of an abstractor (Frege 1891).[4] (We shall come back to this point later on when we describe the rôle of the labels in our framework.)

By means of a suitable harmony between, on the one hand, a functional calculus with abstractors and, on the other hand, a logical calculus with connectives and quantifiers, we want to show that the labelling device is more than just an extra syntactical ornament. It is a useful device to 'put back into the object language' the (lost) capability of handling names (and eventual dependencies), e.g. names of instances of formulas, names of individuals, names of collections of formulas, etc., which is kept at the meta-level in plain logic presentation systems. The lack of this capability makes it difficult (or at least 'unnatural') for the latter systems to handle a class of logics called 'resource' logics where non-declarative features such as 'how many times a formula was used to obtain another formula', 'which order certain formulas were used in order to obtain another formula as a conclusion', etc., do have a crucial rôle to play. The lack of a naming device such as the labelling mechanism, we shall contend, also obscures the connections with some fundamental theorems given by the model-theoretic interpretations.

Moreover, by using labels one can keep track of all proof steps. This feature of labelled natural deduction systems helps to recover the 'local' control virtually lost in plain natural deduction systems. The 'global' character of the so-called *improper* inference rules (i.e. those rules which not only manipulate premises and conclusions, but also involve assumptions, such as, e.g., $\rightarrow$-*introduction*, $\vee$-*elimination*, $\exists$-*elimination*, etc.) is made 'local' by turning the 'discharge functions' into appropriate disciplines for variable-binding via the device of 'abstractors'. As a matter of fact, the use of abstractors to make arbitrary names lose their identity is not a new device: it was already used by Frege in his

[4]See, e.g., the partial English translation of *Grundgesetze I* (Furth 1964). See also Peano's (1889) original device for functional notation.

Grundgesetze I (1893). (Think of the step from 'assuming $x : A$ and arriving at $b(x) : B$' to 'assert $\lambda x.b(x) : A \to B$' as making the arbitrary name 'x' lose the identity it had in $b(x)$ as a name. The binding transforms the name into a mere place-marker for arguments in substitutions.) From this perspective, critical remarks of the sort "Natural Deduction uses global rules (...) which apply to whole deductions, in contrast to rules like elimination rule for $\to$, which apply to formulas" (Girard 1989) (page 35) will have much less impact than they have with respect to 'plain' natural deduction systems.

In what follows we demonstrate how labelled natural deduction presentation systems based on our 'Fregean' functional interpretation (resulting from a generalisation of the Curry–Howard interpretation) can help us provide a more intuitive and more general alternative to most logic presentation systems in the Gentzen tradition. We shall discuss how some 'problematic' aspects of plain natural deduction can be overcome by the use of labels alongside formulas.

Lambda calculus and the rule of application (β-reduction). As far as the semantics of Alonzo Church's λ-calculus is concerned, which he introduces in (1936) 'to propose a definition of effective calculability', the actual meaning-giving rule is the rule of application (so-called β-reduction), which essentially describes, via a process of substitution, the role and the purpose of the λ-symbol binding a variable and working as an 'abstraction' operator, cf. Rule II below:

> I. To replace any part $\lambda x[M]$ of a formula by $\lambda y[\mathrm{S}^x_y M|]$, where y is a variable which does not occur in M.
>
> II. To replace any part $\{\lambda x[M]\}(N)$ of a formula by $\mathrm{S}^x_N M|$, provided that the bound variables in M are distinct both from x and from the free variables in N.
>
> III. To replace any part $\mathrm{S}^x_N M|$ (not immediately following λ) of a formula by $\{\lambda x[M]\}(N)$, provided that the bound variables in M are distinct both from x and from the free variables in N. (Church 1936, p.347)

Rules I and III were given by Church, respectively, to formalise the change in bound variable and to give the inverse of rule II, in order to make the conversion relation (sequences of applications of rules I, II and III) an equivalence relation. This way, a λ-term represents a function 'as a rule', rather than the usual set-theoretic representation as a set of tuples whose last component is the result of applying the function to the previous components of the tuple.

As a matter of historical as well as technical record, it is appropriate to bring here the work of Frege in the formalisation of 'function as a rule' via a notation which finds direct parallels to Church's. In spite of a clear emphasis on formulating precisely the notion of the 'extension' of a concept, Frege's *Grundgesetze* calculus of expressions of the form '$\acute{e}f(e)$', which goes back to his (1891) 'Function and Concept' *Werthverlauf*, brought up a number of operational aspects of what today is seen as the theory of 'functions as rules' as defined by Church with his λ-calculus.

> Wenn ich allgemein sage:
>
> es bedeute
>
> ,$\acute{\varepsilon}\Phi(\varepsilon)$'
>
> den Werthverlauf der Function $\Phi(\xi)$
>
> so bedarf dies ebenso einer Ergängzung wie oben unsere Erklärung von ,$\forall\alpha\Phi(\alpha)$' (*Grundgesetze* I, §9, p. 15)

which was translated as

> If I say in general:
>
> let
>
> '$\acute{\varepsilon}\Phi(\varepsilon)$'
>
> refer to the value-range of the function $\Phi(\xi)$,
>
> then this too requires supplementation, just like our explanation of $\forall\alpha\Phi(\alpha)$ above.

The device of variable-binding, and the idea of having terms representing incomplete 'objects' whenever they contain free variables, were both introduced in a systematic way by Frege in his *Grundgesetze.* As early as 1893 Frege developed in his *Grundgesetze I* what can be seen as the early origins of the notions of *abstraction* and *application*, when showing techniques for transforming functions (expressions with free variables) into value-range terms (expressions with no free variables) by means of an 'introductory' operator of abstraction producing the *Werthverlauf* expression, e.g., '$\acute{\varepsilon}f(\varepsilon)$', and the effect of its corresponding 'eliminatory' operator '$\frown$' on a *value-range* expression.

Expressing how important he considered the introduction of a variable-binding device for the functional calculus (recall that the variable-binding device for the logical calculus (i.e., the universal quantifier) had been introduced earlier in *Begriffsschrift*), Frege says:

> The introduction of a notation for value-ranges seems to me to be one of the most consequential additions to my concept-script that I made since my first publication on this matter. (*Grundgesetze I*, §9, p. 15f.)

In order to make sure that a value-range term (or the term resulting of an application of this to another term) would not end up without a denotation, Frege introduces the operator '\':

> (...) in the case in which $\Phi(\xi)$ was a concept under which one and only one object fell, "$\grave{\varepsilon}\Phi(\varepsilon)$" would designate this object. Now of course this is not possible, because that equation could not be sustained in its general form, but we can serve our purpose by introducing the function
>
> $$\backslash\xi$$
>
> with the specification to distinguish two cases:
> 1. If, for the argument, there is an object Δ such that $\grave{\varepsilon}(\Delta = \varepsilon)$ is the argument, then the value of the function $\backslash\xi$ is to be Δ itself;
> 2. if, for the argument, there is no object Δ such that $\grave{\varepsilon}(\Delta = \varepsilon)$ is the argument, then the argument itself is to be the value of the function $\backslash\xi$. (§11, p. 19)

By doing this, Frege makes sure that "$\backslash\grave{\varepsilon}\Phi(\varepsilon)$" always has a reference. Now, it may be worth noticing the similarity between the two notations

$$\text{Frege's } \backslash\grave{\varepsilon}\Phi(\varepsilon) \qquad \text{and} \qquad \text{Church's } \lambda\varepsilon\Phi(\varepsilon)$$

which are both terms representing a function as a rule. Of course, in Church's case, there was no need to make sure the term always had a reference since the λ-calculus was meant to formalise *partial* functions.

In order to define the use, the purpose of value-range terms, Frege defines the operation of function-application as (Vol. I, §34, pages 52ff), (translated by M. Furth):

> "(...) it is a matter only of designating the value of the function $\Phi(\xi)$ for the argument Δ, i.e., $\Phi(\Delta)$, by means of "Δ" and "$\grave{\varepsilon}\Phi(\varepsilon)$". I do so in this way:
>
> $$\text{“}\Delta \frown \grave{\varepsilon}\Phi(\varepsilon)\text{”}$$
>
> which is to mean the same as "$\Phi(\Delta)$"."

Note the similarity to the rule of functional *application* (Rule II of Church 1936), where 'Δ' is the argument, '$\grave{\varepsilon}\Phi(\varepsilon)$' is the function, and '$\frown$' is the application operator.

Having settled the notation and the meaning of the symbol for the formalisation of 'function as a rule', Frege goes on to show how to use this formal apparatus to define 'numeral' (as opposed to 'number') as a term in the calculus containing value-ranges, much in the same way Church shows how to formalise effective calculability in Arithmetic in his λ-calculus.

Arising out of Church's work of formulating mathematically the notion of 'computable function' in order to prove that Hilbert's Entscheidungsproblem was unsolvable, the rules of the calculus would not prevent a function from being applied to itself; self-application would therefore be 'legal', and this led to the first formalisation of the notion of 'partial function' in the literature. But this meant that the 'set-theoretic' meaning of this formal counterpart to partial function had to be elaborated in a more sophisticated way, leading to substantial work on what is called denotational semantics, of which Dana Scott's domain theory is the best known example. Thus the search for a mathematical 'understanding' of a λ-term took two main directions: operational semantics and denotational semantics, the latter being an approach to the 'meaning' of a λ-term via an interpretation into the universe of (ordered, topologically structured) sets. As to the former, the foundational work was done by G. Plotkin (1974), which helped to consolidate the theory of programming languages, even if the first use of the term 'operational semantics' is credited to D. Scott in a seminal paper of 1970. In any case, it was Church/Frege who prepared the grounds for the study of 'function as a rule' by stating the rule of reduction of a λ-term, thereby making explicit its meaning as use, purpose, consequences, i.e., as a general rule which is to be applied to any input, resulting in a new term where such input replaces the bound variable occurring in the original term. That is the most basic settlement of meaning for the λ-calculus, which comes first and foremost before any (most legitimate!) approach to associate a term to a function between natural numbers, whether via the application of a sequence of rules of reduction or via an interpretation into the realm of (ordered, topological) sets.

Predicate logic and rules of (proof-)reduction. Out of H. Curry's (1934) discovery of parallels between the axioms of (intuitionistic) implicational logic and the types of the combinators of Combinatory Logic, extended by W. Howard's (1969) to full (intuitionistic) predicate logic, there arose a formalisation of Errett Bishop's style of constructive mathematics which was then called intuitionistic type theory by its main author, Per Martin-Löf. The idea behind what is called the 'Curry–Howard' interpretation of predicate logic, leading to the slogan 'propositions are types (of their proofs)', is to associate typing rules with proof rules in the style of Natural Deduction. This association finds a direct counterpart in the so-called Brouwer–Heyting–Kolmogorov (BHK) interpretation of intuitionistic logic, where proofs, not truth-values, determine the meaning of propositions.

$$\begin{array}{rcll} \bot & \text{as} & \varnothing & \text{(empty type)} \\ \alpha \wedge \beta & \text{as} & \alpha \times \beta & \text{(product)} \\ \alpha \vee \beta & \text{as} & \alpha + \beta & \text{(sum)} \\ \forall x\gamma & \text{as} & \Pi x\gamma & \text{(dependent product)} \\ \exists x\gamma & \text{as} & \Sigma x\gamma & \text{(dependent sum)} \end{array}$$

Here is where one can start looking for where meaning as use, purpose, consequences is supposed to show up. For a start, it may be easier to build on the previous discussion of λ-terms, given that not only is type theory considered to be a formalisation of typed λ-calculus, but also that, according to the BHK-interpretation, a proof of an implication $A \to B$ is a function which transforms proofs of A into proofs of B. In other words, formal proofs of $A \to B$ are λ-terms. And we have already discussed where the 'meaning as use, purpose' of λ-terms becomes clear: in the β-reduction rules. By 'lifting' this up to the type-rules for $A \to B$, we have:

$\to$-β-reduction

$$\frac{a : A \qquad \dfrac{\dfrac{[x : A]}{b(x) : B}}{\lambda x.b(x) : A \to B}\ \to\textit{-intr}}{\mathtt{APP}(\lambda x.b(x), a) : B}\ \to\textit{-elim} \quad \beta\text{-reduces to} \quad \frac{a : A}{b(a/x) : B}$$

Associated rewriting:
$\mathtt{APP}(\lambda x.b(x), a) =_\beta b(a/x)$

Taking the rules of *reduction* between proofs in a system of Natural Deduction enriched with terms alongside formulas, such as in the so-called Curry–Howard interpretation (Howard (1980)), of which Martin-Löf's type theory may be seen as an instance, and isolating the terms corresponding to the derivations, one can draw the following parallels. Looking at the conclusion of reduction inference rules, one can take the destructor as being the Attack (or 'Nature', its counterpart in the terminology of Hintikka's Game-Theoretical Semantics, or 'Interpreter' as in Peirce's formulation), and the constructor as being the Assertion (or Hintikka's 'Myself', or Peirce's 'Utterer'). In this way the game-theoretic explanations of logical connectives find a direct counterpart in the functional interpretation, under the view that meaning is use, purpose, application, consequences:

∧-β-reduction

$$\cfrac{\cfrac{a:A \qquad b:B}{\langle a,b\rangle : A\wedge B}\wedge\textit{-intr}}{\mathtt{FST}(\langle a,b\rangle):A}\wedge\textit{-elim} \qquad \beta\text{-reduces to} \qquad a:A$$

$$\cfrac{\cfrac{a:A \qquad b:B}{\langle a,b\rangle : A\wedge B}\wedge\textit{-intr}}{\mathtt{SND}(\langle a,b\rangle):B}\wedge\textit{-elim} \qquad \beta\text{-reduces to} \qquad b:B$$

Associated rewritings:

$\mathtt{FST}(\langle a,b\rangle) =_\beta a$

$\mathtt{SND}(\langle a,b\rangle) =_\beta b$

<u>Assertion</u>/*Introd.*	<u>Attack</u>/*Elim.*	β-reduces to	<u>Defense</u>
Conjunction ('∧'):			
$A\wedge B$	$L?$		A
$A\wedge B$	$R?$		B
$\langle a,b\rangle : A\wedge B$	$\mathtt{FST}(\langle a,b\rangle)$	β-reduces to	$a:A$
$\langle a,b\rangle : A\wedge B$	$\mathtt{SND}(\langle a,b\rangle)$	β-reduces to	$b:B$

∨-β-reduction

$$\cfrac{\cfrac{a:A}{\mathtt{inl}(a):A\vee B}\vee\textit{-intr} \qquad \cfrac{[x:A]}{f(x):C} \qquad \cfrac{[y:B]}{g(y):C}}{\mathtt{CASE}(\mathtt{inl}(a),\upsilon x.f(x),\upsilon y.g(y)):C}\vee\textit{-elim} \quad \beta\text{-reduces to} \quad \cfrac{a:A}{f(a/x):C}$$

$$\cfrac{\cfrac{b:B}{\mathtt{inr}(b):A\vee B}\vee\textit{-intr} \qquad \cfrac{[x:A]}{f(x):C} \qquad \cfrac{[y:B]}{g(y):C}}{\mathtt{CASE}(\mathtt{inr}(b),\upsilon x.f(x),\upsilon y.g(y)):C}\vee\textit{-elim} \quad \beta\text{-reduces to} \quad \cfrac{b:B}{g(b/y):C}$$

Associated rewritings:

$\mathtt{CASE}(\mathtt{inl}(a),\upsilon x.f(x),\upsilon y.g(y)) =_\beta f(a/x)$

$\mathtt{CASE}(\mathtt{inr}(b),\upsilon x.f(x),\upsilon y.g(y)) =_\beta g(b/y)$

<u>Assertion</u>/*Introd.*	<u>Attack</u>/*Elim.*	β-reduces to	<u>Defense</u>
Disjunction ('∨'):			
$A\vee B$	?		A
$A\vee B$	?		B
$\mathtt{inl}(a):A\vee B$	$\mathtt{CASE}(\mathtt{inl}(a),\upsilon x.f(x),\upsilon y.g(y))$	β-reduces to	$a:A,\ f(a/x):C$
$\mathtt{inr}(b):A\vee B$	$\mathtt{CASE}(\mathtt{inr}(b),\upsilon x.f(x),\upsilon y.g(y))$	β-reduces to	$b:B,\ g(b/y):C$

Whilst in both $\texttt{FST}(\langle a,b\rangle)$ and $\texttt{SND}(\langle a,b\rangle)$ the destructors allow access to either of the conjuncts, in $\texttt{CASE}(\texttt{inl}(a), \upsilon x.f(x), \upsilon y.g(y))$ and $\texttt{CASE}(\texttt{inr}(b), \upsilon x.f(x), \upsilon y.g(y))$ the destructor is not given access to either disjunct, but must ask for whichever disjunct comes from the introduction.

$\rightarrow$-β-*reduction*

$$\dfrac{a : A \qquad \dfrac{\dfrac{[x : A]}{b(x) : B}}{\lambda x.b(x) : A \rightarrow B} \rightarrow\textit{-intr}}{\texttt{APP}(\lambda x.b(x), a) : B} \rightarrow\textit{-elim} \quad \beta\text{-reduces to} \quad \dfrac{a : A}{b(a/x) : B}$$

Associated rewriting:

$\texttt{APP}(\lambda x.b(x), a) =_\beta b(a/x)$

<u>Assertion</u>/*Introd.*	<u>Attack</u>/*Elim.*	β-reduces to	<u>Defense</u>
Implication ('$\rightarrow$'):			
$A \rightarrow B$	A ?		B
$a : A$; $\lambda x.b(x) : A \rightarrow B$	$\texttt{APP}(\lambda x.b(x), a)$	β-reduces to	$b(a/x) : B$

$\forall$-β-*reduction*

$$\dfrac{t : D \qquad \dfrac{\dfrac{[x : D]}{g(x) : P(x)}}{\Lambda x.g(x) : \forall x^D.P(x)} \forall\textit{-intr}}{\texttt{EXTR}(\Lambda x.g(x), t) : P(t)} \forall\textit{-elim} \quad \beta\text{-reduces to} \quad \dfrac{t : D}{g(t/x) : P(t)}$$

Associated rewriting:

$\texttt{EXTR}(\Lambda x.g(x), t) =_\beta g(t/x)$

Here it may be helpful to quote Wittgenstein on the meaning of 'all' (and of '$\forall x P(x)$', for that matter):

> Man könnte sagen: Man lernt die Bedeutung von "alle", indem man lernt, daß aus $(x).fx$ fa folgt. – D.h., die Übungen die den Gebrauch dieses Wortes einüben, lehren, laufen || gehen immer darauf hinaus, daß keine Ausnahme gemacht werden darf. || , die den Gebrauch dieses Wortes einüben, – seine Bedeutung lehren, || – zielen immer dahin, daß eine Ausnahme nicht gemacht werden darf.
>
> "Aus 'alle', wenn es so gemeint ist muß doch das folgen." –Wenn es wie gemeint ist? Überlege es Dir, wie meinst Du es? Da schwebt

> Dir etwa noch ein Bild vor – & mehr hast Du nicht. – Nein, es muß nicht, – aber es folgt: Wir vollziehen diesen Übergang.
>
> Und wir sagen: Wenn es || das nicht folgt, dann waren es eben nicht alle! – – und das zeigt nur, wie wir mit Worten in so einer Situation reagieren. –
>
> Wir könnten es auch so sagen: Es kommt uns vor, daß, wenn aus $(x).fx$ nicht mehr fa folgen soll, außer dem Gebrauch des Wortes "alle" noch etwas anderes sich geändert hat || haben muß, etwas, was dem Worte unmittelbar || selbst anhängt.[5]

In the same Ms, we find a remark in the same spirit:

> Und man macht den Sinn von "$(x).fx$" klar, indem man darauf dringt, daß aus ihm "fa" folgt.[6]

As we can observe from the rules of the existential quantifier (below), one cannot explain the immediate consequences of a sentence such as $\exists x^D.P(x)$ by presenting a specific witness, since it does not state that any witness can be used to instantiate the predicate $P(-)$. That is why the rule of $\exists$-*elimination* has to bring in a new local assumption of the kind 'suppose the witness is t', and then somehow get rid of the special status of such a witness. Special conditions

---

[5]which can be translated as:

> You could say: You learn the meaning of 'all' by learning that fa follows from $(x).fx$. - In other words, the exercises that practise, teach, run || the use of this word always result in the fact that no exception may be made. || , which practise the use of this word, - teach its meaning, || - always aim at the fact that an exception must not be made.
>
> 'From 'all', if it is meant that way, this must follow.' - If it is meant how? Think about it, what do you mean? You have another image in mind - & that's all you have. - No, it doesn't have to - but it follows: We make this transition.
>
> And we say: If it doesn't follow, then it wasn't all of them! - and that just shows how we react with words in such a situation. -
>
> We could also put it this way: It seems to us that if fa is no longer to follow from $(x).fx$, something else must have changed besides the use of the word 'all', something that is directly attached to the word itself.

Wittgenstein (2016), Ms-117,12 (1937) (Ms-117: XIII, Philosophische Bemerkungen (WL)) (Accessed 16 Jan 2025)

[6]which can be translated as:

> And we make the meaning of '$(x).fx$' clear by insisting that 'fa' follows from it.

Wittgenstein (2016), Ms-117,97(1937) (Ms-117: XIII, Philosophische Bemerkungen (WL)) (Accessed 16 Jan 2025)

A similar remark from 24.11.14 which was quoted above already appears in the *WW1 Notebooks 1914–16* (Ms-102,37r)

therefore apply when handling the deduction from such a new (local) assumption to the final conclusion C. In the dialogical/game-theoretical approaches, such an issue is dealt with by means of the dichotomy Attacker/Defender, Nature/Myself, Interpreter/Utterer, and by shifting the choice of the witness to the Attacker (or Nature, or Interpreter).

∃-β-reduction

$$\cfrac{\cfrac{s : D \quad g(s) : P(s)}{\varepsilon x.(g(x), s) : \exists x^D.P(x)}\exists\text{-}intr \qquad \cfrac{[t : D, h(t) : P(t)]}{d(h,t) : C}}{\texttt{INST}(\varepsilon x.(g(x), s), \sigma h.\sigma t.d(h,t)) : C}\exists\text{-}elim \quad \beta\text{-reduces to} \quad \cfrac{s : D, g(s) : P(s)}{d(g/h, s/t) : C}$$

Associated rewriting:

$\texttt{INST}(\varepsilon x.(g(x), s), \sigma h.\sigma t.d(h,t)) =_\beta d(g/h, s/t)$

Assertion/*Introd.*	Attack/*Elim.*	β-reduces to	Defense
Universal Quantifier ('∀'):			
$\forall x^D.P(x)$	$a : D$?		$P(a)$
$\Lambda x.g(x) : \forall x^D.P(x)$	$\texttt{EXTR}(\Lambda x.g(x), t)$	β-reduces to	$g(t/x) : P(t)$
Existential Quantifier ('∃'):			
$\exists x^D.P(x)$	?		$s : D, P(s)$
$\varepsilon x.(g(x), s) : \exists x^D.P(x)$	$\texttt{INST}(\varepsilon x.(g(x), s), \sigma h.\sigma t.d(h(t), t))$	β-reduces to	$d(g(s)/h(t), s/t) : C$

Whilst in $\texttt{EXTR}(\Lambda x.g(x), t)$ the destructor has access to the witness (and can use a generic element) a, in $\texttt{INST}(\varepsilon x.(g(x), s), \sigma h.\sigma t.d(h(t), t))$ the only option is to eliminate over the witness s which is inside the term $\varepsilon x.(g(x), s)$ built with the constructor because it was chosen by the introduction.

Id-β-reduction

$$\cfrac{\cfrac{a =_r b : A}{r(a,b) : Id_A(a,b)}Id\text{-}intr \qquad \cfrac{[a =_t b : A]}{d(t) : C}}{\texttt{REWR}(r(a,b), \sigma t.d(t)) : C}Id\text{-}elim \quad \beta\text{-reduces to} \quad \cfrac{a =_r b : A}{d(r/t) : C}$$

Associated rewriting:

$\texttt{REWR}(r(a,b), \sigma t.d(t)) =_\beta d(r/t)$

Assertion/*Introd.*	Attack/*Elim.*	β-reduces to	Defense
Propositional Equality ('$Id_A(a,b)$'):			
$Id_A(a,b)$	?		$a =_r b : A$
$r(a,b) : Id_A(a,b)$	$\texttt{REWR}(r(a,b), \sigma t.d(t))$	β-reduces to	$d(r/t) : C$

In $\mathtt{REWR}(r(a,b), \sigma t.d(t))$ the destructor has no choice regarding the 'reason' for a being equal to b, since r will have been chosen by the time of the assertion, i.e. the application of the introduction rule.

The reference to these parallels is intended to exhibit a connection with the "pragmatist"/"dialogical" approaches to meaning as revealed in Peirce's writings on the interaction between the *Interpreter* and the *Utterer*, which seem to pertain to the "game"/"dialogue" approaches to meaning (Lorenzen, Hintikka, Ehrenfeucht-Fraïssé).

In any case, it seems fair to say that it is due to the pioneers in the introduction of 'dialogical/game-theoretic/pragmatic' aspects, namely Peirce, Lorenzen, Hintikka, Ehrenfeucht-Fraíssé, that the grounds were prepared for the study of 'semantics' of predicate logic by stating the rules of 'attack-defense'/'Nature-Myself'/'Utterer-Interpreter'/elimination-introduction (reduction) for each logical connective and quantifier of a formula, thereby making explicit its meaning as use, purpose, consequences, i.e., as a general rule which is to translate into precise 'rules of interaction' when it comes to sentences and their meaning, use, purpose, consequences. That would be, essentially, the most basic settlement of meaning for predicate logic, which comes first and foremost before any (most legitimate!) approach to associate a sentence with a relation of satisfaction within a model in the realm of set theory.

Arthur F. Ramos
Ruy J. G. B. de Queiroz
Anjolina G. de Oliveira
Dov M. Gabbay
March 2026

Contents

Preface vii

I Foundations: The Rewrite System 1

1 Introduction and Motivation 3
1.1 The Curry–Howard Correspondence and Beyond 3
1.2 Labelled Deduction and Proof-Term Equality 6
1.2.1 The Three Judgement Forms 6
1.2.2 Harmony Between Two Calculi 8
1.2.3 From Proof-Term Equality to Computational Paths . . 8
1.3 Computational Paths: The Core Idea 9
1.3.1 The Path Type 9
1.3.2 Fundamental Operations 10
1.3.3 Rewrite Rules on Paths 11
1.4 On the Two-Level Structure: Path vs. Eq 12
1.4.1 Why Two Levels? 12
1.4.2 Relationship to Other Two-Level Theories 14
1.4.3 Weak Groupoid, Strict Groupoid, and the Quotient . . 14
1.5 Overview of Main Results 15
1.5.1 The Rewrite System 15
1.5.2 The Higher Groupoid Structure 16
1.5.3 Connections to Type Theory and Topology 17
1.5.4 The Lean Formalization 18
1.6 Reading Guide and Prerequisites 18

2 Labelled Natural Deduction and Proof-Terms 23
2.1 The LND Framework 23
2.1.1 Context and Judgement Forms 23
2.1.2 Historical Development 25
2.2 Proof-Terms for Propositional Logic 27

2.2.1 Types and Formulas 27
2.2.2 Terms and Proof-Terms 28
2.2.3 Typing Rules 29
2.2.4 The Abstraction Discipline and Substructural Logics . 30
2.3 β-Reduction and η-Expansion 31
2.3.1 The β-Reduction Rules 31
2.3.2 Dialogical Reading of the β-Rules 32
2.3.3 The η-Expansion Rules 33
2.4 Rewrite Steps and Subject Reduction 34
2.4.1 Rewrite Steps and Positions 34
2.4.2 Reduction Sequences as Rewrite Reasons 35
2.4.3 Subject Reduction 35
2.5 Church–Rosser Property for the Base System 37
2.5.1 The Diamond Property and Confluence 37
2.5.2 Proof via Parallel Reduction 38
2.5.3 Consequences of Confluence 39
2.6 From Reduction Sequences to Computational Paths 41
2.6.1 Operations on Reduction Sequences 41
2.6.2 Preview: The Path Constructors 41
2.6.3 The Central Question 42
2.7 Worked Examples 42
2.7.1 Basic Derivations 43
2.7.2 Reduction Examples 43
2.7.3 The Curry–Howard Correspondence in Detail 44
2.8 Notes and Further Reading 46

3 Computational Paths — Definition and Properties 49
3.1 Steps, Paths, and Path Expressions 50
3.1.1 Atomic Rewrite Steps 50
3.1.2 Operations on Steps 52
3.1.3 The Path Type 52
3.1.4 Path Grammar 54
3.1.5 Typing Rules for Path Constructors 56
3.1.6 Derived Forms and Sugar 57
3.1.7 The Step List as Computational Content 58
3.2 Fundamental Operations: refl, trans, symm, congrArg 59
3.2.1 Reflexivity: The Identity Path 59
3.2.2 Transitivity: Path Composition 59
3.2.3 Symmetry: Path Inversion 60
3.2.4 Congruence: Lifting Paths Through Functions 61
3.2.5 Function Congruence 63

3.2.6 Transport 63
3.3 Proof-Relevance and the Failure of UIP 64
3.3.1 The UIP Principle 64
3.3.2 Failure of UIP for Computational Paths 65
3.3.3 The Canonical Counterexample 65
3.3.4 Explicit Counterexample: Diamond Reduction 67
3.3.5 Step Counting as an Invariant 68
3.3.6 Significance of Non-UIP 69
3.4 Path Algebra: First Properties 69
3.4.1 The Groupoid Laws 69
3.4.2 Proof of the Inverse Laws 72
3.4.3 Functoriality of Congruence 72
3.4.4 Cancellation Laws 74
3.4.5 Preview: Mac Lane Coherence 74
3.4.6 Preview: Triangle Identity 75
3.4.7 Preview: Hexagon Identity for Inverses 75
3.4.8 Preview: Naturality 75
3.4.9 Preview of 2-Paths 75
3.5 Examples and Calculations 76
3.5.1 Building Paths 76
3.5.2 Verifying Algebraic Identities 77
3.5.3 Non-Trivial Loop Spaces 79
3.5.4 Computing with Dependent Paths 80
3.5.5 The Fundamental Groupoid Preview 80
3.5.6 Key Definitions 81
3.5.7 The Two-Level Structure 82
3.5.8 Connection to Earlier Work 82
3.5.9 The Road Ahead 83

4 The Rewrite System on Paths **85**
4.1 The 77-Rule Rewrite System 87
4.1.1 Notation and Conventions 87
4.1.2 Category A: Groupoid Core (Rules 1–8) 89
4.1.3 Category B: Type Constructor Rules (Rules 9–25) . . 92
4.1.4 Category C: Transport Rules (Rules 26–32) 96
4.1.5 Category D: Context Rules (Rules 33–48) 98
4.1.6 Category E: Dependent Context Rules (Rules 49–60) . 103
4.1.7 Category F: Binary Context Rules (Rules 61–68) . . . 106
4.1.8 Category G: Partial Map Rules (Rules 69–72) 108
4.1.9 Category H: Structural Completion (Rules 73–77) . . 109
4.2 Rule Categories and Structure 111

4.2.1 Design Principles . . . 111
4.2.2 The Layered Architecture . . . 112
4.2.3 Dependencies Between Categories . . . 113
4.2.4 Comparison with Related Systems . . . 114
4.3 Rewrite Relations and Strategies . . . 115
4.3.1 The One-Step Rewrite Relation . . . 115
4.3.2 The Reflexive-Transitive Closure . . . 116
4.3.3 Rewrite Equivalence . . . 116
4.3.4 Rewrite Strategies . . . 117
4.3.5 Examples of Rewrite Sequences . . . 118
4.4 Normalization and Implementation . . . 120
4.4.1 Normal Forms . . . 120
4.4.2 The Normalization Function . . . 121
4.4.3 Termination Measures . . . 122
4.4.4 The Path Algebra Quotient . . . 123
4.4.5 Decidability of Path Equivalence . . . 124

5 Confluence and Termination 127
5.1 Termination . . . 128
5.1.1 The Termination Problem . . . 128
5.1.2 The Weight Function . . . 128
5.1.3 The Left-Weight Function . . . 130
5.1.4 The Termination Measure . . . 131
5.1.5 Precedence Orderings and Strict Normalization Data . 132
5.2 Local Confluence . . . 134
5.2.1 Critical Pairs . . . 134
5.2.2 Critical Pair Categories . . . 135
5.2.3 Critical Pair Analysis . . . 137
5.3 The Diamond Lemma and Global Confluence . . . 138
5.3.1 From Local to Global: Newman's Lemma . . . 138
5.3.2 The Diamond Property . . . 139
5.3.3 The Church–Rosser Theorem . . . 139
5.4 Normal Forms and Decidability . . . 140
5.4.1 Characterization of Normal Forms . . . 140
5.4.2 The Word Problem and Decidability . . . 141
5.4.3 Complexity Considerations . . . 142
5.4.4 Semantic Interpretation . . . 143
5.4.5 Connection to Free Groups . . . 144

II Higher Structure: The ω-Groupoid 147

6 The Groupoid of Computational Paths 149

6.1 Groupoids: Definition and Basic Properties 150
6.1.1 First Examples 151
6.1.2 Morphisms of Groupoids 152
6.1.3 Weak versus Strict Groupoids 153
6.2 The Path Groupoid 154
6.2.1 The Data 154
6.2.2 The Weak Path Groupoid 155
6.2.3 Additional Laws 156
6.2.4 Functoriality as a Groupoid Functor 157
6.3 The Fundamental Groupoid 158
6.3.1 The Quotient Construction 158
6.3.2 The Fundamental Group 160
6.3.3 Basepoint Change 161
6.3.4 Comparison with Topological Groupoids 162
6.4 Groupoid Laws Verified 163
6.4.1 The Eight Laws as Groupoid Axioms 163
6.4.2 Detailed Verifications 164
6.4.3 Beyond the Core: Cancellation and Coherence 166
6.4.4 The Free Groupoid Perspective 166

7 Two-Cells — Derivations Between Paths 171

7.1 From Paths to 2-Paths 172
7.1.1 The Idea of a 2-Cell 173
7.1.2 Formal Definition 173
7.1.3 The Vertical Groupoid Structure 175
7.2 The 2-Category Structure 177
7.2.1 Whiskering: Composing Paths with 2-Cells 177
7.2.2 Horizontal Composition 180
7.2.3 The Bicategory of Computational Paths 182
7.3 2-Cells as Path Equalities 183
7.3.1 The RwEq Relation Revisited 184
7.3.2 The Equivalence of RwEq and Derivation$_2$ 184
7.3.3 From Normal Forms to 2-Cells 185
7.3.4 The Free Groupoid Perspective 186
7.4 Coherence at Level 2 187
7.4.1 3-Cells: Paths Between Derivations 187
7.4.2 The Interchange Law 188
7.4.3 The Pentagon Identity 191
7.4.4 The Triangle Identity 193

7.4.5 Naturality of Structural 2-Cells 193
7.4.6 Summary: The Coherence Package 195

8 Three-Cells and Coherence **199**
8.1 3-Cells as Third-Order Derivations 200
8.1.1 The Idea of a 3-Cell 200
8.1.2 Globular Boundaries at Dimension 3 201
8.1.3 Formal Definition 201
8.1.4 Vertical Groupoid at Dimension 3 203
8.2 Three Composition Operations on 3-Cells 204
8.2.1 $\circ_2$: Composition Along 2-Cell Boundaries 205
8.2.2 $\circ_1$: Composition Induced by Vertical Composition of 2-Cells 205
8.2.3 $\circ_0$: Composition Induced by Horizontal Composition of 2-Cells 207
8.3 Interchange Laws in Dimension 3 208
8.4 Coherence Conditions at the 3-Cell Level 210
8.4.1 Named Structural 3-Cells 210
8.4.2 Coherence Among Structural 3-Cells 211
8.5 Connection with Mac Lane Coherence 212
8.5.1 Canonical 3-Cells Between Reassociation Derivations 212
8.6 Examples from the Rewrite System 214
8.6.1 Commuting Independent Rewrites 214
8.6.2 Pentagon Witness as a Concrete 3-Cell 215
8.6.3 Triangle Witness and Unit Elimination Order 216
8.6.4 A Mixed Example Using Inverses 216
8.7 Squier-Style Diamond Fillers 217
8.8 Contractibility Phenomena at Level 3 218

9 The Weak ω-Groupoid of Computational Paths **223**
9.1 Weak ω-Groupoids: Globular Data and Basic Axioms 225
9.1.1 Globular Sets and Boundary Maps 225
9.1.2 Identities and Inverses in All Dimensions 227
9.1.3 Weak Compositions and Higher Laws 227
9.1.4 A concrete $n = 4$ boundary pattern 229
9.1.5 Strict and Weak Viewpoints on the Same Globular Data 231
9.2 Composition Operations in Dimension n 232
9.2.1 A uniform schema for $\circ_k$ 232
9.2.2 Concrete composition atlas for $n = 4$ 234
9.2.3 Recovering the $n = 2$ operations of Chapter 7 235
9.2.4 Recovering the $n = 3$ operations of Chapter 8 235
9.2.5 General n pattern: from principal to deep composition 236

9.3 Main Theorem: Computational Paths Form a Weak ω-Groupoid 238
9.3.1 The Tower of Computational Higher Cells 238
9.3.2 Higher-Dimensional Termination and Confluence . . . 240
9.3.3 The $n = 4$ instance in detail 244
9.3.4 Canonical Comparison Cells from Normalization . . . 245
9.3.5 Finite truncations and compatibility in n 246
9.4 Coherence in All Dimensions 247
9.4.1 Interchange Families in Arbitrary Dimension 247
9.4.2 Associativity and Unit Coherence at All Levels 247
9.4.3 Structural Higher Cells and Naturality 248
9.4.4 The All-Level Coherence Package 249
9.4.5 Coherence as Higher Contractibility 249
9.4.6 Explicit coherence bookkeeping at $n = 4$ 251
9.4.7 General n: coherence growth and control 252
9.5 Connection with Grothendieck's Homotopy Hypothesis 253
9.5.1 The Truncation Ladder as an Internal Postnikov-Style Tower 254
9.5.2 A Precise Comparison Template for the Hypothesis . . 255
9.5.3 Concrete low-dimensional readout and the $n = 4$ bridge 256
9.6 Connection with Identity Types in Homotopy Type Theory . . 257
9.6.1 Compatibility with the Identity Eliminator 258
9.6.2 A concrete $n = 4$ identity-type pattern 259
9.6.3 Transport, Dependent Action, and Whiskering 259
9.7 The Fundamental ω-Groupoid of a Type 260
9.7.1 Homotopy groups inside $\Pi_\omega(A)$ 262
9.7.2 Dimension-4 loop coherence and its meaning 263
9.7.3 General n invariants extracted from the tower 264
9.7.4 Extended Examples 264

10 The Eckmann–Hilton Argument 269
10.1 The Classical Eckmann–Hilton Argument 270
10.1.1 Two Operations and One Unit 270
10.1.2 The Classical Proof 271
10.1.3 A Geometric Reading 272
10.2 Application to Computational Paths 272
10.2.1 Doubly-degenerate 2-cells 273
10.2.2 Vertical and Horizontal Operations 273
10.2.3 Interchange in the Doubly-Degenerate Context 274
10.2.4 Eckmann–Hilton in Computational Paths 275
10.3 Commutativity of π_2 276
10.3.1 The Quotient Relation at Level 2 276

10.3.2 Group Structure 277
10.3.3 A Precise Framework Statement 278
10.4 Higher Eckmann–Hilton and Stabilisation 278
10.4.1 Iterated Degeneracy 279
10.4.2 Higher Eckmann–Hilton Principle 279
10.4.3 Connection with Chapter 9 280
10.5 Computational Interpretation 281
10.5.1 What Do Loop 2-Cells Represent? 281
10.5.2 A Symmetric Rewrite Example from the 77-Rule System 281
10.5.3 Relation with Confluence 282

III Connections: Type Theory, Homotopy, and Beyond 287

11 Path Induction and the J-Eliminator 289
11.1 The Based Path Space 290
11.1.1 Definition and Basic Properties 290
11.1.2 Projections and Structure 291
11.1.3 The Semantic vs. Syntactic Distinction 292
11.1.4 The Free Path Space 293
11.2 Contractibility of the Based Path Space 293
11.2.1 Contractibility: Definition and Significance 293
11.2.2 The Fundamental Contraction Theorem 294
11.2.3 Non-Contractibility at the Syntactic Level 295
11.2.4 Contractibility of the Free Path Space 296
11.3 Derivation of J and Its Computation Rule 296
11.3.1 The J-Eliminator: Statement 296
11.3.2 Derivation from Contractibility 297
11.3.3 Direct Definition via Pattern Matching 298
11.3.4 Path-Level J 298
11.3.5 The Paulin-Mohring Variant 299
11.3.6 Free Path Induction 300
11.4 Transport and Dependent ap 300
11.4.1 Transport Derived from J 301
11.4.2 Dependent Application (apd) 302
11.4.3 Derived Operations 302
11.4.4 Transport in Path Spaces 303
11.4.5 Leibniz's Principle 304
11.5 Uniqueness of the J-Eliminator 304
11.5.1 The Uniqueness Theorem 304
11.5.2 Uniqueness for Free Path Induction 305

11.5.3 Equivalence with Contractibility 305
11.5.4 The Structure Identity Principle 305
11.6 Higher J and Nested Induction 306
11.6.1 Double Path Induction 306
11.6.2 Triple Path Induction 307
11.7 K, UIP, and Their Failure 307
11.7.1 K at the Semantic Level 307
11.7.2 K Fails at the Syntactic Level 308
11.7.3 Two-Level Summary 308
11.8 The Encode-Decode Method 309
11.8.1 The Method 309
11.9 Historical and Conceptual Notes 312
11.9.1 Origins of Path Induction 312
11.9.2 The Contractibility Perspective 313
11.9.3 Paulin-Mohring's Formulation 313
11.9.4 J Without K 313
11.9.5 Computational Paths and Two-Level Type Theory . . . 313

12 Failure of UIP and Proof-Relevance 315
12.1 The Non-UIP Theorem 316
12.1.1 UIP Revisited in the Computational-Path Setting . . . 316
12.1.2 What UIP Would Force Algebraically 316
12.1.3 Canonical Non-UIP Witness 317
12.1.4 Relation to Chapter 3 320
12.2 Failure of Streicher's K 320
12.2.1 K as Loop Collapse 320
12.2.2 Appending a Step Always Changes the Path 321
12.2.3 K Failure as a Structural Phenomenon 321
12.2.4 K Failure as Non-Contractibility 322
12.3 Step-Counting as a Path Invariant 322
12.3.1 From Witness to Invariant 322
12.3.2 Length Filtration of Loop Spaces 324
12.3.3 Worked Calculations 324
12.3.4 Step Count and the Normalization Perspective 325
12.3.5 Infinite Families of Loops 326
12.3.6 Further Stability Properties 327
12.3.7 How Far Does Step Count Go? 328
12.3.8 Stratified Reasoning with Step Count 328
12.4 The Proof-Relevance / Proof-Irrelevance Dialectic 330
12.4.1 Projection-Level UIP 330
12.4.2 Transport and Substitution 331

12.4.3 The Two-Level Architecture 331
12.4.4 Design Principles from the Dialectic 332
12.4.5 A Final Structural Statement 332

13 Fundamental Groupoids of Combinatorial Spaces 335
13.1 Combinatorial Spaces via Generators and Relations 336
13.1.1 The Idea of Cell Attachments 336
13.1.2 Formal Definition 337
13.1.3 The General Strategy 338
13.2 The Circle 339
13.2.1 The Presentation 339
13.2.2 Loop Expressions and Winding Numbers 339
13.2.3 The Isomorphism 341
13.3 The Torus 342
13.3.1 The Polygon Model 342
13.3.2 The Presentation 343
13.3.3 Normal Forms and the Group Structure 343
13.4 The Klein Bottle 345
13.4.1 The Polygon Model 345
13.4.2 The Presentation and Group Structure 346
13.5 Real Projective Plane 347
13.5.1 The Presentation 348
13.5.2 Normal Forms and the Group 348
13.6 Bouquets and Free Groups 349
13.6.1 Definition 349
13.6.2 The Free Group 349
13.6.3 Normal Forms via Word Reduction 350
13.7 The Figure-Eight and Seifert–van Kampen 351
13.7.1 The Seifert–van Kampen Theorem 351
13.7.2 Application to the Figure-Eight 351
13.7.3 Computational Interpretation 352
13.8 Lens Spaces 353
13.8.1 Definition 353
13.8.2 Combinatorial Presentation 353
13.8.3 The Fundamental Group 353

14 Connections to HoTT 357
14.1 Translation Dictionary: HoTT $\leftrightarrow$ Computational Paths 358
14.1.1 Basic Concepts 359
14.1.2 The J-Eliminator: A Point of Contact 359
14.1.3 Groupoid Laws: Same Theorems, Different Proofs . . 360
14.1.4 The Functor Laws for Congruence 361

14.1.5 Higher Structure: 2-Cells and Beyond 361
14.1.6 The ω-Groupoid Structure 362
14.2 What Computational Paths Add 363
14.2.1 Explicit Rewrite Witnesses 363
14.2.2 Decidable Rewrite Equivalence 364
14.2.3 The Two-Level Structure 365
14.2.4 Concrete Computation Examples 365
14.2.5 Computational Meaning of Groupoid Laws 366
14.2.6 Summary: Computational Paths Advantages 366
14.3 What HoTT Adds: Univalence and Higher Inductive Types . . 367
14.3.1 The Univalence Axiom 367
14.3.2 Why Univalence Lies Outside the Present Framework . 368
14.3.3 Higher Inductive Types 369
14.3.4 HITs and Computational Paths: A Comparison 370
14.3.5 Summary: HoTT Advantages 371
14.4 Cubical Type Theory and Computational Paths 372
14.4.1 What Is Cubical Type Theory? 372
14.4.2 Comparison: Cubical vs. Computational Paths 372
14.4.3 The Interval: Same Name, Different Objects 373
14.4.4 How Cubical TT Achieves Computational Univalence 373
14.4.5 What Computational Paths Cannot Achieve (and Why) 374
14.4.6 Cubical Ideas That Could Transfer 375
14.5 Open Problem: Can Computational Paths Model Univalence? . 375
14.5.1 The Challenge 375
14.5.2 Partial Univalence Outside the Present Framework . . 376
14.5.3 Alternative Approaches 376
14.5.4 What Would Be Needed 377
14.5.5 Current Status 378
14.5.6 A Philosophical Remark 378

15 Connections to Category Theory and Higher Algebra 383
15.1 The Path Groupoid as a Free Groupoid on a Computad 384
15.1.1 Graphs, Categories, and Groupoids 385
15.1.2 Computads (1-polygraphs) 386
15.1.3 Computational Paths as a Computad 387
15.1.4 Consequences of the Free Groupoid Perspective 388
15.2 Enrichment and 2-Groupoid Structure 389
15.2.1 Enriched Categories 390
15.2.2 The 2-Groupoid of Computational Paths 391
15.2.3 Strict vs. Weak 2-Groupoids 392
15.2.4 The Gray Tensor Product 392

15.3 Relation to Squier's Theorem 393
15.3.1 Finite Derivation Type 394
15.3.2 Homological Consequences 395
15.3.3 Computational Paths and Squier's Theorem 395
15.3.4 The Homotopy Basis 396
15.3.5 Homological Shadow of FP_3 397
15.4 Double Categories and Double Groupoids 397
15.4.1 Double Categories 397
15.4.2 The Double Groupoid of Computational Paths 398
15.4.3 Connection to Crossed Modules 399
15.4.4 Higher Double Structures 400
15.5 Sketch: Path ∞-Groupoids and ∞-Cosmoi 400
15.5.1 The Homotopy Hypothesis 400
15.5.2 (∞, n)-Categories 401
15.5.3 Quasicategories and the Joyal Model 402
15.5.4 ∞-cosmoi 402
15.5.5 Directed Computational Paths 403
15.5.6 Comparison with HoTT 403
15.6 Historical Notes and Further Reading 407

16 Perspectives and Open Problems 409
16.1 Extension to Dependent Types and Universes 409
16.1.1 The Challenge of Dependency 410
16.1.2 Universes and Type-Level Paths 410
16.1.3 Stratified Universes and Consistency 411
16.2 The Univalence Question 412
16.2.1 Why Full Univalence Fails 412
16.2.2 Partial Univalence for 0-Types 414
16.2.3 Towards a Univalent Extension 415
16.2.4 The Decidability–Univalence Spectrum 417
16.2.5 Why the Univalence Question Matters 418
16.3 Directed Computational Paths 419
16.3.1 From Groupoids to Categories 419
16.3.2 Directed Type Theory 420
16.3.3 Mixed Structures 420
16.4 Applications to Program Verification 421
16.4.1 Refactoring as Path Construction 422
16.4.2 Program Equivalence Proofs 422
16.4.3 Differential Reasoning 423
16.4.4 Security Applications 423
16.5 The Homotopy Theory of Formal Systems 424

16.5.1 Formal Systems as Spaces 424
16.5.2 Homotopy Invariants of Formal Systems 425
16.5.3 Morphisms Between Formal Systems 426
16.5.4 A Research Program 426
16.5.5 The Synthetic Perspective 427
16.6 Conclusion 429

A Term Rewriting Systems **431**
A.1 Basic Definitions 432
A.1.1 Signatures and Terms 432
A.1.2 Substitution and Matching 433
A.1.3 Rewrite Rules and Reduction 434
A.2 Termination 435
A.2.1 Well-Founded Orderings 436
A.2.2 Recursive Path Orderings (RPO) 436
A.2.3 Knuth-Bendix Ordering (KBO) 437
A.2.4 Polynomial Interpretations 438
A.2.5 Lexicographic Combinations 439
A.3 Confluence 439
A.3.1 The Church-Rosser Property 439
A.3.2 Local Confluence 440
A.3.3 Critical Pairs 441
A.3.4 The Knuth-Bendix Completion Procedure 443
A.4 Normalization 444
A.4.1 Normal Forms 444
A.4.2 Unique Normal Forms 445
A.4.3 Decidability of Word Problems 445
A.4.4 Rewriting Strategies 446
A.5 Connection to Computational Paths 446
A.5.1 The Path Rewrite System as a TRS 446
A.5.2 The 77 Rules as a TRS 447
A.5.3 Why Confluence and Termination Matter 448
A.5.4 Techniques Used in the Confluence Proof 448
A.5.5 The Path Algebra as a Completed System 449
A.5.6 Comparison with Other Equational Theories 450
A.6 Further Reading 450

B Groupoids and Higher Groupoids **453**
B.1 Categories and Groupoids 454
B.1.1 Basic Definitions 454
B.1.2 Examples of Groupoids 456
B.1.3 Functors and Natural Transformations 458

B.2 2-Categories and 2-Groupoids . . . 459
B.2.1 2-Categories . . . 459
B.2.2 Strict 2-Groupoids . . . 460
B.2.3 Bicategories: Weak 2-Categories . . . 460
B.3 n-Groupoids and ω-Groupoids . . . 462
B.3.1 The Pattern . . . 462
B.3.2 Globular Sets . . . 463
B.3.3 Strict ω-Categories . . . 463
B.3.4 Weak ω-Categories and Weak ω-Groupoids . . . 464
B.3.5 Truncation . . . 465
B.4 Coherence . . . 466
B.4.1 Mac Lane's Coherence Theorem . . . 466
B.4.2 Stasheff Associahedra . . . 467
B.4.3 Coherence for Bicategories . . . 467
B.4.4 Coherence for ω-Groupoids . . . 468
B.5 Connection to Computational Paths . . . 468
B.5.1 Computational Paths as a Weak ω-Groupoid . . . 468
B.5.2 Stabilisation Above Dimension 3 . . . 469
B.5.3 Comparison with Topological ω-Groupoids . . . 470
B.5.4 The Fundamental Groupoid . . . 470
B.5.5 Summary . . . 471

C The Lean 4 Formalization **473**
C.1 Getting Started . . . 474
C.1.1 Companion repository and reproducibility . . . 474
C.1.2 Accessing the companion repository . . . 474
C.1.3 Project Structure Overview . . . 474
C.2 Core Architecture . . . 475
C.2.1 Path.lean: The Path Type and Basic Operations . . . 475
C.2.2 Step.lean: The Rewrite Relation . . . 477
C.2.3 RwEq.lean: Rewrite Equivalence . . . 478
C.2.4 OmegaGroupoid.lean: Higher Cells and Coherence . . 479
C.3 Key Design Decisions . . . 480
C.3.1 Why Path Uses UIP (Intentionally Not HoTT) . . . 481
C.3.2 Type-Valued 2-Cells . . . 481
C.3.3 Stabilisation and Contractibility above Level 3 . . . 482
C.3.4 MetaStep_3 Constructors for Coherences . . . 484
C.4 Navigating the Proofs . . . 485
C.4.1 Module Hierarchy . . . 485
C.4.2 Finding Specific Theorems . . . 487
C.4.3 Key Lemmas and Where They Live . . . 488

C.5 Extending the Formalization 489
C.5.1 Adding New Rewrite Rules 489
C.5.2 Proving New Coherence Results 490
C.5.3 Contributing Guidelines 491

D Complete List of Named Rewrite Clauses **493**
D.1 Rule Format and Notation 493
D.1.1 Notation Summary 494
D.1.2 Type Conventions 494
D.2 Category A: Groupoid Core (Rules 1–8) 494
D.3 Category B: Type Constructor Rules (Rules 9–25) 495
D.3.1 Binary Congruence and Products (Rules 9–15) 495
D.3.2 Sigma Types (Rules 16–19) 496
D.3.3 Sum Types (Rules 20–21) 496
D.3.4 Function Types (Rules 22–25) 497
D.4 Category C: Transport Rules (Rules 26–32) 497
D.5 Category D: Context Rules (Rules 33–48) 498
D.5.1 Basic Context Schemata (Rules 33–36) 498
D.5.2 Whisker Introduction and Conversion (Rules 37–41) . 498
D.5.3 Whisker Simplification (Rules 42–48) 499
D.6 Category E: Dependent Context Rules (Rules 49–60) 499
D.7 Category F: Binary Context Rules (Rules 61–68) 501
D.7.1 Dependent Binary Context (Rules 61–64) 501
D.7.2 Non-Dependent Binary Context (Rules 65–68) 501
D.8 Category G: Partial Map Rules (Rules 69–72) 502
D.9 Category H: Structural Completion (Rules 73–77) 502
D.9.1 Congruence Rules (Rules 73–75) 502
D.9.2 Cancellation Rules (Rules 76–77) 503
D.10 Quick Reference: Groupoid Rules 504
D.11 Quick Reference: Type Constructor Rules 505
D.11.1 Products 505
D.11.2 Sigma Types 505
D.11.3 Function Types 505
D.12 Quick Reference: Context and Whisker Rules 506
D.12.1 Non-Dependent Context 506
D.12.2 Dependent Context 506
D.13 Quick Reference: Cancellation Rules 506
D.14 Index of Rules by Name 507

Part I

Foundations: The Rewrite System

Chapter 1

Introduction and Motivation

The concept of a computational path *is to type theory what the concept of a path is to topology: a witness to connectivity that carries information not merely about* whether *two objects are connected, but about* how *the connection is made.*

The theory of computational paths begins from a question that may be stated simply: if proofs are programs, and proof simplification is computation, then what, exactly, are the *reasons* for two proofs to be equal? The standard answer—that two proofs are equal when they $\beta\eta$-reduce to a common normal form—is correct as far as it goes; but it tells us only *whether* two proofs are equal, not *how* or *why*. The present book takes those reasons themselves as first-class mathematical objects—*computational paths*. From that starting point it develops two intertwined themes: an explicit rewrite theory of equality and the higher algebra carried by its proof-relevant comparison data. Later chapters show that this structure supports groupoid operations, higher comparison cells, and ultimately a weak ω-groupoid.

The present chapter has three tasks. First, it places the theory of computational paths within the historical tradition of the Curry–Howard correspondence and labelled natural deduction (§1.1–§1.2). Second, we present the central definitions informally and explain the key design decisions of the framework (§1.3–§1.4). Third, we survey the main results of the book and explain how the later chapters fit together (§1.5–§1.6).

1.1 The Curry–Howard Correspondence and Beyond

The functional interpretation of logical connectives is concerned with a certain harmony between, on the one hand, a functional calculus on the expressions built up from the recording of the deduction steps (the *labels*), and, on the other hand, a logical calculus on the formulas. It has been associated with Curry's early discovery of the correspondence between the axioms of intuitionistic implicational logic and the type schemes of the so-called 'combinators' of Combinatory Logic [25], and has been referred to as the *formulae-as-types* interpretation.

Howard's [87] extension of the formulae-as-types paradigm to full intuitionistic first-order predicate logic is what we now call the 'Curry–Howard' functional interpretation. De Bruijn's [27] independent discovery in the context of AUTOMATH further underscored the robustness of the correspondence. Seen chronologically, the progression Curry (1934) $\to$ Howard (1969/1980) $\to$ de Bruijn (1968), followed by Martin-Löf's formulation of intuitionistic type theory [109, 111], marks the transition from isolated proof-theoretic correspondences to a full foundational framework for constructive mathematics. Although Heyting's intuitionistic logic [80, 81] fits naturally into the formulae-as-types paradigm, Tait's intensional interpretations [146, 147] of Gödel's *Dialectica* system of functionals of finite type [75] already suggested that the framework reaches beyond intuitionism. In that broader perspective, the foundations of a *functional* approach to formal logic lie in Frege's *Begriffsschrift* [66] and *Grundgesetze* [67, 68], not in Curry, Howard, or indeed Heyting (see also [36] for a discussion of the relationship between *Grundgesetze* and *Begriffsschrift*).

Remark 1.1 (Frege and Church on "function as rule")**.** A striking historical parallel appears when one compares Church's presentation of λ-conversion with Frege's value-range notation. In Church's 1936 formulation [21], Rules I–III isolate renaming of bound variables, application, and inverse expansion:

I. $\lambda x[M] \to \lambda y[M[y/x]]$ (with y fresh).
II. $\{\lambda x[M]\}(N) \to M[N/x]$.
III. $M[N/x] \to \{\lambda x[M]\}(N)$ (inverse of Rule II).

Frege had already introduced a corresponding abstraction–application discipline in *Grundgesetze* [67], where a function-expression is turned into a value-range term $\acute{\varepsilon}\Phi(\varepsilon)$ and then applied by an explicit operator. As Frege writes in § 9 (English translation in [69]): "Wenn ich allgemein sage: $\acute{\varepsilon}\Phi(\varepsilon)$ den Werthverlauf der Function $\Phi(\xi)$ …" (§ 9), i.e. "If I say in general: let $\acute{\varepsilon}\Phi(\varepsilon)$ refer to the value-range of the function $\Phi(\xi)$, then this too requires supplementation." In this sense, the notations

$$\text{Frege: } \backslash\acute{\varepsilon}\Phi(\varepsilon) \qquad \text{and} \qquad \text{Church: } \lambda\varepsilon.\Phi(\varepsilon)$$

both represent a function by an explicit rule of use.

In its most basic form, the Curry–Howard correspondence establishes a dictionary between two prima facie distinct disciplines:

Logic		**Type Theory**
Propositions	$\leftrightarrow$	Types
Proofs	$\leftrightarrow$	Terms (Programs)
Proof simplification	$\leftrightarrow$	Computation (β-reduction)
Proof normalisation	$\leftrightarrow$	Termination

This dictionary has been enormously fruitful. It forms the conceptual foundation of modern proof assistants such as Coq [149], Agda [118], and Lean [28], and has profoundly influenced both the theory and the practice of formal verification.

However, the standard Curry–Howard dictionary, taken at face value, leaves a fundamental question unanswered. If proofs correspond to programs, and proof simplification corresponds to computation, then what corresponds to *equality of computations*?

Consider two computations of the same result. Both begin with the same expression and end at the same normal form, but the intermediate steps may differ: one might reduce the outermost redex first, the other an inner redex. The Church–Rosser theorem [22] tells us that the two reduction sequences *can* be brought to a common result—we prove the analogous confluence result for our path rewrite system in Chapter 5—but it says nothing about the sequences themselves. Are these sequences—these witnesses to equality—mathematically meaningful in their own right?

The present work answers this question affirmatively: the sequences of rewrite steps that transform one term into another are reified as formal objects called *computational paths*, and these paths are themselves subject to a rich algebraic theory. By extending the Curry–Howard dictionary one level further, we obtain:

Logic (extended)		**Type Theory (extended)**
Equality of proofs	$\leftrightarrow$	Path between terms
Reason for equality	$\leftrightarrow$	Computational path
Equality of reasons	$\leftrightarrow$	Path between paths (2-cell)

It is at this third row that the theory of computational paths begins. The insight that "reasons for equality" can themselves be compared and found equal or distinct opens the door to an infinite hierarchy of higher-dimensional structure—precisely the structure of a weak ω-groupoid [77, 11, 99] (see Chapter 9).

Remark 1.2 (Historical note)**.** The idea that proof-term reduction is not merely a technical device but a semantically significant operation goes back at least to Prawitz's *Natural Deduction* [121], where normalisation was shown to correspond to cut-elimination. The further idea that the *paths* of reduction might be worth studying in their own right was advanced by de Queiroz in the late 1980s [32, 34], building on earlier insights by Lambek and Scott [97], who wrote:

> "In writing $f : A \to B$ we think of f as the 'reason' why A entails B."

The present work may be seen as the full elaboration of the programme implicit in this remark: to make the reasons for entailment—and the reasons for equality of reasons—explicit as structured mathematical objects.

The idea that the identity type can be understood as the type of computational paths was developed in [54, 127], and the full elaboration of the theory of explicit computational paths in type theory was carried out in [124].

1.2 Labelled Natural Deduction and Proof-Term Equality

The framework within which computational paths arise naturally is that of *labelled natural deduction* (LND), developed by de Queiroz and Gabbay [71], and subsequently extended by de Queiroz and de Oliveira [44, 45, 30] over several decades; we present the framework in full in Chapter 2. In this framework, every formula in a deduction carries a *label*—a term from an associated functional calculus that records the structure of the proof. The approach should be seen in the broader context of Gabbay's programme of *Labelled Deductive Systems* [70], where the declarative unit of a logical system is not a bare formula A but a labelled formula $t : A$, read "t labels A" or "t is evidence for A."

From this perspective, a logical system is not simply a calculus of logical deductions on formulas, but a harmonious combination of a functional calculus on the labels and a logical calculus on the formulas. A logic is then defined according to the meta-level features of the conceptual norm that the logic is supposed to formalise: the allowable logical moves are 'controlled' by appropriate constraints on what has been done so far (whether the assumption has been used at all; whether the assumptions have been used in a certain order; whether the assumption has been used more than once; and so on).

1.2.1 The Three Judgement Forms

The key innovation of LND, compared to the standard Curry–Howard setup, is that it makes explicit not only the *proof-terms* but also the *reasons for equality between proof-terms*. While the Curry–Howard correspondence employs two basic judgement forms:

(i) $\Gamma \vdash t : A$ *("t proves A under assumptions Γ")*

(ii) $\Gamma \vdash t = u : A$ *("t and u are equal proofs of A")*

the LND framework introduces a third, richer form:

(iii) $\Gamma \vdash t =_s u : A$ *("t and u are equal proofs of A, **by reason** s")*

The subscript s is the crucial novelty: it records the *rewrite reason*—an explicit, structured witness to the equality. This witness is not merely an annotation or a metalinguistic comment; it is a term in a calculus, subject to its own rules of formation and transformation.

Definition 1.3 (Labelled equality judgement)**.** A *labelled equality judgement* is an expression of the form

$$\Gamma \vdash t =_s u : A$$

where Γ is a context, t and u are proof-terms of type A, and s is a *rewrite reason*—a term in the language of paths denoting a specific sequence of rewrite steps transforming t into u.

Example 1.4 (A simple rewrite reason)**.** Consider the λ-terms:

$$t \equiv (\lambda x.\, x)\, 3 \qquad\qquad u \equiv 3$$

These terms are β-equal. In the standard Curry–Howard setup, we write simply $t = u$ or $t =_\beta u$ and leave it at that. In LND, we write:

$$(\lambda x.\, x)\, 3 \ =_\beta\ 3 \ :\ \mathbb{N}$$

The subscript β is an *atomic step*—an elementary rewrite reason recording the application of the β-reduction rule at the outermost position. When we compose multiple rewrites, we build up more complex reasons. For instance:

$$(\lambda x.\, (\lambda y.\, y)\, x)\, 3 \ =_{\mathsf{trans}(\beta,\,\beta)}\ 3 \ :\ \mathbb{N}$$

Here $\mathsf{trans}(\beta, \beta)$ records two successive β-reductions: first reducing the outermost redex to $(\lambda y.\, y)\, 3$, then reducing the remaining redex to 3.

1.2.2 Harmony Between Two Calculi

The architecture of labelled natural deduction can be summarised in a slogan: *Grundgesetze alongside Begriffsschrift*—the theory of functions of Frege's later work put together with the theory of predicates of his earlier work. The result is that a formula is valid if and only if a deduction of it can be constructed whose label is a *closed* term (i.e., a term containing no free variables), meaning that the truth of the formula relies on no unresolved assumptions.

This two-dimensional architecture—terms alongside formulas—has a number of significant features:

(a) It resumes Gentzen's programme of analysing logical deduction via an analysis of connectives, by introducing the extra dimension (the label) that handles deduction in a more direct fashion than the sequent calculus.

(b) It replaces the *subformula principle* by the *subdeduction principle*: the measure of proof complexity is not the size of the formulas appearing in the proof but the structure of the labels.

(c) It recovers the 'continuation' aspect of classical logic—through a generalised form of *reductio ad absurdum* involving controlled self-application in the labels [71].

(d) It is *resource-aware*: disciplines of abstraction on label variables reflect the disciplines of assumption withdrawal peculiar to the logic being considered.

1.2.3 From Proof-Term Equality to Computational Paths

The crucial observation that leads from labelled deduction to the theory of computational paths is the following. Rewrite reasons—the terms s appearing in the judgement $t =_s u : A$—are themselves structured objects. They can be composed (when two successive rewrites are performed), reversed (when we regard a reduction "backwards" as an expansion), and compared—the formal calculus of these operations is developed in Chapter 3. Two different rewrite reasons might both transform t into u, but via entirely different routes through the space of terms.

Example 1.5 (Two distinct rewrite reasons)**.** Consider the term:

$$t \equiv (\lambda x.\,(\lambda y.\,y\,x)\,z)\,w$$

There are two natural ways to reduce t to the normal form $z\,w$:

Reason s_1 (outer-first):

$$(\lambda x.\,(\lambda y.\,y\,x)\,z)\,w \xrightarrow{\beta} (\lambda y.\,y\,w)\,z \xrightarrow{\beta} z\,w$$

Reason s_2 (inner-first):

$$(\lambda x.\,(\lambda y.\,y\,x)\,z)\,w \xrightarrow{\beta} (\lambda x.\,z\,x)\,w \xrightarrow{\beta} z\,w$$

Both reasons witness the same equality $t = z\,w$, but they are different objects: they differ in the intermediate term and in the order of the reductions performed. The theory of computational paths takes this difference seriously and asks: when should two rewrite reasons be considered equivalent?

This question—the question of equality *among* rewrite reasons—is the starting point for the entire theory. The rewrite reasons of LND are precisely the computational paths of the title. The study of their algebraic structure, the rewrite system that governs their equivalence, and the higher groupoid that they form, constitutes the subject matter of this book.

1.3 Computational Paths: The Core Idea

We now present the central construction informally, deferring the complete formal development to Chapter 3. The key idea is to package together a *trace* (a list of rewrite steps) and a *proof* (a propositional equality) into a single structured object.

1.3.1 The Path Type

Definition 1.6 (Computational path, informal). Let A be a type and let $a, b : A$. At an intuitive level—and in the concrete semantic model used repeatedly in this chapter—a *computational path* from a to b may be represented by a term $p : \mathsf{Path}(a, b)$ with two components:

(i) A *trace*: a finite list of rewrite steps $[\sigma_1, \sigma_2, \ldots, \sigma_n]$, each recording an atomic rewrite applied at a specific position.

(ii) A *proof*: a propositional equality witness $a = b$, certifying that the composition of the steps does indeed take a to b.

This guiding model separates two concerns that are usually conflated. The *proof* component certifies the *fact* that a equals b; it is proof-irrelevant in the sense that the type of this component lives in Prop. The *trace* component records *how* the equality is witnessed; it is proof-relevant and carries the computational content. Later chapters distinguish more sharply between raw path expressions, their rewrite classes, and this semantic trace-based representation.

The analogy with topology is deliberate and illuminating. In topology, a path in a space X from a point x to a point y is a continuous function $\gamma : [0,1] \to X$ with $\gamma(0) = x$ and $\gamma(1) = y$. The path records not merely that x and y are in the same connected component but *how* one can travel from one to the other. Similarly, a computational path records not merely that a and b are rewrite-equivalent but the specific route through the space of terms by which the rewriting proceeds. This analogy is made precise in Chapter 13, where we compute fundamental groups of combinatorial spaces using computational paths.

1.3.2 Fundamental Operations

The path type comes equipped with four fundamental operations, corresponding to the basic ways in which rewrite reasons can be constructed (see Section 3.2 for the rigorous treatment).

Definition 1.7 (Fundamental path operations)**.** Let A and B be types. The following operations are defined on paths:

(i) **Reflexivity.** For any $a : A$, there is a path

$$\mathsf{refl}_a : \mathsf{Path}(a, a)$$

with an empty trace and the trivial equality proof $\mathsf{rfl} : a = a$. This is the path that "stays in place"—the zero-step rewrite.

(ii) **Symmetry.** Given $p : \mathsf{Path}(a, b)$, there is a path

$$\mathsf{symm}(p) : \mathsf{Path}(b, a)$$

obtained by reversing the trace of p and applying the symmetry of propositional equality to the proof component.

(iii) **Transitivity.** Given $p : \mathsf{Path}(a, b)$ and $q : \mathsf{Path}(b, c)$, there is a path

$$\mathsf{trans}(p, q) : \mathsf{Path}(a, c)$$

obtained by concatenating the traces of p and q and composing the proof components by transitivity of propositional equality.

(iv) **Congruence.** Given $f : A \to B$ and $p : \mathsf{Path}(a, b)$, there is a path

$$\mathsf{congrArg}(f, p) : \mathsf{Path}(f(a), f(b))$$

obtained by lifting each step in the trace of p through the function f and applying congruence to the proof component.

We shall frequently use the following notational shorthands:

$$\begin{array}{lll} p \cdot q & \text{for} & \mathsf{trans}(p, q) \\ p^{-1} & \text{for} & \mathsf{symm}(p) \\ f_*(p) & \text{for} & \mathsf{congrArg}(f, p) \end{array}$$

1.3.3 Rewrite Rules on Paths

The operations defined above satisfy a collection of algebraic laws, expressed as *rewrite rules on paths*. We list the most fundamental ones here; the complete system of 77 rules is given in Chapter 4 (with a full listing in Appendix D).

Definition 1.8 (Core rewrite rules)**.** We write $p \rhd q$ to indicate that the path expression p rewrites to the path expression q. The core rules are:

Reflexivity and symmetry:

$$\mathsf{symm}(\mathsf{refl}_a) \rhd \mathsf{refl}_a \qquad \text{(R1: symm\_refl)}$$

$$\mathsf{symm}(\mathsf{symm}(p)) \rhd p \qquad \text{(R2: symm\_symm)}$$

Transitivity and identity:

$$\mathsf{trans}(\mathsf{refl}_a,\, p) \rhd p \qquad \text{(R3: trans\_refl\_left)}$$

$$\mathsf{trans}(p,\, \mathsf{refl}_b) \rhd p \qquad \text{(R4: trans\_refl\_right)}$$

Inverse laws:

$$\mathsf{trans}(p,\, \mathsf{symm}(p)) \rhd \mathsf{refl}_a \qquad \text{(R5: trans\_symm)}$$

$$\mathsf{trans}(\mathsf{symm}(p),\, p) \rhd \mathsf{refl}_b \qquad \text{(R6: symm\_trans)}$$

Symmetry distributes over transitivity:

$$\mathsf{symm}(\mathsf{trans}(p,\, q)) \rhd \mathsf{trans}(\mathsf{symm}(q),\, \mathsf{symm}(p)) \qquad \text{(R7: symm\_trans\_congr)}$$

Associativity:

$$\mathsf{trans}(\mathsf{trans}(p,\, q),\, r) \rhd \mathsf{trans}(p,\, \mathsf{trans}(q,\, r)) \qquad \text{(R8: trans\_assoc)}$$

Reading these rules categorically, refl is the identity morphism, trans is composition, and symm is the inverse. These are the rules of a *groupoid* (Chapter 6). Rules R3–R4 assert that refl is a two-sided identity for trans; rules R5–R6 assert that $\mathsf{symm}(p)$ is a two-sided inverse of p; rule R7 asserts that taking inverses is an anti-homomorphism with respect to composition; and rule R8 asserts associativity.

Remark 1.9 (Weak vs. strict)**.** The rewrite rules R3–R8 hold only up to rewriting ($\triangleright$), not as definitional equalities. This is the hallmark of a *weak* algebraic structure: the laws hold "up to coherent higher cells" rather than "on the nose." The distinction between weak and strict is essential to the theory; the coherence conditions are analysed systematically in Chapter 8 and are discussed further in §1.4. A concrete semantic model may collapse some of these laws to strict equalities of representatives, but the rewrite theory of Chapter 4 lives at the level of raw expressions, where the weakness is genuine.

Example 1.10 (Verifying a compound identity)**.** Let $p : \mathsf{Path}(a, b)$. We verify that $\mathsf{trans}(p,\ \mathsf{trans}(\mathsf{symm}(p),\ p)) \ \triangleright^* \ p$ by a chain of rewrites:

$$\begin{aligned} \mathsf{trans}(p,\ \mathsf{trans}(\mathsf{symm}(p),\ p)) &\xleftarrow{\text{R8}^{-1}} \mathsf{trans}(\mathsf{trans}(p,\ \mathsf{symm}(p)),\ p) \\ &\xrightarrow{\text{R5}} \mathsf{trans}(\mathsf{refl}_a,\ p) \\ &\xrightarrow{\text{R3}} p \end{aligned}$$

The first step applies the inverse of the associativity rule (which is legitimate since we work with rewrite *equivalence*, the symmetric closure of the rewrite relation); the second applies the right inverse law; the third applies the left identity law. The three-step derivation is itself a *2-path*—a witness to the equivalence of two 1-paths (Chapter 7)—and it is precisely such higher witnesses that give rise to the ω-groupoid structure.

1.4 On the Two-Level Structure: Path vs. Eq

A distinctive and deliberate feature of our framework is its *two-level structure*. We maintain a careful distinction between two notions of equality:

(1) **Computational paths** (Path). These are *proof-relevant*: two different paths from a to b are genuinely different mathematical objects, even when they have the same endpoints. Paths carry the rewrite trace and form the substrate of the weak ω-groupoid structure.

(2) **Propositional equality** (Eq). This is *proof-irrelevant*: if $\mathsf{Eq}(a, b)$ holds, then a and b are "the same" for all practical purposes, and any two proofs of $\mathsf{Eq}(a, b)$ are themselves equal. The type Eq satisfies the *Uniqueness of Identity Proofs* (UIP) principle. Crucially, the Path type does *not* satisfy UIP—a fact proved in Chapter 12.

1.4.1 Why Two Levels?

The need for a two-level structure can be motivated from several angles.

From proof theory. In Gentzen's and Prawitz's tradition, a *proof* of a proposition A is a structured object—a derivation tree—and two derivations are equal when one can be transformed into the other by a sequence of normalisation steps. But the sequences of normalisation steps are themselves structured objects, and they can be compared. This naturally leads to a hierarchy:

Level 0:	terms	(proof-terms, inhabitants of types)
Level 1:	paths	(rewrite reasons between terms)
Level 2:	2-paths	(rewrite reasons between paths)
	⋮	

At each level, the objects are proof-relevant: they carry information about *how* an equality is witnessed. To obtain a proof-irrelevant notion of equality, one must quotient by the appropriate equivalence relation at that level.

From type theory.

Remark 1.11 (Equality as a criterion of use)**.** In his 2022 lecture on correctness of assertion and validity of inference, Martin-Löf stresses that equality is "the most important concept in all of mathematics" [113]. For our purposes, this is not a purely rhetorical claim: the distinction between the *fact* of equality and its *witness* is precisely what makes the path level mathematically productive.

A related Wittgensteinian formulation, in the same inferential spirit, is:

> Und man macht den Sinn von "$(x).fx$" klar, indem man darauf dringt, daß aus ihm "fa" folgt.

("And we make the meaning of '$(x).fx$' clear by insisting that 'fa' follows from it.") Read as a criterion of use, this supports the proof-theoretic stance adopted in this book [35, 37].

In Martin-Löf's intensional type theory [111, 109], the identity type $\mathsf{Id}_A(a, b)$ was introduced to express propositional equality; we derive the associated path induction principle (the J-eliminator) from computational paths in Chapter 11. Work by Hofmann and Streicher [86] showed that this type need not be proof-irrelevant: their groupoid model—which we revisit from our perspective in Section 6.1—demonstrated that there can be distinct terms of type $\mathsf{Id}_A(a, a)$, representing distinct "self-identifications" of a. The theory of computational paths makes this phenomenon completely explicit: distinct paths $p, q : \mathsf{Path}(a, a)$ arise from distinct sequences of rewrites.

From the Lean formalisation. In our formalisation, the Path type is defined as a structure in Type (the universe of data-carrying types), while the Eq type lives in Prop (the universe of proof-irrelevant propositions). The proof component of a path is a term of type $\mathsf{Eq}(a, b)$, but the path itself—with its trace—lives in Type. This arrangement is not an accident of the implementation; it reflects the fundamental distinction between the fact of equality and the manner of its witnessing.

1.4.2 Relationship to Other Two-Level Theories

Remark 1.12 (Comparison with HoTT identity types)**.** It is important to emphasise what our path type is *not*. The Path type of this book is *not* the identity type of homotopy type theory (HoTT) [150]; the precise relationship between our framework and HoTT is explored in Chapter 14. The HoTT identity type $\mathsf{Id}_A(a, b)$ is itself proof-relevant and satisfies the univalence axiom; in particular, UIP fails for Id. Our Eq type, by contrast, *does* satisfy UIP—it is the ordinary propositional equality of classical type theory, living in Prop. The proof-relevant structure in our framework comes from the trace component of Path, not from any failure of UIP at the level of Eq.

This design is intentional. By maintaining UIP at the Eq level, we obtain a strict equality that can serve as a foundation for equational reasoning, while the Path level provides the richer, proof-relevant structure. The resulting two-level architecture allows us to study the phenomena of higher-dimensional type theory *within* a setting that remains compatible with classical foundations.

Remark 1.13 (Comparison with two-level type theory)**.** Our two-level structure bears a family resemblance to Voevodsky's two-level type theory (2LTT) [157, 5], which also distinguishes between a "strict" and a "fibrant" layer. However, our approach arrives at this structure from the direction of proof theory and term rewriting rather than from model-theoretic considerations. In 2LTT, the strict level is used primarily as a metatheoretic tool; in our framework, the two levels both have direct proof-theoretic significance.

1.4.3 Weak Groupoid, Strict Groupoid, and the Quotient

The algebraic consequences of the two-level structure can be summarised as follows.

Definition 1.14 (Weak category of paths)**.** The *weak category of paths* over a type A has:

- *Objects*: terms $a, b, c, \ldots$ of type A.
- *Morphisms*: paths $p : \mathsf{Path}(a, b)$.

- *Composition*: $\mathsf{trans}(p, q) : \mathsf{Path}(a, c)$, for $p : \mathsf{Path}(a, b)$ and $q : \mathsf{Path}(b, c)$.
- *Identity*: $\mathsf{refl}_a : \mathsf{Path}(a, a)$.

The category laws—associativity, left identity, right identity—hold *up to rewrite equivalence* ($\sim_{\mathsf{rw}}$), not definitionally.

Definition 1.15 (Weak groupoid of paths). A weak category of paths is a *weak groupoid* when equipped with the inverse operation:

- *Inverse*: $\mathsf{symm}(p) : \mathsf{Path}(b, a)$ for $p : \mathsf{Path}(a, b)$.

The inverse laws hold up to rewrite equivalence (as established rigorously in Chapter 6):

$$\mathsf{trans}(p, \mathsf{symm}(p)) \sim_{\mathsf{rw}} \mathsf{refl}_a \qquad \text{and} \qquad \mathsf{trans}(\mathsf{symm}(p), p) \sim_{\mathsf{rw}} \mathsf{refl}_b.$$

Proposition 1.16 (Strict groupoid by quotient). *Let $\sim_{\mathsf{rw}}$ denote the rewrite equivalence on paths (the reflexive, symmetric, transitive closure of the one-step rewrite relation generated by the 77 rules). The quotient*

$$\mathcal{P}(a, b) \;=\; \mathsf{Path}(a, b) \,/\, \sim_{\mathsf{rw}}$$

forms a strict groupoid*: the groupoid laws hold definitionally (i.e., as equalities of equivalence classes, not merely up to further rewriting).*

This passage from weak to strict is typical of higher category theory (see Chapter 9 for the full construction and Chapter 15 for the categorical perspective). The weak structure is more fundamental and carries more information; the strict structure is obtained by quotienting and is easier to work with when the additional information is not needed.

1.5 Overview of Main Results

We now survey the principal results established in this book. Each theorem is stated here in a self-contained form; the full proofs are given in the chapters indicated.

1.5.1 The Rewrite System

Theorem 1.17 (Confluence & termination; Chapters 4–5). *The 77-rule rewrite system on computational paths is:*

(a) Terminating*: every sequence of rewrites on paths is finite.*

(b) Confluent: *if a path p can be rewritten to both q_1 and q_2, then there exists a path r to which both q_1 and q_2 can be rewritten.*

Consequently, every path has a unique normal form, and rewrite equivalence is decidable: $p \sim_{\mathsf{rw}} q$ *if and only if* $\mathsf{nf}(p) = \mathsf{nf}(q)$.

The decidability of path equivalence is essential for the computational content of the theory. It means that the question "do these two sequences of rewrites amount to the same thing?" can be answered algorithmically, by reducing both sequences to their normal forms and comparing.

Theorem 1.18 (Classification of rewrite rules; Chapter 4). *At the conceptual level, the 77 rewrite rules fall into four natural families. Chapter 4 reorganizes the same system into eight operational categories for the detailed rule-by-rule presentation:*

Category	*Count*	*Role*
Structural	*18*	*Groupoid laws (identity, inverse, associativity)*
Functorial	*24*	*Interaction of* congrArg *with path operations*
Congruence	*20*	*Interchange, whiskering, 2-categorical coherence*
Transport	*15*	*Extension to dependent types*

1.5.2 The Higher Groupoid Structure

Theorem 1.19 (The ω-Groupoid Theorem; Chapters 6–9). *The computational paths over any type A, together with the higher paths between paths (2-cells), paths between 2-cells (3-cells), and so on, form a weak ω-groupoid in the sense of Batanin [11] and Leinster [99]. The non-trivial coherence data extend through dimension 3, and the tower stabilises from dimension 4 onward.*

This is the central structural result of the book. It establishes that the algebraic structure arising from the rewrite system on proof-terms is precisely the structure that higher category theory identifies as a weak ω-groupoid. The same structure was shown by Lumsdaine [105] and van den Berg–Garner [153] to be present in the identity types of Martin-Löf type theory; our result shows that it arises independently from term rewriting.

This hierarchy of higher cells is intimately connected to a fundamental observation made by Voevodsky [158]: the notion of equality itself must be stratified. As Voevodsky noted, equality is an appropriate notion for *elements*—individuals—but fails for collections; isomorphism is appropriate for collections but fails for collections of collections. This leads to a theory of *iterated n-equivalences* which are the correct replacements at each level. The ω-groupoid structure that emerges from our computational paths construction provides a concrete, syntactic realisation of precisely this hierarchy: 1-cells capture equality of terms, 2-cells capture equivalence of proofs, 3-cells capture coherence of equivalences, and so on through all finite levels.

The construction proceeds level by level:

(1) *1-cells*: paths $p : \mathsf{Path}(a, b)$ between terms.

(2) *2-cells*: rewrite derivations $\alpha : p \sim_{\mathsf{rw}} q$ between paths. These are constructed from the 77 rewrite rules.

(3) *3-cells*: derivations between derivations, constructed from the coherence conditions on the rewrite rules.

(4) *n-cells* for all n: the construction iterates, with each level governed by coherence conditions inherited from the previous level.

Theorem 1.20 (Eckmann–Hilton; Chapter 10). *In the double loop space*

$$\Omega^2(A, a) = \mathsf{Path}_2(\mathsf{refl}_a, \mathsf{refl}_a),$$

horizontal and vertical composition of 2-cells coincide and are commutative. This is witnessed by an explicit 3-cell constructed from the rewrite rules.

The Eckmann–Hilton theorem [64] is a cornerstone of higher algebra: in any 2-category (or bicategory), the endomorphisms of the identity 1-cell form a commutative monoid. Our proof provides explicit computational content for this classical result, constructing the witnessing 3-cell as a specific sequence of rewrites.

1.5.3 Connections to Type Theory and Topology

Theorem 1.21 (Derivation of the J-eliminator; Chapter 11). *The path induction principle (Martin-Löf's J-eliminator) is derivable from the structure of computational paths: the based path space $\sum_{b:A} \mathsf{Path}(a, b)$ is contractible.*

This theorem connects our framework to standard type-theoretic practice. The J-eliminator [111] is the fundamental reasoning principle for identity types, and the fact that it is *derivable*—rather than postulated as an axiom—demonstrates the expressive power of the computational paths approach.

Theorem 1.22 (Failure of UIP; Chapter 12). *The* Path *type does not satisfy the* Uniqueness of Identity Proofs *principle. That is, there exist types A, terms $a, b : A$, and paths $p, q : \mathsf{Path}(a, b)$ such that $p \neq q$.*

The failure of UIP for Path is not a deficiency but a feature: it is precisely what allows the higher-dimensional structure to exist. If all paths between the same endpoints were equal, the entire tower of higher cells would collapse, and we would have a mere set rather than a groupoid.

Theorem 1.23 (Fundamental group computations; Chapter 13)**.** *Using computational paths over combinatorially presented spaces, one can compute:*

$$\pi_1(S^1) \cong \mathbb{Z}$$
$$\pi_1(T^2) \cong \mathbb{Z} \times \mathbb{Z}$$
$$\pi_1(\mathbb{RP}^2) \cong \mathbb{Z}/2\mathbb{Z}$$

and verify the Seifert–van Kampen theorem [138, 154, 15] for combinatorial spaces.

These computations demonstrate that computational paths are not merely a proof-theoretic curiosity but a tool with genuine mathematical content, capable of reproducing classical results in algebraic topology [79].

1.5.4 The Lean Formalization

The central structural results stated above, together with a substantial core of the surrounding development, have been formalised in the Lean 4 proof assistant [28]. The formalisation comprises over 10,000 theorems and lemmas, all machine-checked. The code is available at:

`https://github.com/Arthur742Ramos/ComputationalPathsLean`

The formalisation is not merely a post hoc verification of results obtained on paper. In several cases, the process of formalisation led to the discovery of new results, the correction of errors in earlier published proofs, and a deeper understanding of the combinatorial structure of the rewrite system. We comment on the formalisation at relevant points throughout the text; Appendix C provides a detailed guide to the code. The appendix focuses on the mathematical architecture of the formalisation; detailed build and repository-navigation instructions are left to the companion repository itself.

1.6 Reading Guide and Prerequisites

This book is organised in three parts. The main line of argument runs through Parts I and II, where the rewrite-theoretic and higher-structural core is developed. Part III is best read as a collection of interfaces and extensions built on that core. With that in mind, we suggest several reading paths according to the reader's background and interests.

Structure of the Book

Part I: Foundations — The Rewrite System. Chapters 1–5 develop the core theory. Chapter 2 presents the labelled natural deduction framework. Chapter 3 introduces computational paths formally and establishes their basic properties. Chapter 4 gives the complete 77-rule rewrite system. Chapter 5 proves confluence and termination.

Part II: Higher Structure — The ω-Groupoid. Chapters 6–10 develop the higher-dimensional structure. Chapter 6 establishes the weak groupoid structure. Chapters 7–8 construct the 2-cells and 3-cells explicitly. Chapter 9 assembles the weak ω-groupoid and its stabilisation above dimension 3. Chapter 10 proves the Eckmann–Hilton theorem.

Part III: Connections and Applications. Chapters 11–16 connect the theory to neighbouring frameworks and applications. Chapter 11 derives path induction. Chapter 12 analyses the failure of UIP. Chapter 13 computes fundamental groupoids of combinatorial spaces. Chapter 14 relates our work to homotopy type theory. Chapter 15 develops the categorical perspective. Chapter 16 discusses open problems and future directions.

Suggested Reading Paths

For type theorists and proof assistant users. Focus on Parts I and II for the foundations and the ω-groupoid structure, then read Chapter 11 on path induction and Chapter 12 on the failure of UIP. The categorical material in Part III can be skimmed on a first reading.

For algebraic topologists and homotopy theorists. Begin with this chapter's overview, then proceed to Chapter 13 (fundamental groupoids) and Chapter 14 (connections to HoTT) to see how the framework interfaces with familiar homotopical questions. Afterwards, return to Parts I and II to understand the machinery that produces these results.

For term rewriting specialists. The core technical content is in Chapters 4 and 5 (the rewrite system and the confluence proof). The categorical interpretation in Part II shows how the rewrite system gives rise to higher algebraic structure, which may suggest new questions in rewriting theory.

Regarding formalisation. Appendix C describes the Lean 4 formalization in detail. Readers primarily interested in the mathematics can postpone it until after

Parts I and II; readers interested in the implementation can use it as a companion reference, since the correspondence between the two is deliberately close.

Prerequisites

We assume the reader is familiar with:

- Basic logic and set theory.
- Elementary λ-calculus [10, 21, 83], both untyped and simply typed.
- Natural deduction in the Gentzen–Prawitz style [72, 121], including the notions of introduction and elimination rules, normalisation, and the subformula property.

The following topics are helpful but not strictly required; we introduce the necessary background as it arises:

- Term rewriting systems [8, 148] (summarised in Appendix A).
- Elementary category theory [98], in particular the notions of functor, natural transformation, and groupoid (see also Appendix B).
- Dependent type theory [111, 117] and Martin-Löf's identity type (introduced in Chapter 14).
- Algebraic topology, specifically fundamental groups and covering spaces (Chapter 13 is largely self-contained).

Notational Conventions

Throughout the book, we use the following conventions:

- A, B, C range over types; a, b, c range over terms; f, g, h range over functions.
- p, q, r, s range over 1-paths; α, β, γ range over 2-paths; Γ, Δ (and occasionally Φ, Ψ) range over 3-paths.
- $\triangleright$ denotes one-step rewriting on paths; $\triangleright^*$ its reflexive-transitive closure; $\sim_{\mathsf{rw}}$ the symmetric closure of $\triangleright^*$ (rewrite equivalence).
- $\equiv$ denotes definitional equality (syntactic).
- $=$ denotes propositional equality (Eq).

- $\mathsf{Path}(a, b)$ is the type of computational paths from a to b; $\mathsf{Eq}(a, b)$ (or simply $a = b$) is propositional equality.

We now turn to the technical development, beginning in Chapter 2 with the framework of labelled natural deduction that underlies the entire theory.

Chapter 2

Labelled Natural Deduction and Proof-Terms

> *"Logicians should note that a deductive system is concerned not just with unlabelled entailments or sequents $A \to B$ (as in Gentzen's proof theory), but with deductions or proofs of such entailments. In writing $f : A \to B$ we think of f as the 'reason' why A entails B."*
>
> —Lambek & Scott [97]

The framework of *labelled natural deduction* (LND) is the setting in which the theory of computational paths first becomes visible as a general proof-theoretic programme. The point of LND is to make explicit the functional content of deduction. A label does not merely accompany a formula; it records the reason why that formula follows from the assumptions. In that sense, labelled natural deduction adds to ordinary natural deduction an extra dimension—the *label*—through which one can track not merely *that* a proposition follows, but *how* and *why* it follows.

The present chapter develops the LND system for propositional logic and fixes the notational and proof-theoretic machinery used throughout the book. What is at stake is not a mere notational embellishment, nor a syntactical ornament, but a shift in perspective: proofs are treated from the start as structured objects whose internal organisation can be carried along by the deduction itself.

2.1 The LND Framework

2.1.1 Context and Judgement Forms

The central idea of labelled natural deduction is to make explicit the *functional content* of logical deductions. Every formula in a deduction carries a *label*—a term that records the computational structure of the proof. Where traditional natural deduction writes simply "A" for an assumption or conclusion, LND writes "$t : A$", making the witnessing proof-term t an explicit part of the formal system.

Definition 2.1 (Context). A *context* Γ is a finite sequence of declarations of the form $x : A$, where x is a variable and A is a formula. We write:

$$\Gamma = x_1 : A_1, x_2 : A_2, \ldots, x_n : A_n$$

The variables $x_1, \ldots, x_n$ must be pairwise distinct. We write $\cdot$ or simply leave blank for the empty context. When $x : A$ appears in Γ, we say that x is *declared* in Γ with type A.

The basic judgement form of LND is:

$$\Gamma \vdash t : A$$

which asserts that the term t is a proof of formula A under the assumptions recorded in the context Γ. This judgement combines two levels of information:

1. **Logical level:** The formula A is derivable from the assumptions $A_1, \ldots, A_n$.
2. **Computational level:** The term t witnesses this derivability, recording the exact sequence of inference rules used.

Definition 2.2 (Judgement forms). The LND system uses three basic judgement forms:

1. $\Gamma \vdash t : A$ ("t proves A under assumptions Γ")
2. $\Gamma \vdash t = u : A$ ("t and u are equal proofs of A")
3. $\Gamma \vdash t =_s u : A$ ("t and u are equal via reason s")

The third form is the distinctive contribution of LND to proof theory: the subscript s records the *rewrite reason* that witnesses the equality. This is the seed from which computational paths grow.

In the present chapter we concentrate on the first judgement form and on the β-reduction fragment that generates concrete proof transformations. The richer equality judgements $\Gamma \vdash t = u : A$ and $\Gamma \vdash t =_s u : A$ become central in Chapters 3 and 4, where reasons for equality are promoted to first-class path data.

The third judgement form deserves special attention. In standard type theory or proof theory, when we write $t = u$ (or $t =_\beta u$, or $t =_{\beta\eta} u$), we assert merely that an equality holds. The LND approach refines this by making the *reason* for equality—the sequence of reduction steps, the chain of rewrite rules applied—into a first-class mathematical object. Just as we moved from "A is provable" to "t proves A", we now move from "t equals u" to "s witnesses that t equals u".

Remark 2.3 (The two-dimensional structure)**.** The LND framework introduces what we might call a *two-dimensional* logical system:

- The *horizontal dimension* consists of the formulas and the logical calculus of connectives and quantifiers.
- The *vertical dimension* consists of the labels (proof-terms) and the functional calculus of abstraction, application, pairing, etc.

A well-formed judgement requires harmony between these two dimensions: the term must be well-typed according to the formula. This two-dimensional perspective—*Grundgesetze* (terms) alongside *Begriffsschrift* (formulas), to use Fregean terminology—is the architectural principle underlying our approach.

2.1.2 Historical Development

The idea of labelling formulas with proof-terms has a rich history that predates the explicit formulation of the Curry–Howard correspondence.

Curry's theory of functionality (1934). The origins of the formulae-as-types interpretation can be traced to Curry's investigations on combinatory logic [25]. Curry observed that the type-schemes of the basic combinators correspond precisely to axiom schemes of implicational logic:

$$
\begin{aligned}
&\mathsf{I} : A \to A && \text{(identity/reflexivity)}\\
&\mathsf{K} : A \to (B \to A) && \text{(weakening)}\\
&\mathsf{S} : (A \to (B \to C)) \to \\
&\qquad ((A \to B) \to (A \to C)) && \text{(distribution)}\\
&\mathsf{B} : (B \to C) \to ((A \to B) \to (A \to C)) && \text{(transitivity)}\\
&\mathsf{C} : (A \to (B \to C)) \to (B \to (A \to C)) && \text{(permutation)}
\end{aligned}
$$

This was not merely a notational coincidence but a deep structural correspondence. As Curry recognised, the functional character of a combinator—how it transforms its inputs—determines its logical content.

Howard's extension (1969/1980). Howard [87] extended the correspondence to full intuitionistic predicate logic, establishing the now-famous *propositions-as-types* and *proofs-as-programs* identifications:

Logic		**Type Theory**
Proposition A	$\longleftrightarrow$	Type A
Proof of A	$\longleftrightarrow$	Term $t : A$
Implication $A \to B$	$\longleftrightarrow$	Function type $A \to B$
Conjunction $A \wedge B$	$\longleftrightarrow$	Product type $A \times B$
Disjunction $A \vee B$	$\longleftrightarrow$	Sum type $A + B$
Proof simplification	$\longleftrightarrow$	Computation (β-reduction)

Howard's work made precise the slogan that "proofs are programs" and "propositions are specifications."

Tait's intensional interpretation (1965–1967). While Curry and Howard focused on the proof-term correspondence, Tait [146, 147] emphasised a different point: the role of *conversion* (normalisation) as the key semantic notion. In Tait's intensional interpretation, a type is determined not by its canonical elements but by its *rules of computation*:

> "Instead of assuming functionals of higher type, we may regard the definitional schemata for the primitive recursive functionals simply as rules of computation, i.e. for transforming symbols. On this view, Gödel's result may be interpreted as a consistency proof relative to the quantifier-free theory of primitive recursive functionals." — Tait [146]

This shift from *assertability conditions* (what counts as a proof) to *convertibility* (how proofs compute) is crucial for our development. It suggests that the interpretation need not be restricted to intuitionistic logic but can encompass any system with well-behaved computation rules. This view of meaning as determined by the rules of use ("grammar") together with their consequences was further developed by de Queiroz [35, 38], drawing on Wittgensteinian themes [33, 37] (see also Dummett [63] for the philosophical basis of intuitionistic logic and Troelstra–van Dalen [151] for a comprehensive treatment of constructivism in mathematics).

Gabbay's Labelled Deductive Systems (1989–). The framework of *Labelled Deductive Systems* (LDS), developed by Gabbay [70], provides the immediate context for our work. The key insight of LDS is that the declarative unit of a logical system should not be a bare formula A but a *labelled formula* $t : A$. The label t carries meta-level information about the formula:

- In modal logic, t might be a possible-world index.
- In substructural logics, t records which assumptions were used.
- In the Curry–Howard interpretation, t is the proof-term.

From this perspective, a logic is defined by the interplay between a *functional calculus on labels* and a *logical calculus on formulas*. The allowable logical moves are "controlled" by appropriate constraints on the labels. The functional interpretation was extended to the existential quantifier in [48] and to modal necessity in [49].

De Queiroz's rewrite reasons (1988–). The specific innovation that leads to computational paths is de Queiroz's introduction of the judgement form $t =_s u : A$, where s is a term denoting a sequence of rewrites [31, 32, 44, 34]. This makes the "reason for equality" into a first-class object that can itself be manipulated, compared, and reasoned about. The present work develops this idea to its logical conclusion: if reasons for equality are mathematical objects, what are the reasons for equality *between* reasons? The answer leads us to computational paths and their higher structure.

2.2 Proof-Terms for Propositional Logic

We now present the simply-typed λ-calculus, which corresponds to intuitionistic propositional logic under the Curry–Howard correspondence. This system will serve as our base calculus; computational paths will arise from studying the equalities between its terms.

2.2.1 Types and Formulas

Definition 2.4 (Types/Formulas). The set of *types* (equivalently, *formulas*) is generated by the following grammar:

$$A, B ::= \alpha \mid A \to B \mid A \times B \mid A + B \mid \mathbf{0} \mid \mathbf{1}$$

where:

- $\alpha, \beta, \gamma, \ldots$ range over a countably infinite set of *atomic types* (or *propositional variables*).
- $A \to B$ is the *function type* (logically: implication).
- $A \times B$ is the *product type* (logically: conjunction).

- $A + B$ is the *sum type* (logically: disjunction).
- $\mathbf{0}$ is the *empty type* (logically: falsity).
- $\mathbf{1}$ is the *unit type* (logically: truth).

Remark 2.5 (Precedence conventions)**.** We adopt the standard conventions: $\to$ associates to the right, so $A \to B \to C$ means $A \to (B \to C)$. The binary connectives $\times$ and $+$ bind more tightly than $\to$.

The type constructors correspond to the logical connectives of intuitionistic propositional logic. This correspondence is not merely syntactic: as we shall see, the computational behaviour of terms of each type mirrors the proof-theoretic properties of the corresponding connective.

2.2.2 Terms and Proof-Terms

Definition 2.6 (Terms)**.** The set of *terms* (equivalently, *proof-terms* or *λ-terms*) is generated by the following grammar:

$$
\begin{array}{rll}
t, u ::= & x & \text{(variable)} \\
| & \lambda x.\, t & \text{(abstraction)} \\
| & t\, u & \text{(application)} \\
| & \langle t, u \rangle & \text{(pair)} \\
| & \pi_1(t) \mid \pi_2(t) & \text{(projections)} \\
| & \mathsf{inl}(t) \mid \mathsf{inr}(t) & \text{(injections)} \\
| & \mathsf{case}(t, x.u_1, y.u_2) & \text{(case analysis)} \\
| & \mathsf{abort}(t) & \text{(ex falso)} \\
| & \star & \text{(unit element)}
\end{array}
$$

Remark 2.7 (Terminology: term vs. proof-term)**.** From this point onward we use *proof-term* as the primary name for the expressions generated by the grammar of Definition 2.6, since it emphasises the Curry–Howard reading (a term *is* a proof). We write simply "term" when no ambiguity arises. The two words are interchangeable throughout the book.

The term constructors fall into two categories that mirror the logical structure:

- **Introduction forms:** $\lambda x.\, t$ (for $\to$), $\langle t, u \rangle$ (for $\times$), $\mathsf{inl}(t)$ and $\mathsf{inr}(t)$ (for $+$), $\star$ (for $\mathbf{1}$). These correspond to methods of *constructing* a proof of a compound proposition.

- **Elimination forms:** $t\,u$ (for $\to$), $\pi_1(t)$ and $\pi_2(t)$ (for $\times$). For sums we use $\mathsf{case}(t, x.u_1, y.u_2)$, and for **0** we use $\mathsf{abort}(t)$. These correspond to methods of *using* a proof of a compound proposition.

Definition 2.8 (Free and bound variables)**.** The variable x is *bound* by the abstraction $\lambda x.\,t$ and by the bindings in $\mathsf{case}(t, x.u_1, y.u_2)$. A variable that is not bound is *free*. We write $\mathsf{FV}(t)$ for the set of free variables of t. A term with no free variables is *closed*.

Definition 2.9 (Substitution)**.** We write $t[u/x]$ for the result of substituting u for all free occurrences of x in t, with the usual capture-avoiding convention: bound variables are renamed as necessary to avoid capture of free variables in u.

2.2.3 Typing Rules

The typing judgement $\Gamma \vdash t : A$ is defined by the following inference rules. Each rule has two faces: read top-down it is a *construction principle*, telling us how to build new proof-terms from old; read bottom-up it is a *reasoning principle*, telling us which logical moves are legitimate. The introduction rules create proofs of compound propositions, while the elimination rules decompose them—and the resulting β-reduction rules (Section 2.12) arise precisely from the interaction between the two.

Definition 2.10 (Typing rules)**. Variables and structural rules.**

$$\frac{}{\Gamma, x : A, \Delta \vdash x : A}\ [\textsc{Var}]$$

A variable is typed by looking it up in the context. This corresponds to the logical principle that an assumption may be used as a premise.

Implication ($\to$)**.**

$$\frac{\Gamma, x : A \vdash t : B}{\Gamma \vdash \lambda x.\,t : A \to B}\ [\to\text{-I}] \qquad \frac{\Gamma \vdash t : A \to B \quad \Gamma \vdash u : A}{\Gamma \vdash t\,u : B}\ [\to\text{-E}]$$

The introduction rule ($\to$-I) corresponds to the logical principle of conditional proof: to prove $A \to B$, assume A and derive B. The λ-abstraction records this assumption discharge.

The elimination rule ($\to$-E) corresponds to *modus ponens*: from $A \to B$ and A, conclude B. Application of a function to an argument realises this inference.

Conjunction ($\times$)**.**

$$\frac{\Gamma \vdash t : A \quad \Gamma \vdash u : B}{\Gamma \vdash \langle t, u \rangle : A \times B}\ [\times\text{-I}]$$

$$\frac{\Gamma \vdash t : A \times B}{\Gamma \vdash \pi_1(t) : A}\ [\times\text{-E}_1] \qquad \frac{\Gamma \vdash t : A \times B}{\Gamma \vdash \pi_2(t) : B}\ [\times\text{-E}_2]$$

A conjunction $A \times B$ is proved by proving both conjuncts. A conjunction is used by extracting either conjunct.

Disjunction $(+)$**.**

$$\frac{\Gamma \vdash t : A}{\Gamma \vdash \mathsf{inl}(t) : A + B}\ [+\text{-I}_1] \qquad \frac{\Gamma \vdash t : B}{\Gamma \vdash \mathsf{inr}(t) : A + B}\ [+\text{-I}_2]$$

$$\frac{\Gamma \vdash t : A + B \quad \Gamma, x : A \vdash u_1 : C \quad \Gamma, y : B \vdash u_2 : C}{\Gamma \vdash \mathsf{case}(t, x.u_1, y.u_2) : C}\ [+\text{-E}]$$

A disjunction $A + B$ is proved by proving either disjunct. A disjunction is used by case analysis: to derive C from $A + B$, show that C follows from A and that C follows from B.

Falsity $(\mathbf{0})$ **and Truth** $(\mathbf{1})$**.**

$$\frac{\Gamma \vdash t : \mathbf{0}}{\Gamma \vdash \mathsf{abort}(t) : A}\ [\mathbf{0}\text{-E}] \qquad \frac{}{\Gamma \vdash \star : \mathbf{1}}\ [\mathbf{1}\text{-I}]$$

From falsity, anything follows (*ex falso quodlibet*). Truth is trivially provable.

Remark 2.11 (Harmony between introduction and elimination)**.** The introduction and elimination rules for each connective stand in a precise relationship that Prawitz [121] called *harmony*. The elimination rules specify what can be derived from a proof of a compound formula, and this is exactly what the introduction rules put into the proof. This harmony is the source of the computational content: when an introduction is immediately followed by an elimination of the same connective, a *redex* is formed that can be simplified.

2.2.4 The Abstraction Discipline and Substructural Logics

A crucial observation, emphasized in Gabbay's LDS framework, is that the λ-abstraction rule admits variation. When forming $\lambda x.\, t$ from t, several possibilities arise depending on how x occurs in t:

(a) **Vacuous abstraction:** $x \notin \mathsf{FV}(t)$. The assumption was not used at all.

(b) **Linear abstraction:** x occurs exactly once in t. The assumption was used exactly once.

(c) **Affine abstraction:** x occurs at most once in t. The assumption was used at most once.

(d) **Relevant abstraction:** x occurs at least once in t. The assumption was actually used.

(e) **Unrestricted abstraction:** x may occur any number of times. This is standard intuitionistic logic.

By restricting which abstractions are permitted, one obtains substructural logics:

Restriction on λ-abstraction	**Resulting logic**
Exactly one use (no vacuous, no contraction)	Linear logic
At least one use (no vacuous abstraction)	Relevant logic
At most one use (no contraction)	Affine logic
No restriction	Intuitionistic logic

This observation is significant because it shows that the computational interpretation is not limited to intuitionistic logic. By varying the abstraction discipline, the same framework accommodates resource-sensitive logics. The common thread is that *modus ponens* (application) remains unchanged—all implications share this elimination rule—while the introduction rules differ in their assumptions about resource use.

2.3 β-Reduction and η-Expansion

The computational content of proofs is expressed through *reduction rules*. These rules describe how a proof that first constructs and then destructs a compound formula can be simplified—a "detour" through an unnecessary complexity can be eliminated. Each β-step will become an atomic step in a computational path (Definition 3.1 in Chapter 3), and chains of such steps give rise to the path algebra developed in Chapters 3–4.

2.3.1 The β-Reduction Rules

Definition 2.12 (β-reduction rules). The *β-reduction rules* are:

$$(\lambda x.\, t)\, u \to_\beta t[u/x] \qquad (\beta_\to)$$

$$\pi_1(\langle t, u\rangle) \to_\beta t \qquad (\beta_{\times,1})$$

$$\pi_2(\langle t, u\rangle) \to_\beta u \qquad (\beta_{\times,2})$$

$$\mathsf{case}(\mathsf{inl}(t), x.u_1, y.u_2) \to_\beta u_1[t/x] \qquad (\beta_{+,1})$$

$$\mathsf{case}(\mathsf{inr}(t), x.u_1, y.u_2) \to_\beta u_2[t/y] \qquad (\beta_{+,2})$$

where $t[u/x]$ denotes capture-avoiding substitution of u for x in t.

Each β-rule captures a specific computational intuition:

- $(\beta_{\rightarrow})$: A function $\lambda x.\, t$ applied to an argument u computes by substituting u for the parameter x throughout the body t. This is the fundamental rule of functional computation.
- $(\beta_{\times,1})$ and $(\beta_{\times,2})$: Projecting from a pair extracts the corresponding component. The pair $\langle t, u\rangle$ "remembers" both components; projection selects one.
- $(\beta_{+,1})$ and $(\beta_{+,2})$: Case analysis on a tagged value selects the appropriate branch and substitutes the payload. If the scrutinee is $\mathsf{inl}(t)$, we take the first branch with t substituted for x; if $\mathsf{inr}(t)$, we take the second branch.

Remark 2.13 (Prawitz's reduction rules)**.** The β-rules correspond exactly to Prawitz's [121] *reduction rules* for natural deduction proofs. Prawitz showed that when an introduction rule is immediately followed by an elimination rule for the same connective, the resulting "detour" can be eliminated. Our β-rules are the term-level realisation of this proof transformation. For example, the rule $(\beta_{\rightarrow})$ corresponds to the elimination of an implication detour:

$$\text{From} \quad \dfrac{\dfrac{\vdots}{A \rightarrow B}\,[\rightarrow\text{-I}] \quad A}{B}\,[\rightarrow\text{-E}] \quad \text{reduce to} \quad \dfrac{\vdots}{B}$$

where the derivation of A is substituted for uses of the assumption A.

2.3.2 Dialogical Reading of the β-Rules

The same β-reductions admit a dialogical reading: introduction rules play the role of a *Defender*'s assertion, elimination rules play the role of an *Attacker*'s challenge, and the reductum gives the continuation (or defense) of the interaction. This is the proof-theoretic counterpart of game-theoretic semantics in the tradition of Lorenzen/Hintikka and of Peirce's Utterer/Interpreter contrast.

Conn.	Defender (intro)	Attacker (elim)	Defense / continuation
$\wedge$ / $\times$	assert $\langle t, u\rangle : A \times B$	ask for π_1 or π_2	return the requested component (t or u), i.e. $\beta_{\times,1/2}$
$\vee$ / $+$	assert $\mathsf{inl}(t)$ or $\mathsf{inr}(u)$	demand case analysis	continue in the chosen branch with substitution ($\beta_{+,1/2}$)
$\rightarrow$	assert $\lambda x.\, t : A \rightarrow B$	provide an input $u : A$	compute $t[u/x]$ ($\beta_{\rightarrow}$)
$\forall$	assert a uniform rule $x \mapsto g(x)$	choose an instance a	return the instance $g(a)$ (first-order extension)
$\exists$	assert witness pair (a, p) with $p : P(a)$	impose a local "suppose x is the witness" challenge	answer by instantiating the continuation (first-order extension)

Read this way, β-reduction makes explicit the "meaning as use" principle: the meaning of each connective is fixed by the admissible challenges and by how those challenges are answered in canonical interactions [35, 38, 37, 39]. The corresponding identity-type case is discussed in Chapter 11 (Remark 11.30).

2.3.3 The η-Expansion Rules

While β-rules express that "introduction followed by elimination" can be simplified, η-rules express a dual principle: a term is equivalent to its "canonical expansion."

Definition 2.14 (η-expansion rules). The *η-expansion rules* (or *η-laws*) are:

$$t =_\eta \lambda x.\, t\, x \quad \text{where } x \notin \mathsf{FV}(t) \tag{$\eta_\rightarrow$}$$

$$t =_\eta \langle \pi_1(t), \pi_2(t)\rangle \tag{$\eta_\times$}$$

$$t =_\eta \mathsf{case}(t, x.\mathsf{inl}(x), y.\mathsf{inr}(y)) \tag{η_+}$$

$$t =_\eta \star \quad \text{when } \Gamma \vdash t : \mathbf{1} \tag{η_1}$$

The η-laws express *extensionality* principles:

- ($\eta_\rightarrow$): Two functions are equal if they give the same result on all inputs. The term t of type $A \rightarrow B$ is equivalent to the η-expansion $\lambda x.\, t\, x$, which makes the functional nature explicit.
- ($\eta_\times$): Two pairs are equal if their components are equal. Any term t of product type is equivalent to the pair of its projections.
- (η_+): Any term of sum type is equivalent to case analysis that reconstructs the tagging. This is less commonly used but maintains the symmetry.

- (η_1): All terms of unit type are equal to $\star$. There is only one element of type $\mathbf{1}$.

Remark 2.15 (Directionality). While β-rules are naturally oriented as *reductions* (from more complex to simpler), η-laws are typically treated as bidirectional *equalities*. In some contexts, η is used as an expansion (left-to-right); in others, as a contraction (right-to-left). The choice affects confluence properties and the structure of normal forms. For our purposes, we shall primarily be concerned with the equational theory generated by $\beta\eta$, though the directed rewriting theory is also important.

Definition 2.16 ($\beta\eta$-equivalence). The relation $=_{\beta\eta}$ is the smallest equivalence relation on well-typed terms containing all instances of the β-rules and η-laws and closed under the congruence rules:

- If $t =_{\beta\eta} t'$, then $\lambda x.\, t =_{\beta\eta} \lambda x.\, t'$.
- If $t =_{\beta\eta} t'$ and $u =_{\beta\eta} u'$, then $t\, u =_{\beta\eta} t'\, u'$.
- Similarly for pair formation, projections, injections, case analysis.

The relation $=_{\beta\eta}$ captures when two proof-terms should be considered "the same proof." However, this is precisely where the theory of computational paths begins: we shall argue that the *way* two terms are shown equal—the specific sequence of β and η steps—is itself significant.

2.4 Rewrite Steps and Subject Reduction

We now formalise what we mean by a *rewrite step* and establish the fundamental property that reduction preserves types.

2.4.1 Rewrite Steps and Positions

Definition 2.17 (Position). A *position* in a term is a finite sequence of positive integers specifying a path from the root to a subterm. We write $t|_p$ for the subterm of t at position p, and $t[u]_p$ for the result of replacing the subterm at position p with u.

Definition 2.18 (Rewrite step). A *rewrite step* is, in this chapter, an instance of a β-rule applied at some position within a term. Formally, if $\ell \to_\beta r$ is a β-rule instance and p is a position in t such that $t|_p = \ell$, then:

$$t \to_\beta t[r]_p$$

is a rewrite step. We write $t \to_{\beta}^{(p)} u$ when the step occurs at position p. Later chapters enlarge the palette of atomic path steps to include symmetry, congruence, transport, and higher structural witnesses.

Example 2.19. Consider the term $t \equiv (\lambda f.\, f\,(\pi_1(\langle a, b\rangle)))\, g$, where a, b, g are variables or closed terms of appropriate types. There are two positions where β-reductions can occur:

1. At the root (position ϵ), by $(\beta_{\to})$:

$$t \to_{\beta} g\,(\pi_1(\langle a, b\rangle))$$

2. At position 1.1 (inside the body of the λ), by $(\beta_{\times,1})$:

$$t \to_{\beta} (\lambda f.\, f\, a)\, g$$

These are two different one-step reductions from the same term.

2.4.2 Reduction Sequences as Rewrite Reasons

Definition 2.20 (Reduction sequence)**.** A *reduction sequence* (or *rewrite reason*) from t to u is a finite sequence of rewrite steps:

$$t = t_0 \to_{\beta} t_1 \to_{\beta} \cdots \to_{\beta} t_n = u$$

We denote such a sequence by $s : t \rightsquigarrow u$ and call n the *length* of s. When $n = 0$, we have the *empty sequence* (or *identity*) at t.

The crucial observation that motivates the theory of computational paths is that reduction sequences can be recorded as structured terms, not merely as the assertion that a reduction exists. The judgement form $t =_s u : A$ makes this recording explicit.

Definition 2.21 (Multi-step reduction)**.** We write $t \to_{\beta}^{*} u$ ("t reduces to u in zero or more steps") if there exists a reduction sequence from t to u. We write $t \to_{\beta}^{+} u$ ("t reduces to u in one or more steps") if there exists a reduction sequence of length at least 1.

2.4.3 Subject Reduction

The most fundamental property of typed reduction is that types are preserved: reducing a well-typed term yields another well-typed term of the same type.

Theorem 2.22 (Subject reduction)**.** *If $\Gamma \vdash t : A$ and $t \to_{\beta} u$, then $\Gamma \vdash u : A$.*

Proof. By induction on the derivation of $\Gamma \vdash t : A$, with a case analysis on where the reduction step occurs. We consider the key cases.

Case $(\beta_{\rightarrow})$: Suppose the reduction is $(\lambda x.\, t')\, u' \rightarrow_{\beta} t'[u'/x]$ occurring at the root. From the typing derivation $\Gamma \vdash (\lambda x.\, t')\, u' : A$, we have by inversion:

- $\Gamma \vdash \lambda x.\, t' : B \rightarrow A$ for some B
- $\Gamma \vdash u' : B$

From the first, by inversion of ($\rightarrow$ -I):

- $\Gamma, x : B \vdash t' : A$

By the substitution lemma (which states that substituting a well-typed term for a variable preserves typing), we conclude:

$$\Gamma \vdash t'[u'/x] : A$$

Case $(\beta_{\times,1})$: Suppose the reduction is $\pi_1(\langle t', u' \rangle) \rightarrow_{\beta} t'$ occurring at the root. From $\Gamma \vdash \pi_1(\langle t', u' \rangle) : A$, by inversion of ($\times$-E$_1$):

- $\Gamma \vdash \langle t', u' \rangle : A \times B$ for some B

By inversion of ($\times$-I):

- $\Gamma \vdash t' : A$

which is the desired conclusion.

Case $(\beta_{+,1})$: Suppose the reduction

$$\mathsf{case}(\mathsf{inl}(t'), x.u_1, y.u_2) \rightarrow_{\beta} u_1[t'/x]$$

occurs at the root. From the typing of the redex, by inversions:

- $\Gamma \vdash \mathsf{inl}(t') : B + C$ for some B, C
- $\Gamma, x : B \vdash u_1 : A$
- $\Gamma \vdash t' : B$

By the substitution lemma:

$$\Gamma \vdash u_1[t'/x] : A$$

Congruence cases: If the reduction occurs in a subterm, we apply the induction hypothesis to that subterm and use the appropriate typing rule to reconstruct the overall typing. □

Lemma 2.23 (Substitution). *If* $\Gamma, x : A, \Delta \vdash t : B$ *and* $\Gamma \vdash u : A$, *then* $\Gamma, \Delta \vdash t[u/x] : B$.

The proof is a standard induction on the derivation of $\Gamma, x : A, \Delta \vdash t : B$. The key case is the application rule: if $t = t_1\, t_2$ with $\Gamma, x : A, \Delta \vdash t_1 : C \to B$ and $\Gamma, x : A, \Delta \vdash t_2 : C$, then by the induction hypothesis $\Gamma, \Delta \vdash t_1[u/x] : C \to B$ and $\Gamma, \Delta \vdash t_2[u/x] : C$, and a single application of ($\to$-E) yields $\Gamma, \Delta \vdash (t_1\, t_2)[u/x] : B$. The abstraction case requires renaming the bound variable to avoid capture, after which the induction hypothesis applies to the body. The remaining cases (variables, pairs, projections, injections, case analysis) follow the same pattern.

Corollary 2.24. *If* $\Gamma \vdash t : A$ *and* $t \to_\beta^* u$, *then* $\Gamma \vdash u : A$.

Proof. By induction on the length of the reduction sequence, applying Theorem 2.22 at each step. □

Subject reduction tells us that the type system is "sound" with respect to computation: we never "go wrong" by reducing a well-typed term. This is the type-theoretic analogue of the fact that valid logical inferences preserve truth.

2.5 Church–Rosser Property for the Base System

A fundamental property of the λ-calculus, and of our proof-term system, is *confluence*: if a term can reduce in two different ways, the results can always be brought back together. The Church–Rosser property for the base system established here should be compared with the stronger confluence result for the 77-rule path rewrite system proved in Chapter 5; see also Appendix A for the general theory of confluence in term rewriting systems.

2.5.1 The Diamond Property and Confluence

Definition 2.25 (Diamond property). A relation $\to$ has the *diamond property* if whenever $t \to u_1$ and $t \to u_2$, there exists v such that $u_1 \to v$ and $u_2 \to v$.

Definition 2.26 (Confluence). A relation $\to$ is *confluent* if its reflexive-transitive closure $\to^*$ has the diamond property. Equivalently, $\to$ is confluent if whenever $t \to^* u_1$ and $t \to^* u_2$, there exists v such that $u_1 \to^* v$ and $u_2 \to^* v$.

The diamond property for $\to_\beta$ itself does *not* hold in general. Consider:

$$(\lambda x.\, x\, x)\, ((\lambda y.\, y)\, z)$$

This term has two one-step reducts, and the diamond does not close in one step. However, confluence holds for the reflexive-transitive closure.

Theorem 2.27 (Church–Rosser). *The β-reduction relation $\to_\beta$ is confluent. That is, if $t \to_\beta^* u_1$ and $t \to_\beta^* u_2$, then there exists v such that $u_1 \to_\beta^* v$ and $u_2 \to_\beta^* v$.*

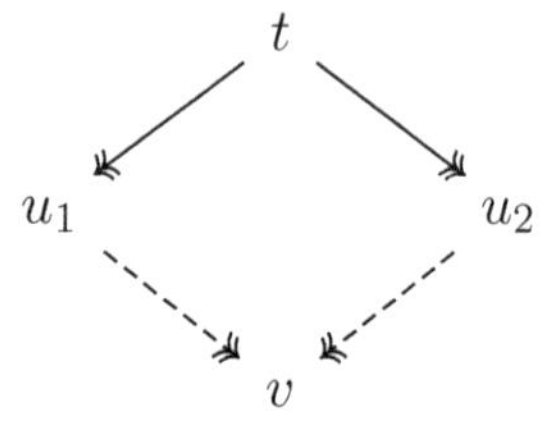

2.5.2 Proof via Parallel Reduction

Why does confluence matter for computational paths? Because the question "do these two reduction sequences amount to the same thing?" is, at its core, the question that the path rewrite system of Chapter 5 must answer algorithmically. The parallel reduction technique introduced below provides the cleanest route to the diamond property: instead of tracking individual redex contractions one at a time, we allow *all* redexes to fire simultaneously, and then show that simultaneous reduction has the diamond property directly. This idea—reducing the general multi-step problem to a single-step property of a coarsened relation—will reappear in the confluence proof for the 77-rule system in Chapter 5.

The classical proof, due to Tait and Martin-Löf, uses the technique of *parallel reduction*.

Definition 2.28 (Parallel reduction). The *parallel reduction* relation $t \Rightarrow u$ is defined inductively as follows:

1. $x \Rightarrow x$ for any variable x.
2. If $t \Rightarrow t'$, then $\lambda x.\, t \Rightarrow \lambda x.\, t'$.
3. If $t \Rightarrow t'$ and $u \Rightarrow u'$, then $t\, u \Rightarrow t'\, u'$.
4. If $t \Rightarrow t'$ and $u \Rightarrow u'$, then $(\lambda x.\, t)\, u \Rightarrow t'[u'/x]$.
5. Similar rules for pairs, projections, injections, and case analysis:
   - $\langle t, u\rangle \Rightarrow \langle t', u'\rangle$ if $t \Rightarrow t'$ and $u \Rightarrow u'$.
   - $\pi_i(\langle t, u\rangle) \Rightarrow t_i'$ if $t \Rightarrow t'$ and $u \Rightarrow u'$ (for $i = 1, 2$).
   - $\mathsf{case}(\mathsf{inl}(t), x.u_1, y.u_2) \Rightarrow u_1'[t'/x]$ if $t \Rightarrow t'$ and $u_1 \Rightarrow u_1'$.

Parallel reduction allows any subset (possibly empty, possibly all) of the redexes in a term to be reduced simultaneously.

Lemma 2.29. $\to_\beta \subseteq \Rightarrow \subseteq \to_\beta^*$

Proof. For the first inclusion: any single β-step is a parallel reduction (reduce one redex, leave others unchanged). For the second: any parallel reduction can be simulated by a sequence of single steps (reduce the redexes one at a time). □

Definition 2.30 (Complete development)**.** The *complete development* t^* of a term t is obtained by simultaneously reducing all redexes in t:

$$\begin{aligned} x^* &= x \\ (\lambda x.\, t)^* &= \lambda x.\, t^* \\ ((\lambda x.\, t)\, u)^* &= t^*[u^*/x] \\ (t\, u)^* &= t^*\, u^* \quad \text{(if } t \text{ is not a } \lambda\text{-abstraction)} \\ \langle t, u \rangle^* &= \langle t^*, u^* \rangle \\ \pi_i(\langle t, u \rangle)^* &= (t_i)^* \quad \text{for } i = 1, 2 \\ \pi_i(t)^* &= \pi_i(t^*) \quad \text{(if } t \text{ is not a pair)} \\ &\vdots \quad \text{(and similarly for sums)} \end{aligned}$$

Lemma 2.31. *If $t \Rightarrow u$, then $u \Rightarrow t^*$.*

Proof. By induction on the derivation of $t \Rightarrow u$, showing that u parallel-reduces to t^*. The key point is that t^* reduces "more" than any other single parallel reduction step from t. □

Lemma 2.32 (Diamond property for parallel reduction)**.** *If $t \Rightarrow u_1$ and $t \Rightarrow u_2$, then there exists v such that $u_1 \Rightarrow v$ and $u_2 \Rightarrow v$.*

Proof. Take $v = t^*$. By Lemma 2.31, $u_1 \Rightarrow t^*$ and $u_2 \Rightarrow t^*$. □

Proof of Theorem 2.27. From Lemma 2.29, the reflexive-transitive closures satisfy $(\to_\beta)^* = (\Rightarrow)^*$. From Lemma 2.32, $\Rightarrow$ has the diamond property. A relation with the diamond property has a confluent reflexive-transitive closure. Therefore $(\to_\beta)^*$ is confluent. □

2.5.3 Consequences of Confluence

Definition 2.33 (Normal form)**.** A term t is in *β-normal form* (or simply *normal form*) if there is no u such that $t \to_\beta u$. We say u is *a normal form of* t if $t \to_\beta^* u$ and u is in normal form.

Corollary 2.34 (Uniqueness of normal forms)**.** *If t has normal forms u_1 and u_2, then $u_1 = u_2$.*

Proof. By confluence, there exists v with $u_1 \to_\beta^* v$ and $u_2 \to_\beta^* v$. Since u_1 and u_2 are normal forms, these reductions must have length 0, so $u_1 = v = u_2$. □

Corollary 2.35 (Consistency)**.** *There is no closed term of type* **0**.

Proof. Suppose $\vdash t : \mathbf{0}$. By strong normalisation (not yet proved, but true for simply-typed λ-calculus), t has a normal form u. By subject reduction, $\vdash u : \mathbf{0}$. But a normal closed term of type **0** would have to begin with an introduction rule for **0**, and no such rule exists. Contradiction. □

The uniqueness of normal forms tells us that if two reduction sequences from t both lead to normal forms, they lead to the *same* normal form. But this does *not* tell us that the paths themselves are equal. Distinct reduction sequences are genuinely different objects, even if they lead to the same destination.

Example 2.36 (Distinct paths to the same destination)**.** Consider the term:

$$t \equiv (\lambda x.\, (\lambda y.\, y)\, x)\, ((\lambda z.\, z)\, w)$$

where w is a variable. There are (at least) three different reduction paths to the normal form w:

Path 1 (outer-first):

$$\begin{array}{lll} & (\lambda x.\, (\lambda y.\, y)\, x)\, ((\lambda z.\, z)\, w) & \\ \to_\beta & (\lambda y.\, y)\, ((\lambda z.\, z)\, w) & \text{(by } \beta_\to \text{ at root)} \\ \to_\beta & (\lambda z.\, z)\, w & \text{(by } \beta_\to \text{ at root)} \\ \to_\beta & w & \text{(by } \beta_\to \text{ at root)} \end{array}$$

Path 2 (inner-first):

$$\begin{array}{lll} & (\lambda x.\, (\lambda y.\, y)\, x)\, ((\lambda z.\, z)\, w) & \\ \to_\beta & (\lambda x.\, (\lambda y.\, y)\, x)\, w & \text{(by } \beta_\to \text{ in argument)} \\ \to_\beta & (\lambda y.\, y)\, w & \text{(by } \beta_\to \text{ at root)} \\ \to_\beta & w & \text{(by } \beta_\to \text{ at root)} \end{array}$$

Path 3 (body-first):

$$\begin{array}{lll} & (\lambda x.\, (\lambda y.\, y)\, x)\, ((\lambda z.\, z)\, w) & \\ \to_\beta & (\lambda x.\, x)\, ((\lambda z.\, z)\, w) & \text{(by } \beta_\to \text{ in body)} \\ \to_\beta & (\lambda z.\, z)\, w & \text{(by } \beta_\to \text{ at root)} \\ \to_\beta & w & \text{(by } \beta_\to \text{ at root)} \end{array}$$

All three paths have length 3 and reach the same normal form w. But they are different mathematical objects—they record different sequences of computational steps. The theory of computational paths makes this difference precise and asks: when should two such paths be considered "the same"?

2.6 From Reduction Sequences to Computational Paths

We are now ready to make the conceptual transition from informal "reduction sequences" to formal *computational paths*. The key insight is that the operations we naturally perform on reduction sequences can be given a precise algebraic structure.

2.6.1 Operations on Reduction Sequences

Given reduction sequences, we naturally want to:

1. **Compose** them: If $p : t \rightsquigarrow u$ and $q : u \rightsquigarrow v$, we can form the composite sequence $p \cdot q : t \rightsquigarrow v$.
2. **Invert** them: If $p : t \rightsquigarrow u$ is a sequence of β-steps, we can consider running it "backwards" to get a sequence $p^{-1} : u \rightsquigarrow t$ (though this requires interpreting β-reduction as a bidirectional equality rather than a directed rewrite).
3. **Apply congruence**: If $p : t \rightsquigarrow t'$ is a sequence and f is a term with a hole, we can reduce "under f" to get $f[p] : f[t] \rightsquigarrow f[t']$.
4. **Do nothing**: There is a trivial sequence $\mathsf{refl}_t : t \rightsquigarrow t$ of length 0.

These operations satisfy evident laws:

- Composition is associative: $(p \cdot q) \cdot r = p \cdot (q \cdot r)$.
- The reflexivity sequence is a unit: $\mathsf{refl}_t \cdot p = p$ and $p \cdot \mathsf{refl}_u = p$.
- Inversion gives an "inverse": $p \cdot p^{-1} = \mathsf{refl}_t$ and $p^{-1} \cdot p = \mathsf{refl}_u$ (up to appropriate equivalence).

This is precisely the structure of a *groupoid* (see Chapter 6). But there is more: the congruence operation satisfies functoriality laws, and the interactions among these operations give rise to additional structure.

2.6.2 Preview: The Path Constructors

The full theory of computational paths (developed in Chapter 3) formalises this structure. The basic path constructors are:

Definition 2.37 (Path constructors (preview)).

$$
\begin{array}{rll}
p, q ::= & \mathsf{refl}_t & \text{(identity path at } t\text{)} \\
| & \beta(\ell, r) & \text{(}\beta\text{-step from redex } \ell \text{ to reductum } r\text{)} \\
| & \mathsf{symm}(p) & \text{(reversal/inverse)} \\
| & \mathsf{trans}(p, q) & \text{(composition)} \\
| & \mathsf{congrArg}(f, p) & \text{(congruence: apply } f \text{ to path)}
\end{array}
$$

with typing rules:

$$
\frac{}{\mathsf{refl}_t : t \rightsquigarrow t} \qquad \frac{p : t \rightsquigarrow u}{\mathsf{symm}(p) : u \rightsquigarrow t} \qquad \frac{p : t \rightsquigarrow u \quad q : u \rightsquigarrow v}{\mathsf{trans}(p, q) : t \rightsquigarrow v}
$$

The β-constructor creates a path for each β-reduction step. The refl, symm, and trans constructors give the groupoid structure (Chapter 6). The $\mathsf{congrArg}$ constructor allows paths to be lifted through term formation.

2.6.3 The Central Question

With these path constructors, we can *name* the different reduction sequences from Example 2.36. But now a fundamental question arises:

> *When should two paths be considered equal?*

Consider the composition $\mathsf{trans}(\mathsf{refl}_t, p)$. Intuitively, doing nothing and then doing p should be the same as just doing p. So we want:

$$\mathsf{trans}(\mathsf{refl}_t, p) = p$$

But this is now an equality between *paths*. If we take the LND philosophy seriously, this equality should itself have a witness—a reason or evidence for the equality. This leads us to *2-cells* (Chapter 7) or *derivations between paths*, and the beginnings of higher categorical structure.

The theory of computational paths, developed in the following chapters, provides a precise answer: paths are related by a system of rewrite rules that capture all the expected coherences (groupoid laws, functoriality laws, and more; see Chapter 8). The key insight is that these rewrite rules generate a confluent rewriting system (Chapter 5), so that two paths are equal if and only if they reduce to a common normal form.

2.7 Worked Examples

To consolidate the material of this chapter, we present several worked examples that illustrate the LND framework in action. These examples demonstrate both the typing derivations and the reduction behaviour of proof-terms.

2.7.1 Basic Derivations

Example 2.38 (Identity combinator)**.** The simplest non-trivial proof-term is the identity $\lambda x.\, x : A \to A$, which proves that every proposition implies itself. The derivation is:

$$\dfrac{\dfrac{}{x : A \vdash x : A}\,[\textsc{Var}]}{\vdash \lambda x.\, x : A \to A}\,[\to\text{-I}]$$

This corresponds to the combinator I in combinatory logic.

Example 2.39 (Constant combinator)**.** The term $\lambda x.\, \lambda y.\, x : A \to (B \to A)$ proves the *weakening* axiom. The derivation is:

$$\dfrac{\dfrac{\dfrac{}{x : A, y : B \vdash x : A}\,[\textsc{Var}]}{x : A \vdash \lambda y.\, x : B \to A}\,[\to\text{-I}]}{\vdash \lambda x.\, \lambda y.\, x : A \to (B \to A)}\,[\to\text{-I}]$$

This corresponds to the combinator K. Note that y does not occur in the body x; this is a vacuous abstraction, which is allowed in intuitionistic logic but forbidden in relevant logic.

Example 2.40 (Composition combinator)**.** The term $\lambda f.\, \lambda g.\, \lambda x.\, f\,(g\,x)$ of type $(B \to C) \to ((A \to B) \to (A \to C))$ witnesses the transitivity of implication. This corresponds to the combinator B.

Example 2.41 (S combinator)**.** The term

$$\lambda f.\, \lambda g.\, \lambda x.\, f\,x\,(g\,x) : (A \to (B \to C)) \to ((A \to B) \to (A \to C))$$

corresponds to the combinator S and witnesses the distribution axiom. Note that x occurs twice in the body—once as argument to f and once as argument to g. This use of *contraction* (using an assumption multiple times) distinguishes intuitionistic from linear logic.

2.7.2 Reduction Examples

Example 2.42 (A simple β-reduction)**.** Consider the identity at the function type $A \to A$ applied to the identity on A:

$$\lambda x.\, x : (A \to A) \to (A \to A), \qquad \lambda y.\, y : A \to A.$$

Hence

$$((\lambda x.\, x)\,(\lambda y.\, y)) : A \to A.$$

This reduces in one step:

$$(\lambda x.\, x)\,(\lambda y.\, y) \to_\beta \lambda y.\, y$$

by substituting $\lambda y.\, y$ for x in the body x.

Example 2.43 (Projection from a pair)**.** Let $t : A \times B$ be a term of product type. Then:

$$\pi_1(\langle \pi_1(t), \pi_2(t) \rangle) \to_\beta \pi_1(t)$$
$$\pi_2(\langle \pi_1(t), \pi_2(t) \rangle) \to_\beta \pi_2(t)$$

These show the β-rules for products in action. Note that $\langle \pi_1(t), \pi_2(t) \rangle$ is the η-expansion of t, so these reductions "undo" the expansion.

Example 2.44 (A non-trivial Church numeral computation)**.** The Church numeral for 2 is:

$$\mathsf{c}_2 \equiv \lambda f.\, \lambda x.\, f\,(f\,x) : (A \to A) \to (A \to A)$$

To obtain a typable non-trivial computation in the simply typed setting, instantiate A by $B \to B$, and let

$$\mathsf{d} \equiv \lambda g.\, \lambda y.\, g\,(g\,y) : (B \to B) \to (B \to B).$$

Applying c_2 to d yields:

$$\begin{aligned} \mathsf{c}_2\,\mathsf{d} &= (\lambda f.\, \lambda x.\, f\,(f\,x))\,(\lambda g.\, \lambda y.\, g\,(g\,y)) \\ &\to_\beta \lambda x.\,(\lambda g.\, \lambda y.\, g\,(g\,y))\,((\lambda g.\, \lambda y.\, g\,(g\,y))\,x) \\ &\to_\beta^* \lambda x.\, \lambda y.\, x\,(x\,(x\,(x\,y))) \end{aligned}$$

which is c_4 at base type B. In other words, applying the doubling operator twice produces four iterations of the input function.

2.7.3 The Curry–Howard Correspondence in Detail

The following table summarises the Curry–Howard correspondence for the connectives we have introduced:

Logic	Types	Terms
Proposition A	Type A	Proof $t : A$
$A \supset B$	$A \to B$	$\lambda x.\, t$
Modus ponens	Application	$t\, u$
$A \wedge B$	$A \times B$	$\langle t, u \rangle$
Left projection	π_1	$\pi_1(t)$
Right projection	π_2	$\pi_2(t)$
$A \vee B$	$A + B$	$\mathsf{inl}(t)$, $\mathsf{inr}(t)$
Disjunction elim.	Case	$\mathsf{case}(t, x.u_1, y.u_2)$
$\bot$ (falsity)	$\mathbf{0}$	(no introduction)
Ex falso	Abort	$\mathsf{abort}(t)$
$\top$ (truth)	$\mathbf{1}$	$\star$
Proof simplification	β-reduction	$(\lambda x.\, t)\, u \to_\beta t[u/x]$
Proof identity	η-expansion	$t =_\eta \lambda x.\, t\, x$

This correspondence is not merely formal; it reflects deep structural parallels. The harmony between introduction and elimination rules corresponds to β-reduction. The extensionality principles (two proofs are equal if they behave the same way) correspond to the η-laws. And the principle that proofs normalise corresponds to the consistency of the logic.

Summary

The purpose of this chapter has been to fix the labelled setting in which the whole book moves. Labelled natural deduction (LND), together with the simply-typed λ-calculus with products, sums, unit, and empty types, provides the base system for the theory of computational paths. What matters in LND is that every judgement $\Gamma \vdash t : A$ carries a proof-term t recording *how* the conclusion follows from the assumptions, not merely *that* it does. That extra dimension is what makes it possible to compare different proofs of the same proposition and, eventually, to study the space of such comparisons.

The computational content of the proof-terms is expressed through β-reduction and η-expansion rules for each type constructor. We established subject reduction (Theorem 2.22), ensuring that types are preserved under rewriting, and the Church–Rosser property (Theorem 2.27), guaranteeing that distinct reduction sequences from the same term can always be brought together. These results are not merely technical preliminaries. Subject reduction ensures that the rewrite steps are *well-typed paths*, and confluence ensures that the resulting algebra of paths is mathematically stable rather than accidental. The chapter closed by introducing the path constructors—refl, symm, trans, congrArg, and β-steps—which become the basic material of computational paths in the chapters that follow.

2.8 Notes and Further Reading

Historical sources. The references gathered here mark the main proof-theoretic and type-theoretic background of the chapter. The Curry–Howard correspondence is named after Haskell Curry's work on combinatory logic [25, 26]. It also reflects William Howard's foundational paper [87], written in 1969 and only published in 1980. Tait's work on intensional interpretations [146, 147] provided crucial insights into the computational content of proofs.

The LND framework and the idea of tracking rewrite reasons were developed by de Queiroz and collaborators [32, 44, 35, 71]. Gabbay's Labelled Deductive Systems [70] provides the broader context for labelled proof systems. The connections between proof theory and computer programming were explored in [51], with further work on abstract data types and type theory in [52].

Textbooks. For background on the simply-typed λ-calculus, see Barendregt [9]. Another standard reference is Hindley–Seldin [82, 83]. The proof-theoretic perspective is developed in Prawitz [121, 122] and Gentzen's original work [72].

For type theory and the propositions-as-types correspondence, Martin-Löf's papers [110, 111] are foundational. Accessible introductions include Girard–Lafont–Taylor [74] and the textbook by Sørensen and Urzyczyn [140].

Substructural logics. The connection between restricted λ-abstraction and substructural logics was explored by many authors, including Lambek [96], and is developed systematically in the context of linear logic [73]. See also van Benthem [152] for the "categorial hierarchy" of logics.

Confluence. The Church–Rosser theorem for the λ-calculus was first proved by Church and Rosser [22]. The parallel reduction proof technique is due to Tait and Martin-Löf; Takahashi gives a particularly clean presentation. For a comprehensive treatment of confluence in term rewriting, see Baader and Nipkow [8].

We have established in this chapter the foundational framework of LND: the types/formulas, the terms/proof-terms, the $\beta\eta$-theory, and the basic properties of type preservation and confluence. The next chapter develops the full theory of computational paths, introducing all the path constructors and stating the rewrite system (Chapter 4) that governs path equality. The J-eliminator derived in Chapter 11 will later show how path induction connects to the equality type structure introduced here.

Formalization Notes. The equality-specific infrastructure used later in this chapter is formalised in the following files of the accompanying Lean development (`https://github.com/Arthur742Ramos/ComputationalPathsLean`):

- `Path/Basic/Context.lean` — context operations and substitution-style actions on paths
- `Path/Basic/Congruence.lean` — congruence principles for functions and dependent families
- `Path/Basic/PathAlgebra.lean` — algebraic operations on paths and their basic laws
- `Path/Rewrite/LNDEQ.lean` — the labelled-natural-deduction equality presentation used by the later rewrite chapters

Chapter 3

Computational Paths — Definition and Properties

"The clarification of the notion of normal form for equality reasoning took an important step with the work of Statman in the late 1970s. The concept of 'direct computation' was instrumental in the development of Statman's approach."

—de Queiroz, de Oliveira & Ramos

We now come to the central mathematical objects of this book: *computational paths*. Whereas Chapter 2 established the framework of labelled natural deduction and proof-terms, the present chapter introduces the formal structure that captures what we have been calling a "reason for equality"—an object which witnesses not merely *that* two terms are equal, but records *how* and *why* they are equal.

The fundamental point is this: in the labelled natural deduction judgement $t =_s u : A$, the subscript s is not a notational decoration but a first-class mathematical entity. We call s a *computational path from t to u*. Just as proof-terms are programs that compute values, computational paths are programs that compute equalities. And just as we can reason about the equality of programs, we can reason about the equality of paths—leading to a hierarchy of structure that extends infinitely upward.

This chapter develops the foundational definitions and elementary properties of computational paths. We begin with atomic rewrite steps and build up to the full grammar of path expressions. We then examine in detail the fundamental operations—reflexivity, transitivity, symmetry, and congruence—that allow paths to be combined and manipulated. A key section addresses the failure of the Uniqueness of Identity Proofs (UIP) principle: computational paths are genuinely *proof-relevant*, and this proof-relevance is the source of the higher structure to come. We conclude with a preview of the algebraic laws that paths satisfy, setting the stage for the complete rewrite system of Chapter 4.

3.1 Steps, Paths, and Path Expressions

The formal definition and properties of computational paths presented in this chapter follow the development in [54, 128, 127, 130, 56, 57].

The construction of computational paths proceeds in two stages. First, we identify the *atomic steps*—the primitive rewrite operations that transform one term into another in a single move. Second, we define the *path expressions*—compound terms built from atomic steps using algebraic operations like composition and inversion.

3.1.1 Atomic Rewrite Steps

An atomic step represents a single application of a definitional equality at some position within a term. In the context of the simply-typed λ-calculus, the basic steps correspond to the fundamental computational rules: β-reduction, η-expansion, and the various congruence principles.

Definition 3.1 (Step structure). An *elementary rewrite step* between elements of a type A is a triple (a, b, h) consisting of:

- A *source* term $a : A$
- A *target* term $b : A$
- A *justification* h recording one designated atomic equality step from a to b in the ambient theory

We write $\mathsf{Step}(A)$ for the type of all such steps in A, and we write $a \xrightarrow{h} b$ or simply $a \to b$ when the justification is clear from context.

Remark 3.2 (Directionality of steps). A step has an inherent direction: it goes *from a to b*. This directionality is crucial. Although the underlying definitional equality is symmetric—if a equals b, then b equals a—the computational content of going from a to b is distinct from that of going from b to a. The former might represent a reduction, the latter an expansion. We will introduce explicit symmetry operations to reverse steps when needed.

For the simply-typed λ-calculus of Chapter 2, the atomic steps are classified according to the reduction rule or congruence principle they employ.

Definition 3.3 (Classification of atomic steps)**.** The atomic steps for the simply-typed λ-calculus fall into two natural families. The first family comprises the *computational steps*—the rules that actually perform calculation. The most fundamental of these is the β-step,

$$(\lambda x.\, t)\, u \xrightarrow{\beta} t[u/x],$$

which replaces a redex (reducible expression) with its contractum. Alongside it stands the η-step,

$$\lambda x.\, (f\, x) \xrightarrow{\eta} f \qquad (x \notin \mathsf{FV}(f)),$$

which collapses a vacuous abstraction, enforcing extensionality for functions. Products and sums have their own analogous computation rules: projection from a pair ($\pi_1(\langle t, u\rangle) \xrightarrow{\beta} t$, $\pi_2(\langle t, u\rangle) \xrightarrow{\beta} u$), surjective pairing ($\langle \pi_1(t), \pi_2(t)\rangle \xrightarrow{\eta} t$), and case analysis on injections ($\mathsf{case}(\mathsf{inl}(t), x.u_1, y.u_2) \xrightarrow{\beta} u_1[t/x]$, $\mathsf{case}(\mathsf{inr}(t), x.u_1, y.u_2) \xrightarrow{\beta} u_2[t/y]$).

The second family comprises the *structural steps*—the congruence rules that propagate computation through compound terms. A rewrite inside the body of a λ-abstraction is recorded by a ξ-step ($\lambda x.\, t \xrightarrow{\xi} \lambda x.\, t'$ when $t \to t'$); a rewrite in the function position of an application by a μ-step ($f\, u \xrightarrow{\mu} f'\, u$ when $f \to f'$); and a rewrite in the argument position by a ν-step ($f\, u \xrightarrow{\nu} f\, u'$ when $u \to u'$). Product and sum constructors and eliminators admit analogous congruence steps.

The congruence steps express a fundamental closure principle: definitional equality is preserved by *every* term-forming operation. If subterms are equal, then the compound terms built from them are also equal. Together the two families ensure that every single-step transformation in the calculus is recorded as a named atomic step—a building block from which computational paths will be assembled.

Example 3.4 (A chain of steps)**.** Consider the term

$$t = (\lambda x.\, x)\, ((\lambda y.\, y)\, z).$$

There are several possible reduction sequences to the normal form z:

Sequence 1 (inner reduction first):

$$\begin{aligned}(\lambda x.\, x)\, ((\lambda y.\, y)\, z) &\xrightarrow{\nu} (\lambda x.\, x)\, z && \text{(by } \beta \text{ in argument)}\\ &\xrightarrow{\beta} z\end{aligned}$$

Sequence 2 (outer reduction first):

$$\begin{aligned}(\lambda x.\, x)\,((\lambda y.\, y)\, z) &\xrightarrow{\beta} (\lambda y.\, y)\, z \\ &\xrightarrow{\beta} z\end{aligned}$$

Both sequences arrive at the same result, but through different routes. The computational path formalism will distinguish these routes as different mathematical objects.

3.1.2 Operations on Steps

Before constructing paths, we define basic operations on individual steps.

Definition 3.5 (Step reversal)**.** Given a step $s : \mathsf{Step}(A)$ with $s = (a, b, h)$, the *reversed step* $\bar{s} : \mathsf{Step}(A)$ is defined by:

$$\bar{s} = (b, a, h^{-1})$$

where h^{-1} is the symmetric witness derived from h. Reversing a step twice returns the original step: $\bar{\bar{s}} = s$.

Definition 3.6 (Step mapping)**.** Given a function $f : A \to B$ and a step $s = (a, b, h)$ in A, the *image step* $f_*(s) : \mathsf{Step}(B)$ is defined by:

$$f_*(s) = (f(a), f(b), \mathsf{congrArg}(f, h))$$

This operation satisfies the expected functoriality properties:

$$\begin{aligned}\mathsf{id}_*(s) &= s \\ (g \circ f)_*(s) &= g_*(f_*(s)) \\ f_*(\bar{s}) &= \overline{f_*(s)}\end{aligned}$$

3.1.3 The Path Type

Individual steps capture the atomic moves of computation. But a computation rarely consists of a single move. When we reduce the term $(\lambda x.\, x)((\lambda y.\, y)\, z)$ to its normal form z, we perform *two* steps—and the order in which we perform them matters at the level of computational content, even though it does not affect the final result. What we need, therefore, is a way to *compose* atomic steps into structured sequences, recording not only the start and end points but the entire intermediate history. This is exactly what the path type provides: it packages a sequence of steps together with a proof that they form a valid chain, yielding a single first-class mathematical object that witnesses equality while remembering how that equality was established.

Before giving the formal definition, it is worth pausing to consider what kind of mathematical object we are about to introduce. In standard type theory, the identity type $\mathsf{Id}_A(a, b)$ asserts that a and b are equal, but—at least in proof-irrelevant foundations—it carries no information about *why* they are equal. Our path type $\mathsf{Path}(a, b)$ is designed to fill this gap: it records the entire computational history of the identification. Different sequences of rewrite steps produce different paths, even when they connect the same endpoints. The path type is therefore *proof-relevant by construction*, and this proof-relevance is the wellspring of the higher-dimensional structure that occupies the remainder of the book.

Remark 3.7 (Paths versus identity types)**.** The path type $\mathsf{Path}(a, b)$ should not be confused with the identity type $\mathsf{Id}_A(a, b)$ of intensional Martin-Löf type theory [109]. While both express a notion of equality, they differ in a crucial way:

- In standard proof-irrelevant foundations (for example extensional proof-assistant kernels or classical set theory), the identity type is *proof-irrelevant*: any two proofs of $a = b$ are equal. The type $\mathsf{Id}_A(a, b)$ is either empty or a singleton.
- The path type $\mathsf{Path}(a, b)$ is *proof-relevant* by design: different step traces are distinguished, so $\mathsf{Path}(a, b)$ can contain many distinct inhabitants.

In Homotopy Type Theory (HoTT) [150] (see Chapter 14), the identity type itself is proof-relevant. Our path type can be seen as a way to *introduce* proof-relevant identity into foundations where it is not primitive.

We are now ready for the central definition of this chapter—and the most important definition in the entire book. Everything that follows—the groupoid structure of Chapter 6, the 2-cells and 3-cells of Chapters 7–8, the ω-groupoid of Chapter 9, the derivation of path induction and the failure of UIP—depends on this single construction. The path type takes the informal idea of a "reason for equality" and makes it fully precise: a concrete data structure consisting of a sequence of atomic rewrite steps together with a proof that those steps compose correctly.

Definition 3.8 (Computational path)**.** A *computational path* from a to b in a type A is represented semantically by a pair (s, h) where:

- s is a list of elementary steps $[s_1, s_2, \ldots, s_n]$, each $s_i : \mathsf{Step}(A)$, where for each $1 \leq i < n$ the target of s_i coincides with the source of s_{i+1}

- $h : a = b$ is the ambient equality induced by composing the justifications carried by the steps of s (with $h := \mathsf{refl}_a$ when s is empty)

We write $\mathsf{Path}_A(a, b)$ or simply $\mathsf{Path}(a, b)$ for the type of all computational paths from a to b in A.

The step list s is the *computational trace*: it records which rewrite operations were performed and in what order. The propositional equality h is the *semantic content*: it witnesses in the ambient theory that the trace really connects a to b. The two components serve different purposes:

- The semantic component h ensures *well-formedness*: we cannot construct a path from a to b unless a truly equals b.
- The trace component s preserves *intensional information*: different traces distinguish different paths, even when they have the same endpoints.

This concrete pair/list representation should be distinguished from the raw path syntax studied in Chapter 4. At the raw syntactic level, groupoid laws hold only up to rewriting; in the present semantic model, some of those laws collapse to strict equalities of representatives. We will signal explicitly which level is in use.

3.1.4 Path Grammar

The path type admits a rich grammar of constructors. We list the fundamental ones, which suffice for all basic constructions.

Definition 3.9 (Path constructors)**.** The following operations construct computational paths:

Reflexivity. For any term $a : A$, there is a *reflexivity path*:

$$\mathsf{refl}_a : \mathsf{Path}(a, a)$$

defined by $\mathsf{refl}_a = ([\,], \mathsf{refl})$—the empty step list paired with the trivial equality proof. This represents "staying in place": a is equal to itself via no computation at all.

Single step. For any elementary step $s : a \to b$, there is a corresponding path:

$$\mathsf{step}(s) : \mathsf{Path}(a, b)$$

defined by $\mathsf{step}(s) = ([s], h_s)$ where h_s is the equality witness from s. This embeds atomic steps into paths.

Symmetry. For any path $p : \mathsf{Path}(a, b)$, there is an *inverse path*:

$$\mathsf{symm}(p) : \mathsf{Path}(b, a)$$

If $p = (s, h)$ where $s = [s_1, \ldots, s_n]$, then:

$$\mathsf{symm}(p) = ([\bar{s}_n, \ldots, \bar{s}_1], h^{-1})$$

The step list is reversed and each step is inverted; the semantic equality is symmetrized.

Transitivity (composition). For paths $p : \mathsf{Path}(a, b)$ and $q : \mathsf{Path}(b, c)$, there is a *composite path*:

$$\mathsf{trans}(p, q) : \mathsf{Path}(a, c)$$

If $p = (s_p, h_p)$ and $q = (s_q, h_q)$, then:

$$\mathsf{trans}(p, q) = (s_p \mathbin{+\!\!+} s_q, h_p \cdot h_q)$$

where $\mathbin{+\!\!+}$ denotes list concatenation and $\cdot$ denotes transitivity of equality.

Congruence. For a function $f : A \to B$ and a path $p : \mathsf{Path}_A(a, b)$, there is a *congruence path*:

$$\mathsf{congrArg}(f, p) : \mathsf{Path}_B(f(a), f(b))$$

If $p = (s, h)$ where $s = [s_1, \ldots, s_n]$, then:

$$\mathsf{congrArg}(f, p) = ([f_*(s_1), \ldots, f_*(s_n)], \mathsf{congrArg}(f, h))$$

Each step is mapped through f, and the semantic equality is lifted by the built-in congruence principle.

Remark 3.10 (Infix and postfix conventions)**.** We adopt the following notational conventions throughout:

- **Composition:** We write $p \cdot q$ or $p \bullet q$ for $\mathsf{trans}(p, q)$. Composition is read left-to-right: p is followed by q.
- **Inversion:** We write p^{-1} for $\mathsf{symm}(p)$.
- **Action on paths:** We write $f_*(p)$ for $\mathsf{congrArg}(f, p)$, emphasizing the functorial nature of the operation.
- **Reflexivity:** We often write $\mathbf{1}_a$ or simply $\mathbf{1}$ for refl_a.

With this notation, many path equations take familiar algebraic forms. For instance, the inverse law becomes $p \cdot p^{-1} = \mathbf{1}$.

3.1.5 Typing Rules for Path Constructors

We present the formal typing rules for paths in a natural deduction style, making the judgement forms precise.

Definition 3.11 (Path judgements)**.** The basic judgement form for computational paths is:

$$\Gamma \vdash p : \mathsf{Path}_A(a, b)$$

which asserts that p is a computational path from a to b in type A, under the assumptions recorded in context Γ. This presupposes that $\Gamma \vdash a : A$ and $\Gamma \vdash b : A$.

The inference rules for constructing paths are as follows:

Formation rule.

$$\frac{\Gamma \vdash A : \mathsf{Type} \qquad \Gamma \vdash a : A \qquad \Gamma \vdash b : A}{\Gamma \vdash \mathsf{Path}_A(a, b) : \mathsf{Type}}\ [\textsc{Path-Form}]$$

Reflexivity.

$$\frac{\Gamma \vdash a : A}{\Gamma \vdash \mathsf{refl}_a : \mathsf{Path}_A(a, a)}\ [\textsc{Path-Refl}]$$

Symmetry.

$$\frac{\Gamma \vdash p : \mathsf{Path}_A(a, b)}{\Gamma \vdash \mathsf{symm}(p) : \mathsf{Path}_A(b, a)}\ [\textsc{Path-Symm}]$$

Transitivity.

$$\frac{\Gamma \vdash p : \mathsf{Path}_A(a, b) \qquad \Gamma \vdash q : \mathsf{Path}_A(b, c)}{\Gamma \vdash \mathsf{trans}(p, q) : \mathsf{Path}_A(a, c)}\ [\textsc{Path-Trans}]$$

Argument congruence.

$$\frac{\Gamma \vdash f : A \to B \qquad \Gamma \vdash p : \mathsf{Path}_A(a, a')}{\Gamma \vdash \mathsf{congrArg}(f, p) : \mathsf{Path}_B(f\,a, f\,a')}\ [\textsc{Path-CongrArg}]$$

Function congruence.

$$\frac{\Gamma \vdash p : \mathsf{Path}_{A \to B}(f, g) \qquad \Gamma \vdash a : A}{\Gamma \vdash \mathsf{congrFun}(p, a) : \mathsf{Path}_B(f\,a, g\,a)}\ [\textsc{Path-CongrFun}]$$

Single step.

$$\frac{\Gamma \vdash a : A \qquad \Gamma \vdash b : A \qquad \Gamma \vdash h : a = b}{\Gamma \vdash \mathsf{step}(a, b, h) : \mathsf{Path}_A(a, b)} \text{[PATH-STEP]}$$

These rules define the *canonical forms* of paths. Every path is either reflexivity, a single step, or built from smaller paths using symmetry, transitivity, or congruence.

3.1.6 Derived Forms and Sugar

Several useful path constructors can be derived from the primitive ones.

Definition 3.12 (Path from propositional equality)**.** Given a propositional equality $h : a = b$, we can form a path:

$$\mathsf{ofEq}(h) = \mathsf{step}(a, b, h) : \mathsf{Path}(a, b)$$

This is the canonical embedding of propositional equalities into paths. Later chapters sometimes denote the same constructor by $\mathsf{sc}(h)$ when emphasising its role as a single semantic step.

Definition 3.13 (Binary congruence)**.** For a binary function $f : A \to B \to C$ and paths $p : \mathsf{Path}(a, a')$ and $q : \mathsf{Path}(b, b')$, define:

$$\mathsf{congr}_2(f, p, q) : \mathsf{Path}(f\,a\,b, f\,a'\,b')$$

by:

$$\mathsf{congr}_2(f, p, q) = \mathsf{trans}(\mathsf{congrArg}(f\,a, q), \mathsf{congrArg}(\lambda x.\, f\,x\,b', p))$$

This first applies $f\,a$ to the path q (getting from $f\,a\,b$ to $f\,a\,b'$), then varies a to a' while holding the second argument fixed at b'.

Definition 3.14 (Product paths)**.** For paths $p : \mathsf{Path}(a, a')$ and $q : \mathsf{Path}(b, b')$ in types A and B respectively, define:

$$\mathsf{prodMk}(p, q) : \mathsf{Path}_{A \times B}((a, b), (a', b'))$$

by applying binary congruence to the pairing constructor.

Definition 3.15 (Sigma paths)**.** For a type family $B : A \to \mathsf{Type}$, a path $p : \mathsf{Path}_A(a, a')$, and a dependent path $q : \mathsf{Path}_{B(a')}(\mathsf{transport}(p, b), b')$, define:

$$\mathsf{sigmaMk}(p, q) : \mathsf{Path}_{\Sigma B}(\langle a, b\rangle, \langle a', b'\rangle)$$

This constructs a path in a sigma type from its components.

3.1.7 The Step List as Computational Content

The step list of a path is its distinguishing feature—the component that makes paths proof-relevant. Let us examine this more closely.

Definition 3.16 (Step extraction). For a path $p = (s, h) : \mathsf{Path}(a, b)$, we write:

- $\mathsf{steps}(p) = s$ for the step list
- $\mathsf{toEq}(p) = h$ for the semantic equality

Lemma 3.17 (Step list properties). *The following hold for step lists:*

1. $\mathsf{steps}(\mathsf{refl}_a) = []$ *(the empty list)*
2. $\mathsf{steps}(\mathsf{step}(s)) = [s]$ *(a singleton list)*
3. $\mathsf{steps}(\mathsf{trans}(p, q)) = \mathsf{steps}(p) +\!\!+ \mathsf{steps}(q)$
4. $\mathsf{steps}(\mathsf{symm}(p)) = \mathsf{rev}(\mathsf{map}(\mathsf{inv}, \mathsf{steps}(p)))$
5. $\mathsf{steps}(\mathsf{congrArg}(f, p)) = \mathsf{map}(f_*, \mathsf{steps}(p))$

where rev *reverses a list and* map *applies a function elementwise.*

Proof. Immediate from the definitions of the path operations. □

Proposition 3.18 (Path equality via step lists). *Two paths* $p, q : \mathsf{Path}(a, b)$ *are equal (as elements of the path type) if and only if they have equal step lists:*

$$p = q \iff \mathsf{steps}(p) = \mathsf{steps}(q)$$

The ambient equality component $\mathsf{toEq}(p) : \mathsf{Eq}(a, b)$ *is fixed once the endpoints are fixed, because* $\mathsf{Eq}(a, b)$ *is proof-irrelevant. Thus the step list is the only proof-relevant datum carried by a path.*

Proof. By construction, a path from a to b consists of a step list together with an inhabitant of the ambient equality type $\mathsf{Eq}(a, b)$. Since $\mathsf{Eq}(a, b)$ lives in the proof-irrelevant ambient equality layer, any two such witnesses are equal. Therefore two paths with the same endpoints are equal exactly when their step lists agree. □

This proposition is fundamental: it tells us that the entire mathematical content of a path lies in its step list. Two paths from a to b are the same exactly when they consist of the same sequence of elementary operations.

3.2 Fundamental Operations: refl, trans, symm, congrArg

Having defined paths and their constructors, we now examine the fundamental operations in detail. Each operation has a precise type, a computational behaviour, and algebraic properties that we will establish. Together, these operations endow the collection of paths on a type with the structure of a groupoid, as we shall prove in Chapter 6.

3.2.1 Reflexivity: The Identity Path

Definition 3.19 (Reflexivity). For any term $a : A$, the *reflexivity path* is:

$$\mathsf{refl}_a : \mathsf{Path}(a, a)$$

Its step list is empty, and its semantic content is the trivial equality $\mathsf{refl} : a = a$.

Typing rule.

$$\frac{\Gamma \vdash a : A}{\Gamma \vdash \mathsf{refl}_a : \mathsf{Path}_A(a, a)}$$

Interpretation. The reflexivity path represents the "trivial" witness to equality: a is equal to itself by virtue of being the same term. No computation is performed; we simply observe the identity.

In groupoid terms, refl_a is the *identity morphism* at the object a. In topological terms, it is the *constant path* that stays at the point a for all time. In categorical terms, it is the identity 1-cell in the fundamental groupoid.

Key properties. Reflexivity will serve as the identity element for path composition:

$$\mathsf{trans}(\mathsf{refl}_a, p) \sim_{\mathsf{rw}} p \qquad \text{(left identity)}$$

$$\mathsf{trans}(p, \mathsf{refl}_b) \sim_{\mathsf{rw}} p \qquad \text{(right identity)}$$

where $\sim_{\mathsf{rw}}$ denotes rewrite equivalence (to be defined precisely in Chapter 4). These properties make paths form a groupoid-like structure.

3.2.2 Transitivity: Path Composition

Definition 3.20 (Transitivity). For paths $p : \mathsf{Path}(a, b)$ and $q : \mathsf{Path}(b, c)$, the *composite path* is:

$$\mathsf{trans}(p, q) : \mathsf{Path}(a, c)$$

Its step list is the concatenation of the component step lists, and its semantic content is the transitivity of the component equalities.

Typing rule.

$$\frac{\Gamma \vdash p : \mathsf{Path}(a, b) \qquad \Gamma \vdash q : \mathsf{Path}(b, c)}{\Gamma \vdash \mathsf{trans}(p, q) : \mathsf{Path}(a, c)}$$

Interpretation. If p witnesses that a rewrites to b, and q witnesses that b rewrites to c, then $\mathsf{trans}(p, q)$ witnesses that a rewrites to c by first following p and then following q. The composite path contains all the steps of p followed by all the steps of q.

In groupoid terms, $\mathsf{trans}(p, q)$ is the *composition* of morphisms. In topological terms, it is *path concatenation*. The concatenation is performed at the intermediate point b: the path p ends where q begins.

Key property: associativity. Composition of paths is associative:

$$\mathsf{trans}(\mathsf{trans}(p, q), r) \sim_{\mathsf{rw}} \mathsf{trans}(p, \mathsf{trans}(q, r))$$

This is not a definitional equality but a *rewrite equivalence*: there is an explicit 2-path (a path between paths) witnessing this equation. The reason is that step-list concatenation is definitionally associative:

$$(s_p \mathbin{+\!\!+} s_q) \mathbin{+\!\!+} s_r = s_p \mathbin{+\!\!+} (s_q \mathbin{+\!\!+} s_r)$$

so the two sides have identical step lists.

Proposition 3.21 (Fourfold associativity). *For four composable paths p, q, r, s, any two parenthesizations give the same result:*

$$\mathsf{trans}(\mathsf{trans}(\mathsf{trans}(p, q), r), s) \sim_{\mathsf{rw}} \mathsf{trans}(p, \mathsf{trans}(q, \mathsf{trans}(r, s)))$$

More generally, the grouping of any finite composition is irrelevant.

Proof. By induction, using the binary associativity law and the fact that list concatenation is strictly associative. □

3.2.3 Symmetry: Path Inversion

Definition 3.22 (Symmetry). For a path $p : \mathsf{Path}(a, b)$, the *inverse path* is:

$$\mathsf{symm}(p) : \mathsf{Path}(b, a)$$

Its step list is obtained by reversing the original list and inverting each step. Its semantic content is the symmetric equality.

Typing rule.

$$\frac{\Gamma \vdash p : \mathsf{Path}(a, b)}{\Gamma \vdash \mathsf{symm}(p) : \mathsf{Path}(b, a)}$$

Interpretation. If p witnesses that a rewrites to b, then $\mathsf{symm}(p)$ witnesses that b rewrites back to a by traversing the same path in reverse. Each forward step becomes a backward step; each reduction becomes an expansion.

In groupoid terms, $\mathsf{symm}(p)$ is the *inverse morphism*. The existence of inverses is what distinguishes groupoids from mere categories. In topological terms, it is the *reversed path*: the same trajectory travelled backward.

Key properties.

Proposition 3.23 (Inverse laws). *For any path p : $\mathsf{Path}(a, b)$:*

1. $\mathsf{trans}(p, \mathsf{symm}(p)) \sim_{\mathsf{rw}} \mathsf{refl}_a$ *(right inverse)*
2. $\mathsf{trans}(\mathsf{symm}(p), p) \sim_{\mathsf{rw}} \mathsf{refl}_b$ *(left inverse)*

These laws assert that following a path and then its reverse brings you back to where you started (up to rewrite equivalence).

Proposition 3.24 (Involution). *For any path p : $\mathsf{Path}(a, b)$:*

$$\mathsf{symm}(\mathsf{symm}(p)) \sim_{\mathsf{rw}} p$$

This says that reversing a path twice returns the original path.

Proposition 3.25 (Contravariance). *For composable paths p and q:*

$$\mathsf{symm}(\mathsf{trans}(p, q)) \sim_{\mathsf{rw}} \mathsf{trans}(\mathsf{symm}(q), \mathsf{symm}(p))$$

Symmetry is *contravariant*: the inverse of a composition is the composition of the inverses in *reverse order*. This is familiar from group theory: $(ab)^{-1} = b^{-1}a^{-1}$.

Proposition 3.26 (Symmetry of reflexivity). *For any term a:*

$$\mathsf{symm}(\mathsf{refl}_a) \sim_{\mathsf{rw}} \mathsf{refl}_a$$

The inverse of the identity path is itself. This is immediate: staying in place and then reversing still leaves you in place.

3.2.4 Congruence: Lifting Paths Through Functions

Definition 3.27 (Argument congruence). For a function $f : A \to B$ and a path $p : \mathsf{Path}_A(a, a')$, the *congruence path* is:

$$\mathsf{congrArg}(f, p) : \mathsf{Path}_B(f(a), f(a'))$$

Each step in p is mapped through f, and the semantic equality is lifted by the standard congruence principle.

Typing rule.

$$\frac{\Gamma \vdash f : A \to B \qquad \Gamma \vdash p : \mathsf{Path}_A(a, a')}{\Gamma \vdash \mathsf{congrArg}(f, p) : \mathsf{Path}_B(f(a), f(a'))}$$

Interpretation. If p witnesses that a equals a' (via a sequence of rewrites), then $\mathsf{congrArg}(f, p)$ witnesses that $f(a)$ equals $f(a')$ by applying f to each step of the rewrite sequence. This captures the fundamental principle that *equals can be substituted for equals*: if two arguments are equal, then the results of applying any function to them are also equal.

In categorical terms, $\mathsf{congrArg}(f, -)$ is a *functor*: it maps morphisms (paths) in A to morphisms in B in a structure-preserving way. This functorial character is established formally as Theorem 6.20 in Chapter 6.

Functoriality properties.

Proposition 3.28 (Preservation of identity). *For any function $f : A \to B$ and term $a : A$:*

$$\mathsf{congrArg}(f, \mathsf{refl}_a) \sim_{\mathsf{rw}} \mathsf{refl}_{f(a)}$$

Applying f to the trivial path at a gives the trivial path at $f(a)$.

Proposition 3.29 (Preservation of composition). *For any function $f : A \to B$ and paths $p : \mathsf{Path}(a, b)$, $q : \mathsf{Path}(b, c)$:*

$$\mathsf{congrArg}(f, \mathsf{trans}(p, q)) \sim_{\mathsf{rw}} \mathsf{trans}(\mathsf{congrArg}(f, p), \mathsf{congrArg}(f, q))$$

Applying f to a composite path gives the composite of the applied paths.

Proposition 3.30 (Preservation of inverses). *For any function $f : A \to B$ and path $p : \mathsf{Path}(a, b)$:*

$$\mathsf{congrArg}(f, \mathsf{symm}(p)) \sim_{\mathsf{rw}} \mathsf{symm}(\mathsf{congrArg}(f, p))$$

Applying f to an inverse path gives the inverse of the applied path.

Proposition 3.31 (Composition of congruences). *For functions $f : A \to B$ and $g : B \to C$, and path $p : \mathsf{Path}_A(a, a')$:*

$$\mathsf{congrArg}(g, \mathsf{congrArg}(f, p)) \sim_{\mathsf{rw}} \mathsf{congrArg}(g \circ f, p)$$

Proposition 3.32 (Identity function). *For any path $p : \mathsf{Path}(a, b)$:*

$$\mathsf{congrArg}(\mathsf{id}, p) \sim_{\mathsf{rw}} p$$

These properties establish that congrArg is a genuine functor: it preserves identities and composition up to rewrite equivalence.

3.2.5 Function Congruence

For completeness, we also define congruence in the function position.

Definition 3.33 (Function congruence)**.** For a path $p : \mathsf{Path}_{A \to B}(f, g)$ between functions and a term $a : A$, the *function congruence path* is:

$$\mathsf{congrFun}(p, a) : \mathsf{Path}_B(f(a), g(a))$$

Typing rule.

$$\frac{\Gamma \vdash p : \mathsf{Path}(f, g) \qquad \Gamma \vdash a : A}{\Gamma \vdash \mathsf{congrFun}(p, a) : \mathsf{Path}(f(a), g(a))}$$

This operation lifts a path between functions to a path between their values at a given argument. It will be particularly important when we consider higher path algebra and the coherence conditions (Chapter 8) for function types.

3.2.6 Transport

A fundamental operation derived from paths is *transport*: the ability to move elements along a path in a dependent type.

Definition 3.34 (Transport)**.** Given a type family $D : A \to \mathsf{Type}$, a path $p : \mathsf{Path}_A(a, b)$, and an element $x : D(a)$, the *transport of x along p* is:

$$\mathsf{transport}(p, x) : D(b)$$

This operation uses the semantic content of p to coerce x from $D(a)$ to $D(b)$.

Transport satisfies expected coherence properties:

Proposition 3.35 (Transport laws)**.**

1. $\mathsf{transport}(\mathsf{refl}_a, x) = x$
2. $\mathsf{transport}(\mathsf{trans}(p, q), x) = \mathsf{transport}(q, \mathsf{transport}(p, x))$
3. $\mathsf{transport}(\mathsf{symm}(p), \mathsf{transport}(p, x)) = x$
4. $\mathsf{transport}(p, \mathsf{transport}(\mathsf{symm}(p), y)) = y$

These show that transport along reflexivity is the identity, transport composes correctly, and transport along inverse paths cancels.

3.3 Proof-Relevance and the Failure of UIP

The computational path framework is fundamentally *proof-relevant*: distinct paths with the same endpoints are genuinely different mathematical objects. This stands in sharp contrast to *proof-irrelevant* approaches where any two proofs of the same equality are identified. The failure of the Uniqueness of Identity Proofs (UIP) principle is not a bug but a feature—it is precisely the source of the higher structure we seek.

Scope note. This section introduces UIP, its companion axiom K, and the simplest counterexample showing their failure for computational paths. The treatment here is deliberately concise: it establishes the key definitions and the foundational `stepLoop` witness that the remaining chapters build on. A comprehensive, self-contained analysis of UIP failure—including stronger counterexamples, trace-based invariants, and the relationship to proof irrelevance—is deferred to Chapter 12.

3.3.1 The UIP Principle

Definition 3.36 (Uniqueness of Identity Proofs). The *Uniqueness of Identity Proofs* (UIP) principle for a type A asserts that for any $a, b : A$ and any two proofs p, q of $a = b$, the proofs are themselves equal:

$$\mathsf{UIP}_A : \prod_{a,b:A} \prod_{p,q:\mathsf{Id}_A(a,b)} \mathsf{Id}_{\mathsf{Id}_A(a,b)}(p, q)$$

Equivalently: the type $\mathsf{Id}_A(a, b)$ is a *proposition* (has at most one element).

Definition 3.37 (Streicher's K principle [86]). Streicher's axiom K asserts that every self-identity is reflexivity:

$$\mathsf{K}_A : \prod_{a:A} \prod_{p:\mathsf{Id}_A(a,a)} p = \mathsf{refl}_a$$

The K principle is equivalent to UIP in the presence of the standard rules for identity types.

In classical foundations (set theory, or type theory with proof-irrelevant equality), UIP holds automatically: the type $a = b$ is either empty (if $a \neq b$) or, when inhabited, has a unique witness up to proof irrelevance. But this is because the equality type carries no computational structure.

3.3.2 Failure of UIP for Computational Paths

Theorem 3.38 (UIP fails for paths). *The type* $\mathsf{Path}(a, b)$ *does not satisfy UIP. For any nonempty type* A*, there exist terms and paths such that two paths with the same endpoints are not equal:*

$$\exists a : A,\ \exists p, q : \mathsf{Path}(a, a), \quad p \neq q$$

Proof. Let a be any term in a nonempty type A. Consider two paths:

1. $p = \mathsf{refl}_a$, the reflexivity path with empty step list.
2. $q = \mathsf{step}((a, a, \mathsf{refl}))$, a path with a single trivial step.

Both paths have source and target equal to a. However:

$$\mathsf{steps}(p) = [] \neq [(a, a, \mathsf{refl})] = \mathsf{steps}(q)$$

Since paths with different step lists are unequal (Proposition 3.18), we have $p \neq q$. This first counterexample is deliberately minimal: it shows that UIP already fails at the level of raw path representations, even before one exploits genuinely different computational histories. Chapter 12 develops stronger non-UIP phenomena using nontrivial trace invariants. □

Corollary 3.39 (K fails for paths). *Streicher's K principle fails for* Path*: not every loop (path from* a *to itself) equals* refl_a.

Proof. The path q in the proof above is a loop at a but $q \neq \mathsf{refl}_a$. □

3.3.3 The Canonical Counterexample

We now present the simplest and most illuminating counterexample to UIP, which serves as the foundational witness for the non-triviality of path spaces.

Example 3.40 (The stepLoop counterexample). Fix any type A and any element $a : A$. We construct two paths $p, q : \mathsf{Path}(a, a)$ and prove $p \neq q$.

Construction:

$$p := \mathsf{refl}_a = \mathsf{Path.mk}\ []\ \mathsf{rfl} \qquad \text{(empty step list)}$$

$$q := \mathsf{stepLoop}(a) = \mathsf{Path.mk}\ [(a, a, \mathsf{rfl})]\ \mathsf{rfl} \qquad \text{(singleton step list)}$$

Both p and q are loops at a—they have identical source (a), target (a), and underlying equality proof (rfl). The *only* difference is their step lists.

Proof of non-identity: Suppose for contradiction that $p = q$. Then by congruence:

$$\mathsf{steps}(p) = \mathsf{steps}(q)$$

But:

$$[] = [(a, a, \mathsf{rfl})]$$

since the empty list and a singleton list have different lengths. This contradicts our assumption, so $p \neq q$.

Remark 3.41 (Why this counterexample is fundamental)**.** The stepLoop example reveals the essential mechanism of UIP failure:

1. **Semantic agreement**: Both paths witness the same propositional equality ($a = a$ via rfl).
2. **Syntactic disagreement**: The paths differ in their *computational record*—the sequence of rewrite steps.
3. **Step counting**: The invariant $|p| = 0 \neq 1 = |q|$ provides a simple, decidable witness of non-identity.
4. **Universality**: This construction works for *any* type A with *any* element a—there is no escape from UIP failure.

The key insight is that while p and q represent the "same" equality in the proof-irrelevant sense, they encode *different computational histories*. The path p says "a equals a by doing nothing," while q says "a equals a by explicitly recording the reflexivity step." Chapter 12 revisits this phenomenon in depth and develops it into the non-UIP theorem for all nonempty types. The step-counting invariant introduced below is extended to an additive functional on paths in Section 12.3 of that chapter.

Corollary 3.42 (Infinite loop spaces)**.** *For any type A and element $a : A$, the loop space $\mathsf{Path}(a, a)$ is infinite. Specifically, define:*

$$p_n := \underbrace{\mathsf{stepLoop}(a) \cdot \mathsf{stepLoop}(a) \cdot \ldots \cdot \mathsf{stepLoop}(a)}_{n \text{ times}}$$

with $p_0 = \mathsf{refl}_a$. Then $|p_n| = n$, and for $m \neq n$, we have $p_m \neq p_n$.

Proof. By the step-count properties established later in this section, $|p_n| = n$. If $p_m = p_n$, then $m = |p_m| = |p_n| = n$. Contrapositive gives $m \neq n \Rightarrow p_m \neq p_n$. □

This shows that every point in a type carries not just "the point itself" but an entire countably infinite space of self-identifications. This is a first glimpse of the higher categorical structure inherent in computational paths. The Eckmann–Hilton argument (Chapter 10) shows that the composition of 2-dimensional loops is abelian, mirroring the classical result that π_n is abelian for $n \geq 2$.

3.3.4 Explicit Counterexample: Diamond Reduction

A complementary counterexample arises from the *diamond property* of λ-calculus reduction. Consider a term with two redexes that can be reduced in either order.

Example 3.43 (Diamond counterexample)**.** Let $t = (\lambda x.\,(\lambda y.\,y)\,x)\,z$, a term with an inner and an outer redex. There are two reduction paths to the normal form z:

Path p (outer first):

$$\begin{aligned} t &= (\lambda x.\,(\lambda y.\,y)\,x)\,z \\ &\xrightarrow{\beta_1} (\lambda y.\,y)\,z && \text{(reduce outer redex)} \\ &\xrightarrow{\beta_2} z && \text{(reduce remaining redex)} \end{aligned}$$

Path q (inner first):

$$\begin{aligned} t &= (\lambda x.\,(\lambda y.\,y)\,x)\,z \\ &\xrightarrow{\xi} (\lambda x.\,x)\,z && \text{(reduce under } \lambda) \\ &\xrightarrow{\beta_3} z && \text{(reduce outer redex)} \end{aligned}$$

Both paths start at t and end at z, but they go through different intermediate terms:

- Path p goes through $(\lambda y.\,y)\,z$
- Path q goes through $(\lambda x.\,x)\,z$

The step lists are:

$$\begin{aligned} \mathsf{steps}(p) &= [\beta_1, \beta_2] \\ \mathsf{steps}(q) &= [\xi, \beta_3] \end{aligned}$$

These are different sequences of steps, so $p \neq q$ as paths.

This example illustrates a deep phenomenon: in any system with confluent but non-trivial rewriting, there are typically multiple paths between terms. The Church–Rosser theorem [22] guarantees these paths reach the same destination, but the paths themselves are distinct.

3.3.5 Step Counting as an Invariant

The step count is the simplest tool for proving that two paths differ: it assigns to each path a natural number recording its length, and paths with different lengths are necessarily distinct. This invariant is coarse—it cannot distinguish all pairs of unequal paths—but its simplicity makes it the right instrument for the foundational non-UIP results of this section and Chapter 12.

Definition 3.44 (Step count). For a path $p : \mathsf{Path}(a, b)$, the *step count* is:

$$|p| = \mathsf{length}(\mathsf{steps}(p))$$

the number of elementary steps in the path.

Lemma 3.45 (Step count properties). *1.* $|\mathsf{refl}_a| = 0$

2. $|\mathsf{step}(s)| = 1$

3. $|\mathsf{trans}(p, q)| = |p| + |q|$

4. $|\mathsf{symm}(p)| = |p|$

5. $|\mathsf{congrArg}(f, p)| = |p|$

Proof. Immediate from the definitions: reflexivity has no steps, single steps have one step, composition concatenates step lists, symmetry preserves length (it reverses but doesn't add or remove), and congruence maps step-by-step. □

Corollary 3.46 (Paths with different step counts are unequal). *If* $|p| \neq |q|$*, then* $p \neq q$.

This gives a simple test: paths with different step counts are necessarily distinct. The invariant cannot distinguish all pairs of distinct paths (two paths may share the same step count yet differ in their individual steps), but it suffices for many purposes.

Example 3.47 (Infinitely many loops). At any point a, there are infinitely many distinct loops:

$$\begin{aligned} p_0 &= \mathsf{refl}_a & |p_0| &= 0 \\ p_1 &= \mathsf{step}((a, a, \mathsf{refl})) & |p_1| &= 1 \\ p_2 &= \mathsf{trans}(p_1, p_1) & |p_2| &= 2 \\ p_n &= \mathsf{trans}(p_{n-1}, p_1) & |p_n| &= n \end{aligned}$$

These form an infinite family of loops at a, all distinct because they have different step counts. The "loop space" $\mathsf{Path}(a, a)$ is thus infinite (in fact, countably infinite with this construction).

3.3.6 Significance of Non-UIP

The failure of UIP has profound consequences:

1. Paths carry genuine computational information. A path is not just a certificate that two terms are equal; it records *how* the equality is established. Different computational routes are preserved as different mathematical objects.

2. Loop spaces are non-trivial. The type $\mathsf{Path}(a, a)$ of self-paths at a is not necessarily trivial. It can contain many elements, and these elements can themselves be related by higher paths. This is the beginning of the higher structure.

3. Higher structure emerges naturally. If paths can be different, we can ask: when are two paths "the same" in a meaningful sense? The answer is given by *2-paths*—paths between paths. And since 2-paths can also differ, we get 3-paths, and so on. This infinite tower of structure is exactly what's captured by the notion of a weak ω-groupoid.

4. Connection to homotopy type theory. In Homotopy Type Theory (HoTT; see Chapter 14), the identity type $\mathsf{Id}_A(a, b)$ is inherently proof-relevant—UIP fails in general. Our computational paths provide a concrete, syntactic model of this behaviour. The paths between terms correspond to homotopies between points; the higher paths correspond to higher homotopies.

3.4 Path Algebra: First Properties

Having established the basic definitions and the non-triviality of paths, we now develop their algebraic theory. The paths on any type form a structure satisfying the laws of a *groupoid*—a category where every morphism has an inverse. But more is true: there are also paths between paths, and paths between those, and the entire structure forms what is known as a *weak ω-groupoid* (Chapter 9).

Here we preview the main algebraic identities; their complete development, including all 77 rewrite rules, occupies Chapter 4.

3.4.1 The Groupoid Laws

A groupoid is a category in which every morphism is invertible. The collection of paths on a type A forms a groupoid with:

- Objects: the terms of type A

- Morphisms from a to b: the paths $p : \mathsf{Path}(a, b)$
- Identity at a: the reflexivity path refl_a
- Composition: transitivity $\mathsf{trans}(p, q)$
- Inverse: symmetry $\mathsf{symm}(p)$

However, the groupoid laws hold only *up to rewrite equivalence*, not as strict equalities. We denote rewrite equivalence by $\sim_{\mathsf{rw}}$ and develop it formally in the next chapter.

Proposition 3.48 (Groupoid laws for paths). *The following identities hold up to rewrite equivalence:*

Identity laws.

$$\mathsf{trans}(\mathsf{refl}_a, p) \sim_{\mathsf{rw}} p \qquad \text{(G1: left identity)}$$
$$\mathsf{trans}(p, \mathsf{refl}_b) \sim_{\mathsf{rw}} p \qquad \text{(G2: right identity)}$$

Inverse laws.

$$\mathsf{trans}(p, \mathsf{symm}(p)) \sim_{\mathsf{rw}} \mathsf{refl}_a \qquad \text{(G3: right inverse)}$$
$$\mathsf{trans}(\mathsf{symm}(p), p) \sim_{\mathsf{rw}} \mathsf{refl}_b \qquad \text{(G4: left inverse)}$$

Associativity.

$$\mathsf{trans}(\mathsf{trans}(p, q), r) \sim_{\mathsf{rw}} \mathsf{trans}(p, \mathsf{trans}(q, r)) \qquad \text{(G5: associativity)}$$

Involution.

$$\mathsf{symm}(\mathsf{symm}(p)) \sim_{\mathsf{rw}} p \qquad \text{(G6: involution)}$$

Contravariance of inverses.

$$\mathsf{symm}(\mathsf{trans}(p, q)) \sim_{\mathsf{rw}} \mathsf{trans}(\mathsf{symm}(q), \mathsf{symm}(p)) \qquad \text{(G7: contravariance)}$$

Symmetry of reflexivity.

$$\mathsf{symm}(\mathsf{refl}_a) \sim_{\mathsf{rw}} \mathsf{refl}_a \qquad \text{(G8: symm of refl)}$$

Proof. **(G1) Left identity.** By definition:

$$\mathsf{trans}(\mathsf{refl}_a, p) = ([\,] \mathbin{+\!\!+} \mathsf{steps}(p), h) = (\mathsf{steps}(p), h) = p$$

since $[\,] \mathbin{+\!\!+} s = s$ for every list s.

(G2) Right identity. Similarly:

$$\mathsf{trans}(p, \mathsf{refl}_b) = (\mathsf{steps}(p) \mathbin{+\!\!+} [\,], h) = (\mathsf{steps}(p), h) = p$$

(G3) Right inverse. The composition $\mathsf{trans}(p, \mathsf{symm}(p))$ has step list $\mathsf{steps}(p) \mathbin{+\!\!+} \mathsf{rev}(\mathsf{map}(\mathsf{inv}, \mathsf{steps}(p)))$. Chapter 4 introduces the inverse rewrite rule sending this expression to refl_a. Hence $\mathsf{trans}(p, \mathsf{symm}(p)) \sim_{\mathsf{rw}} \mathsf{refl}_a$.

(G4) Left inverse. Dually, the corresponding inverse rewrite rule in Chapter 4 gives $\mathsf{trans}(\mathsf{symm}(p), p) \sim_{\mathsf{rw}} \mathsf{refl}_b$.

(G5) Associativity. By strict equality of step lists:

$$(\mathsf{steps}(p) \mathbin{+\!\!+} \mathsf{steps}(q)) \mathbin{+\!\!+} \mathsf{steps}(r) = \mathsf{steps}(p) \mathbin{+\!\!+} (\mathsf{steps}(q) \mathbin{+\!\!+} \mathsf{steps}(r))$$

since list concatenation is associative.

(G6) Involution. We compute:

$$\begin{aligned}
&\mathsf{steps}(\mathsf{symm}(\mathsf{symm}(p))) \\
&\quad = \mathsf{rev}(\mathsf{map}(\mathsf{inv}, \\
&\qquad\quad \mathsf{rev}(\mathsf{map}(\mathsf{inv}, \mathsf{steps}(p))))) \\
&\quad = \mathsf{map}(\mathsf{inv}, \mathsf{map}(\mathsf{inv}, \mathsf{steps}(p))) \\
&\quad = \mathsf{map}(\mathsf{inv} \circ \mathsf{inv}, \mathsf{steps}(p)) \\
&\quad = \mathsf{steps}(p)
\end{aligned}$$

using the fact that $\mathsf{rev} \circ \mathsf{rev} = \mathsf{id}$ and $\mathsf{inv} \circ \mathsf{inv} = \mathsf{id}$.

(G7) Contravariance. Writing $s_p = \mathsf{steps}(p)$ and $s_q = \mathsf{steps}(q)$,

$$\begin{aligned}
\mathsf{steps}(\mathsf{symm}(\mathsf{trans}(p, q))) &= \mathsf{rev}(\mathsf{map}(\mathsf{inv}, s_p \mathbin{+\!\!+} s_q)) \\
&= \mathsf{rev}(\mathsf{map}(\mathsf{inv}, s_q)) \mathbin{+\!\!+} \mathsf{rev}(\mathsf{map}(\mathsf{inv}, s_p)) \\
&= \mathsf{steps}(\mathsf{trans}(\mathsf{symm}(q), \mathsf{symm}(p))).
\end{aligned}$$

(G8) Symmetry of reflexivity. We have:

$$\mathsf{steps}(\mathsf{symm}(\mathsf{refl}_a)) = \mathsf{rev}(\mathsf{map}(\mathsf{inv}, [\,])) = [\,] = \mathsf{steps}(\mathsf{refl}_a)$$

□

These eight laws are the core of groupoid structure. They will be witnessed by explicit 2-paths (paths between paths) when we develop the full rewrite system.

3.4.2 Proof of the Inverse Laws

The inverse laws (G3) and (G4) deserve a more detailed treatment, as they are central to the groupoid structure and illustrate the role of the cancellation rewrite rules.

Lemma 3.49 (Left-right cancellation). *For any path* $p : \mathsf{Path}(a, b)$ *and path* $q : \mathsf{Path}(a, c)$*:*

$$\mathsf{trans}(p, \mathsf{trans}(\mathsf{symm}(p), q)) \sim_{\mathsf{rw}} q$$

Similarly, for $p : \mathsf{Path}(a, b)$ *and* $q : \mathsf{Path}(b, c)$*:*

$$\mathsf{trans}(\mathsf{symm}(p), \mathsf{trans}(p, q)) \sim_{\mathsf{rw}} q$$

Proof. The first equation follows from:

$$\begin{aligned}\mathsf{trans}(p, \mathsf{trans}(\mathsf{symm}(p), q)) &\sim_{\mathsf{rw}} \mathsf{trans}(\mathsf{trans}(p, \mathsf{symm}(p)), q) \\ &\qquad \text{(by associativity)} \\ &\sim_{\mathsf{rw}} \mathsf{trans}(\mathsf{refl}_a, q) \quad \text{(by right inverse)} \\ &\sim_{\mathsf{rw}} q \quad \text{(by left identity)}\end{aligned}$$

The second equation is proved similarly using the left inverse law. □

These cancellation laws will appear as primitive rewrite rules (R76–R77) in Chapter 4, avoiding the need to go through the intermediate steps.

3.4.3 Functoriality of Congruence

The congruence operation $\mathsf{congrArg}(f, -)$ is a *functor* from the groupoid of paths in A to the groupoid of paths in B.

Proposition 3.50 (Functoriality laws). *For any function* $f : A \to B$*:*

Preservation of identity.

$$\mathsf{congrArg}(f, \mathsf{refl}_a) \sim_{\mathsf{rw}} \mathsf{refl}_{f(a)} \qquad \text{(F1: preserves identity)}$$

Preservation of composition.

$$\mathsf{congrArg}(f, \mathsf{trans}(p, q)) \sim_{\mathsf{rw}} \mathsf{trans}(\mathsf{congrArg}(f, p), \mathsf{congrArg}(f, q)) \qquad \text{(F2: preserves comp)}$$

Preservation of inverses.

$$\mathsf{congrArg}(f, \mathsf{symm}(p)) \sim_{\mathsf{rw}} \mathsf{symm}(\mathsf{congrArg}(f, p)) \qquad \text{(F3: preserves inv)}$$

Proof. **(F1)** By definition:

$$\mathsf{steps}(\mathsf{congrArg}(f, \mathsf{refl}_a)) = \mathsf{map}(f_*, []) = [] = \mathsf{steps}(\mathsf{refl}_{f(a)})$$

(F2) We compute:

$$\begin{aligned}\mathsf{steps}(\mathsf{congrArg}(f, \mathsf{trans}(p, q))) &= \mathsf{map}(f_*, \mathsf{steps}(p) +\!\!+ \mathsf{steps}(q)) \\ &= \mathsf{map}(f_*, \mathsf{steps}(p)) \\ &\quad +\!\!+ \mathsf{map}(f_*, \mathsf{steps}(q)) \\ &= \mathsf{steps}(\mathsf{congrArg}(f, p)) \\ &\quad +\!\!+ \mathsf{steps}(\mathsf{congrArg}(f, q)) \\ &= \mathsf{steps}(\mathsf{trans}(\mathsf{congrArg}(f, p), \\ &\qquad \mathsf{congrArg}(f, q)))\end{aligned}$$

(F3) We have:

$$\begin{aligned}&\mathsf{steps}(\mathsf{congrArg}(f, \mathsf{symm}(p))) \\ &\quad = \mathsf{map}(f_*, \mathsf{rev}(\mathsf{map}(\mathsf{inv}, \mathsf{steps}(p)))) \\ &\quad = \mathsf{rev}(\mathsf{map}(f_*, \mathsf{map}(\mathsf{inv}, \mathsf{steps}(p)))) \\ &\quad = \mathsf{rev}(\mathsf{map}(f_* \circ \mathsf{inv}, \mathsf{steps}(p))) \\ &\quad = \mathsf{rev}(\mathsf{map}(\mathsf{inv} \circ f_*, \mathsf{steps}(p))) \\ &\qquad \text{(by naturality of inversion)} \\ &\quad = \mathsf{steps}(\mathsf{symm}(\mathsf{congrArg}(f, p)))\end{aligned}$$

where we use the fact that $f_*(\bar{s}) = \overline{f_*(s)}$. □

Proposition 3.51 (Congruence composition). *For functions $f : A \to B$ and $g : B \to C$:*

$$\mathsf{congrArg}(g, \mathsf{congrArg}(f, p)) \sim_{\mathsf{rw}} \mathsf{congrArg}(g \circ f, p) \qquad \text{(F4: functor comp)}$$

And for the identity function:

$$\mathsf{congrArg}(\mathsf{id}, p) \sim_{\mathsf{rw}} p \qquad \text{(F5: identity functor)}$$

Proof. **(F4)** By computation on step lists:

$$\begin{aligned}\mathsf{steps}(\mathsf{congrArg}(g, \mathsf{congrArg}(f, p))) &= \mathsf{map}(g_*, \mathsf{map}(f_*, \mathsf{steps}(p))) \\ &= \mathsf{map}(g_* \circ f_*, \mathsf{steps}(p)) \\ &= \mathsf{map}((g \circ f)_*, \mathsf{steps}(p)) \\ &= \mathsf{steps}(\mathsf{congrArg}(g \circ f, p))\end{aligned}$$

(F5) Since $\mathsf{id}_*(s) = s$ for any step s:

$$\mathsf{steps}(\mathsf{congrArg}(\mathsf{id}, p)) = \mathsf{map}(\mathsf{id}_*, \mathsf{steps}(p)) = \mathsf{steps}(p)$$

□

These laws say that congrArg behaves exactly as a functor should: it preserves all the categorical structure up to equivalence.

3.4.4 Cancellation Laws

An important family of identities involves *cancellation*: composing a path with its inverse yields the identity, but there are also partial cancellation laws when inverses appear in compositions.

Proposition 3.52 (Cancellation laws).

$$\mathsf{trans}(p, \mathsf{trans}(\mathsf{symm}(p), q)) \sim_{\mathsf{rw}} q \qquad \text{(C1: left cancellation)}$$
$$\mathsf{trans}(\mathsf{symm}(p), \mathsf{trans}(p, q)) \sim_{\mathsf{rw}} q \qquad \text{(C2: right cancellation)}$$

These laws provide a "shortcut" through the inverse laws plus associativity. They express that p and $\mathsf{symm}(p)$ can be cancelled when they appear adjacent (after reassociation).

3.4.5 Preview: Mac Lane Coherence

When working with multiple compositions, different association patterns must be coherent. The higher cells witnessing this coherence will only be introduced in Chapter 7; for now we record the shape of the first constraint those witnesses must satisfy.

Remark 3.53 (Pentagon identity, preview). For four composable paths p, q, r, s, the two canonical ways of reassociating from $((pq)r)s$ to $p(q(rs))$ are related by a canonical 2-cell in the higher theory:

$$\alpha_{p,q,r} \cdot (\mathsf{id}_p \cdot \alpha_{q,r,s}) = \alpha_{p,qr,s} \cdot \alpha_{p,q,rs}$$

where $\alpha_{p,q,r} : (pq)r \sim_{\mathsf{rw}} p(qr)$ is the associator 2-path.

This is a first instance of *higher coherence*: not only do the groupoid laws hold, but the witnesses to those laws satisfy their own coherence conditions. This pattern continues at all levels, giving rise to the full structure of a weak ω-groupoid.

3.4.6 Preview: Triangle Identity

The triangle identity relates associativity and the identity laws. As with the pentagon, the precise witness belongs to the 2-cell calculus developed later; here we only record the phenomenon.

Remark 3.54 (Triangle identity, preview)**.** For composable paths p and q:

$$\mathsf{trans}(\mathsf{trans}(p, \mathsf{refl}), q) \sim_{\mathsf{rw}} \mathsf{trans}(p, q)$$

and this is witnessed by composing the right identity law with the trivial 2-path on q.

3.4.7 Preview: Hexagon Identity for Inverses

The distribution of symmetry over composition satisfies a hexagon-like coherence. Again, the full statement lives one categorical level up and will be formalized only after 2-cells are available.

Remark 3.55 (Hexagon identity, preview)**.** For three composable paths p, q, r:

$$\mathsf{symm}(\mathsf{trans}(\mathsf{trans}(p, q), r)) \sim_{\mathsf{rw}} \mathsf{trans}\big(\mathsf{symm}(r), \mathsf{trans}(\mathsf{symm}(q), \mathsf{symm}(p))\big)$$

This can be derived by iterating the contravariance law (G7).

3.4.8 Preview: Naturality

The algebraic operations on paths satisfy *naturality* conditions with respect to congruence. We state the guiding pattern now and give the actual 2-cell construction in Chapter 7.

Remark 3.56 (Naturality of associativity, preview)**.** For composable paths p, q, r and a function f:

$$\mathsf{congrArg}(f, \alpha_{p,q,r}) = \alpha_{f_*(p), f_*(q), f_*(r)}$$

where α denotes the associator 2-path. Similar naturality holds for all the groupoid isomorphisms.

3.4.9 Preview of 2-Paths

The equations above are stated as "rewrite equivalences" $\sim_{\mathsf{rw}}$. But in the computational path framework, every equivalence is *witnessed* by an explicit higher path.

Definition 3.57 (2-paths)**.** A *2-path* from p to q, where $p, q : \mathsf{Path}(a, b)$, is an element:

$$\alpha : \mathsf{Path}_2(p, q)$$

of the *2-path type*. This witnesses that p and q are rewrite- equivalent.

Each groupoid law in Proposition 3.48 is witnessed by a specific 2-path:

- $\lambda_p : \mathsf{Path}_2(\mathsf{trans}(\mathsf{refl}, p), p)$ witnesses left identity
- $\rho_p : \mathsf{Path}_2(\mathsf{trans}(p, \mathsf{refl}), p)$ witnesses right identity
- $\varepsilon_p : \mathsf{Path}_2(\mathsf{trans}(p, \mathsf{symm}(p)), \mathsf{refl})$ witnesses right inverse
- $\eta_p : \mathsf{Path}_2(\mathsf{trans}(\mathsf{symm}(p), p), \mathsf{refl})$ witnesses left inverse
- $\alpha_{p,q,r} : \mathsf{Path}_2(\mathsf{trans}(\mathsf{trans}(p, q), r), \mathsf{trans}(p, \mathsf{trans}(q, r)))$ witnesses associativity

These 2-paths are not arbitrary: they are generated by specific rewrite rules and satisfy their own coherence conditions. The full development is given in subsequent chapters.

3.5 Examples and Calculations

We conclude this chapter with concrete examples that illustrate the definitions and concepts introduced.

3.5.1 Building Paths

Example 3.58 (Simple β-reduction path)**.** Consider the terms $t = (\lambda x.\, x)\, y$ and $u = y$. These are related by a β-reduction: the redex $(\lambda x.\, x)\, y$ contracts to y.

The computational path witnessing this is:

$$p = \mathsf{step}(\beta) : \mathsf{Path}((\lambda x.\, x)\, y, y)$$

where β is the elementary step corresponding to this specific β-reduction. The step list is $[\beta]$, a singleton.

Example 3.59 (A composite path)**.** Consider $t = (\lambda x.\, x)\, ((\lambda y.\, y)\, z)$ and $u = z$. We can build a path in stages:

First, a path for the inner reduction:

$$p_1 : \mathsf{Path}((\lambda y.\, y)\, z, z)$$

with $p_1 = \mathsf{step}(\beta_1)$.

Then, lift this through the outer application using congruence:

$$p_2 = \mathsf{congrArg}((\lambda t.\, (\lambda x.\, x)\, t), p_1) : \mathsf{Path}((\lambda x.\, x)\, ((\lambda y.\, y)\, z), (\lambda x.\, x)\, z)$$

Finally, add the outer reduction:

$$p_3 : \mathsf{Path}((\lambda x.\, x)\, z, z)$$

with $p_3 = \mathsf{step}(\beta_2)$.

The complete path is the composition:

$$p = \mathsf{trans}(p_2, p_3) : \mathsf{Path}(t, z)$$

with step list $[\nu(\beta_1), \beta_2]$ (where ν indicates the step is in argument position).

Example 3.60 (A loop constructed from cancellation)**.** Let $p : \mathsf{Path}(a, b)$ be any path. The composition:

$$q = \mathsf{trans}(p, \mathsf{symm}(p)) : \mathsf{Path}(a, a)$$

is a loop at a. By the right inverse law, $q \sim_{\mathsf{rw}} \mathsf{refl}_a$, meaning there is a 2-path witnessing their rewrite equivalence.

However, $q \neq \mathsf{refl}_a$ as elements of $\mathsf{Path}(a, a)$: if p has n steps, then q has $2n$ steps, while refl_a has 0 steps. They are equivalent but not equal.

3.5.2 Verifying Algebraic Identities

Example 3.61 (Verifying left identity)**.** Let $p : \mathsf{Path}(a, b)$ with step list $s = [s_1, \ldots, s_n]$.

The left-hand side is:

$$\mathsf{trans}(\mathsf{refl}_a, p) = ([\,] \mathbin{+\!\!+} s, h) = (s, h) = p$$

Observe that this is in fact a strict equality, not merely a rewrite equivalence.

Because list concatenation satisfies $[\,] \mathbin{+\!\!+} s = s$ definitionally, we have $\mathsf{trans}(\mathsf{refl}_a, p) = p$ *strictly* in this model. The left identity law holds as an equality, not merely as a rewrite equivalence.

However, the right identity is more subtle:

$$\mathsf{trans}(p, \mathsf{refl}_b) = (s \mathbin{+\!\!+} [\,], h) = (s, h) = p$$

This also holds strictly because $s \mathbin{+\!\!+} [\,] = s$.

The situation changes when we consider the *rewrite rules*: the identity laws appear in the rewrite system because the general framework doesn't assume these strict equalities. In a more abstract setting (such as the categorical semantics), the laws hold only up to isomorphism.

Example 3.62 (Verifying associativity). Let $p : \mathsf{Path}(a, b)$, $q : \mathsf{Path}(b, c)$, $r : \mathsf{Path}(c, d)$ with step lists s_p, s_q, s_r respectively.

Left-hand side:

$$\mathsf{trans}(\mathsf{trans}(p, q), r) = \big((s_p +\!\!+ s_q) +\!\!+ s_r,\ h_{pqr}\big)$$

Right-hand side:

$$\mathsf{trans}(p,\ \mathsf{trans}(q, r)) = \big(s_p +\!\!+ (s_q +\!\!+ s_r),\ h_{pqr}\big)$$

Since list concatenation is associative:

$$(s_p +\!\!+ s_q) +\!\!+ s_r = s_p +\!\!+ (s_q +\!\!+ s_r)$$

the two sides are strictly equal:

$$\mathsf{trans}(\mathsf{trans}(p, q), r) = \mathsf{trans}(p, \mathsf{trans}(q, r)).$$

Again, associativity holds strictly in this concrete model, while the rewrite system treats it as a non-trivial rule for generality.

Example 3.63 (Verifying contravariance of symmetry). Let $p : \mathsf{Path}(a, b)$ with step list $s_p = [s_1, \ldots, s_m]$ and $q : \mathsf{Path}(b, c)$ with step list $s_q = [t_1, \ldots, t_n]$.

The left-hand side $\mathsf{symm}(\mathsf{trans}(p, q))$ has:

$$\begin{aligned}
&\mathsf{steps}(\mathsf{symm}(\mathsf{trans}(p, q)))\\
&\quad = \mathsf{rev}(\mathsf{map}(\mathsf{inv},\ s_p +\!\!+ s_q))\\
&\quad = \mathsf{rev}([\bar{s}_1, \ldots, \bar{s}_m, \bar{t}_1, \ldots, \bar{t}_n])\\
&\quad = [\bar{t}_n, \ldots, \bar{t}_1, \bar{s}_m, \ldots, \bar{s}_1]
\end{aligned}$$

The right-hand side $\mathsf{trans}(\mathsf{symm}(q), \mathsf{symm}(p))$ has:

$$\begin{aligned}
&\mathsf{steps}(\mathsf{trans}(\mathsf{symm}(q), \mathsf{symm}(p)))\\
&\quad = \mathsf{steps}(\mathsf{symm}(q)) +\!\!+ \mathsf{steps}(\mathsf{symm}(p))\\
&\quad = [\bar{t}_n, \ldots, \bar{t}_1] +\!\!+ [\bar{s}_m, \ldots, \bar{s}_1]\\
&\quad = [\bar{t}_n, \ldots, \bar{t}_1, \bar{s}_m, \ldots, \bar{s}_1]
\end{aligned}$$

The step lists are identical, so:

$$\mathsf{symm}(\mathsf{trans}(p, q)) = \mathsf{trans}(\mathsf{symm}(q), \mathsf{symm}(p))$$

holds strictly. The contravariance law (G7) is a definitional equality in our model.

3.5.3 Non-Trivial Loop Spaces

The examples below illustrate a single underlying point: *every inhabited type, no matter how elementary, carries an infinite loop space once we track computational content.* The richness is not an artefact of exotic type constructions; it is a direct consequence of proof relevance.

Example 3.64 (Loop space of the integers)**.** Consider the type $\mathbb{Z}$ of integers with its usual operations. The loop space $\mathsf{Path}_{\mathbb{Z}}(0, 0)$ contains all computational paths from 0 to itself.

One family of such loops arises from the equation $n + (-n) = 0$:

$$p_n = \mathsf{step}(\text{“}n + (-n) = 0\text{”}) : \mathsf{Path}(0, 0)$$

Another family from associativity and commutativity manipulations:

$$q = \mathsf{step}(\text{“}0 + 0 = 0\text{”}) : \mathsf{Path}(0, 0)$$

These form distinct loops, demonstrating that even simple types can have rich loop spaces when we track computational content.

Example 3.65 (Loop space of the booleans)**.** Consider the type $\mathsf{Bool} = \{\mathsf{true}, \mathsf{false}\}$. At each point, there are infinitely many loops:

$$\begin{aligned} \mathsf{refl}_{\mathsf{true}} &: \mathsf{Path}(\mathsf{true}, \mathsf{true}) && \text{(0 steps)} \\ \mathsf{step}((\mathsf{true}, \mathsf{true}, \mathsf{refl})) &: \mathsf{Path}(\mathsf{true}, \mathsf{true}) && \text{(1 step)} \end{aligned}$$

and similarly for false. The loop spaces $\mathsf{Path}(\mathsf{true}, \mathsf{true})$ and $\mathsf{Path}(\mathsf{false}, \mathsf{false})$ are both infinite.

However, $\mathsf{Path}(\mathsf{true}, \mathsf{false})$ is empty: there is no computational path from true to false because these values are definitionally distinct.

Example 3.66 (Paths in function types)**.** Let $f, g : A \to B$ be functions. A path $p : \mathsf{Path}(f, g)$ witnesses that f and g are equal. In the simply-typed λ-calculus, such a path typically arises from:

- An η-expansion/contraction: $\lambda x.\, f\, x \overset{\eta}{\leftrightarrow} f$
- A pointwise equality lifted through λ: if $\forall x.\, f\, x = g\, x$, then $f = g$
- A chain of ξ-steps under the λ-binder

For example, if $f = \lambda x.\, (\lambda y.\, y)\, x$ and $g = \lambda x.\, x$, then there is a path:

$$p : \mathsf{Path}(f, g)$$

consisting of a ξ-step that reduces the inner redex under the outer λ.

3.5.4 Computing with Dependent Paths

Example 3.67 (Transport along a path). Let $D : A \to \mathsf{Type}$ be a type family, and let $p : \mathsf{Path}(a, b)$ in A. If we have $x : D(a)$, we can transport it along p to get $\mathsf{transport}(p, x) : D(b)$.

Concretely, suppose $A = \mathbb{N}$ and $D(n) = \mathsf{Vec}(n, B)$, the type of vectors of length n with entries in B. If $p : \mathsf{Path}(n, m)$ witnesses that $n = m$ (e.g., through arithmetic simplification), then:

$$\mathsf{transport}(p, v) : \mathsf{Vec}(m, B)$$

for any $v : \mathsf{Vec}(n, B)$. The transport "moves" the vector from one length index to another.

Example 3.68 (Dependent paths in sigma types). Consider a sigma type $\Sigma_{x:A} B(x)$. A path between two elements $\langle a, b \rangle$ and $\langle a', b' \rangle$ consists of:

1. A path $p : \mathsf{Path}_A(a, a')$ in the first component
2. A path $q : \mathsf{Path}_{B(a')}(\mathsf{transport}(p, b), b')$ in the second component

This is more subtle than for simple pairs because the second component lives in a *different* type after transport. The path p must first move b from $B(a)$ to $B(a')$, and only then can we compare it to b'.

For example, in the type $\Sigma_{n:\mathbb{N}} \mathsf{Vec}(n, B)$ of length-indexed vectors, a path between $\langle n, v \rangle$ and $\langle m, w \rangle$ requires:

- A path $p : \mathsf{Path}(n, m)$ witnessing $n = m$
- A path $q : \mathsf{Path}(\mathsf{transport}(p, v), w)$ in $\mathsf{Vec}(m, B)$

3.5.5 The Fundamental Groupoid Preview

Example 3.69 (Paths form a groupoid). For any type A, the collection of paths organizes into a *groupoid*:

- **Objects:** The terms $a, b, c, \ldots : A$
- **Morphisms:** The paths $p : \mathsf{Path}(a, b)$
- **Identity:** The reflexivity path $\mathsf{refl}_a : \mathsf{Path}(a, a)$
- **Composition:** The transitivity $\mathsf{trans}(p, q)$
- **Inverse:** The symmetry $\mathsf{symm}(p)$

The groupoid laws hold (up to rewrite equivalence):

$$\mathsf{trans}(\mathsf{refl}, p) \sim_{\mathsf{rw}} p \sim_{\mathsf{rw}} \mathsf{trans}(p, \mathsf{refl}) \qquad \text{(identity)}$$
$$\mathsf{trans}(p, \mathsf{symm}(p)) \sim_{\mathsf{rw}} \mathsf{refl} \sim_{\mathsf{rw}} \mathsf{trans}(\mathsf{symm}(p), p) \qquad \text{(inverses)}$$
$$\mathsf{trans}(\mathsf{trans}(p, q), r) \sim_{\mathsf{rw}} \mathsf{trans}(p, \mathsf{trans}(q, r)) \qquad \text{(associativity)}$$

This is the *fundamental groupoid* of A (see Chapter 6 for the general construction and Chapter 13 for later applications). Unlike an ordinary category, every morphism has an inverse, making it a groupoid.

Example 3.70 (Higher structure preview)**.** The groupoid laws hold "up to rewrite equivalence," which means there are *2-paths* (paths between paths) witnessing them:

$$\lambda_p : \mathsf{Path}_2(\mathsf{trans}(\mathsf{refl}, p), p)$$

These 2-paths themselves satisfy coherence conditions (Chapter 8), witnessed by *3-paths*, and so on. The full structure is a *weak ω-groupoid* (Chapter 9).

For instance, the pentagon identity asserts that two different ways of reassociating four paths give the same 2-path. Starting from $((pq)r)s$, we can reach $p(q(rs))$ via two routes:

- *Upper route:* Apply α to get $(pq)(rs)$, then α again to get $p(q(rs))$.
- *Lower route:* Apply $\alpha \cdot 1$ to get $(p(qr))s$, then α to get $p((qr)s)$, then $1 \cdot \alpha$ to get $p(q(rs))$.

The pentagon coherence states that these two routes, viewed as composite 2-paths, are equal—witnessed by a 3-path.

Summary

This chapter has introduced the central objects of the book: computational paths. What emerges is a notion of equality witness that records not only truth but also derivational history.

3.5.6 Key Definitions

1. **Steps** are atomic rewrite operations—single applications of β-reduction, η-expansion, or congruence rules. Each step records a source term, a target term, and a justification for their equality.
2. **Paths** are structured witnesses to equality, consisting of a step list (the computational trace) and a semantic equality (the propositional content). The type $\mathsf{Path}(a, b)$ captures paths from a to b.

3. **Fundamental operations**—reflexivity (refl), transitivity (trans), symmetry (symm), and congruence (congrArg)—allow paths to be constructed and combined.

4. **UIP fails**: distinct paths with the same endpoints are genuinely different, as witnessed by different step lists. This proof-relevance is the source of higher structure.

5. **Groupoid laws** hold up to rewrite equivalence: paths satisfy identity, inverse, and associativity laws, and congruence is functorial. But these laws are witnessed by explicit 2-paths.

3.5.7 The Two-Level Structure

The key insight is that computational paths occupy a middle ground between intensional and extensional approaches to equality:

- **Intensional aspect**: Different paths are distinguished; the computational trace is preserved. Two paths with the same endpoints but different step lists are genuinely different objects.

- **Extensional aspect**: The semantic equality (the toEq projection) depends only on endpoints, not on the trace. All paths from a to b witness the same propositional equality.

This two-level structure—intensional on paths, extensional on their semantic content—is the key to reconciling computational content with mathematical identity.

3.5.8 Connection to Earlier Work

The concepts introduced in this chapter have deep roots in the literature:

Statman's direct computation (1977–1978). The notion that equality proofs themselves can be objects of study, subject to normalisation, originates in Statman's work on the equational calculus E [143, 144]. Statman introduced "computations" as objects witnessing equalities, and studied their normal forms.

Le Chenadec's equational logic (1989). Le Chenadec [20] developed the LE system with explicit proof terms for equality, including a normalisation procedure. This work anticipated many themes of computational paths.

Martin-Löf's identity types (1975–1984). The intensional identity type of Martin-Löf type theory [109] is a precursor to our path type. The distinction between intensional and extensional equality that we navigate here was first articulated in Martin-Löf's foundational work.

De Queiroz's rewrite reasons (1988–). The specific formulation of $t =_s u : A$ with an explicit "rewrite reason" s is due to de Queiroz [32, 44, 34]. This innovation makes the reason for equality a first-class object, which is exactly what computational paths formalise. The treatment of equality as "the missing entity" in natural deduction was developed in [46], while a normalisation procedure for the equational fragment was given in [29].

Homotopy Type Theory (2010–). The connection between identity types and homotopy theory, developed in HoTT [150], provides a topological interpretation of our algebraic structures. Paths correspond to homotopies, and higher paths correspond to higher homotopies.

3.5.9 The Road Ahead

The definitions in this chapter are foundational, but their full force is still ahead of us:

1. **Chapter 4** will introduce the complete *rewrite system* on computational paths: the 77 rules that govern how paths can be transformed and simplified, and the formal notion of *rewrite equivalence*.
2. **Part II** will establish the *groupoid structure* in full: the weak category and weak groupoid formed by paths, and the strictification process that produces an honest groupoid. It will then develop the *higher structure*: paths between paths (2-paths), paths between those (3-paths), and the full weak ω-groupoid that emerges.
3. **Part III** will connect the theory to *type theory* and *homotopy*: path induction, the failure of UIP, fundamental groupoids of combinatorial spaces, and the relationship to homotopy type theory and category theory.

The computational-path framework thus provides a concrete, syntactic approach to proof-relevant equality. Rather than postulating higher structure abstractly, we let it arise from the computational behaviour of proof terms themselves. The theoretical gain is that coherence comes with explicit witnesses; the practical gain is that the same framework supports implementation and automation.

The next chapter makes this promise precise by introducing the rewrite system itself.

Formalization Notes. The core path operations introduced in this chapter are formalised in the following files of the accompanying Lean development (`https://github.com/Arthur742Ramos/ComputationalPathsLean`):

- `Path/Basic/Core.lean` — the basic `Path` structure, symmetry, transitivity, transport, and projection to propositional equality
- `Path/Basic/Context.lean` — context application and substitution-style operations on paths
- `Path/Basic/Congruence.lean` — congruence constructions such as functional action on paths

Chapter 4

The Rewrite System on Paths

> *"The word problem for finitely presented algebraic structures is, in general, undecidable. The method of rewriting—reducing expressions to canonical forms by means of directed equations—provides a powerful technique for those cases where decidability can be achieved."*
>
> —Baader & Nipkow [8]

How to read this chapter. This chapter serves two purposes: as an exposition to be read linearly on a first encounter, and as a reference to be consulted when later chapters invoke specific rules. On a first reading, focus on Category A (the groupoid core, §4.1.2), which contains the heart of the path algebra in just eight rules, and skim the remaining categories to absorb their general shape. Later, when Chapter 5 analyses critical pairs or Chapter 7 interprets rewrite steps as 2-cells, return to the relevant rule cluster for the precise statement. The worked examples at the end of each category are designed to build intuition and can be read selectively.

The computational paths introduced in Chapter 3 carry a rich algebraic structure: they can be composed, inverted, and transformed in ways that mirror the topology of identity proofs. The next question is therefore unavoidable: when are two syntactically different paths to be considered *equivalent*? When do two witnesses of the same equality carry the same computational content?

In classical term rewriting [8, 148], these questions are addressed by a *rewrite system*: a collection of directed equations that reduce complex expressions to simpler, canonical forms. In this chapter, we introduce the complete rewrite system for computational paths—a system organised around 77 named rewrite clauses and schemata governing how paths can be simplified and transformed. These rules are organised into eight categories, each corresponding to a distinct layer of the path algebra:

1. **Groupoid core** (Rules 1–8): the fundamental algebraic laws of path composition, inversion, and identity.
2. **Type constructor rules** (Rules 9–25): how paths interact with products, sigma types, sums, and function types.
3. **Transport rules** (Rules 26–32): the behaviour of paths under transport in dependent type families.
4. **Context rules** (Rules 33–48): how paths propagate through one-hole contexts and substitution.
5. **Dependent context rules** (Rules 49–60): the dependent counterpart of context propagation.
6. **Binary context rules** (Rules 61–68): congruence for two-argument contexts, both dependent and non-dependent.
7. **Partial map rules** (Rules 69–72): interactions between left-partial and right-partial congruences.
8. **Structural completion** (Rules 73–77): closure under path constructors and cancellation principles.

The design of this system is driven by two imperatives. First, *completeness*: the rules must capture all the algebraic laws that computational paths satisfy, so that rewrite equivalence coincides with semantic equality. Second, *effectiveness*: the system must be confluent and terminating, so that every path reduces to a unique normal form and path equivalence becomes decidable. The first imperative is addressed in this chapter; the second is established in Chapter 5.

The 77-clause catalogue may at first seem daunting, but it has a clear internal logic—and the number 77 is not arbitrary. The system is driven by a small set of design principles (Section 4.2), and once those principles are understood, the rule count becomes inevitable rather than mysterious.

The first eight rules establish the groupoid structure that was previewed in Chapter 3: they are the directed versions of the standard groupoid axioms, and any rewrite system for paths must include them or their equivalents. The next seventeen rules (Categories B and C) extend this core to the type constructors and transport operations of dependent type theory. These rules express, at the path level, the same β-η discipline that governs ordinary type theory one level down. Categories D through G (thirty-six rules) ensure that the rewrite relation is a *congruence*: that rewrites can be performed inside any context, dependent or binary, without breaking the algebraic structure. The final five rules (Category H) close the system by providing the compatibility and cancellation laws needed for the confluence proof.

In short, each rule exists because some combination of the design principles demands it, and removing any rule would either break completeness (some valid path equation would become unprovable) or confluence (some path would have two distinct normal forms).

4.1 The 77-Rule Rewrite System

We now present the complete rewrite system. Each displayed clause is written in the form $\ell \to r$, meaning that the left-hand side ℓ may be replaced by the right-hand side r in any context. The rules are numbered consecutively from 1 to 77 and organised into eight categories (A through H). Some entries are primitive contractions, while others are compatibility schemata generating context-closure instances. A compact reference table of all 77 named clauses is provided in Appendix D.

The reader encountering this system for the first time should begin with Category A (the groupoid core, §4.1.2) and then proceed through the remaining categories at whatever depth suits their needs. Categories B–C (§§4.1.3–4.1.4) extend the groupoid to type constructors and transport; Categories D–G (§§4.1.5–4.1.8) handle contexts and binary operations; and Category H (§4.1.9) closes the system. The design principles in Section 4.2 explain *why* each category has the shape it does.

4.1.1 Notation and Conventions

Before presenting the rules, we establish notation for the operations that appear in them. The fundamental path operations were defined in Chapter 3; here we also introduce several auxiliary operations that arise from the interaction of paths with type-theoretic structure.

Path operations (from Chapter 3).

- $\mathsf{refl}_a : \mathsf{Path}(a, a)$ — the reflexivity (identity) path at a.
- $\mathsf{symm}(p) : \mathsf{Path}(b, a)$ — the symmetry (inverse) of $p : \mathsf{Path}(a, b)$.
- $\mathsf{trans}(p, q) : \mathsf{Path}(a, c)$ — the transitivity (composition) of $p : \mathsf{Path}(a, b)$ and $q : \mathsf{Path}(b, c)$.

- $\mathsf{congrArg}(f, p) : \mathsf{Path}(f(a), f(b))$ — the congruence of $f : A \to B$ applied to $p : \mathsf{Path}(a, b)$.
- $\mathsf{congrFun}(p, a) : \mathsf{Path}(f(a), g(a))$ — the function congruence of $p : \mathsf{Path}(f, g)$ evaluated at $a : A$.
- $\mathsf{transport}(p, x) : B(b)$ — the transport of $x : B(a)$ along $p : \mathsf{Path}(a, b)$ in a type family B.

Type constructor operations.

- $\mathsf{prodMk}(p, q) : \mathsf{Path}((a_1, b_1), (a_2, b_2))$ — the product path from $p : \mathsf{Path}(a_1, a_2)$ and $q : \mathsf{Path}(b_1, b_2)$.
- $\mathsf{sigmaMk}(p, q)$ — the sigma path from component paths.
- $\mathsf{sigmaFst}(p)$, $\mathsf{sigmaSnd}(p)$ — the first and second projections of a sigma path.
- $\mathsf{lamCongr}(p)$ — the function extensionality path constructed from a family of paths $p : \prod_{x:A} \mathsf{Path}(f(x), g(x))$.

Context operations. A *context* C on a type A is a function $C : A \to B$ equipped with a lifting operation on paths. Given a path $p : \mathsf{Path}(a_1, a_2)$, we write $C[p]$ for the lifted path $\mathsf{congrArg}(C, p) : \mathsf{Path}(C(a_1), C(a_2))$.

- $L_C(r, p)$ — the *left whisker*: given $r : \mathsf{Path}(x, C(a_1))$ and $p : \mathsf{Path}(a_1, a_2)$, this composes r with $C[p]$.
- $R_C(p, t)$ — the *right whisker*: given $p : \mathsf{Path}(a_1, a_2)$ and $t : \mathsf{Path}(C(a_2), y)$, this composes $C[p]$ with t.

A *dependent context* DC operates similarly but for a dependent function $C : (a : A) \to B(a)$, where the codomain varies with the input. The transport map $T_p : B(a_1) \to B(a_2)$ plays the role of the identity on the codomain.

Binary context operations. A *binary context* $K : A \to B \to C$ gives rise to partial maps:

- $\mathsf{mapLeft}_K(p, b) : \mathsf{Path}(K(a_1, b), K(a_2, b))$ — congruence in the first argument, holding the second fixed.
- $\mathsf{mapRight}_K(a, q) : \mathsf{Path}(K(a, b_1), K(a, b_2))$ — congruence in the second argument, holding the first fixed.
- $\mathsf{map}_2(K, p, q) : \mathsf{Path}(K(a_1, b_1), K(a_2, b_2))$ — binary congruence, varying both arguments simultaneously.

Semantic operations.

- $\mathsf{sc}(h) : \mathsf{Path}(a, b)$ — the *step constructor*, which lifts a propositional equality $h : a = b$ to a computational path. This is the canonical embedding of ordinary equalities into the path type, introduced in Chapter 3 as $\mathsf{ofEq}(h)$.
- $\mathsf{toEq}(p) : a = b$ — the semantic projection, extracting the underlying equality from a path $p : \mathsf{Path}(a, b)$.

Dependent application.

- $\mathsf{apd}_f(p) : \mathsf{Path}(f(a), f(b))$ — the *dependent application* of f along p, a dependent generalisation of $\mathsf{congrArg}$.

Throughout the rules, we use the following conventions:

- p, q, r denote paths of appropriate types.
- a, b, c denote terms.
- f, g, h denote functions.
- C denotes a context, DC a dependent context, and K a binary context.
- The symbol $\to$ denotes the one-step rewrite relation.

4.1.2 Category A: Groupoid Core (Rules 1–8)

Every algebraic structure has a skeleton—a minimal set of laws without which the structure ceases to exist. For groupoids, that skeleton consists of the identity, inverse, and associativity laws. The eight rules of Category A are exactly these laws, oriented as directed rewrite rules.

Without these rules, the rewrite system would be unable to simplify even the most basic path expressions: a composition with the identity path would remain unreduced, an inverted inverse would stay doubly wrapped, and two reductions of the same expression via different strategies could never be shown equivalent. In short, Category A is the engine that makes path normalisation possible. Everything that follows—type constructors, transport, contexts—is an extension of this core.

The first eight rules express the fundamental algebraic laws of computational paths as a groupoid. These are the directed versions of the groupoid identities previewed in Chapter 3.

Definition 4.1 (Groupoid core rules)**.** The *groupoid core* consists of eight rewrite rules, naturally organised into four pairs.

The first pair governs the interaction of symmetry with itself. Inverting the identity path produces the identity path, and inverting twice recovers the original:

Symmetry of reflexivity.

$$\mathsf{symm}(\mathsf{refl}_a) \to \mathsf{refl}_a \tag{R1}$$

Involution of symmetry.

$$\mathsf{symm}(\mathsf{symm}(p)) \to p \tag{R2}$$

The second pair establishes that refl is a two-sided identity for composition—the path-level analogue of the monoid identity law:

Left identity.

$$\mathsf{trans}(\mathsf{refl}_a, p) \to p \tag{R3}$$

Right identity.

$$\mathsf{trans}(p, \mathsf{refl}_b) \to p \tag{R4}$$

The third pair asserts that every path has a two-sided inverse. These are the rules that make the structure a *groupoid* rather than merely a category:

Right inverse.

$$\mathsf{trans}(p, \mathsf{symm}(p)) \to \mathsf{refl}_a \tag{R5}$$

Left inverse.

$$\mathsf{trans}(\mathsf{symm}(p), p) \to \mathsf{refl}_b \tag{R6}$$

The final pair describes how symm distributes over trans and how trans associates. Rule R7 says that inversion is an anti-homomorphism; Rule R8 chooses right-association as the canonical bracketing:

Contravariance of inversion.

$$\mathsf{symm}(\mathsf{trans}(p, q)) \to \mathsf{trans}(\mathsf{symm}(q), \mathsf{symm}(p)) \tag{R7}$$

Associativity of composition.

$$\mathsf{trans}(\mathsf{trans}(p,q),r) \to \mathsf{trans}(p,\mathsf{trans}(q,r)) \tag{R8}$$

Remark 4.2 (Directedness of groupoid rules)**.** Each of these eight equations appears in the groupoid literature as a *symmetric* identity: for instance, associativity is usually stated as $\mathsf{trans}(\mathsf{trans}(p,q),r) = \mathsf{trans}(p,\mathsf{trans}(q,r))$. In a rewrite system, however, each equation is given a *direction.* The choice of direction is guided by the goal of termination: the right-hand side should be "simpler" than the left in some well-founded measure. For Rules R3–R6, the right-hand side is strictly smaller in size. For Rule R8, we choose right-association as the normal form by convention; this is a standard choice in the term-rewriting literature [8]. For Rule R2, the double-negation elimination reduces nesting depth. For Rule R7, the right-hand side distributes symm to the leaves, which facilitates further simplification via Rules R1 and R2.

In Chapter 7, these rewrite steps will themselves be interpreted as 2-cells—paths between paths—giving rise to the bicategorical structure of computational paths.

Example 4.3 (Reduction of a composite path)**.** Consider the path expression $\mathsf{trans}(\mathsf{symm}(\mathsf{refl}_a),\mathsf{trans}(\mathsf{refl}_a,p))$. We reduce it step by step using the groupoid rules:

$$\begin{aligned}
\mathsf{trans}(\mathsf{symm}(\mathsf{refl}_a),\mathsf{trans}(\mathsf{refl}_a,p)) &\to \mathsf{trans}(\mathsf{refl}_a,\mathsf{trans}(\mathsf{refl}_a,p)) && \text{by R1}\\
&\to \mathsf{trans}(\mathsf{refl}_a,p) && \text{by R3}\\
&\to p && \text{by R3}
\end{aligned}$$

Three applications of the groupoid rules reduce a complex expression to the underlying path p.

Example 4.4 (Inverse cancellation)**.** The expression

$$\mathsf{trans}(\mathsf{trans}(p,q),\mathsf{symm}(\mathsf{trans}(p,q)))$$

reduces to refl_a by a single application of Rule R5. Alternatively, we can first distribute the inverse using Rule R7 and then cancel:

$$\mathsf{trans}(\mathsf{trans}(p,q),\mathsf{symm}(\mathsf{trans}(p,q))) \to \mathsf{refl}_a \qquad \text{(by R5)}$$

Or, expanding the inner symm first:

$$\begin{aligned}
&\mathsf{trans}(\mathsf{trans}(p,q),\mathsf{symm}(\mathsf{trans}(p,q)))\\
&\quad\to \mathsf{trans}(\mathsf{trans}(p,q),\mathsf{trans}(\mathsf{symm}(q),\mathsf{symm}(p))) && \text{by R7}\\
&\quad\to \mathsf{trans}(p,\mathsf{trans}(q,\mathsf{trans}(\mathsf{symm}(q),\mathsf{symm}(p)))) && \text{by R8}\\
&\quad\to \mathsf{trans}(p,\mathsf{trans}(\mathsf{refl}_b,\mathsf{symm}(p))) && \text{by R5}\\
&\quad\to \mathsf{trans}(p,\mathsf{symm}(p)) && \text{by R3}\\
&\quad\to \mathsf{refl}_a && \text{by R5}
\end{aligned}$$

Both reduction sequences reach the same normal form. This is an instance of the *confluence* property, proved in Chapter 5.

Example 4.5 (Simplification of a round-trip path)**.** Consider the path expression $\mathsf{trans}(\mathsf{trans}(p, \mathsf{refl}_b),\ \mathsf{symm}(p))$, where $p : \mathsf{Path}(a, b)$. This represents a "round trip": traverse p, pause at b (via refl_b), then return via $\mathsf{symm}(p)$. The rewrite system reduces it to refl_a in three steps:

$$\begin{aligned} &\mathsf{trans}(\mathsf{trans}(p, \mathsf{refl}_b),\ \mathsf{symm}(p)) \\ &\quad \to \mathsf{trans}(p,\ \mathsf{symm}(p)) && \text{(R4)} \\ &\quad \to \mathsf{refl}_a && \text{(R5)} \end{aligned}$$

Observe that Rule R3 (left identity) is not needed here; the right identity R4 fires first because refl_b appears on the right of the inner trans. After that, the resulting expression $\mathsf{trans}(p, \mathsf{symm}(p))$ matches the pattern of Rule R5 directly.

More generally, this example illustrates the design principle behind the groupoid rules: each rule targets a specific form of redundancy, and their combined effect is to collapse every "null-homotopic" path expression—one that traverses and then retraces the same path—to refl.

4.1.3 Category B: Type Constructor Rules (Rules 9–25)

With the groupoid core in place, the rewrite system can simplify any expression built from refl, symm, and trans alone. But paths in dependent type theory do not exist in isolation—they live inside compound types, and the system must be able to look inside these structures. A path in a product type $A \times B$ should decompose into component paths in A and B; a path between functions should be determined by pointwise paths; a path in a sigma type must account for transport between fibres. Without rules governing these interactions, the rewrite system would be blind to the internal structure of dependent type theory: it could simplify $\mathsf{trans}(p, \mathsf{symm}(p))$ to refl by Rule R5, but it could not simplify $\mathsf{congrArg}(\pi_1, \mathsf{prodMk}(p, q))$ to p.

The seventeen rules of Category B fill this gap. They express, at the path level, the same β-η laws that govern ordinary terms one level down. Each type constructor—products, sigma types, coproducts, and function types—contributes its own cluster of rules, and together they ensure that compound paths reduce to their canonical components.

The type constructor rules govern how computational paths interact with the type-forming operations of dependent type theory: products, sigma types, coproducts, and function types. These rules express β-η-like laws at the level of paths.

We begin with the decomposition of binary congruence, which sets the stage for all subsequent type constructor rules. When a function takes two arguments, a path varying both can always be split into two stages: one varying each argument in turn. This decomposition is not merely a convenience; it is the path-level manifestation of the fact that every binary operation factors through partial applications, and it ensures that the partial map rules of Category G (Rules 69–72) can subsequently reduce each factor.

Binary congruence decomposition. For a binary function $f : A \to B \to C$ and paths $p : \mathsf{Path}(a_1, a_2)$, $q : \mathsf{Path}(b_1, b_2)$:

$$\mathsf{map}_2(f, p, q) \to \mathsf{trans}(\mathsf{mapRight}_f(a_1, q),\ \mathsf{mapLeft}_f(p, b_2)) \tag{R9}$$

Binary congruence decomposes into a composition: first vary the second argument (holding the first fixed at a_1), then vary the first argument (holding the second fixed at b_2). This is a naturality condition expressing that the two "directions" of a binary map can be sequenced.

Example 4.6 (Rule R9 in action)**.** Let $f(x, y) = x + y$ on $\mathbb{N}$, and let $p : \mathsf{Path}(2, 3)$, $q : \mathsf{Path}(5, 7)$. Then R9 decomposes the path $\mathsf{map}_2(f, p, q)$ from $2 + 5 = 7$ to $3 + 7 = 10$ into two stages:

$$\mathsf{map}_2(f, p, q) \ \to\ \mathsf{trans}(\underbrace{\mathsf{mapRight}_f(2, q)}_{2+5 \rightsquigarrow 2+7},\ \underbrace{\mathsf{mapLeft}_f(p, 7)}_{2+7 \rightsquigarrow 3+7}).$$

First the second argument moves from 5 to 7 while the first stays at 2; then the first argument moves from 2 to 3 while the second stays at 7.

The next six rules govern how paths interact with product types. Just as a pair (a, b) is built from its components and can be decomposed by projection, a path in a product type can be built from component paths and decomposed by applying projections. Rules R10–R13 provide the path-level β- and η-laws, while R14–R15 express the compatibility of products with inversion and functorial mapping.

Product β-rules. For paths $p : \mathsf{Path}(a_1, a_2)$, $q : \mathsf{Path}(b_1, b_2)$:

$$\mathsf{congrArg}(\pi_1, \mathsf{map}_2(\langle\cdot,\cdot\rangle, p, q)) \to p \tag{R10}$$

$$\mathsf{congrArg}(\pi_2, \mathsf{map}_2(\langle\cdot,\cdot\rangle, p, q)) \to q \tag{R11}$$

Projecting the first (resp. second) component of a product path yields the first (resp. second) component path.

Product recursion. For a function $f : A \to B \to C$:

$$\mathsf{congrArg}(\mathsf{Prod.rec}(f), \mathsf{map}_2(\langle\cdot,\cdot\rangle, p, q)) \to \mathsf{map}_2(f, p, q) \tag{R12}$$

Applying the product recursor to a product path yields binary congruence with respect to the uncurried function.

Product η-rule.

$$\mathsf{prodMk}(\mathsf{congrArg}(\pi_1, p),\ \mathsf{congrArg}(\pi_2, p)) \to p\ . \tag{R13}$$

A product path is determined by its components: reassembling the projections of a path yields the original path.

Product symmetry.

$$\mathsf{symm}(\mathsf{prodMk}(p, q)) \to \mathsf{prodMk}(\mathsf{symm}(p), \mathsf{symm}(q)) \tag{R14}$$

The inverse of a product path is the product of the inverses.

Product functoriality. For functions g and h acting on the respective components:

$$\begin{aligned}&\mathsf{congrArg}(\lambda x.\,(g(x.\mathsf{fst}),\ h(x.\mathsf{snd})),\ \mathsf{prodMk}(p, q))\\ &\quad \to \mathsf{prodMk}(\mathsf{congrArg}(g, p),\ \mathsf{congrArg}(h, q))\end{aligned} \tag{R15}$$

A component-wise function applied to a product path acts independently on each component.

Sigma types introduce a subtlety absent from products: the second component's type depends on the first. Consequently, the projection and inversion rules for sigma paths must account for transport between fibers. Rules R16–R19 encode this dependent structure, with the step constructor sc mediating between the path and propositional levels.

Sigma β-rules. For paths p and q in a sigma type:

$$\mathsf{congrArg}(\Sigma.\mathsf{fst},\ \mathsf{sigmaMk}(p, q)) \to \mathsf{sc}(\mathsf{toEq}(p)) \tag{R16}$$

$$\mathsf{sigmaSnd}(\mathsf{sigmaMk}(p, q)) \to \mathsf{sc}(\mathsf{toEq}(q)) \tag{R17}$$

Projecting from a sigma path yields the semantic content of the component. The appearance of sc reflects the fact that dependent projections interact with transport.

Sigma η-rule.

$$\mathsf{sigmaMk}(\mathsf{sigmaFst}(p),\ \mathsf{sigmaSnd}(p)) \to p \tag{R18}$$

A sigma path is determined by its components.

Sigma symmetry.

$$\mathsf{symm}(\mathsf{sigmaMk}(p, q)) \to \mathsf{sigmaMk}(\mathsf{symm}(p),\ \mathsf{symmSnd}(p, q)) \qquad \text{(R19)}$$

Inverting a sigma path inverts the first component and appropriately adjusts the second component, accounting for the transport that arises in dependent types.

Sum types behave more simply than products or sigma types at the path level, because paths within a coproduct cannot cross between the left and right injections. The following two rules express that the sum eliminator commutes with path lifting in each summand.

Sum computation rules. For functions $f : A \to C$ and $g : B \to C$:

$$\mathsf{congrArg}(\mathsf{Sum.rec}(f, g),\ \mathsf{congrArg}(\mathsf{inl}, p)) \to \mathsf{congrArg}(f, p) \qquad \text{(R20)}$$
$$\mathsf{congrArg}(\mathsf{Sum.rec}(f, g),\ \mathsf{congrArg}(\mathsf{inr}, q)) \to \mathsf{congrArg}(g, q) \qquad \text{(R21)}$$

Sum elimination applied to a left-injected path reduces via f; applied to a right-injected path, it reduces via g.

For function types, the central principle is function extensionality: a path between functions is determined by a family of pointwise paths. Rules R22–R23 express the β- and η-laws for this extensionality principle, while R24 shows that inversion distributes pointwise. Rule R25 then handles dependent application on a trivial base path, completing the coverage of type constructors.

Function β-rule. For a family of paths $p : \prod_{x:A} \mathsf{Path}(f(x), g(x))$ and $a : A$:

$$\mathsf{congrFun}(\mathsf{lamCongr}(p), a) \to p(a) \qquad \text{(R22)}$$

Applying a function extensionality path at a point yields the pointwise path at that point.

Function η-rule. For a path $p : \mathsf{Path}(f, g)$ between functions:

$$\mathsf{lamCongr}(\lambda x.\, \mathsf{congrFun}(p, x)) \to p \qquad \text{(R23)}$$

Reassembling a function path from its pointwise evaluations yields the original path.

Function symmetry.

$$\mathsf{symm}(\mathsf{lamCongr}(p)) \to \mathsf{lamCongr}(\lambda x.\, \mathsf{symm}(p(x))) \qquad \text{(R24)}$$

The inverse of a function extensionality path is constructed pointwise.

Dependent application on reflexivity. For a function $f : A \to B$:

$$\mathsf{apd}_f(\mathsf{refl}_a) \to \mathsf{refl}_{f(a)} \tag{R25}$$

The dependent application of f along the identity path at a yields the identity path at $f(a)$.

Remark 4.7 (Type constructors as functors)**.** Rules R10–R13 and R16–R18 are the path-level analogues of the standard β-η laws for product and sigma types. Just as the β-rule for products states $\pi_1(\langle a, b\rangle) = a$, Rule R10 states that the corresponding path equation holds at the level of computational paths. The η-rules (R13 and R18) express the uniqueness of the decomposition.

Rules R14, R15, R19, and R24 express the compatibility of path inversion and congruence with the type constructors. These can be understood categorically: each type constructor gives rise to a functor on the path groupoid, and these rules express the functoriality conditions. A categorical interpretation of such sequences of rewrites was developed in [125].

4.1.4 Category C: Transport Rules (Rules 26–32)

The next group of rules captures a fundamentally different phenomenon from the type constructor rules. Categories A and B dealt with paths that live *within* a single type; Category C deals with paths that move *between* types in a dependent family.

In dependent type theory, a path $p : \mathsf{Path}_A(a_1, a_2)$ in the base type induces a transport map $\mathsf{transport}(p, -) : B(a_1) \to B(a_2)$ between the fibres. Without rules governing this transport, the rewrite system would be unable to simplify any expression involving dependent types: sigma paths, dependent function paths, and the coherence conditions that connect them would all remain stuck. Rules R26–R32 ensure that transport interacts correctly with the groupoid structure—transport over refl is the identity, transport over a composite is the composite of transports, and transport over a path and its inverse cancel—so that normal forms exist even for the dependent fragments of the path algebra.

The transport rules govern the interaction of computational paths with the transport operation in dependent type families. Recall from Chapter 3 that for a type family $B : A \to \mathsf{Type}$ and a path $p : \mathsf{Path}_A(a_1, a_2)$, transport gives a map $\mathsf{transport}(p, -) : B(a_1) \to B(a_2)$. The following rules describe when transport operations can be simplified at the level of paths.

Transport over reflexivity.

$$\mathsf{sc}(\mathsf{transport\_refl}(a, x)) \to \mathsf{refl}_x \tag{R26}$$

Transporting along the identity path refl_a is the identity operation: the resulting path from $\mathsf{transport}(\mathsf{refl}_a, x)$ to x is the reflexivity path refl_x.

Transport over composition.

$$\mathsf{sc}(\mathsf{transport\_trans}(p, q, x)) \to \iota(h_{27}) \tag{R27}$$

where $\iota(h_{27})$ denotes the canonical path arising from the coherence: transporting along a composite path $\mathsf{trans}(p, q)$ is equivalent to transporting first along p and then along q. The term h_{27} encodes this coherence datum.

Transport and left inverse.

$$\mathsf{sc}(\mathsf{transport\_symm\_left}(p, x)) \to \iota(h_{28}) \tag{R28}$$

Transporting forward along p and then backward along $\mathsf{symm}(p)$ returns to the starting point, up to a canonical coherence path.

Transport and right inverse.

$$\mathsf{sc}(\mathsf{transport\_symm\_right}(p, y)) \to \iota(h_{29}) \tag{R29}$$

The dual: transporting backward along $\mathsf{symm}(p)$ and then forward along p returns to the starting point.

When transport operates within sigma types, it must act on both components of the dependent pair, with the second component's transport depending on the first. The remaining three rules handle these sigma-specific transport interactions.

Transport in sigma types (first component).

$$\mathsf{sc}(\mathsf{transport\_sigmaMk\_fst}(p, q, x)) \to \iota(h_{30}) \tag{R30}$$

Transport in sigma types (dependent component).

$$\mathsf{sc}(\mathsf{transport\_sigmaMk\_dep}(p, q, x)) \to \iota(h_{31}) \tag{R31}$$

Substitution in sigma types (dependent).

$$\mathsf{sc}(\mathsf{subst\_sigmaMk\_dep}(p, q, x)) \to \iota(h_{32}) \tag{R32}$$

Remark 4.8 (The role of coherence paths)**.** Rules R27–R32 involve canonical *coherence paths* $\iota(h)$. These are paths constructed from the underlying propositional equalities that witness the relevant transport coherence conditions. The notation $\iota(h)$ indicates a path in normal form—specifically, a step constructor sc applied to a derived equality. The explicit form of each h_i involves composition of the basic transport laws and is determined by the type-theoretic structure. The key point is that each left-hand side reduces to a *canonical* right-hand side, ensuring that transport coherences have unique normal forms. For the normal-form analysis below, the crucial point is that each $\iota(h_i)$ is still atomic from the perspective of the rewrite system: it is a canonical step term rather than a composite path expression.

Rules R26–R32 extend the rewrite system to handle the full complexity of dependent type theory. Without these rules, paths involving transport would not reduce, and the normal form theorem would fail for dependent types.

4.1.5 Category D: Context Rules (Rules 33–48)

Categories A–C equipped the rewrite system with the means to simplify paths at the top level of an expression: groupoid identities, type constructor laws, and transport coherences. But in practice, the subexpression that needs simplifying is rarely at the top—it is typically buried inside a function application, a composition, or a dependent context. Category D ensures that the rewrite relation can reach inside such contexts.

The key principle is that the rewrite relation must be a *congruence*: if p rewrites to q, then $C[p]$ rewrites to $C[q]$ for any context C. Without this property, the system could simplify a path in isolation but would be unable to propagate that simplification into the surrounding term.

The context rules are among the most technically important in the entire system. They govern how paths interact with *contexts*—functions $C : A \to B$ through which paths can be propagated. The central operation is $C[p] = \mathsf{congrArg}(C, p)$: the action of a context on a path. The auxiliary operations L_C (left whisker) and R_C (right whisker) manage the composition of context-lifted paths with ambient paths.

Context congruence (meta-rule).

$$p \to q \quad \Longrightarrow \quad C[p] \to C[q] \tag{R33}$$

If p rewrites to q, then $C[p]$ rewrites to $C[q]$ for any context C. This is a *compatibility* rule: it ensures that the rewrite relation is closed under context application. Technically, it is a rule scheme that generates one rule instance for each rewrite step and each context.

Symmetry through contexts.

$$\mathsf{symm}(C[p]) \to C[\mathsf{symm}(p)] \tag{R34}$$

Inverting a context-lifted path is the same as lifting the inverted path. This expresses the *functoriality* of context application with respect to path inversion.

Context absorption (left).

$$\mathsf{trans}(C[p],\ \mathsf{trans}(C[\mathsf{symm}(p)],\ v)) \to \mathsf{trans}(C[\mathsf{trans}(p, \mathsf{symm}(p))],\ v) \tag{R35}$$

When a context-lifted path and its inverse appear adjacent in a composition, they can be absorbed into a single context application.

Context absorption (right).

$$\mathsf{trans}(\mathsf{trans}(v,\ C[p]),\ C[\mathsf{symm}(p)]) \to \mathsf{trans}(v,\ C[\mathsf{trans}(p, \mathsf{symm}(p))]) \tag{R36}$$

The dual of Rule R35, absorbing on the right.

The next five rules introduce and manipulate the *whisker operations* L_C and R_C, which arise whenever a context-lifted path $C[p]$ is composed with an ambient path. These operations serve as intermediate normal forms that track the interaction between context structure and path composition, enabling further simplification.

Left whisker introduction.

$$\mathsf{trans}(r, C[p]) \to L_C(r, p) \tag{R37}$$

When $r : \mathsf{Path}(x, C(a_1))$ is composed on the left with $C[p] : \mathsf{Path}(C(a_1), C(a_2))$, the result is recorded as a left whisker $L_C(r, p)$. This operation tracks the context structure, enabling further simplification.

Example 4.9 (Rule R37: recording context structure)**.** Let $C = \mathsf{succ}$, $p : \mathsf{Path}(3, 5)$, and $r : \mathsf{Path}(0, \mathsf{succ}(3))$. The composition $\mathsf{trans}(r, \mathsf{congrArg}(\mathsf{succ}, p))$ composes an arbitrary path r with a context-lifted path. Rule R37 rewrites this to $L_{\mathsf{succ}}(r, p)$, which records that the right factor arose from the succ context applied to p. This intermediate form enables Rules R39 and R42–R46 to simplify further.

Left whisker from right whisker.

$$\mathsf{trans}(r, R_C(p, \mathsf{refl}_{C(a_2)})) \to L_C(r, p) \tag{R38}$$

A right whisker with a trivial tail reduces to a left whisker.

Left-to-right whisker conversion.

$$\mathsf{trans}(L_C(r,p),t) \to \mathsf{trans}(r,R_C(p,t)) \tag{R39}$$

Composing a left whisker with a tail converts to a right whisker. This rule, together with R37, allows flexible manipulation of compositions involving context-lifted paths.

Right whisker introduction.

$$\mathsf{trans}(C[p],t) \to R_C(p,t) \tag{R40}$$

When a context-lifted path $C[p]$ is composed on the right with $t : \mathsf{Path}(C(a_2),y)$, the result is a right whisker $R_C(p,t)$.

Right whisker associativity.

$$\mathsf{trans}(R_C(p,t),u) \to R_C(p,\mathsf{trans}(t,u)) \tag{R41}$$

Composing a right whisker with a further path absorbs the composition into the tail of the whisker.

When a whisker involves a reflexivity path—either as the lifted path or as the ambient path—the whisker reduces to a simpler form. The following four rules handle these identity cases, eliminating trivial whisker structure.

Left whisker with reflexivity.

$$L_C(r,\mathsf{refl}_a) \to r \tag{R42}$$

Left-whiskering by the identity path is the identity operation.

Left whisker with reflexivity prefix.

$$L_C(\mathsf{refl}_{C(a_1)},p) \to C[p] \tag{R43}$$

Left-whiskering by the identity at $C(a_1)$ yields the context application directly.

Right whisker with reflexivity.

$$R_C(\mathsf{refl}_a,r) \to r \tag{R44}$$

Right-whiskering by the identity path is the identity operation.

Right whisker with reflexivity suffix.

$$R_C(p, \mathsf{refl}_{C(a_2)}) \to C[p] \tag{R45}$$

Right-whiskering with a trivial tail yields the context application directly.

The final three context rules eliminate redundant nesting of whisker operations. When an intermediate whisker involves a reflexivity path, the nested structure can be collapsed to a single whisker.

Left whisker idempotence.

$$L_C(L_C(r, \mathsf{refl}_{a_1}), p) \to L_C(r, p) \tag{R46}$$

An intermediate left whisker by reflexivity can be absorbed.

Right whisker idempotence (prefix).

$$R_C(p, R_C(\mathsf{refl}_{a_2}, t)) \to R_C(p, t) \tag{R47}$$

An intermediate right whisker by reflexivity can be absorbed.

Right whisker idempotence (identity prefix).

$$R_C(\mathsf{refl}_{a_1}, R_C(p, t)) \to R_C(p, t) \tag{R48}$$

A redundant reflexivity prefix on a nested right whisker is removed.

Remark 4.10 (Whisker operations and 2-categorical structure)**.** The whisker operations L_C and R_C are analogous to the *whiskering* operations in a 2-category. In a 2-category, given a 2-cell $\alpha : f \Rightarrow g$ and a 1-cell h, the left whisker $h \circ \alpha$ and right whisker $\alpha \circ h$ compose the 2-cell with the 1-cell. Rules R37–R48 ensure that these operations interact correctly with composition, identity, and each other.

The proliferation of whisker rules (sixteen in total) reflects the complexity of managing compositions in a context-aware rewrite system. Each rule eliminates a specific form of redundancy, and together they guarantee that normal forms are free of extraneous whisker structure.

Example 4.11 (Context simplification)**.** Let $C = \mathsf{succ} : \mathbb{N} \to \mathbb{N}$ be the successor context, and consider the path expression $\mathsf{trans}(C[\mathsf{refl}_n], C[p])$ for some $p : \mathsf{Path}(n, m)$. We reduce:

$$\begin{aligned}
&\mathsf{trans}(C[\mathsf{refl}_n], C[p]) \\
&\quad = \mathsf{trans}(\mathsf{congrArg}(\mathsf{succ}, \mathsf{refl}_n),\ \mathsf{congrArg}(\mathsf{succ}, p)) \\
&\quad \to \mathsf{trans}(\mathsf{refl}_{\mathsf{succ}(n)},\ \mathsf{congrArg}(\mathsf{succ}, p)) && \text{by R25}^{\dagger} \\
&\quad \to \mathsf{congrArg}(\mathsf{succ}, p) && \text{by R3} \\
&= C[p]
\end{aligned}$$

[†]For non-dependent f, apd_f coincides with $\mathsf{congrArg}(f, -)$, making R25 applicable. The reflexivity application is absorbed, leaving only the essential context action.

Example 4.12 (Congruence propagation: from $a = b$ to $f(a) = f(b)$). Let $f : \mathbb{N} \to \mathbb{N}$ and suppose $p : \mathsf{Path}(a, b)$. We want to derive a path from $f(a)$ to $f(b)$ and then compose it with its inverse, showing full reduction. Taking $C = f$ as the context:

$$\begin{aligned}
&\mathsf{trans}(C[p],\ \mathsf{symm}(C[p])) \\
&\quad \to \mathsf{trans}(C[p],\ C[\mathsf{symm}(p)]) && \text{(R34)} \\
&\quad \to R_C(p,\ C[\mathsf{symm}(p)]) && \text{(R40)} \\
&\quad \to R_C(p,\ R_C(\mathsf{symm}(p),\ \mathsf{refl}_{C(a)})) && \text{(R40)}
\end{aligned}$$

Alternatively, we can use the context absorption rule directly:

$$\begin{aligned}
&\mathsf{trans}(\mathsf{trans}(\mathsf{refl}_{f(a)},\ C[p]),\ C[\mathsf{symm}(p)]) \\
&\quad \to \mathsf{trans}(\mathsf{refl}_{f(a)},\ C[\mathsf{trans}(p, \mathsf{symm}(p))]) && \text{(R36)} \\
&\quad \to \mathsf{trans}(\mathsf{refl}_{f(a)},\ C[\mathsf{refl}_a]) && \text{(R33, R5)} \\
&\quad \to \mathsf{trans}(\mathsf{refl}_{f(a)},\ \mathsf{refl}_{f(a)}) && \text{(R25)} \\
&\quad \to \mathsf{refl}_{f(a)} && \text{(R3)}
\end{aligned}$$

The absorption rules R35 and R36 are designed precisely for this pattern: they detect adjacent context-lifted paths that cancel and fold them into a single context application, which then simplifies via the groupoid rules.

Example 4.13 (Nested context: simplifying $g(f(a)) = g(f(b))$). Let $f : A \to B$ and $g : B \to D$, and let $p : \mathsf{Path}(a, b)$. The composite context $C = g \circ f$ lifts p to a path from $g(f(a))$ to $g(f(b))$. We can build this in two ways and show they reduce to the same form.

Method 1: Direct. Apply the composite context:

$$(g \circ f)[p] \;=\; \mathsf{congrArg}(g \circ f,\ p) \;:\; \mathsf{Path}(g(f(a)),\ g(f(b)))$$

Method 2: Two-step lifting. First lift through f, then through g:

$$\begin{aligned}
&\mathsf{congrArg}(g,\ \mathsf{congrArg}(f,\ p)) \\
&\quad = g[f[p]]
\end{aligned}$$

Now consider the expression $\mathsf{trans}(g[f[\mathsf{refl}_a]],\ g[f[p]])$. We reduce step by step:

$$\begin{aligned}
\mathsf{trans}(g[f[\mathsf{refl}_a]],\ g[f[p]]) &\to \mathsf{trans}(g[\mathsf{refl}_{f(a)}],\ g[f[p]]) && \text{(R33)} \\
&\to \mathsf{trans}(\mathsf{refl}_{g(f(a))},\ g[f[p]]) && \text{(R33)} \\
&\to g[f[p]] && \text{(R3)}
\end{aligned}$$

This illustrates how the congruence meta-rule R33 propagates rewrites through nested contexts, and the groupoid rules then clean up the resulting identity compositions.

4.1.6 Category E: Dependent Context Rules (Rules 49–60)

Category D ensured that rewrites propagate through non-dependent contexts, where source and target live in the same type. Dependent type theory, however, abounds with situations where the codomain varies with the input: a dependent function $C : (a : A) \to B(a)$ maps the endpoints of a path into *different* types $B(a_1)$ and $B(a_2)$. Composing paths in this setting requires transport to bridge the gap between fibres, and the whisker operations must be adjusted accordingly.

Category E provides the dependent counterparts of the context rules. Its twelve rules mirror the non-dependent ones but incorporate the transport maps that the dependent structure demands.

The dependent context rules mirror the non-dependent context rules (Category D), but for dependent functions $C : (a : A) \to B(a)$ where the codomain varies with the argument. The key difference is that composing paths in a dependent setting requires *transport* to align the types. We write $T_p : B(a_1) \to B(a_2)$ for the transport map along $p : \mathsf{Path}(a_1, a_2)$, and $\mathsf{DC}[p]$ for the dependent context application.

The dependent whisker operations are:

- $DL_C(r, p)$: given $r : \mathsf{Path}(x, C(a_1))$ and $p : \mathsf{Path}(a_1, a_2)$, this composes r with the dependent action $\mathsf{DC}[p]$, incorporating the necessary transport.
- $DR_C(p, t)$: given $p : \mathsf{Path}(a_1, a_2)$ and $t : \mathsf{Path}(C(a_2), y)$ (after transport adjustment), this composes $\mathsf{DC}[p]$ with t.

Dependent context congruence (meta-rule).

$$p \to q \quad \Longrightarrow \quad \mathsf{DC}[p] \to \mathsf{DC}[q] \tag{R49}$$

Dependent symmetry.

$$\mathsf{symm}(\mathsf{DC}[p]) \to \mathsf{DC.symmMap}(p) \tag{R50}$$

Inverting a dependent context application involves the symmetry map, which accounts for the transport adjustment required in the dependent case.

The next four rules define the dependent whisker operations DL_C and DR_C and establish their basic interaction with composition. These parallel the non-dependent whisker rules R37–R41 but additionally account for the transport maps that arise in the dependent setting.

Dependent left whisker introduction.

$$\mathsf{trans}(T_p(r), \mathsf{DC}[p]) \to DL_C(r, p) \tag{R51}$$

Here $T_p(r)$ is the transport of r along p, composed with the dependent context application.

Dependent left-to-right conversion.

$$\mathsf{trans}(DL_C(r, p), t) \to \mathsf{trans}(T_p(r), DR_C(p, t)) \tag{R52}$$

Dependent right whisker introduction.

$$\mathsf{trans}(\mathsf{DC}[p], t) \to DR_C(p, t) \tag{R53}$$

Dependent right whisker associativity.

$$\mathsf{trans}(DR_C(p, t), u) \to DR_C(p, \mathsf{trans}(t, u)) \tag{R54}$$

When a dependent whisker involves a reflexivity path, the whisker simplifies or vanishes entirely. These four rules are the dependent analogues of Rules R42–R45.

Dependent left whisker with reflexivity.

$$DL_C(r, \mathsf{refl}_a) \to r \tag{R55}$$

Dependent left whisker with reflexivity prefix.

$$DL_C(\mathsf{refl}_{C(a_1)}, p) \to \mathsf{DC}[p] \tag{R56}$$

Dependent right whisker with reflexivity.

$$DR_C(\mathsf{refl}_a, r) \to r \tag{R57}$$

Dependent right whisker with reflexivity suffix.

$$DR_C(p, \mathsf{refl}_{C(a_2)}) \to \mathsf{DC}[p] \tag{R58}$$

The final two dependent context rules eliminate redundant nesting, paralleling the non-dependent idempotence rules R46–R48.

Dependent left whisker idempotence.

$$DL_C(DL_C(r, \mathsf{refl}_{a_1}), p) \to DL_C(r, p) \tag{R59}$$

Dependent right whisker idempotence.

$$DR_C(p, DR_C(\mathsf{refl}_{a_2}, t)) \to DR_C(p, t) \tag{R60}$$

Remark 4.14 (Parallel structure of Categories D and E)**.** Categories D and E exhibit a striking parallel. Twelve of the sixteen non-dependent rules (R33–R48) have direct dependent counterparts in R49–R60. The correspondence is:

Operation	**Non-dependent**	**Dependent**
Congruence	R33	R49
Symmetry	R34	R50
Left whisker introduction	R37	R51
Left-to-right conversion	R39	R52
Right whisker introduction	R40	R53
Right whisker associativity	R41	R54
Left whisker refl	R42	R55
Left whisker refl prefix	R43	R56
Right whisker refl	R44	R57
Right whisker refl suffix	R45	R58
Left whisker idempotence	R46	R59
Right whisker idempotence	R47	R60

The dependent rules are necessarily more complex because they must account for transport adjustments, but their logical role is identical. The non-dependent rules R35, R36, R38, and R48 have no direct dependent counterparts; these involve cancellation patterns that are subsumed in the dependent case by the transport coherences of Category C.

Example 4.15 (Dependent pair paths and transport interaction)**.** Let $B : A \to \mathsf{Type}$ be a type family, and consider a dependent pair type $\Sigma_{a:A} B(a)$. Suppose we have a path $p : \mathsf{Path}_A(a_1, a_2)$ in the base and a dependent path q in the fiber. The sigma path $\mathsf{sigmaMk}(p, q)$ represents a path between dependent pairs.

We show how the rewrite rules simplify expressions involving sigma paths and transport.

(a) Projecting a sigma path. Applying the β-rules to $\mathsf{sigmaMk}(p, q)$:

$$\mathsf{congrArg}(\Sigma.\mathsf{fst},\ \mathsf{sigmaMk}(p, q)) \to \mathsf{sc}(\mathsf{toEq}(p)) \quad \text{by R16}$$
$$\mathsf{sigmaSnd}(\mathsf{sigmaMk}(p, q)) \to \mathsf{sc}(\mathsf{toEq}(q)) \quad \text{by R17}$$

The sigma path decomposes cleanly into its base and fiber components. The appearance of $\mathsf{sc} \circ \mathsf{toEq}$ reflects the passage through propositional equality that is characteristic of dependent projections.

(b) Round-trip on sigma paths. Consider inverting and composing a sigma path:

$$\mathsf{symm}(\mathsf{sigmaMk}(p, q)) \to \mathsf{sigmaMk}(\mathsf{symm}(p),\ \mathsf{symmSnd}(p, q)) \qquad \text{by R19}$$

Note the asymmetry: while the first component simply inverts via $\mathsf{symm}(p)$, the second component requires the adjusted operation $\mathsf{symmSnd}(p, q)$ because the fiber $B(a_2)$ over the target of p differs from $B(a_1)$. The transport map $T_{\mathsf{symm}(p)}$ is implicit in symmSnd.

(c) Transport along reflexivity in a sigma type. When the base path is refl_a, transport becomes trivial:

$$\mathsf{sc}(\mathsf{transport\_refl}(a, x)) \to \mathsf{refl}_x \qquad \text{by R26}$$

This means that for $\mathsf{sigmaMk}(\mathsf{refl}_a, q)$, the transport coherences in rules R30–R32 collapse:

$$\begin{aligned}&\mathsf{sigmaMk}(\mathsf{sigmaFst}(\mathsf{sigmaMk}(\mathsf{refl}_a, q)),\\ &\quad \mathsf{sigmaSnd}(\mathsf{sigmaMk}(\mathsf{refl}_a, q)))\\ &\quad \to \mathsf{sigmaMk}(\mathsf{refl}_a, q) \qquad \text{(R16, R17, R18)}\end{aligned}$$

The η-rule R18 guarantees that decomposing a sigma path into its components and reassembling recovers the original, even in the dependent setting. This is the sigma-type analogue of the product η-rule (Example 4.17).

4.1.7 Category F: Binary Context Rules (Rules 61–68)

Categories D and E handled unary contexts; Category F extends the congruence property to binary operations. Since many type-theoretic constructions take two arguments—$\mathsf{trans}(p, q)$, $\mathsf{prodMk}(p, q)$, $\mathsf{map}_2(f, p, q)$—the rewrite relation must also propagate through each argument position of these binary operations.

The eight rules below follow a systematic four-fold pattern: left argument vs. right argument, dependent vs. non-dependent. Each rule asserts that a rewrite in one argument of a binary context propagates to the whole expression.

The binary context rules handle congruence for functions of two arguments. For a binary context $K : A \to B \to C$, we have the dependent and non-dependent variants:

- DBC — dependent binary context (where K may have dependent codomain).
- BC — non-dependent binary context.

Dependent binary context: left congruence.

$$p \to q \quad \Longrightarrow \quad \mathsf{DBC.mapLeft}_K(p, b) \to \mathsf{DBC.mapLeft}_K(q, b) \qquad \text{(R61)}$$

Dependent binary context: right congruence.

$$p \to q \quad \Longrightarrow \quad \mathsf{DBC.mapRight}_K(a, p) \to \mathsf{DBC.mapRight}_K(a, q) \qquad \text{(R62)}$$

Dependent binary context: binary congruence (left).

$$p \to q \quad \Longrightarrow \quad \mathsf{DBC.map}_2(K, p, r) \to \mathsf{DBC.map}_2(K, q, r) \qquad \text{(R63)}$$

Dependent binary context: binary congruence (right).

$$q \to r \quad \Longrightarrow \quad \mathsf{DBC.map}_2(K, p, q) \to \mathsf{DBC.map}_2(K, p, r) \qquad \text{(R64)}$$

The following four rules are the non-dependent counterparts of R61–R64. They ensure that the rewrite relation is also preserved by the non-dependent binary context operations $\mathsf{BC.mapLeft}$, $\mathsf{BC.mapRight}$, and $\mathsf{BC.map}_2$.

Non-dependent binary context: left congruence.

$$p \to q \quad \Longrightarrow \quad \mathsf{BC.mapLeft}_K(p, b) \to \mathsf{BC.mapLeft}_K(q, b) \qquad \text{(R65)}$$

Non-dependent binary context: right congruence.

$$p \to q \quad \Longrightarrow \quad \mathsf{BC.mapRight}_K(a, p) \to \mathsf{BC.mapRight}_K(a, q) \qquad \text{(R66)}$$

Non-dependent binary context: binary congruence (left).

$$p \to q \quad \Longrightarrow \quad \mathsf{BC.map}_2(K, p, r) \to \mathsf{BC.map}_2(K, q, r) \qquad \text{(R67)}$$

Non-dependent binary context: binary congruence (right).

$$q \to r \quad \Longrightarrow \quad \mathsf{BC.map}_2(K, p, q) \to \mathsf{BC.map}_2(K, p, r) \qquad \text{(R68)}$$

Remark 4.16 (Binary context rules as compatibility conditions). The eight rules of Category F are all *compatibility* rules (also called *congruence rules* in the rewriting literature). They state that the rewrite relation is preserved by the binary context operations. These rules do not simplify expressions; rather, they ensure that rewrites can be performed under binary context constructors. Without them, the rewrite relation would not be a congruence on the full path algebra, and the quotient construction of Section 4.4 would fail.

The four-fold repetition (left/right, dependent/non-dependent, for both unary and binary positions) reflects the systematic nature of the rule design: every path-forming operation must be covered by a compatibility rule.

Example 4.17 (Product path composition and simplification)**.** Let p : $\mathsf{Path}_A(a, a')$ and $q : \mathsf{Path}_B(b, b')$, and consider the product path $\mathsf{prodMk}(p, q)$: $\mathsf{Path}_{A\times B}((a, b),\ (a', b'))$. We verify that projecting and then reassembling recovers the original path, and show how composition of product paths simplifies.

(a) Projection and reassembly. Applying the product β-rules to the product path:

$$\begin{aligned} &\mathsf{congrArg}(\pi_1,\ \mathsf{prodMk}(p, q)) \to p && \text{by R10} \\ &\mathsf{congrArg}(\pi_2,\ \mathsf{prodMk}(p, q)) \to q && \text{by R11} \end{aligned}$$

Now reassembling via the η-rule:

$$\begin{aligned} &\mathsf{prodMk}(\mathsf{congrArg}(\pi_1,\ \mathsf{prodMk}(p, q)), \\ &\quad \mathsf{congrArg}(\pi_2,\ \mathsf{prodMk}(p, q))) \\ &\quad \to \mathsf{prodMk}(p, q) && \text{(R10, R11, R13)} \end{aligned}$$

(b) Inversion of a product path. The inverse of $\mathsf{prodMk}(p, q)$ distributes to the components:

$$\mathsf{symm}(\mathsf{prodMk}(p, q)) \to \mathsf{prodMk}(\mathsf{symm}(p),\ \mathsf{symm}(q)) \qquad \text{by R14}$$

Composing a product path with its inverse, we obtain:

$$\begin{aligned} &\mathsf{trans}(\mathsf{prodMk}(p, q),\ \mathsf{symm}(\mathsf{prodMk}(p, q))) \\ &\quad \to \mathsf{trans}(\mathsf{prodMk}(p, q),\ \mathsf{prodMk}(\mathsf{symm}(p), \mathsf{symm}(q))) && \text{by R14} \end{aligned}$$

Projecting the first component of this composition via R10 and then applying the groupoid rule R5 yields refl_a, and similarly the second component yields refl_b, confirming that the composition reduces to $\mathsf{refl}_{(a,b)}$ at the product level.

4.1.8 Category G: Partial Map Rules (Rules 69–72)

Rule R9 decomposes binary congruence into a composition of partial maps—one varying each argument in turn. The four rules of Category G complete this pipeline: two compatibility rules (R69–R70) propagate rewrites through partial maps, and two computation rules (R71–R72) reduce a partial map applied to a step constructor to a single step.

When a binary function $f : A \to B \to C$ is applied to paths in both arguments, Rule R9 decomposes the result into a composition of partial maps.

Left map congruence.

$$p \to q \quad \Longrightarrow \quad \mathsf{mapLeft}_f(p, b) \to \mathsf{mapLeft}_f(q, b) \qquad \text{(R69)}$$

Right map congruence.

$$p \to q \quad \Longrightarrow \quad \mathsf{mapRight}_f(a, p) \to \mathsf{mapRight}_f(a, q) \tag{R70}$$

Left map computation.

$$\mathsf{mapLeft}_f(\mathsf{sc}(h), b) \to \mathsf{sc}(\mathsf{congrArg}(\lambda x.\, f(x, b),\ h)) \tag{R71}$$

A left map applied to a step constructor reduces to a step constructor with the congruence applied to the underlying equality.

Right map computation.

$$\mathsf{mapRight}_f(a, \mathsf{sc}(h)) \to \mathsf{sc}(\mathsf{congrArg}(f(a, \cdot),\ h)) \tag{R72}$$

Similarly, a right map applied to a step constructor reduces by pushing the partial application through.

Remark 4.18 (Interaction with binary congruence)**.** Rules R69–R72 complement the binary congruence decomposition of Rule R9. Once $\mathsf{map}_2(f, p, q)$ has been decomposed into a composition of $\mathsf{mapRight}$ and $\mathsf{mapLeft}$ by Rule R9, the partial map computation rules R71–R72 allow each component to be further simplified when the argument path is a step constructor. The congruence rules R69–R70 ensure that simplifications within the argument paths propagate outward.

4.1.9 Category H: Structural Completion (Rules 73–77)

The preceding seven categories have established the groupoid algebra, the type-constructor laws, and the congruence properties. Category H closes the circle with the last five rules needed for the system to be self-contained.

Three of these rules (R73–R75) are compatibility rules for symm and trans themselves—without them, a rewrite deep inside a symm or trans expression would be invisible to the surrounding context. The remaining two (R76–R77) are extended cancellation laws that handle a common but tricky pattern: inverse cancellation occurring in the *middle* of a longer composition. Without R76 and R77, reducing $\mathsf{trans}(p, \mathsf{trans}(\mathsf{symm}(p), q))$ would require a detour through associativity (R8) followed by inverse cancellation (R5) and then identity elimination (R3)—a three-step process that complicates the termination argument. By including the cancellation as a primitive rule, we obtain a cleaner reduction and a simpler termination proof.

The final five rules close the rewrite system under the basic path constructors and provide the cancellation laws needed for normalisation.

Symmetry congruence.

$$p \to q \quad \Longrightarrow \quad \mathsf{symm}(p) \to \mathsf{symm}(q) \tag{R73}$$

The rewrite relation is preserved under path inversion.

Transitivity congruence (left).

$$p \to q \quad \Longrightarrow \quad \mathsf{trans}(p, r) \to \mathsf{trans}(q, r) \tag{R74}$$

Rewrites in the left component of a composition propagate to the whole.

Transitivity congruence (right).

$$q \to r \quad \Longrightarrow \quad \mathsf{trans}(p, q) \to \mathsf{trans}(p, r) \tag{R75}$$

Rewrites in the right component of a composition propagate to the whole.

The final two rules are *extended cancellation laws* that handle inverse cancellation occurring in the middle of a longer composition. They complement the basic inverse laws R5 and R6, which only apply when the cancellation produces a standalone refl.

Left cancellation.

$$\mathsf{trans}(p, \mathsf{trans}(\mathsf{symm}(p), q)) \to q \tag{R76}$$

A path followed by its inverse, followed by another path, cancels to just the third path.

Right cancellation.

$$\mathsf{trans}(\mathsf{symm}(p), \mathsf{trans}(p, q)) \to q \tag{R77}$$

The inverse of a path followed by the path itself, followed by another path, cancels to just the third path.

Remark 4.19 (Role of the cancellation rules)**.** Rules R76 and R77 extend the basic inverse laws R5 and R6. Where R5 states $\mathsf{trans}(p, \mathsf{symm}(p)) \to \mathsf{refl}$, Rule R76 handles the case where the cancellation occurs in the *middle* of a longer composition: $\mathsf{trans}(p, \mathsf{trans}(\mathsf{symm}(p), q))$. Without these rules, reducing such expressions would require an intermediate step through associativity (R8), potentially complicating the termination argument. By including them as primitive rules, we obtain a more direct reduction path.

Example 4.20 (Rule R76: mid-composition cancellation)**.** Let $p : \mathsf{Path}(a, b)$ and $q : \mathsf{Path}(a, c)$. The expression $\mathsf{trans}(p, \mathsf{trans}(\mathsf{symm}(p), q))$ represents: go from a to b via p, return from b to a via $\mathsf{symm}(p)$, then continue from a to c via q. The round trip p followed by $\mathsf{symm}(p)$ is redundant. Rule R76 eliminates it in a single step:

$$\mathsf{trans}(p, \mathsf{trans}(\mathsf{symm}(p), q)) \;\to\; q.$$

Without R76, the same result would require three steps: reassociate by R8, cancel by R5, then eliminate the identity by R3.

Remark 4.21 (Closure under path constructors)**.** Rules R33, R49, R61–R70, and R73–R75 are all *compatibility rules*: they state that if $p \to q$, then $F(p) \to F(q)$ for various path constructors F. Together, these rules ensure that the one-step rewrite relation $\to$ is closed under all path-forming operations. This is essential for the relation to be a *congruence* on the term algebra of paths.

The full set of compatibility rules covers:

- Unary context application: R33 (non-dependent), R49 (dependent)
- Binary context: R61–R68 (dependent and non-dependent, left and right, unary and binary positions)
- Partial maps: R69–R70
- Path constructors: R73 (symm), R74–R75 (trans, left and right)

4.2 Rule Categories and Structure

Having presented the complete system of 77 rules, we now step back to examine its structure. Why these particular rules? Why this particular organization? What principles guided the design?

4.2.1 Design Principles

The rewrite system is shaped by several interlocking design principles.

Principle 1: Semantic completeness. The system must be *complete* with respect to semantic equality: two paths p and q should be rewrite-equivalent if and only if they denote the same underlying equality. More precisely:

$$p \sim_{\mathsf{rw}} q \quad\iff\quad \mathsf{toEq}(p) = \mathsf{toEq}(q)$$

This principle dictates which rules must be present: any identity that holds semantically must be derivable by the rewrite system. The 77 rules are precisely those needed to achieve this completeness for the full type-theoretic path algebra.

Principle 2: Termination. Every rule must reduce the "size" of its input according to some well-founded measure. This ensures that the rewrite process always terminates. The choice of which direction to orient each equation (left-to-right or right-to-left) is determined by the termination measure. For instance, associativity (R8) is oriented to produce right-associated terms, and contravariance (R7) is oriented to distribute symm inward.

Principle 3: Confluence. The system must be *confluent* [116]: if a path p can be rewritten in two different ways, the results must be joinable—that is, they must have a common reduct. Confluence, combined with termination, yields *unique normal forms*. The proof of confluence is given in Chapter 5.

Principle 4: Systematic coverage. Every path-forming operation must be accounted for. For each type constructor (products, sums, sigma types, function types), there must be rules governing how paths interact with the constructor's introduction and elimination forms. For each structural operation (symm, trans, congrArg), there must be compatibility rules ensuring that rewrites propagate through.

Principle 5: Minimality. Subject to the above requirements, the rule set should be as small as possible. Redundant rules complicate the confluence proof and increase the risk of non-termination from overlapping rewrites. The 77 rules represent a careful balance: enough for completeness, few enough for manageability.

4.2.2 The Layered Architecture

The eight categories form a natural hierarchy, with each layer building on the previous ones:

Layer	Category	Rules	Role
1	A: Groupoid core	1–8	Paths form a groupoid
2	B: Type constructors	9–25	Products, sums, Σ, functions
3	C: Transport	26–32	Dependent type families
4	D: Contexts	33–48	Compatibility with substitution
5	E: Dependent contexts	49–60	Dependent counterpart of Layer 4
6	F: Binary contexts	61–68	Binary operations
7	G: Partial maps	69–72	Partial application
8	H: Structural completion	73–77	Closure and cancellation

Layer 1 (Groupoid core). This is the foundation. The eight rules of Category A capture the algebraic essence of path composition and inversion. Any rewrite system for paths must include these rules or their equivalents. They correspond to the axioms of a *free groupoid*—the groupoid generated by a set of atomic steps, subject to the standard groupoid laws.

Layer 2 (Type constructors). This layer extends the groupoid structure to account for the type-forming operations of dependent type theory. Each type constructor introduces new path-forming operations (e.g., prodMk for products, $\mathsf{sigmaMk}$ for sigma types, $\mathsf{lamCongr}$ for functions), and the rules of Category B govern how these interact with the groupoid operations. The β-η rules for paths (R10–R13, R16–R18, R22–R23) are particularly important: they ensure that paths in compound types are determined by their components.

Layer 3 (Transport). Transport is the mechanism by which paths in the base type A act on the fibers of a dependent type family $B : A \to \mathsf{Type}$. The transport rules (R26–R32) ensure that this action is compatible with the groupoid structure established in Layer 1.

Layers 4–7 (Context and binary context rules). These layers ensure that the rewrite system is a *congruence*: the rewrite relation is preserved by all term-forming operations. The context rules (Layers 4–5) handle unary contexts, and the binary context rules (Layers 6–7) handle binary contexts. Together, they guarantee that normalisation can proceed under any subterm.

Layer 8 (Structural completion). The final layer closes the system. The compatibility rules (R73–R75) ensure closure under symm and trans, while the cancellation rules (R76–R77) provide the "extended inverse laws" needed for complete normalisation.

4.2.3 Dependencies Between Categories

The categories are not independent; there are important dependencies and interactions between them.

1. **Category A is self-contained.** The groupoid core rules reference only the basic path operations (refl, symm, trans) and can be studied independently of all other categories.
2. **Category B depends on Category A.** The type constructor rules reference groupoid operations (e.g., R14 uses symm, R9 uses trans). The normal

forms of type-constructor paths are expressed in terms of the groupoid normal forms.

3. **Category C depends on Categories A and B.** The transport rules reference both groupoid operations and type constructor operations (e.g., sigma type transport in R30–R32).

4. **Categories D–G depend on Categories A–C.** The context rules reference all previously defined operations. The whisker operations L_C and R_C compose context-lifted paths with arbitrary paths, which may involve type constructors and transport.

5. **Category H depends on Category A.** The cancellation rules R76 and R77 reference only groupoid operations, but the compatibility rules R73–R75 apply to the entire system.

Example 4.22 (Cross-category reduction)**.** Consider the path

$$\mathsf{symm}(\mathsf{prodMk}(\mathsf{refl}_a, q))$$

in a product type. This requires rules from two categories:

$$\begin{aligned}
&\mathsf{symm}(\mathsf{prodMk}(\mathsf{refl}_a, q)) \\
&\quad \to \mathsf{prodMk}(\mathsf{symm}(\mathsf{refl}_a), \mathsf{symm}(q)) && \text{(R14)} \\
&\quad \to \mathsf{prodMk}(\mathsf{refl}_a, \mathsf{symm}(q)) && \text{(R1)}
\end{aligned}$$

The product symmetry rule (Cat. B) distributes symm to the components, and then the groupoid rule (Cat. A) simplifies $\mathsf{symm}(\mathsf{refl}_a)$.

4.2.4 Comparison with Related Systems

The 77-rule rewrite system occupies a distinctive position in the landscape of term rewriting systems and type-theoretic equational theories.

Comparison with free groupoid presentations. A standard presentation of the free groupoid on a set of generators uses essentially the eight rules of Category A. Our system extends this to handle the rich structure of dependent type theory. The additional 69 rules (Categories B–H) can be understood as expressing the *functoriality* and *compatibility* conditions that arise when a groupoid is embedded in a type-theoretic setting.

Squier–Guiraud–Malbos comparison. The work of Squier [141] and its extensions by Guiraud and Malbos connect term rewriting systems to higher-dimensional algebra. In their framework, a confluent and terminating rewrite system gives rise to a *coherent presentation* of the corresponding algebraic structure. Our 77-clause rewrite package can be seen as providing such a coherent presentation for the path algebra. The critical pairs of the system (analysed in Chapter 5) correspond to the 2-cells of the coherent presentation.

Comparison with HoTT path algebra. In homotopy type theory [150], the identity type satisfies groupoid laws that hold only up to higher identity proofs. The path algebra of HoTT is *presented* by rules analogous to our Category A, but the "rules" are witnessed by identity proofs rather than rewrite steps. Our rewrite system provides a *computational* counterpart to the HoTT path algebra: instead of axiomatizing the laws, we compute with them.

Comparison with classical term rewriting. The standard references on term rewriting [8, 148] develop the theory of abstract rewrite systems, including confluence and termination. Our system is a concrete instance of this theory, but with the additional structure of type-theoretic contexts and dependent types. The context and binary context rules (Categories D–G) are specific to our setting and have no direct analogue in classical term rewriting.

4.3 Rewrite Relations and Strategies

With the 77 displayed clauses in hand, we now define the formal rewrite relations that they generate and discuss strategies for applying them.

4.3.1 The One-Step Rewrite Relation

Definition 4.23 (One-step rewrite)**.** The *one-step rewrite relation* $\rightarrow$ on paths is the least relation satisfying:

1. **Displayed clauses:** If $\ell \rightarrow r$ is one of the displayed rule patterns (R1–R77), interpreted with its schematic variables and admissibility conditions, and σ is a substitution matching the rule variables to concrete paths, then $\sigma(\ell) \rightarrow \sigma(r)$.

2. **Compatibility:** If $p \rightarrow q$, then $F(p) \rightarrow F(q)$ for every path-forming operation F. This is ensured by the compatibility rules R33, R49, R61–R70, R73–R75.

We write $p \rightarrow q$ and say that *p rewrites to q* in one step.

The one-step relation is finitely branching: from any given path, there are only finitely many possible one-step rewrites. This is because each path has finitely many subexpressions, and each subexpression matches at most finitely many rule left-hand sides.

Proposition 4.24 (Soundness of one-step rewrite). *If $p \to q$, then* $\mathsf{toEq}(p) = \mathsf{toEq}(q)$.

Proof. By induction on the derivation of $p \to q$. For each base rule $\ell \to r$ (R1–R77), one verifies that the left-hand side and right-hand side denote the same underlying equality. This is straightforward for the groupoid rules (Cat. A): for instance, $\mathsf{toEq}(\mathsf{symm}(\mathsf{refl}_a)) = \mathsf{toEq}(\mathsf{refl}_a)$ because both denote the reflexivity proof of $a = a$. For the type constructor rules (Cat. B), soundness follows from the corresponding β-η laws for the type constructors. For the compatibility rules, soundness follows by induction on the context. □

4.3.2 The Reflexive-Transitive Closure

Definition 4.25 (Multi-step rewrite). The *multi-step rewrite relation* $\twoheadrightarrow$ (also written $\to^*$) is the reflexive-transitive closure of $\to$:

$$p \twoheadrightarrow q \iff \exists\, p_0, p_1, \ldots, p_n : p = p_0 \to p_1 \to \cdots \to p_n = q$$

We say that p *reduces to* q if $p \twoheadrightarrow q$.

Corollary 4.26 (Soundness of multi-step rewrite). *If $p \twoheadrightarrow q$, then* $\mathsf{toEq}(p) = \mathsf{toEq}(q)$.

Proof. By transitivity of equality, from Proposition 4.24. □

4.3.3 Rewrite Equivalence

Definition 4.27 (Rewrite equivalence). *Rewrite equivalence*, denoted $p \sim_{\mathsf{rw}} q$, is the reflexive-symmetric-transitive closure of $\to$:

$$p \sim_{\mathsf{rw}} q \iff \exists\, p_0, \ldots, p_n : p = p_0 \leftrightarrow p_1 \leftrightarrow \cdots \leftrightarrow p_n = q$$

where each $\leftrightarrow$ is either $\to$ or $\leftarrow$.

Proposition 4.28 (Rewrite equivalence is a congruence). *Rewrite equivalence is a* congruence *on the algebra of path expressions. That is, it is preserved by all path-forming operations:*

1. *If $p \sim_{\mathsf{rw}} p'$, then* $\mathsf{symm}(p) \sim_{\mathsf{rw}} \mathsf{symm}(p')$.
2. *If $p \sim_{\mathsf{rw}} p'$ and $q \sim_{\mathsf{rw}} q'$, then* $\mathsf{trans}(p, q) \sim_{\mathsf{rw}} \mathsf{trans}(p', q')$.

3. *If* $p \sim_{\mathsf{rw}} p'$, *then* $\mathsf{congrArg}(f, p) \sim_{\mathsf{rw}} \mathsf{congrArg}(f, p')$ *for any* f.

4. *If* $p \sim_{\mathsf{rw}} p'$, *then* $C[p] \sim_{\mathsf{rw}} C[p']$ *for any context* C.

5. *Similarly for all other path constructors.*

Proof. The congruence property follows from the compatibility rules. For (1): if $p \to p'$, then $\mathsf{symm}(p) \to \mathsf{symm}(p')$ by R73. The extension to $\sim_{\mathsf{rw}}$ follows by induction on the length of the equivalence chain. Properties (2)–(5) follow analogously from R74–R75, R33, and R49–R70, respectively. □

Proposition 4.29 (Completeness of rewrite equivalence)**.** *Rewrite equivalence coincides with semantic equality:*

$$p \sim_{\mathsf{rw}} q \quad \Longleftrightarrow \quad \mathsf{toEq}(p) = \mathsf{toEq}(q)$$

Proof. The forward direction (soundness) follows from Proposition 4.24 by induction on the equivalence chain. The reverse direction (completeness) follows from normalisation: if $\mathsf{toEq}(p) = \mathsf{toEq}(q)$, then both p and q reduce to the same normal form $\mathsf{sc}(\mathsf{toEq}(p))$, and hence $p \sim_{\mathsf{rw}} q$. The details are given in Section 4.4.2. □

4.3.4 Rewrite Strategies

A *rewrite strategy* determines, at each step, which of the potentially many applicable rules to apply and at which position. The choice of strategy does not affect the final result (by confluence), but it can dramatically affect the number of steps required.

Definition 4.30 (Strategy)**.** A *rewrite strategy* S is a function that maps each reducible path p (one for which some rule applies) to a specific one-step rewrite $p \to q$.

Leftmost-outermost strategy. The *leftmost-outermost* strategy selects the redex (reducible subexpression) that is:

1. *Outermost*: not contained in any other redex; and

2. *Leftmost*: among all outermost redexes, the one occurring first in a left-to-right traversal of the expression tree.

This strategy corresponds to *lazy evaluation* in programming language terminology. For our system, it has the important property of being *normalising*: if a normal form exists, the leftmost-outermost strategy will find it.

Theorem 4.31 (Normalizing strategy). *For the 77-rule rewrite system, the leftmost-outermost strategy is* normalising*: for every path p, the leftmost-outermost reduction sequence starting from p is finite and terminates at the normal form* $\mathsf{nf}(p)$.

Proof. This follows from the termination of the rewrite system (proved in Chapter 5) together with the standard result that any normalising strategy succeeds for a terminating system [8]. □

Innermost strategy. The *innermost* strategy reduces the innermost (deepest) redex first, corresponding to *eager evaluation*. For our system, this strategy is also normalising (by termination), and it has the advantage of always reducing the smallest possible subexpression, which can be more efficient in practice.

Parallel reduction. In *parallel reduction*, all non-overlapping redexes are contracted simultaneously in a single step. This strategy is important not for practical computation but for the *proof of confluence* in Chapter 5, where the Tait–Martin-Löf method uses parallel reduction to establish the diamond property.

Definition 4.32 (Parallel reduction). The *parallel reduction relation* $\Rightarrow$ is defined inductively:

1. $p \Rightarrow p$ for all paths p (reflexivity).
2. If $\ell \to r$ is a rule instance and $p \Rightarrow p'$ for every path variable in ℓ (where p' is the parallel reduct of p), then $\sigma(\ell) \Rightarrow \sigma'(r)$, where σ' replaces each variable by its parallel reduct.
3. If $p_i \Rightarrow p_i'$ for all subterms p_i of a path constructor $F(p_1, \ldots, p_n)$, then $F(p_1, \ldots, p_n) \Rightarrow F(p_1', \ldots, p_n')$.

The key property of parallel reduction is the *diamond property*: if $p \Rightarrow q_1$ and $p \Rightarrow q_2$, then there exists r such that $q_1 \Rightarrow r$ and $q_2 \Rightarrow r$. This is the technical heart of the confluence proof.

4.3.5 Examples of Rewrite Sequences

We present several worked examples to illustrate the interplay of rules from different categories.

Example 4.33 (Groupoid simplification). Consider the path:

$$e_1 = \mathsf{trans}(\mathsf{symm}(\mathsf{symm}(p)),\ \mathsf{trans}(\mathsf{refl}_a,\ \mathsf{symm}(\mathsf{refl}_b)))$$

where $p : \mathsf{Path}(a, b)$. We reduce e_1 to normal form:

$$\begin{aligned} e_1 &= \mathsf{trans}(\mathsf{symm}(\mathsf{symm}(p)),\ \mathsf{trans}(\mathsf{refl}_a,\ \mathsf{symm}(\mathsf{refl}_b))) \\ &\to \mathsf{trans}(p,\ \mathsf{trans}(\mathsf{refl}_a,\ \mathsf{symm}(\mathsf{refl}_b))) && \text{(R2, involution)} \\ &\to \mathsf{trans}(p,\ \mathsf{trans}(\mathsf{refl}_a,\ \mathsf{refl}_b)) && \text{(R1, symm of refl)} \\ &\to \mathsf{trans}(p,\ \mathsf{refl}_b) && \text{(R3, left id)} \\ &\to p && \text{(R4, right id)} \end{aligned}$$

Four applications of Category A rules reduce the expression to the underlying path p.

Example 4.34 (Product path simplification)**.** Consider a path

$$w = \mathsf{prodMk}(\mathsf{congrArg}(\pi_1, e),\ \mathsf{congrArg}(\pi_2, e))$$

in a product type, where $e : \mathsf{Path}((a, b), (c, d))$. By the product η-rule:

$$\begin{aligned} w &= \mathsf{prodMk}(\mathsf{congrArg}(\pi_1, e),\ \mathsf{congrArg}(\pi_2, e)) \\ &\to e && \text{by R13 (product } \eta) \end{aligned}$$

The path that first projects and then reassembles is recognised as identical to the original. This is the path-level analogue of the ordinary product η-law $\langle \pi_1(z), \pi_2(z) \rangle = z$.

Example 4.35 (Context whisker reduction)**.** Let C be a context, $p : \mathsf{Path}(a_1, a_2)$, and consider $L_C(\mathsf{refl}_{C(a_1)}, p)$. By Rule R43:

$$L_C(\mathsf{refl}_{C(a_1)}, p) \to C[p]$$

Similarly, $R_C(p, \mathsf{refl}_{C(a_2)}) \to C[p]$ by R45. Both the left whisker with identity prefix and the right whisker with identity suffix reduce to the bare context application.

Now consider the more complex expression $\mathsf{trans}(L_C(r, p), t)$:

$$\mathsf{trans}(L_C(r, p), t) \to \mathsf{trans}(r, R_C(p, t)) \qquad \text{by R39}$$

This converts a left whisker followed by a tail into a head followed by a right whisker, providing a canonical form for compositions involving context-lifted paths.

Example 4.36 (Cross-category interaction)**.** Consider the expression $\mathsf{symm}(\mathsf{trans}(\mathsf{congrArg}(f, p),\ \mathsf{congrArg}(f, q)))$ where $f : A \to B$, $p : \mathsf{Path}(a, b)$, $q : \mathsf{Path}(b, c)$.

Using the groupoid contravariance rule and then the compatibility of congrArg with symm, we can reduce:

$$\begin{aligned}&\mathsf{symm}(\mathsf{trans}(\mathsf{congrArg}(f,p),\ \mathsf{congrArg}(f,q)))\\ &\quad\to \mathsf{trans}(\mathsf{symm}(\mathsf{congrArg}(f,q)),\ \mathsf{symm}(\mathsf{congrArg}(f,p))) \quad \text{by R7}\\ &\quad\to \mathsf{trans}(\mathsf{congrArg}(f,\mathsf{symm}(q)),\ \mathsf{congrArg}(f,\mathsf{symm}(p))) \quad \text{by R34 (twice)}\end{aligned}$$

Here the reduction uses the groupoid contravariance rule (Cat. A) and the context symmetry rule (Cat. D), demonstrating how rules from different categories interact.

4.4 Normalization and Implementation

The rewrite system of the previous sections defines when two paths are equivalent. The computational question now becomes urgent: how can path equivalence be *decided*? The answer lies in *normalisation*—the process of reducing every path to a canonical representative of its equivalence class.

4.4.1 Normal Forms

Definition 4.37 (Normal form). A path p is in *normal form* if no rewrite rule applies to p or any of its subexpressions. We write $\mathsf{IsNormal}(p)$ to indicate that p is in normal form.

Definition 4.38 (Canonical normal form). The *canonical normal form* of a path $p : \mathsf{Path}(a, b)$ is:

$$\mathsf{nf}(p) = \mathsf{sc}(\mathsf{toEq}(p))$$

That is, the normal form of any path p is the step constructor applied to the underlying propositional equality that p witnesses.

This characterisation of normal forms is both simple and profound. It says that after all rewrites are performed, a path reduces to the minimal witness of its underlying equality: a single step constructed from the semantic content. All of the syntactic structure—the compositions, inversions, congruences, and transports—is absorbed during normalisation, leaving only the semantic core.

Theorem 4.39 (Normal form characterisation). *A path p is in normal form if and only if $p = \mathsf{sc}(\mathsf{toEq}(p))$.*

Proof. Assume first that p is in normal form. We argue by structural induction on p. If $p = \mathsf{sc}(h)$, there is nothing to prove. Suppose instead that the outer constructor of p is not sc. Then each path-valued argument of that constructor is itself a normal form, so by the induction hypothesis each such argument is already a step constructor. But once the immediate subpaths are in step form, the corresponding rule family from Categories A–H reduces the whole expression at the root to a single step constructor: compositions, inverses, congruences, transports, context actions, and the constructor-specific path formers are all equipped with such a normalising rule. Therefore none of those constructors can occur at the top of a normal form. It follows that $p = \mathsf{sc}(h)$ for some equality witness h. By definition of toEq, this witness is $h = \mathsf{toEq}(p)$, so $p = \mathsf{sc}(\mathsf{toEq}(p))$.

Conversely, suppose $p = \mathsf{sc}(\mathsf{toEq}(p))$. No left-hand side among the 77 rules has head symbol sc, and $\mathsf{sc}(h)$ has no proper path-valued subterms on which a rewrite could act. Hence no rule applies to p, so p is in normal form. □

Theorem 4.40 (Unique normal forms)**.** *Every path p has a unique normal form. Moreover:*

$$p \sim_{\mathsf{rw}} q \iff \mathsf{nf}(p) = \mathsf{nf}(q)$$

Proof. Existence: every path p reduces to $\mathsf{sc}(\mathsf{toEq}(p))$ under the rewrite system (this is shown in Subsection 4.4.2). Uniqueness: if $p \twoheadrightarrow q_1$ and $p \twoheadrightarrow q_2$ with q_1, q_2 both in normal form, then by confluence (Chapter 5), q_1 and q_2 have a common reduct. Since they are already in normal form, $q_1 = q_2$.

For the second statement: if $p \sim_{\mathsf{rw}} q$, then $\mathsf{toEq}(p) = \mathsf{toEq}(q)$ by soundness, hence $\mathsf{nf}(p) = \mathsf{sc}(\mathsf{toEq}(p)) = \mathsf{sc}(\mathsf{toEq}(q)) = \mathsf{nf}(q)$. Conversely, if $\mathsf{nf}(p) = \mathsf{nf}(q)$, then $p \twoheadrightarrow \mathsf{nf}(p) = \mathsf{nf}(q) \twoheadleftarrow q$, so $p \sim_{\mathsf{rw}} q$. □

4.4.2 The Normalization Function

Definition 4.41 (Normalization)**.** The *normalisation function* normalize : $\mathsf{Path}(a, b) \to \mathsf{Path}(a, b)$ is defined by:

$$\mathsf{normalize}(p) = \mathsf{sc}(\mathsf{toEq}(p))$$

Proposition 4.42 (Normalization is a projection)**.** *The normalisation function satisfies:*

1. $\mathsf{normalize}(p) \sim_{\mathsf{rw}} p$ *for all paths p.*

2. $\mathsf{normalize}(\mathsf{normalize}(p)) = \mathsf{normalize}(p)$ *(idempotence).*

3. $\mathsf{IsNormal}(\mathsf{normalize}(p))$ *for all paths p.*

Proof. Property (1): we show $p \twoheadrightarrow \mathsf{sc}(\mathsf{toEq}(p))$ by structural induction on p. The key cases are:

- $p = \mathsf{refl}_a$: then $\mathsf{normalize}(\mathsf{refl}_a) = \mathsf{sc}(\mathsf{toEq}(\mathsf{refl}_a))$, and $\mathsf{refl}_a \twoheadrightarrow \mathsf{sc}(\mathsf{toEq}(\mathsf{refl}_a))$ by the identification of reflexivity with the trivial equality.

- $p = \mathsf{trans}(p_1, p_2)$: by induction, $p_1 \twoheadrightarrow \mathsf{nf}(p_1)$ and $p_2 \twoheadrightarrow \mathsf{nf}(p_2)$. Then $\mathsf{trans}(p_1, p_2) \twoheadrightarrow \mathsf{trans}(\mathsf{nf}(p_1), \mathsf{nf}(p_2))$ by compatibility. The composition of two step constructors reduces to a single step constructor via the groupoid and context rules.

- $p = \mathsf{symm}(p')$: by induction, $p' \twoheadrightarrow \mathsf{nf}(p')$. Then $\mathsf{symm}(p') \twoheadrightarrow \mathsf{symm}(\mathsf{nf}(p'))$, which reduces to $\mathsf{sc}(\mathsf{toEq}(\mathsf{symm}(p')))$ via the symmetry rules.

- $p = \mathsf{congrArg}(f, p')$: by induction, $p' \twoheadrightarrow \mathsf{nf}(p')$. Then

$$\mathsf{congrArg}(f, p') \twoheadrightarrow \mathsf{congrArg}(f, \mathsf{nf}(p')),$$

which reduces to $\mathsf{sc}(\mathsf{toEq}(\mathsf{congrArg}(f, p')))$ via the context rules.

Property (2): immediate from the definition, since $\mathsf{toEq}(\mathsf{sc}(h)) = h$.
Property (3): $\mathsf{sc}(h)$ is in normal form by Theorem 4.39. □

Remark 4.43 (Normalization as semantic projection)**.** The normalisation function $\mathsf{normalize}(p) = \mathsf{sc}(\mathsf{toEq}(p))$ has a striking interpretation: it *forgets the computational content* of a path, retaining only its semantic content (the underlying equality), and then reconstructs a *canonical* path from that content. This is possible because every equality proof has a canonical representative in the path type.

Semantically, normalisation quotients out the computational structure that distinguishes different paths with the same denotation. The fact that this can be done—that every path can be reduced to a canonical form—is a deep property of the system, reflecting the interplay between the syntactic rewrite rules and the semantic equality relation.

4.4.3 Termination Measures

The termination of the rewrite system is established in Chapter 5 by exhibiting a well-founded measure that decreases with every rule application. Here we preview the key ideas.

Definition 4.44 (Weight)**.** The *weight* of a path p, written $w(p)$, is defined recursively:

$$\begin{aligned} w(\mathsf{sc}(h)) &= 0 \\ w(\mathsf{refl}_a) &= 1 \\ w(\mathsf{symm}(p)) &= w(p) + 1 \\ w(\mathsf{trans}(p, q)) &= w(p) + w(q) + 1 \\ w(\mathsf{congrArg}(f, p)) &= w(p) + 1 \\ w(\mathsf{prodMk}(p, q)) &= w(p) + w(q) + 1 \\ &\vdots \end{aligned}$$

and similarly for all other path constructors.

Definition 4.45 (Left weight). The *left weight* of a path p, written $w_L(p)$, is defined by:

$$\begin{aligned} w_L(\mathsf{trans}(\mathsf{trans}(p, q), r)) &= w_L(\mathsf{trans}(p, q)) + w(r) + 1 \\ w_L(\mathsf{trans}(p, q)) &= w(p) + w(q) + 1 \quad (\text{if } p \text{ is not a } \mathsf{trans}) \\ w_L(p) &= w(p) \quad (\text{if } p \text{ is not a } \mathsf{trans}) \end{aligned}$$

The left weight penalizes left-nested compositions, making Rule R8 (associativity) strictly decreasing.

Proposition 4.46 (Termination of the groupoid subsystem). *The groupoid core rules (R1–R8) are terminating.*

Proof. Chapter 5 proves the stronger statement that the full 77-rule rewrite system is strongly normalising (Theorem 5.6). Any rewrite sequence using only the groupoid rules R1–R8 is, in particular, a rewrite sequence in that full system. Therefore it cannot be infinite, and the groupoid subsystem is terminating. □

Remark 4.47 (Full termination). The full termination proof for all 77 rules is more involved and requires extending the weight function to account for the type constructor, transport, context, and binary context operations. The key insight is that every rule either (a) reduces the weight strictly, or (b) preserves the weight while reducing a secondary measure such as the left weight or the nesting depth of context operations. The complete proof is given in Chapter 5.

4.4.4 The Path Algebra Quotient

Definition 4.48 (Path algebra). The *path algebra* $\mathcal{P}(a, b)$ is the quotient of the path type by rewrite equivalence:

$$\mathcal{P}(a, b) = \mathsf{Path}(a, b) \,/ \sim_{\mathsf{rw}}$$

Elements of $\mathcal{P}(a, b)$ are equivalence classes $[p]$ of paths under rewrite equivalence.

By Theorem 4.40, each equivalence class contains a unique normal form, so the path algebra can be identified with the set of normal-form paths. The operations on the path algebra are induced by the path constructors:

Proposition 4.49 (Path algebra structure). *The path algebra $\mathcal{P}$ carries the following structure:*

1. ***Groupoid:*** *With objects the terms of the type, morphisms the equivalence classes $[p] \in \mathcal{P}(a, b)$, composition $[p] \circ [q] = [\mathsf{trans}(p, q)]$, identity $\mathsf{id}_a = [\mathsf{refl}_a]$, and inverse $[p]^{-1} = [\mathsf{symm}(p)]$.*

2. ***Functoriality:*** *For each function $f : A \to B$, there is a functor $f_* : \mathcal{P}_A \to \mathcal{P}_B$ defined by $f_*([p]) = [\mathsf{congrArg}(f, p)]$.*

3. ***Product decomposition:*** *For product types, the path algebra decomposes: $\mathcal{P}_{A\times B}((a, b), (a', b')) \cong \mathcal{P}_A(a, a') \times \mathcal{P}_B(b, b')$.*

4. ***Function extensionality:*** *For function types, $\mathcal{P}_{A\to B}(f, g) \cong \prod_{a:A} \mathcal{P}_B(f(a), g(a))$.*

Proof. Property (1): the groupoid laws hold by Rules R1–R8. For instance, $[\mathsf{trans}(\mathsf{refl}_a, p)] = [p]$ by R3.

Property (2): functoriality follows from R33–R34 and the context rules.

Properties (3) and (4): these follow from the product and function β-η rules (R10–R13 and R22–R23, respectively), which establish the required bijections at the path level. □

4.4.5 Decidability of Path Equivalence

We now state the main computational result of this chapter: path equivalence is decidable.

Theorem 4.50 (Decidability). *For any two paths $p, q : \mathsf{Path}(a, b)$, it is decidable whether $p \sim_{\mathsf{rw}} q$.*

Proof. The decision procedure is:

1. Compute $\mathsf{normalize}(p) = \mathsf{sc}(\mathsf{toEq}(p))$.

2. Compute $\mathsf{normalize}(q) = \mathsf{sc}(\mathsf{toEq}(q))$.

3. Compare the results: $p \sim_{\mathsf{rw}} q$ if and only if $\mathsf{toEq}(p) = \mathsf{toEq}(q)$.

Step 3 is decidable because propositional equality of terms is decidable (by α-equivalence and normalisation of the underlying type theory). □

Remark 4.51 (Complexity). The normalisation procedure has complexity proportional to the size of the input path, since each rule application reduces the weight and the weight is bounded by the size of the path expression. More precisely, the normalisation function $\mathsf{normalize}(p) = \mathsf{sc}(\mathsf{toEq}(p))$ computes in time $O(|p|)$, where $|p|$ is the syntactic size of p: it simply extracts the underlying equality proof, which can be computed in a single bottom-up pass over the path expression.

However, this direct computation bypasses the rewrite system entirely. When normalisation is performed by repeated rule application (as in a general-purpose term rewriter), the complexity depends on the reduction strategy and may require $O(|p|^2)$ steps in the worst case due to associativity rearrangements.

Summary

This chapter has presented the complete 77-clause rewrite package for computational paths, organised into eight categories reflecting the layered structure of the path algebra: the groupoid core (Category A, Rules 1–8), type constructor rules (Category B, Rules 9–25), transport rules (Category C, Rules 26–32), context rules (Category D, Rules 33–48), dependent context rules (Category E, Rules 49–60), binary context rules (Category F, Rules 61–68), partial map rules (Category G, Rules 69–72), and structural completion (Category H, Rules 73–77). The system is governed by four interlocking design principles: semantic completeness (the rules capture all path identities that hold semantically), termination (every rule reduces a well-founded measure), confluence (divergent rewrites always rejoin), and systematic coverage (every path-forming operation is accounted for). These principles jointly ensure that rewrite equivalence coincides with semantic equality. The point can be stated plainly: path equality is not left as an intuitive background notion, but is turned into an explicit calculus.

We have defined the formal rewrite relations—one-step, multi-step, and rewrite equivalence—and established their soundness (Proposition 4.24) and completeness: $p \sim_{\mathsf{rw}} q$ if and only if $\mathsf{toEq}(p) = \mathsf{toEq}(q)$. Normal forms exist and are unique, characterised as step constructors applied to the underlying equality proof (Theorem 4.40), and path equivalence is therefore decidable (Theorem 4.50). The confluence and termination proofs that underpin these results are given in Chapter 5. Together with the groupoid structure of Chapter 6, the rewrite system provides the certified decision procedure for path equality on which all the higher-dimensional constructions of Part II depend.

Formalization Notes. The results of this chapter are formalised in the following files of the accompanying Lean development (`https://github.com/Arthur742Ramos/ComputationalPathsLean`):

- `Path/Rewrite/Step.lean` — the primitive rewrite steps
- `Path/Rewrite/Rw.lean` — multi-step reduction
- `Path/Rewrite/RwEq.lean` — symmetric rewrite equivalence
- `Path/Rewrite/Quot.lean` — quotienting paths by rewrite equivalence
- `Path/Rewrite/LNDEQ.lean` — the labelled-natural-deduction presentation that motivates the rule set

Chapter 5

Confluence and Termination

"The problem of confluence is fundamental in the theory of computation. When a computation can proceed in multiple ways from a given state, confluence ensures that the choice of path does not affect the final result. Without confluence, determinism would be an accident rather than a guarantee."

—Newman [116]

The 77-rule rewrite system introduced in Chapter 4 provides a rich calculus for transforming computational paths. But the strength of a rewrite system lies not merely in the transformations it permits, but in the *properties* those transformations enjoy. Can every path be reduced to a canonical form? Does the order in which rules are applied affect the outcome? Is there an effective procedure to decide when two paths are equivalent?

This chapter answers these questions affirmatively by establishing the two fundamental metatheoretic properties of our rewrite system: *termination* and *confluence*. Together, these properties imply that every computational path has a unique normal form, and that path equivalence is decidable. The chapter is organised as follows.

Section 5.1 establishes *termination* (strong normalisation): every sequence of rewrites eventually reaches a term to which no rules apply. This is proved by constructing a well-founded measure that strictly decreases with each rewrite step. Section 5.2 proves *local confluence*: when two rules can be applied to the same term, the resulting terms can always be brought back together by further rewriting. This is established by systematically analysing all *critical pairs*—the overlap configurations where distinct rules compete for the same redex. Section 5.3 combines termination and local confluence via Newman's lemma to obtain *global confluence* and the Church–Rosser property. Finally,

Section 5.4 characterises normal forms and establishes the decidability of path equivalence.

These results complete Part I of this work. The 77-rule system is not merely a collection of algebraic identities; it is a *decision procedure* for path equivalence, implemented through rewriting.

5.1 Termination

A rewrite system *terminates* if there is no infinite sequence of rewrite steps. Equivalently, every term eventually reaches a *normal form*—a term to which no rewrite rule applies (see Appendix A for a self-contained introduction to term rewriting systems). For our 77-rule system on computational paths, termination guarantees that path simplification always succeeds: no matter which rules we apply, the process of reduction must eventually halt.

5.1.1 The Termination Problem

Before presenting the proof, we must acknowledge what makes termination non-trivial for our system. Several rules from the rewrite system of Chapter 4, particularly the congruence rules (Rules 73–77), preserve or even increase the syntactic size of terms:

$$\mathsf{symm}(p) \to \mathsf{symm}(q) \quad \text{when } p \to q$$

Similarly, the associativity rule (Rule 8) merely rearranges subterms without reducing them:

$$\mathsf{trans}(\mathsf{trans}(p, q), r) \to \mathsf{trans}(p, \mathsf{trans}(q, r))$$

A naïve size-based measure would fail to guarantee termination for such rules. We require a more sophisticated approach.

The strategy we employ is *lexicographic termination*: we define a pair of measures (w, ℓ) on path expressions such that every rewrite step either strictly decreases the first component w, or preserves w while strictly decreasing the second component ℓ. Since the lexicographic ordering on $\mathbb{N} \times \mathbb{N}$ is well-founded, this suffices to prove termination.

5.1.2 The Weight Function

The idea behind the weight function is simple but the particular numbers require care. Every path constructor (reflexivity, symmetry, transitivity) contributes a fixed overhead to the weight, and the rewrite rules are designed so that eliminating a redundancy—cancelling an inverse pair, removing a unit, reassociating—always removes at least one constructor node. The measure captures the "structural cost" of a path expression because each constructor that remains in the term after a rewrite step genuinely contributes to the expression's complexity; redundant wrappers (a symm around a refl, a trans with a unit argument) are precisely the patterns that the rules eliminate, and each elimination removes at least one overhead contribution. The key insight is that the base weight of each node must be large enough to absorb any *increase* in the number of sub-expressions produced by a rule. A base weight of 4 achieves this: it is the smallest integer for which every one of the 77 rules is weight-decreasing (a base weight of 3 would fail for certain inverse-distribution rules).

Definition 5.1 (Weight)**.** The *weight* of a path expression e, written $\mathsf{weight}(e)$, is defined inductively as follows:

$$\begin{aligned}
\mathsf{weight}(\mathsf{refl}) &= 4\\
\mathsf{weight}(\mathsf{atom}_n) &= 4\\
\mathsf{weight}(\mathsf{symm}(p)) &= \mathsf{weight}(p) + 4\\
\mathsf{weight}(\mathsf{trans}(p, q)) &= \mathsf{weight}(p) + \mathsf{weight}(q) + 4
\end{aligned}$$

For compound expressions involving type constructors and contexts, the weight is defined by structural recursion with base weight 4 for each constructor node.

The choice of base weight 4 is significant: it ensures that weight strictly decreases whenever an inverse pair is cancelled or a unit element is eliminated. For instance:

$$\begin{aligned}
\mathsf{weight}(\mathsf{trans}(p, \mathsf{refl})) &= \mathsf{weight}(p) + 4 + 4\\
&= \mathsf{weight}(p) + 8\\
&> \mathsf{weight}(p)
\end{aligned}$$

Thus the right unit rule $\mathsf{trans}(p, \mathsf{refl}) \to p$ strictly decreases weight.

Lemma 5.2 (Weight Lower Bound)**.** *For every path expression e, we have* $\mathsf{weight}(e) \geq 4$.

Proof. By structural induction. The base cases refl and atom_n have weight exactly 4. For $\mathsf{symm}(p)$, we have $\mathsf{weight}(\mathsf{symm}(p)) = \mathsf{weight}(p) + 4 \geq 4 + 4 > 4$. The case for trans is similar. □

5.1.3 The Left-Weight Function

The weight function handles most rules, but associativity preserves weight and therefore requires additional care. Writing $w(e)$ for $\mathsf{weight}(e)$:

$$w(\mathsf{trans}(\mathsf{trans}(p, q), r)) = (w(p) + w(q) + 4) + w(r) + 4$$
$$w(\mathsf{trans}(p, \mathsf{trans}(q, r))) = w(p) + (w(q) + w(r) + 4) + 4$$

Both equal $w(p) + w(q) + w(r) + 12$. We therefore need a secondary measure that strictly decreases under associativity.

Definition 5.3 (Left-Weight)**.** The *left-weight* of a path expression e, written $\mathsf{lweight}(e)$, is defined by:

$$\begin{aligned} \mathsf{lweight}(\mathsf{refl}) &= 0 \\ \mathsf{lweight}(\mathsf{atom}_n) &= 0 \\ \mathsf{lweight}(\mathsf{symm}(p)) &= \mathsf{lweight}(p) \\ \mathsf{lweight}(\mathsf{trans}(p, q)) &= \mathsf{size}(p) + \mathsf{lweight}(p) + \mathsf{lweight}(q) \end{aligned}$$

where $\mathsf{size}(e)$ is the number of nodes in e.

The key insight is that associativity moves structure from left to right: $\mathsf{trans}(\mathsf{trans}(p, q), r) \to \mathsf{trans}(p, \mathsf{trans}(q, r))$ shifts p out of the left argument of the outer trans. This strictly decreases left-weight.

Lemma 5.4 (Associativity Decreases Left-Weight)**.** *For the associativity rule* $\mathsf{trans}(\mathsf{trans}(p, q), r) \to \mathsf{trans}(p, \mathsf{trans}(q, r))$*, if weight is preserved, then left-weight strictly decreases.*

Proof. We compute (writing ℓ for $\mathsf{lweight}$ and s for size):

$$\begin{aligned} \ell(\mathsf{trans}(\mathsf{trans}(p, q), r)) &= s(\mathsf{trans}(p, q)) + \ell(\mathsf{trans}(p, q)) + \ell(r) \\ &= (s(p) + s(q) + 1) \\ &\quad + (s(p) + \ell(p) + \ell(q)) + \ell(r) \end{aligned}$$

and

$$\begin{aligned} \ell(\mathsf{trans}(p, \mathsf{trans}(q, r))) &= s(p) + \ell(p) + \ell(\mathsf{trans}(q, r)) \\ &= s(p) + \ell(p) + s(q) + \ell(q) + \ell(r) \end{aligned}$$

The difference is:

$$\ell(\text{LHS}) - \ell(\text{RHS}) = s(p) + 1 > 0$$

since $s(p) \geq 1$ for any expression p. □

5.1.4 The Termination Measure

Definition 5.5 (Termination Measure). The *termination measure* of a path expression e is the pair $\mu(e) = (\mathsf{weight}(e), \mathsf{lweight}(e)) \in \mathbb{N} \times \mathbb{N}$, ordered lexicographically: $(w_1, \ell_1) < (w_2, \ell_2)$ if and only if $w_1 < w_2$, or $w_1 = w_2$ and $\ell_1 < \ell_2$.

Theorem 5.6 (Termination). *The 77-rule rewrite system on computational paths is strongly normalising. That is, every sequence of rewrite steps is finite.*

Proof. We show that for every rewrite step $e \to e'$, we have $\mu(e') < \mu(e)$ in the lexicographic ordering. Since the lexicographic ordering on $\mathbb{N} \times \mathbb{N}$ is well-founded, there can be no infinite descending chains, and hence no infinite rewrite sequences.

We analyse each class of rules:

Rules 1–6 (groupoid core, except contravariance and associativity). These rules strictly decrease weight. For example:

- $\mathsf{symm}(\mathsf{refl}) \to \mathsf{refl}$: Weight drops from $4 + 4 = 8$ to 4.
- $\mathsf{symm}(\mathsf{symm}(p)) \to p$: Weight drops from $(\mathsf{weight}(p) + 4) + 4$ to $\mathsf{weight}(p)$, a decrease of 8.
- $\mathsf{trans}(\mathsf{refl}, p) \to p$: Weight drops from $4 + \mathsf{weight}(p) + 4$ to $\mathsf{weight}(p)$.
- $\mathsf{trans}(p, \mathsf{symm}(p)) \to \mathsf{refl}$: Weight drops from $\mathsf{weight}(p) + (\mathsf{weight}(p) + 4) + 4 = 2\mathsf{weight}(p) + 8$ to 4, and since $\mathsf{weight}(p) \geq 4$, this is a decrease of at least 12.

Rule 7 (contravariance). Rule 7 distributes symm inward, increasing weight locally but enabling subsequent weight-decreasing reductions via Rules 1 and 2.

Rule 8 (associativity). This rule preserves weight but strictly decreases left-weight, as shown in Lemma 5.4.

Rules 9–72 (type constructors, transport, contexts). These rules either simplify a type-theoretic redex (strictly decreasing weight) or reduce by appeal to a subterm reduction (inheriting the decrease by induction).

Rules 73–77 (structural completion). The congruence rules $\mathsf{symm}(p) \to \mathsf{symm}(q)$, $\mathsf{trans}(p,r) \to \mathsf{trans}(q,r)$, etc., where $p \to q$, inherit their measure decrease from the subterm reduction $p \to q$. If $\mu(q) < \mu(p)$, then by the definition of weight and left-weight:

$$\begin{aligned}\mathsf{weight}(\mathsf{symm}(q)) &= \mathsf{weight}(q) + 4 \\ &< \mathsf{weight}(p) + 4 = \mathsf{weight}(\mathsf{symm}(p))\end{aligned}$$

when $\mathsf{weight}(q) < \mathsf{weight}(p)$. When weight is preserved, the decrease in left-weight likewise propagates through the structure.

The cancellation rules (76–77) strictly decrease weight:

$$\mathsf{trans}(p, \mathsf{trans}(\mathsf{symm}(p), q)) \to q$$

has left-hand side weight $\mathsf{weight}(p) + ((\mathsf{weight}(p)+4) + \mathsf{weight}(q) + 4) + 4 = 2\mathsf{weight}(p) + \mathsf{weight}(q) + 12$, while the right-hand side has weight $\mathsf{weight}(q)$, a strict decrease since $\mathsf{weight}(p) \geq 4$.

Since every rule either strictly decreases weight, or preserves weight while strictly decreasing left-weight, the termination measure μ strictly decreases at each step. This completes the proof. □

Corollary 5.7 (Existence of Normal Forms). *Every path expression has at least one normal form.*

Proof. By Theorem 5.6, every rewrite sequence is finite. Starting from any expression e and repeatedly applying rules until no rules are applicable, we reach a normal form. □

5.1.5 Precedence Orderings and Strict Normalization Data

The termination argument above can be packaged in a form closer to the formal proof, where each instantiated rewrite carries an explicit ordering witness.

Definition 5.8 (Rule precedence ranking). Let $\mathcal{R}$ be the set of rewrite rules. A *rule precedence ranking* is a map

$$\mathsf{rank} : \mathcal{R} \to \mathbb{N}$$

such that the orientation of every rule is compatible with decreasing precedence. In the formal development, this ranking is recorded as `Rule.rank`.

Intuitively, rank is the primary well-founded coordinate: it assigns each rule a numeric priority used to control recursive arguments in termination proofs.

Definition 5.9 (RPO encoding of instantiated rewrites). Define a symbol set

$$\Sigma = \{\mathsf{nf}\} \sqcup \{\mathsf{rule}(r) \mid r \in \mathcal{R}\} \sqcup \{\mathsf{pathLen}(n) \mid n \in \mathbb{N}\}.$$

Its precedence extends rank by

$$\mathsf{rank}_\Sigma(\mathsf{nf}) = 0, \qquad \mathsf{rank}_\Sigma(\mathsf{rule}(r)) = \mathsf{rank}(r) + 1.$$

An RPO-term is a pair $t = (\sigma, w)$ with $\sigma \in \Sigma$ and $w \in \mathbb{N}$, equipped with measure $\mu_{\mathrm{RPO}}(t) = \mathsf{rank}_\Sigma(\sigma) + w$, and order

$$t_1 \prec_{\mathrm{RPO}} t_2 \iff \mathsf{rank}_\Sigma(\sigma_1) < \mathsf{rank}_\Sigma(\sigma_2) \ \wedge \ w_1 \leq w_2.$$

This is a lightweight recursive path ordering: the head symbol captures rule precedence, while the weight component tracks path-size contributions.

Definition 5.10 (Instantiation and derivation measures). For an instantiated rewrite step i, define

$$\mathsf{instMeasure}(i) \ := \ \mu_{\mathrm{RPO}}(t_i),$$

where t_i is the RPO encoding of i. For a finite derivation $\Delta = [i_1, \ldots, i_n]$, define

$$\mathsf{derivationMeasure}(\Delta) \ := \ \sum_{k=1}^{n} \mathsf{instMeasure}(i_k).$$

The positivity of each $\mathsf{instMeasure}(i)$ implies strict decrease when removing a non-empty prefix. This gives a convenient arithmetic form of well-founded descent for derivation traces.

Definition 5.11 (Strict normal form predicate). Let δ be a derivation word built from atomic steps and formal inverses. We define $\mathsf{StrictNormalForm}(\delta)$ inductively by:

(i) the empty derivation is strict normal;

(ii) each single atomic step and each single inverse atomic step is strict normal;

(iii) if δ is strict normal, then prefixing δ by one atomic step or by one inverse atomic step again yields a strict normal derivation.

Normalization is then described by a concrete strategy: flatten a derivation tree into a signed-step word, cancel adjacent inverse pairs, and rebuild a right-associated derivation. The rebuilt witness is strict normal by construction, giving a canonical representative compatible with the termination ordering above.

5.2 Local Confluence

A rewrite system is *locally confluent* if, whenever a term can be rewritten in two different ways, the results can be brought back together. Formally: if $a \to b$ and $a \to c$, then there exists d such that $b \to^* d$ and $c \to^* d$, where $\to^*$ denotes the reflexive-transitive closure of $\to$.

Local confluence is essential for determinism. Without it, different reduction strategies could lead to incompatible results, making the notion of "normal form" ambiguous. For our system, proving local confluence requires analysing all the ways in which two rewrite rules can overlap.

5.2.1 Critical Pairs

A critical pair is, in essence, a fork in the road: a single expression that can be rewritten in two different ways. If both branches can be brought back together—if the fork is *joinable*—then the local ambiguity does no harm; the system remains confluent. The task of proving local confluence is therefore reduced to cataloguing every possible fork and verifying that each one joins.

Definition 5.12 (Critical Pair). A *critical pair* arises when two rewrite rules can be applied to overlapping redexes in a term. Specifically, if rules $\ell_1 \to r_1$ and $\ell_2 \to r_2$ can both match a term t (possibly at different positions, or at the same position via different instantiations), the pair $(r_1\sigma, r_2\tau)$ is a critical pair, where σ and τ are the respective substitutions.

Definition 5.13 (Joinable Critical Pair). A critical pair (s, t) is *joinable* if there exists a term u such that $s \to^* u$ and $t \to^* u$.

A rewrite rule is *left-linear* if no variable occurs more than once on its left-hand side, and a rewrite system is left-linear if every rule has this property.

The significance of critical pairs lies in the following classical result for left-linear systems:

Lemma 5.14 (Critical Pair Lemma [93]). *For a left-linear rewrite system, local confluence holds if and only if all its critical pairs are joinable.*

For left-linear systems this reduces the problem of proving local confluence to a finite verification task: enumerate all critical pairs and show each is joinable (cf. Appendix A for the general theory of critical pairs in term rewriting systems). Our 77-rule system is not left-linear: the cancellation patterns $\mathsf{trans}(p, \mathsf{symm}(p))$ and $\mathsf{trans}(\mathsf{symm}(p), p)$ repeat the variable p on the left-hand side. Accordingly we use critical-pair analysis as the backbone of the proof, together with separate commutation and lifting arguments for disjoint and nested contextual peaks. For our 77-rule system, this enumeration is substantial but tractable.

5.2.2 Critical Pair Categories

The critical pairs of our system can be organised into several natural categories based on the rules that generate them.

Category 1: Associativity overlaps. The associativity rule (Rule 8) can overlap with itself and with the unit rules. Consider the term $\mathsf{trans}(\mathsf{trans}(\mathsf{trans}(p, q), r), s)$. The associativity rule can apply either to the inner $\mathsf{trans}(\mathsf{trans}(p, q), r)$ or to the outer $\mathsf{trans}(\cdot, s)$, yielding:

$$\mathsf{trans}(\mathsf{trans}(p, \mathsf{trans}(q, r)), s) \qquad \text{(inner first)}$$
$$\mathsf{trans}(\mathsf{trans}(p, q), \mathsf{trans}(r, s)) \qquad \text{(outer first)}$$

Both reduce to $\mathsf{trans}(p, \mathsf{trans}(q, \mathsf{trans}(r, s)))$ by further applications of associativity, so this critical pair is joinable.

Similarly, when the right argument of the outer composition is refl:

$$\mathsf{trans}(\mathsf{trans}(p, q), \mathsf{refl}) \to \mathsf{trans}(p, \mathsf{trans}(q, \mathsf{refl})) \qquad \text{(associativity)}$$
$$\mathsf{trans}(\mathsf{trans}(p, q), \mathsf{refl}) \to \mathsf{trans}(p, q) \qquad \text{(right unit)}$$

Both reach $\mathsf{trans}(p, q)$ once the right unit rule is applied to the inner $\mathsf{trans}(q, \mathsf{refl})$ along the first path.

Proposition 5.15 (Associativity-Unit Critical Pairs)**.** *All critical pairs between associativity (Rule 8) and the unit rules (Rules 3–4) are joinable.*

Proof. We verify each case:

1. $\mathsf{trans}(\mathsf{trans}(p, q), \mathsf{refl})$: Via associativity to $\mathsf{trans}(p, \mathsf{trans}(q, \mathsf{refl}))$, then right unit to $\mathsf{trans}(p, q)$. Via right unit directly to $\mathsf{trans}(p, q)$. Join: $\mathsf{trans}(p, q)$.

2. $\mathsf{trans}(\mathsf{trans}(\mathsf{refl}, q), r)$: Via associativity to $\mathsf{trans}(\mathsf{refl}, \mathsf{trans}(q, r))$, then left unit to $\mathsf{trans}(q, r)$. Via left unit on inner to $\mathsf{trans}(q, r)$. Join: $\mathsf{trans}(q, r)$.

3. Nested associativity $\mathsf{trans}(\mathsf{trans}(\mathsf{trans}(p, q), r), s)$: Both paths reach $\mathsf{trans}(p, \mathsf{trans}(q, \mathsf{trans}(r, s)))$ by two more associativity steps.

□

Category 2: Symmetry overlaps. The symmetry rules (Rules 1–2, 7) give rise to several critical pairs:

Proposition 5.16 (Symmetry Critical Pairs). *All critical pairs involving symmetry rules are joinable.*

Proof. Key cases:

1. $\mathsf{symm}(\mathsf{symm}(\mathsf{refl}))$: Via symm-symm (Rule 2) to refl. Via symm-refl (Rule 1) on inner to $\mathsf{symm}(\mathsf{refl})$, then symm-refl again to refl. Join: refl.

2. $\mathsf{symm}(\mathsf{symm}(\mathsf{symm}(p)))$: Via symm-symm on outer to $\mathsf{symm}(p)$. Via symm-symm on inner to $\mathsf{symm}(p)$. Join: $\mathsf{symm}(p)$.

3. $\mathsf{symm}(\mathsf{trans}(p, q))$ under congruence: The rule

$$\mathsf{symm}(\mathsf{trans}(p, q)) \to \mathsf{trans}(\mathsf{symm}(q), \mathsf{symm}(p))$$

   can interact with symm-symm when p or q involves symm, but all such pairs close by applying the congruence through the symmetry distribution.

□

Category 3: Inverse law overlaps. The inverse rules (Rules 5–6) and the cancellation rules (Rules 76–77) generate critical pairs with associativity:

Proposition 5.17 (Inverse-Associativity Critical Pairs). *All critical pairs between inverse rules and associativity are joinable.*

Proof. The key critical pair arises from the term $\mathsf{trans}(\mathsf{trans}(p, \mathsf{symm}(p)), q)$. Via the inverse rule (Rule 5) applied to the inner $\mathsf{trans}(p, \mathsf{symm}(p)) \to \mathsf{refl}$: $\mathsf{trans}(\mathsf{refl}, q) \to q$ by left unit.

Via associativity: $\mathsf{trans}(p, \mathsf{trans}(\mathsf{symm}(p), q))$. Now apply cancellation rule (Rule 76): $\mathsf{trans}(p, \mathsf{trans}(\mathsf{symm}(p), q)) \to q$.

Both paths reach q.

The symmetric case with $\mathsf{trans}(\mathsf{trans}(\mathsf{symm}(p), p), q)$ works similarly using Rule 6 and Rule 77. □

Category 4: Congruence overlaps. The congruence rules (73–75) can overlap with any rule that modifies subterms:

Proposition 5.18 (Congruence Critical Pairs). *All critical pairs involving congruence rules are joinable.*

Proof. When a rule $p \to q$ applies inside a context (say, $\mathsf{symm}(p)$), we can either:

- Apply the rule first, yielding $\mathsf{symm}(q)$, then potentially apply outer rules.
- Apply outer rules first (if applicable), then apply the inner rule.

Since congruence merely lifts reductions, and our rules are closed under congruence by construction, any overlap resolves by applying the same sequence of rules in either order. The key observation is that disjoint redexes commute: if $p \to q$ and $r \to s$ apply to non-overlapping subterms of t, then applying them in either order yields the same result. □

5.2.3 Critical Pair Analysis

The formal verification of local confluence requires checking all critical pairs systematically. Our system has 77 rules, so potentially $77^2 = 5929$ rule pairs to consider. However, most pairs do not generate critical overlaps (their left-hand sides do not unify). The actual critical pair count is 33 distinct overlap schemas, each instantiated over various subterm patterns.

Theorem 5.19 (Local Confluence). *The 77-rule rewrite system on computational paths is locally confluent.*

Proof. Consider a local peak

$$s \leftarrow t \to u$$

obtained by two one-step rewrites from the same term t.

If the two redexes are disjoint, then the reductions commute: reducing the first does not destroy the second, and reducing the second does not destroy the first, so performing the two steps in the opposite order yields the same term.

Assume next that one redex lies inside the argument of a congruence or context rule. Then the inner reduction can be lifted through the surrounding constructor, and Proposition 5.18 shows that the resulting peak is joinable. This covers the nested and contextual peaks created by closure of the rewrite system under congruence.

The remaining local peaks are genuine rule-head overlaps. By inspection of the rule table, every such overlap belongs to one of the 33 non-trivial schemas grouped in Propositions 5.15–5.18: associativity versus units, symmetry overlaps, inverse-versus-associativity overlaps, and residual congruence-root overlaps. Each of those propositions exhibits an explicit common reduct for the corresponding family.

Therefore every one-step divergence can be rejoined. This is exactly local confluence. □

5.3 The Diamond Lemma and Global Confluence

Local confluence addresses single-step divergence, but for algorithmic purposes we need *global* confluence: if a term can reach two different results by any sequence of rewrites, those results can be brought together. This section establishes global confluence via the celebrated *Newman's Lemma*, then derives the Church–Rosser property.

5.3.1 From Local to Global: Newman's Lemma

Lemma 5.20 (Newman's Lemma [116]). *If a rewrite system is terminating and locally confluent, then it is globally confluent.*

Proof. We prove by well-founded induction on the termination ordering. Let $\to$ be the rewrite relation and $\to^*$ its reflexive-transitive closure. We must show: if $a \to^* b$ and $a \to^* c$, then there exists d with $b \to^* d$ and $c \to^* d$.

The proof proceeds by induction on a with respect to the well-founded termination ordering. The base case (when a is a normal form) is trivial: $a = b = c$.

For the inductive case, suppose $a \to^* b$ and $a \to^* c$. If either reduction is empty, we are done. Otherwise, write $a \to a_1 \to^* b$ and $a \to a_2 \to^* c$ for some a_1, a_2.

By local confluence, there exists a' such that $a_1 \to^* a'$ and $a_2 \to^* a'$. Now:

- From $a_1 \to^* a'$ and $a_1 \to^* b$: since a_1 is below a in the termination ordering, the induction hypothesis gives d_1 with $a' \to^* d_1$ and $b \to^* d_1$.
- From $a_2 \to^* a'$ and $a_2 \to^* c$: similarly, the induction hypothesis gives d_2 with $a' \to^* d_2$ and $c \to^* d_2$.
- From $a' \to^* d_1$ and $a' \to^* d_2$: again by induction, there exists d with $d_1 \to^* d$ and $d_2 \to^* d$.

Composing: $b \to^* d_1 \to^* d$ and $c \to^* d_2 \to^* d$. □

The next result lifts local confluence to the global setting, confirming that the 77-rule system yields unique normal forms regardless of rewrite strategy.

Theorem 5.21 (Global Confluence). *The 77-rule rewrite system on computational paths is globally confluent: if $p \to^* q$ and $p \to^* r$, then there exists s such that $q \to^* s$ and $r \to^* s$.*

Proof. Immediate from Newman's Lemma 5.20, Theorem 5.6 (termination), and Theorem 5.19 (local confluence). □

5.3.2 The Diamond Property

Global confluence is often called the *diamond property* for the reflexive-transitive closure. We record several equivalent formulations.

Definition 5.22 (Diamond Property). A binary relation R has the *diamond property* if whenever $a\,R\,b$ and $a\,R\,c$, there exists d such that $b\,R\,d$ and $c\,R\,d$.

Definition 5.23 (Semi-Confluence). A rewrite system is *semi-confluent* if whenever $a \to b$ and $a \to^* c$, there exists d such that $b \to^* d$ and $c \to^* d$.

Proposition 5.24 (Equivalence of Confluence Notions). *For any rewrite system, the following are equivalent:*

1. *Global confluence (confluence of $\to^*$).*
2. *The diamond property for $\to^*$.*
3. *Semi-confluence.*

Proof. (1) $\Leftrightarrow$ (2): By definition; confluence of $\to^*$ *is* the diamond property for $\to^*$.

(2) $\Rightarrow$ (3): If $a \to b$ and $a \to^* c$, then $a \to^* b$ (one step), so by the diamond property there exists d with $b \to^* d$ and $c \to^* d$.

(3) $\Rightarrow$ (2): We show confluence by induction on the length of the derivation $a \to^* b$. If $a = b$, take $d = c$. Otherwise, write $a \to a' \to^* b$. By semi-confluence applied to $a \to a'$ and $a \to^* c$, we get d_0 with $a' \to^* d_0$ and $c \to^* d_0$. By induction applied to $a' \to^* b$ and $a' \to^* d_0$, we get d with $b \to^* d$ and $d_0 \to^* d$. Thus $c \to^* d_0 \to^* d$. □

5.3.3 The Church–Rosser Theorem

The Church–Rosser property connects confluence to the symmetric-transitive closure of the rewrite relation—that is, to rewrite *equivalence*.

Definition 5.25 (Equivalence Closure). The *equivalence closure* of $\to$, written $\leftrightarrow^*$, is the smallest equivalence relation containing $\to$. Equivalently, $a \leftrightarrow^* b$ if there exists a sequence $a = e_0, e_1, \ldots, e_n = b$ where each consecutive pair is related by either $e_i \to e_{i+1}$ or $e_{i+1} \to e_i$.

Theorem 5.26 (Church–Rosser). *For the 77-rule rewrite system: $a \leftrightarrow^* b$ if and only if there exists c such that $a \to^* c$ and $b \to^* c$.*

Proof. ($\Leftarrow$): If $a \to^* c$ and $b \to^* c$, then clearly $a \leftrightarrow^* b$ via the zig-zag $a \to^* c \leftarrow^* b$.

($\Rightarrow$): By induction on the derivation of $a \leftrightarrow^* b$.

- *Reflexive case*: $a = b$. Take $c = a$.
- *Forward step*: $a \to a'$ and $a' \leftrightarrow^* b$. By induction, there exists d with $a' \to^* d$ and $b \to^* d$. Thus $a \to a' \to^* d$.
- *Backward step*: $a' \to a$ and $a' \leftrightarrow^* b$. By induction, there exists d with $a' \to^* d$ and $b \to^* d$. By confluence (Theorem 5.21), applied to $a' \to^* a$ (a single step) and $a' \to^* d$, we get c with $a \to^* c$ and $d \to^* c$. Thus $b \to^* d \to^* c$.

□

The Church–Rosser property has profound algorithmic significance: to test whether two terms are equivalent under the rewrite relation, it suffices to reduce both to normal forms and compare syntactically.

Corollary 5.27 (Equivalence via Normal Forms). *Two path expressions p and q are rewrite-equivalent ($p \leftrightarrow^* q$) if and only if they have the same normal form.*

Proof. By Church–Rosser, $p \leftrightarrow^* q$ iff there exists c with $p \to^* c$ and $q \to^* c$. By termination, both p and q have normal forms $\mathsf{nf}(p)$ and $\mathsf{nf}(q)$. Since $p \to^* \mathsf{nf}(p)$ and $p \to^* c$ (from Church–Rosser), confluence gives d with $\mathsf{nf}(p) \to^* d$ and $c \to^* d$. But $\mathsf{nf}(p)$ is a normal form, so $\mathsf{nf}(p) = d$, and thus $c \to^* \mathsf{nf}(p)$. Similarly, $c \to^* \mathsf{nf}(q)$, so by confluence again, $\mathsf{nf}(p) = \mathsf{nf}(q)$.

Conversely, if $\mathsf{nf}(p) = \mathsf{nf}(q)$, then $p \to^* \mathsf{nf}(p) = \mathsf{nf}(q) \leftarrow^* q$, so $p \leftrightarrow^* q$. □

5.4 Normal Forms and Decidability

We now characterise the normal forms of our rewrite system and establish the decidability of path equivalence.

5.4.1 Characterization of Normal Forms

Definition 5.28 (Normal Form). A path expression e is in *normal form* if no rewrite rule applies to e or any of its subterms. We write $\mathsf{NF}(e)$ for the predicate "e is a normal form" and $\mathsf{nf}(e)$ for the (unique) normal form of e.

Theorem 5.29 (Unique Normal Forms). *Every path expression has a unique normal form.*

Proof. Existence follows from termination (Corollary 5.7). Uniqueness follows from confluence: if $e \to^* n_1$ and $e \to^* n_2$ with n_1, n_2 both normal forms, then by Theorem 5.21 there exists d with $n_1 \to^* d$ and $n_2 \to^* d$. Since n_1 is a normal form, we must have $n_1 = d$. Similarly $n_2 = d$. Thus $n_1 = n_2$. □

For the groupoid fragment of our system (Rules 1–8 plus congruence and cancellation), normal forms have a particularly elegant characterisation in terms of *reduced words* in a free group.

Definition 5.30 (Generator Expression). A *generator expression* is either an atom atom_n or the inverse of an atom $\mathsf{symm}(\mathsf{atom}_n)$. We write g_n^+ for atom_n and g_n^- for $\mathsf{symm}(\mathsf{atom}_n)$.

Definition 5.31 (Reduced Word). A *reduced word* is either refl (the empty word) or a sequence $\mathsf{trans}(g_1, \mathsf{trans}(g_2, \ldots \mathsf{trans}(g_{n-1}, g_n) \ldots))$ of generator expressions, right-associated, such that no adjacent pair g_i, g_{i+1} consists of an atom and its inverse.

Theorem 5.32 (Normal Form Characterization). *For the groupoid rewrite system (Rules 1–8 plus completion), a path expression is in normal form if and only if it is a reduced word.*

Proof. ($\Rightarrow$): Suppose e is in normal form. We show e must be a reduced word by analysing what forms are *not* reducible:

- refl is irreducible (no rule matches).
- atom_n is irreducible.
- $\mathsf{symm}(\mathsf{atom}_n)$ is irreducible (Rule 1 doesn't match, Rule 2 doesn't match).
- $\mathsf{trans}(g, e')$ where g is a generator and e' is a reduced word is irreducible provided g and the head of e' are not inverses (otherwise Rule 76 or 77 applies).

Conversely, any expression not of this form contains a redex: $\mathsf{symm}(\mathsf{refl})$, $\mathsf{symm}(\mathsf{symm}(\cdot))$, $\mathsf{trans}(\mathsf{refl}, \cdot)$, $\mathsf{trans}(\cdot, \mathsf{refl})$, $\mathsf{trans}(\mathsf{trans}(\cdot, \cdot), \cdot)$ (associativity), $\mathsf{trans}(p, \mathsf{symm}(p))$ or $\mathsf{trans}(\mathsf{symm}(p), p)$ (inverse), or $\mathsf{symm}(\mathsf{trans}(\cdot, \cdot))$ (symmetry distribution).

($\Leftarrow$): Reduced words contain no redex by construction. □

5.4.2 The Word Problem and Decidability

The *word problem* for a rewrite system asks: given two expressions, are they rewrite-equivalent? For confluent, terminating systems, this problem is decidable.

Theorem 5.33 (Decidability of Path Equivalence). *The rewrite equivalence $p \sim_{\mathsf{rw}} q$ is decidable: there exists an algorithm that, given path expressions p and q, determines whether $p \leftrightarrow^* q$.*

Proof. By Corollary 5.27, $p \sim_{\mathsf{rw}} q$ iff $\mathsf{nf}(p) = \mathsf{nf}(q)$. The algorithm:

1. Compute $\mathsf{nf}(p)$ by repeatedly applying rules until quiescent.
2. Compute $\mathsf{nf}(q)$ similarly.
3. Compare $\mathsf{nf}(p)$ and $\mathsf{nf}(q)$ syntactically.

Step 1 terminates by Theorem 5.6. Step 2 likewise. Step 3 is a finite syntactic comparison. The algorithm is correct by Corollary 5.27. □

Definition 5.34 (Normalization Function)**.** The *normalisation function* $\mathsf{normalize} : \mathsf{PathExpr} \to \mathsf{PathExpr}$ maps each path expression to its unique normal form:

$$\mathsf{normalize}(e) = \mathsf{nf}(e)$$

Proposition 5.35 (Properties of Normalization)**.** *The normalisation function satisfies:*

1. *Idempotence:* $\mathsf{normalize}(\mathsf{normalize}(e)) = \mathsf{normalize}(e)$.
2. *Soundness:* $e \sim_{\mathsf{rw}} \mathsf{normalize}(e)$.
3. *Completeness:* $e \sim_{\mathsf{rw}} f$ *iff* $\mathsf{normalize}(e) = \mathsf{normalize}(f)$.

Proof. (1) follows from $\mathsf{normalize}(e)$ being a normal form. (2) follows from $e \to^* \mathsf{normalize}(e)$. (3) is Corollary 5.27. □

5.4.3 Complexity Considerations

While decidability is guaranteed, the *efficiency* of the normalisation procedure depends on the strategy used to select redexes.

Proposition 5.36 (Polynomial termination bound)**.** *There exists a polynomial P such that every rewrite sequence from an expression e to its normal form has length at most $P(\mathsf{size}(e))$.*

Proof. Let

$$M(e) := (\mathsf{weight}(e), \mathsf{lweight}(e))$$

be the lexicographic termination measure used in the proof of Theorem 5.6. Every rewrite step strictly decreases M.

Write $N := \mathsf{size}(e)$. By structural recursion on expressions, each constructor contributes only a constant amount to weight beyond the sum of the weights of its immediate subexpressions, so $\mathsf{weight}(e)$ is bounded above by a linear function of N. Likewise, $\mathsf{lweight}(e)$ counts left-nesting of compositions and is bounded by a quadratic function of N; in particular there is a constant C such that

$$\mathsf{weight}(e) \leq CN, \qquad \mathsf{lweight}(e) \leq CN^2$$

for all expressions of size N.

If a reduction sequence starts at e, then each step either decreases the first component of M or leaves the first component fixed and decreases the second. Hence the total number of steps is at most

$$\mathsf{weight}(e)\big(\mathsf{lweight}(e) + 1\big) + \mathsf{lweight}(e),$$

since for each possible value of weight there can be at most $\mathsf{lweight}(e) + 1$ decreases of the second component before the first component must drop.

Substituting the polynomial bounds for $\mathsf{weight}(e)$ and $\mathsf{lweight}(e)$ gives a polynomial bound in N. Thus there exists a polynomial P such that every rewrite sequence from e to its normal form has length at most $P(\mathsf{size}(e))$. □

In practice, innermost normalisation (reducing the deepest redexes first) often reaches normal form faster than arbitrary strategies, as it avoids duplicating work on subterms.

5.4.4 Semantic Interpretation

The decidability result has a pleasing semantic interpretation. Recall from Chapter 3 that every computational path $p : \mathsf{Path}(a, b)$ projects to an ordinary equality $\mathsf{toEq}(p) : a = b$. The rewrite system preserves this projection.

Lemma 5.37 (Rewriting Preserves Semantics). *If* $p \to q$, *then* $\mathsf{toEq}(p) = \mathsf{toEq}(q)$.

Proof. By case analysis on the rewrite rule. Each rule is an equation that holds in the groupoid interpretation, hence preserves the underlying equality. □

Corollary 5.38 (Soundness of Normal Forms). $\mathsf{toEq}(\mathsf{nf}(p)) = \mathsf{toEq}(p)$.

Proof. By transitivity of equality and Lemma 5.37 applied to each step in $p \to^* \mathsf{nf}(p)$. □

This means that normalisation is a *refinement*: it simplifies the syntactic representation of a path without changing its semantic content. Two paths are rewrite-equivalent precisely when they represent the same equality proof, up to the groupoid laws (see Chapter 6 for the groupoid interpretation of this equivalence).

5.4.5 Connection to Free Groups

The normal form characterisation (Theorem 5.32) reveals a deep connection between path algebra and free group theory. The loop fragment of our system matches the reduction theory of free groups.

Definition 5.39 (Free Group on Generators)**.** The *free group* $F(X)$ on a set X of generators consists of reduced words over the alphabet $X \cup X^{-1}$, with multiplication given by concatenation followed by cancellation of adjacent inverse pairs.

Theorem 5.40 (Loop Path Algebra as Free Group)**.** *The quotient of path expressions by rewrite equivalence, restricted to paths* $\mathsf{Path}(a, a)$ *for a fixed point* a*, is isomorphic to the free group on the atom generators.*

Proof. Let L_a denote the quotient of loop expressions at a by rewrite equivalence. By Theorems 5.29 and 5.32, each class $[p] \in L_a$ contains a unique reduced word normal form. Define

$$\overline{\phi} : L_a \to F(\mathbb{N})$$

by sending $[p]$ to the reduced word represented by that normal form, reading each atom_n as the generator n and each $\mathsf{symm}(\mathsf{atom}_n)$ as n^{-1}. This is well-defined because rewrite-equivalent expressions have the same normal form.

Conversely, if $w = g_1 \cdots g_k$ is a reduced word in the free group, define $\psi(w) \in L_a$ to be the class of the right-associated loop expression obtained by interpreting each positive letter as the corresponding atom and each negative letter as its inverse:

$$\psi(w) = \big[\mathsf{trans}(g_1, \mathsf{trans}(g_2, \ldots, \mathsf{trans}(g_{k-1}, g_k) \ldots))\big],$$

with the empty word sent to $[\mathsf{refl}]$. Because w is already reduced, this expression is already in normal form, so $\overline{\phi}(\psi(w)) = w$.

For the other composite, let $[p] \in L_a$ and let $\mathsf{nf}(p)$ be its unique reduced-word normal form. Then $p \sim_{\mathsf{rw}} \mathsf{nf}(p)$, so

$$\psi(\overline{\phi}([p])) = [\mathsf{nf}(p)] = [p].$$

Hence $\overline{\phi}$ and ψ are inverse bijections.

Finally, composition in L_a is induced by trans, while multiplication in the free group is concatenation followed by reduction. Under the bijection above, both operations are computed by concatenating the two reduced words and then reducing to the unique reduced normal form. The empty word corresponds to $[\mathsf{refl}]$, and inversion of words corresponds to symm. Therefore $\overline{\phi}$ is a group isomorphism. □

This connection explains why our 77-rule system works: it implements the standard algorithm for solving the word problem in free groups via reduction to normal form. The additional rules (9–72) extend this to handle the type-theoretic structure of computational paths.

Summary

This chapter establishes the metatheoretic foundations that make the rewrite theory usable rather than merely suggestive. Once termination and confluence are in place, path equality becomes something that can be computed and checked. More precisely:

1. **Termination** (Section 5.1): Every rewrite sequence is finite, proved via a lexicographic measure on weight and left-weight.
2. **Local Confluence** (Section 5.2): Every critical pair is joinable, verified by systematic analysis of rule overlaps.
3. **Global Confluence** (Section 5.3): By Newman's lemma, termination and local confluence imply global confluence. The Church–Rosser theorem follows: equivalence classes are characterised by shared normal forms.
4. **Decidability** (Section 5.4): Path equivalence is decidable by normalising both sides and comparing. Normal forms correspond to reduced words in a free group.

These results complete Part I of this work. The 77-rule system is not merely a theoretical curiosity; it is a certified decision procedure for path equality. In Part II, we will build on these foundations to explore the higher categorical structure that emerges when paths themselves become objects of study: the groupoid of computational paths (Chapter 6), its higher cells (Chapters 7 and 8), and ultimately the weak ω-groupoid (Chapter 9). The decidability established here also plays a central role in computing fundamental groups of combinatorial spaces in Chapter 13.

Historical Notes.
The references behind this chapter come from two converging traditions: general rewriting theory and the proof-theoretic study of equality. The study of confluence in rewriting systems began with Church and Rosser's 1936 proof that the λ-calculus is confluent [22]. Newman's 1942 lemma [116] showed that local confluence implies global confluence for terminating systems, dramatically simplifying confluence proofs. The critical pair method was introduced by

Knuth and Bendix in their 1970 paper [93], which also gave the completion algorithm for constructing confluent systems.

The connection between path algebra and free groups was observed by Hofmann and Streicher [86] in their groupoid model of type theory. The word problem for free groups was solved by Dehn [60] in 1911; our normalisation procedure is essentially Dehn's algorithm applied to path expressions.

The 77-rule system itself emerged from the work of de Queiroz and collaborators on labelled natural deduction [44, 50, 47], extended to handle the full type-theoretic structure of computational paths. The mechanised verification confirms that no rules are missing for confluence, a non-trivial result given the system's complexity.

Formalization Notes. The results of this chapter are formalised in the following files of the accompanying Lean development (`https://github.com/Arthur742Ramos/ComputationalPathsLean`):

- `Path/Rewrite/NewmanLemma.lean` — abstract diamond, local-confluence, and Newman's-lemma infrastructure
- `Path/Rewrite/DiamondProperties.lean` — diamond-style closure properties used in the confluence proofs
- `Path/Rewrite/ConfluenceProof.lean` — computational-path-specific local-confluence certificates and global confluence arguments
- `Path/Rewrite/Confluence.lean` — the consolidated confluence interface
- `Path/Rewrite/Termination.lean` — termination via the measures developed in Section 5.1

Part II

Higher Structure: The ω-Groupoid

Chapter 6

The Groupoid of Computational Paths

"A groupoid is a category in which every arrow is an isomorphism. This innocent-looking definition conceals a wealth of structure."

—Saunders Mac Lane [98]

Part I developed the theory of computational paths as a term rewriting system. We introduced paths between terms of a type (Chapter 3), organised them by means of 77 rewrite rules (Chapter 4), and proved confluence and termination for the expression layer used by the rewrite semantics (Chapter 5). The eight groupoid laws G1–G8 appeared in Proposition 3.48 as algebraic identities holding *up to rewrite equivalence*. What has not yet been brought fully into view is the structure those laws witness.

The purpose of this chapter is to do precisely that. We show that the collection of computational paths on a type A forms a *groupoid*—a category in which every morphism is invertible [126]. More precisely, the raw paths form a *weak* groupoid [59], where the category axioms hold only up to rewrite equivalence $\sim_{\mathsf{rw}}$; quotienting by $\sim_{\mathsf{rw}}$ yields a *strict* groupoid that we denote $\Pi_1(A)$ and call the *fundamental groupoid* of A. Chapter 3 already previewed the groupoid laws G1–G8 as algebraic identities among path operations (Proposition 3.48); what that chapter stated, the present chapter proves in full, placing the laws inside the categorical framework that gives them their proper meaning.

This is the opening chapter of Part II. Whereas Part I was concerned with the rewrite system as a computational device, Part II studies the *higher algebraic structure* that the rewrite system reveals. Groupoids are merely the first level: in Chapters 7–9 we will see that paths between paths (2-cells) and higher paths organise themselves into a weak ω-groupoid. The present chapter lays the categorical foundations for that development.

The chapter is organised as follows. Section 6.1 recalls the definition of a groupoid and introduces the examples we will need. Section 6.2 constructs the weak path groupoid on any type A and examines the distinction between weak and strict groupoid structure. Section 6.3 passes to the quotient to obtain the fundamental groupoid $\Pi_1(A)$ and its automorphism groups $\pi_1(A, a)$, connecting our type-theoretic development to classical algebraic topology. Section 6.4 revisits the eight groupoid laws G1–G8 from Chapter 3, now understood as the axioms of the groupoid structure, and verifies them using the rewrite system of Chapter 4.

6.1 Groupoids: Definition and Basic Properties

We begin with the abstract definition of a groupoid. Readers familiar with category theory will recognise this as a standard concept (see also Appendix B for a more detailed treatment of groupoids and their higher-dimensional analogues); those who are not should find that the definition is entirely self-contained. A groupoid is, in essence, a "group with many objects": just as a group captures the symmetries of a single object, a groupoid captures the symmetries and identifications among a collection of objects.

Definition 6.1 (Category). A *category* $\mathcal{C}$ consists of the following data:

(i) A collection $\mathrm{Ob}(\mathcal{C})$ of *objects*.

(ii) For each pair of objects a, b, a collection $\mathrm{Hom}_{\mathcal{C}}(a, b)$ of *morphisms* (or *arrows*) from a to b.

(iii) For each object a, an *identity morphism* $\mathrm{id}_a \in \mathrm{Hom}_{\mathcal{C}}(a, a)$.

(iv) For each triple of objects a, b, c, a *composition* operation

$$\circ : \mathrm{Hom}_{\mathcal{C}}(b, c) \times \mathrm{Hom}_{\mathcal{C}}(a, b) \to \mathrm{Hom}_{\mathcal{C}}(a, c)$$

writing $g \circ f$ for the composite of $f : a \to b$ and $g : b \to c$.

These data are required to satisfy:

(C1) **Associativity.** $h \circ (g \circ f) = (h \circ g) \circ f$ for all composable f, g, h.

(C2) **Left identity.** $\mathrm{id}_b \circ f = f$ for all $f : a \to b$.

(C3) **Right identity.** $f \circ \mathrm{id}_a = f$ for all $f : a \to b$.

Definition 6.2 (Groupoid)**.** A *groupoid* is a category $\mathcal{G}$ in which every morphism is invertible: for each $f \in \mathrm{Hom}_{\mathcal{G}}(a, b)$ there exists a morphism $f^{-1} \in \mathrm{Hom}_{\mathcal{G}}(b, a)$ such that

$$f^{-1} \circ f = \mathrm{id}_a, \tag{6.1}$$

$$f \circ f^{-1} = \mathrm{id}_b. \tag{6.2}$$

Remark 6.3 (Uniqueness of inverses)**.** In any groupoid the inverse is unique. If both g and g' satisfy the inverse conditions for f, then

$$g = g \circ \mathrm{id}_b = g \circ (f \circ g') = (g \circ f) \circ g' = \mathrm{id}_a \circ g' = g'.$$

We may therefore speak of *the* inverse f^{-1} without ambiguity.

6.1.1 First Examples

Before connecting groupoids to computational paths, we collect the standard examples. These will serve both as illustrations of the definition and as objects of comparison throughout this chapter.

Example 6.4 (Groups as one-object groupoids)**.** Every group $(G, \cdot, e, (-)^{-1})$ gives rise to a groupoid $\mathbf{B}G$ with a single object $*$ and

$$\mathrm{Hom}_{\mathbf{B}G}(*, *) = G.$$

Composition is the group operation, the identity morphism is the group identity e, and inverses are the group inverses. Conversely, any groupoid with exactly one object is just a group. Thus groupoids generalise groups from one object to many.

Example 6.5 (Discrete groupoid)**.** Given a set X, the *discrete groupoid* $\mathcal{D}(X)$ has $\mathrm{Ob}(\mathcal{D}(X)) = X$ and

$$\mathrm{Hom}_{\mathcal{D}(X)}(a, b) = \begin{cases} \{\mathrm{id}_a\} & \text{if } a = b, \\ \varnothing & \text{otherwise.} \end{cases}$$

The only morphisms are identities. This is the "most trivial" groupoid on a set of objects.

Example 6.6 (Codiscrete (pair) groupoid)**.** Given a set X, the *codiscrete groupoid* (also called the *pair groupoid* or *indiscrete groupoid*) $\mathcal{I}(X)$ has $\mathrm{Ob}(\mathcal{I}(X)) = X$ and exactly one morphism between any two objects: $\mathrm{Hom}_{\mathcal{I}(X)}(a, b) = \{(a, b)\}$. Composition is $(b, c) \circ (a, b) = (a, c)$ and inverses are $(a, b)^{-1} = (b, a)$. This is the "most connected" groupoid: every object is isomorphic to every other.

Example 6.7 (Equivalence relations as groupoids)**.** Given a set X with an equivalence relation $\sim$, define a groupoid with objects X and

$$\mathrm{Hom}(a,b) = \begin{cases} \{(a,b)\} & \text{if } a \sim b, \\ \varnothing & \text{otherwise.} \end{cases}$$

Reflexivity provides identities, symmetry provides inverses, and transitivity provides composition. Thus every equivalence relation determines a groupoid in which the hom-sets have at most one element. Such groupoids are sometimes called *setoids* in the type-theoretic literature.

Example 6.8 (Fundamental groupoid of a topological space)**.** Let X be a topological space. The *fundamental groupoid* $\Pi_1(X)$ has points of X as objects and homotopy classes of paths as morphisms:

$$\mathrm{Hom}_{\Pi_1(X)}(x,y) = \{\gamma : [0,1] \to X \mid \gamma(0) = x,\ \gamma(1) = y\}/\simeq$$

where $\simeq$ denotes homotopy relative to endpoints. Composition is concatenation of paths, the identity at x is the constant path, and inverses are given by reversing the parametrisation. This is the classical construction of algebraic topology [79, 114]. The fundamental group $\pi_1(X, x_0) = \mathrm{Hom}_{\Pi_1(X)}(x_0, x_0)$ is recovered as the automorphism group at a basepoint.

This example is the direct inspiration for our construction: we will build an analogous groupoid from computational paths rather than continuous paths.

6.1.2 Morphisms of Groupoids

To compare groupoids we need the appropriate notion of structure-preserving map.

Definition 6.9 (Functor between groupoids)**.** Let $\mathcal{G}$ and $\mathcal{H}$ be groupoids. A *functor* $F : \mathcal{G} \to \mathcal{H}$ consists of:

(i) A function $F : \mathrm{Ob}(\mathcal{G}) \to \mathrm{Ob}(\mathcal{H})$ on objects.

(ii) For each $a, b \in \mathrm{Ob}(\mathcal{G})$, a function $F_{a,b} : \mathrm{Hom}_{\mathcal{G}}(a,b) \to \mathrm{Hom}_{\mathcal{H}}(F(a), F(b))$.

These must satisfy:

$$F(\mathrm{id}_a) = \mathrm{id}_{F(a)}, \qquad \text{(preserves identity)}$$
$$F(g \circ f) = F(g) \circ F(f). \qquad \text{(preserves composition)}$$

Remark 6.10 (Functors preserve inverses)**.** Any functor between groupoids automatically preserves inverses:

$$F(f^{-1}) = F(f)^{-1}.$$

To see this, observe that $F(f) \circ F(f^{-1}) = F(f \circ f^{-1}) = F(\mathrm{id}_b) = \mathrm{id}_{F(b)}$, and similarly for the other composite. By uniqueness of inverses (Remark 6.3), we conclude $F(f^{-1}) = F(f)^{-1}$.

This observation is significant. The congruence operation $\mathsf{congrArg}(f, -)$ of Chapter 3 is precisely a functor between path groupoids (Proposition 3.50), so its preservation of inverses (law F3) follows from the general theory rather than from an extra axiom.

6.1.3 Weak versus Strict Groupoids

The standard definition of groupoid (Definition 6.2) requires the axioms to hold as *equalities*. In many settings, however, the axioms hold only up to a specified equivalence relation.

Why should we care about this distinction? Consider the computational paths of Chapter 3. We would like to say that $\mathsf{trans}(\mathsf{refl}_a, p)$ "is" p, but in fact these are syntactically different path expressions. What is true is that they are *rewrite-equivalent*: the rewrite system transforms one into the other. The groupoid axioms hold not as syntactic identities but as equivalences witnessed by rewrite sequences. This is the prototypical example of a *weak* groupoid: a structure where the laws are satisfied "up to" a specified notion of equivalence rather than on the nose.

Definition 6.11 (Weak groupoid)**.** A *weak groupoid* consists of the data of Definition 6.2—objects, morphisms, identity, composition, and inverses—but where the axioms (C1)–(C3), (6.1), and (6.2) hold only up to a specified equivalence relation $\sim$ on morphisms. That is:

$$h \circ (g \circ f) \sim (h \circ g) \circ f, \qquad \text{(weak C1)}$$
$$\mathrm{id}_b \circ f \sim f, \qquad \text{(weak C2)}$$
$$f \circ \mathrm{id}_a \sim f, \qquad \text{(weak C3)}$$
$$f^{-1} \circ f \sim \mathrm{id}_a, \qquad \text{(weak inv. left)}$$
$$f \circ f^{-1} \sim \mathrm{id}_b. \qquad \text{(weak inv. right)}$$

Moreover, we require that $\sim$ be a *congruence*: if $f \sim f'$ and $g \sim g'$ (with appropriate types), then $g \circ f \sim g' \circ f'$ and $f^{-1} \sim f'^{-1}$.

Remark 6.12 (From weak to strict by quotient)**.** Given a weak groupoid $(\mathcal{G}, \sim)$, one can always form a strict groupoid by passing to the quotient: define $\overline{\mathrm{Hom}}(a,b) = \mathrm{Hom}(a,b)/{\sim}$, and let composition and inverses act on equivalence classes. The congruence condition ensures this is well-defined. This passage from weak to strict is exactly the construction we will employ for computational paths.

The distinction between weak and strict is fundamental. In homotopy theory, the fundamental groupoid is strict because one works with homotopy *classes* of paths. In intensional type theory, identity proofs are first-class objects, and UIP is generally a non-trivial principle. In the formalisation accompanying this book, a path is explicitly represented as a structure carrying both a proof-irrelevant identity witness and a computational trace recording the rewrite steps. The framework therefore provides both options: the weak groupoid of raw paths (with trace data) and the strict groupoid obtained by quotienting.

6.2 The Path Groupoid

We now construct the central object of this chapter: the groupoid whose objects are terms of a type and whose morphisms are computational paths. The data have already been assembled in Chapter 3; our task here is to verify that they satisfy the groupoid axioms.

The construction is both natural and instructive. In topology, the fundamental groupoid of a space is built from continuous paths modulo homotopy. Here we build the *path groupoid* from computational paths modulo rewrite equivalence. The parallel is exact: reflexivity plays the role of the constant path, transitivity plays the role of path concatenation, and symmetry plays the role of path reversal. The groupoid axioms—associativity, unit laws, and inverse laws—are witnessed by the rewrite rules R1–R8 of Chapter 4.

6.2.1 The Data

Fix a type A. We define a structure $\mathcal{P}(A)$ as follows.

Definition 6.13 (Path groupoid data)**.** The *path groupoid data* on a type A consists of:

Groupoid concept	**Path concept**	**Notation**
Objects	Terms $a : A$	$\mathrm{Ob}(\mathcal{P}(A)) = A$
Morphisms $a \to b$	Paths $p : \mathsf{Path}(a,b)$	$\mathrm{Hom}(a,b) = \mathsf{Path}(a,b)$
Identity at a	Reflexivity refl_a	$\mathrm{id}_a = \mathsf{refl}_a$
Composition	Transitivity $\mathsf{trans}(p,q)$	$q \circ p = \mathsf{trans}(p,q)$
Inverse	Symmetry $\mathsf{symm}(p)$	$p^{-1} = \mathsf{symm}(p)$

Remark 6.14 (Composition order). We write $q \circ p$ for the categorical composite and $\mathsf{trans}(p, q)$ for the path-theoretic composite. These are the same operation, but with a reversal in the order of arguments: $q \circ p = \mathsf{trans}(p, q)$. The path convention lists arguments in the order of traversal (p first, then q); the categorical convention lists them in the order of application (q applied after p). We will freely use both notations, choosing whichever is more natural in context. When discussing the groupoid axioms, we prefer the categorical notation $q \circ p$; when computing with the rewrite system, we prefer $\mathsf{trans}(p, q)$.

6.2.2 The Weak Path Groupoid

Theorem 6.15 (Paths form a weak groupoid). *For any type A, the data $\mathcal{P}(A)$ of Definition 6.13 form a weak groupoid with respect to rewrite equivalence $\sim_{\mathsf{rw}}$.*

Proof. We must verify the five axioms of Definition 6.11. Each is one of the groupoid laws G1–G8 established in Proposition 3.48. We translate from the categorical notation $q \circ p = \mathsf{trans}(p, q)$ to the path notation used in Chapter 3.

Associativity (weak C1). For composable paths $p : \mathsf{Path}(a, b)$, $q : \mathsf{Path}(b, c)$, $r : \mathsf{Path}(c, d)$, we must show $r \circ (q \circ p) \sim_{\mathsf{rw}} (r \circ q) \circ p$, i.e.,

$$\mathsf{trans}(p, \mathsf{trans}(q, r)) \sim_{\mathsf{rw}} \mathsf{trans}(\mathsf{trans}(p, q), r).$$

This is exactly law G5, established in Proposition 3.48.

Left identity (weak C2). For $p : \mathsf{Path}(a, b)$, we must show $\mathrm{id}_b \circ p \sim_{\mathsf{rw}} p$, i.e., $\mathsf{trans}(p, \mathsf{refl}_b) \sim_{\mathsf{rw}} p$. This is law G2.

Right identity (weak C3). For $p : \mathsf{Path}(a, b)$, we must show $p \circ \mathrm{id}_a \sim_{\mathsf{rw}} p$, i.e., $\mathsf{trans}(\mathsf{refl}_a, p) \sim_{\mathsf{rw}} p$. This is law G1.

Left inverse. For $p : \mathsf{Path}(a, b)$, we must show $p^{-1} \circ p \sim_{\mathsf{rw}} \mathrm{id}_a$, i.e., $\mathsf{trans}(p, \mathsf{symm}(p)) \sim_{\mathsf{rw}} \mathsf{refl}_a$. This is law G3.

Right inverse. For $p : \mathsf{Path}(a, b)$, we must show $p \circ p^{-1} \sim_{\mathsf{rw}} \mathrm{id}_b$, i.e., $\mathsf{trans}(\mathsf{symm}(p), p) \sim_{\mathsf{rw}} \mathsf{refl}_b$. This is law G4.

Congruence. It remains to verify that $\sim_{\mathsf{rw}}$ is a congruence for composition and inversion. Suppose $p \sim_{\mathsf{rw}} p'$ and $q \sim_{\mathsf{rw}} q'$ with $p, p' : \mathsf{Path}(a, b)$ and $q, q' : \mathsf{Path}(b, c)$. Then $\mathsf{trans}(p, q) \sim_{\mathsf{rw}} \mathsf{trans}(p', q')$ because the rewrite system respects the structure of trans (see the congruence closure properties established in Chapter 5). Similarly, $\mathsf{symm}(p) \sim_{\mathsf{rw}} \mathsf{symm}(p')$. □

Remark 6.16 (Non-strictness of the path groupoid). The path groupoid $\mathcal{P}(A)$ is genuinely *not* strict in general. For instance, the identity law $\mathsf{trans}(\mathsf{refl}_a, p) \sim_{\mathsf{rw}} p$ holds as a rewrite equivalence, not as a definitional equality, when p is a non-trivial path. The witness to this equivalence is itself a 2-path (a path between paths), and it is precisely these 2-paths that will give rise to the higher groupoid structure in Chapter 7.

This is in sharp contrast with the topological fundamental groupoid, where one works with homotopy *classes* and the axioms hold strictly. The additional structure present in the weak path groupoid—the witnesses to the groupoid laws—is where the "higher" information resides.

Example 6.17 (Composition and inversion in the path groupoid)**.** Let A be a type with terms $a, b, c : A$ and suppose we have elementary paths $s : \mathsf{Path}(a, b)$ and $t : \mathsf{Path}(b, c)$. In the path groupoid $\mathcal{P}(A)$:

- **Composition:** The composite $t \circ s$ in the groupoid is $\mathsf{trans}(s, t) : \mathsf{Path}(a, c)$. This is the path that first follows s from a to b, then follows t from b to c.
- **Inversion:** The inverse s^{-1} in the groupoid is $\mathsf{symm}(s) : \mathsf{Path}(b, a)$. This "reverses" the path, going from b back to a.
- **Round trip:** The composite $s^{-1} \circ s = \mathsf{trans}(s, \mathsf{symm}(s))$ goes from a to b and back. By law G3, this is equivalent to the identity: $\mathsf{trans}(s, \mathsf{symm}(s)) \sim_{\mathsf{rw}} \mathsf{refl}_a$. This equivalence is not a definitional equality—$\mathsf{trans}(s, \mathsf{symm}(s))$ and refl_a are syntactically different—but the rewrite system reduces the former to the latter via rule R5.
- **Detour elimination:** $\mathsf{trans}(\mathsf{trans}(s, \mathsf{symm}(s)), t) \sim_{\mathsf{rw}} t$ by G3 followed by G1. Thus any "round trip" inserted into a path can be eliminated.

6.2.3 Additional Laws

Beyond the five axioms required for a groupoid, the laws G6–G8 provide additional structure that constrains the behaviour of inverses.

Proposition 6.18 (Inverse laws in the path groupoid)**.** *In the weak path groupoid $\mathcal{P}(A)$:*

(a) ***Involution:*** $(p^{-1})^{-1} \sim_{\mathsf{rw}} p$ *for all* $p : \mathsf{Path}(a, b)$.

(b) ***Contravariance:*** $(q \circ p)^{-1} \sim_{\mathsf{rw}} p^{-1} \circ q^{-1}$ *for all composable* p, q.

(c) ***Identity inverse:*** $(\mathrm{id}_a)^{-1} \sim_{\mathsf{rw}} \mathrm{id}_a$ *for all* $a : A$.

Proof. In path notation:

(a) $\mathsf{symm}(\mathsf{symm}(p)) \sim_{\mathsf{rw}} p$ is law G6.

(b) $\mathsf{symm}(\mathsf{trans}(p, q)) \sim_{\mathsf{rw}} \mathsf{trans}(\mathsf{symm}(q), \mathsf{symm}(p))$ is law G7.

(c) $\mathsf{symm}(\mathsf{refl}_a) \sim_{\mathsf{rw}} \mathsf{refl}_a$ is law G8.

□

Remark 6.19 (Derivability of inverse laws). Laws (a) and (b) are, in principle, derivable from the groupoid axioms in any groupoid. For the involution law, note that p^{-1} has inverse p (since $p \circ p^{-1} \sim_{\mathsf{rw}} \mathrm{id}$ and $p^{-1} \circ p \sim_{\mathsf{rw}} \mathrm{id}$), and by uniqueness of inverses (Remark 6.3), $(p^{-1})^{-1} = p$. Similarly, $p^{-1} \circ q^{-1}$ serves as an inverse for $q \circ p$ because:

$$\begin{aligned}(q \circ p) \circ (p^{-1} \circ q^{-1}) &\sim_{\mathsf{rw}} q \circ (p \circ p^{-1}) \circ q^{-1} \\ &\sim_{\mathsf{rw}} q \circ \mathrm{id} \circ q^{-1} \\ &\sim_{\mathsf{rw}} q \circ q^{-1} \\ &\sim_{\mathsf{rw}} \mathrm{id},\end{aligned}$$

and analogously for the other composite. The fact that G6 and G7 are independently derivable from G1–G5 is a consistency check: the rewrite system includes no redundant equalities.

However, in a *weak* groupoid the argument is more subtle because the uniqueness of inverses itself holds only up to $\sim_{\mathsf{rw}}$. The explicit rewrite rules for G6 and G7 provide direct witnesses, bypassing the need for the higher-level uniqueness argument.

6.2.4 Functoriality as a Groupoid Functor

Having established the groupoid structure of paths on a single type, we now ask: how does this structure behave under functions between types? If $f : A \to B$ is a function and $p : \mathsf{Path}_A(a, a')$ is a path in A, then $\mathsf{congrArg}(f, p) : \mathsf{Path}_B(f(a), f(a'))$ is a path in B. The question is whether this mapping respects the groupoid operations—and the answer, as we shall see, is that it constitutes a *functor* between path groupoids.

Theorem 6.20 ($\mathsf{congrArg}$ as a functor). *For any function $f : A \to B$, the operation $\mathsf{congrArg}(f, -)$ defines a functor $\mathcal{P}(A) \to \mathcal{P}(B)$ between path groupoids, sending each object a to $f(a)$ and each morphism $p : \mathsf{Path}(a, b)$ to $\mathsf{congrArg}(f, p) : \mathsf{Path}(f(a), f(b))$. The functor axioms (preservation of identity and composition) are exactly the laws F1 and F2 of Proposition 3.50.*

Proof. Law F1 gives $\mathsf{congrArg}(f, \mathsf{refl}_a) \sim_{\mathsf{rw}} \mathsf{refl}_{f(a)}$, which is preservation of identity. Law F2 gives

$$\mathsf{congrArg}(f, \mathsf{trans}(p, q)) \sim_{\mathsf{rw}} \mathsf{trans}(\mathsf{congrArg}(f, p), \mathsf{congrArg}(f, q)),$$

which is preservation of composition. By Remark 6.10, preservation of inverses (law F3) follows automatically. □

Corollary 6.21. *For composable functions $f : A \to B$ and $g : B \to C$:*

(i) $\mathsf{congrArg}(g, -) \circ \mathsf{congrArg}(f, -) \sim_{\mathsf{rw}} \mathsf{congrArg}(g \circ f, -)$ *(law F4).*

(ii) $\mathsf{congrArg}(\mathrm{id}_A, -) \sim_{\mathsf{rw}} \mathrm{id}_{\mathcal{P}(A)}$ *(law F5).*

Thus the assignment $A \mapsto \mathcal{P}(A)$, $f \mapsto \mathsf{congrArg}(f, -)$ is itself a (weak) functor from the category of types and functions to the category of groupoids and functors.

This perspective reveals a deep principle: *every function between types induces a functor between their path groupoids.* In the language of homotopy type theory, this says that every function is "continuous" with respect to the path structure—it maps paths to paths, preserving all the algebraic structure.

6.3 The Fundamental Groupoid

The weak path groupoid $\mathcal{P}(A)$ retains all the intensional information of computational paths: two paths that are rewrite-equivalent are still distinct as elements of $\mathsf{Path}(a, b)$. To obtain a strict groupoid—one in which the axioms hold as genuine equalities rather than merely up to equivalence—we quotient by the rewrite equivalence. This is the algebraic analogue of the passage from paths to homotopy classes in topology: just as two homotopic paths in a topological space become a single morphism in the fundamental groupoid $\Pi_1(X)$, two rewrite-equivalent computational paths become a single morphism in the fundamental groupoid $\Pi_1(A)$ of a type.

6.3.1 The Quotient Construction

The quotient is made particularly tractable by the confluence theorem for groupoid expressions (Chapter 5): two rewrite-equivalent paths have presentations sharing the same reduced normal form under the free-group interpretation. This gives a concrete algorithmic criterion for deciding when two equivalence classes coincide.

Definition 6.22 (Rewrite equivalence class)**.** For terms $a, b : A$, the *rewrite equivalence class* of a path $p : \mathsf{Path}(a, b)$ is

$$[p] = \{q : \mathsf{Path}(a, b) \mid p \sim_{\mathsf{rw}} q\}.$$

We write $\mathsf{Path}(a, b)/{\sim_{\mathsf{rw}}}$ for the set of all equivalence classes.

The results of Chapter 5 give this quotient a particularly concrete description.

Proposition 6.23 (Normal form representatives). *Choose groupoid-expression presentations e_p, e_q of p, q : $\mathsf{Path}(a, b)$. By the Church–Rosser theorem for expressions (Chapter 5),*

$$[p] = [q] \quad \Longleftrightarrow \quad \mathrm{nf}_{\mathrm{Expr}}(e_p) = \mathrm{nf}_{\mathrm{Expr}}(e_q).$$

Hence each equivalence class is determined by a unique reduced expression normal form (equivalently, a reduced free-group word), even though raw path values may retain different step traces.

The formal confluence theorem operates at the level of free-group expressions via the canonical interpretation map, not directly on raw path values. Confluence on the algebraic expression system lifts to typed paths through the results of Chapter 5.

Definition 6.24 (Fundamental groupoid)**.** The *fundamental groupoid* of a type A, denoted $\Pi_1(A)$, is the groupoid with:

Component	**Definition**
Objects	Terms $a : A$
Morphisms $a \to b$	$\Pi_1(A)(a, b) = \mathsf{Path}(a, b)/\sim_{\mathsf{rw}}$
Identity at a	$[\mathsf{refl}_a]$
Composition	$[q] \circ [p] = [\mathsf{trans}(p, q)]$
Inverse	$[p]^{-1} = [\mathsf{symm}(p)]$

Theorem 6.25 ($\Pi_1(A)$ is a strict groupoid)**.** *The fundamental groupoid $\Pi_1(A)$ is well-defined and is a strict groupoid: all groupoid axioms hold as equalities of equivalence classes.*

Proof. We verify well-definedness and the axioms.

Well-definedness of composition. Suppose $[p] = [p']$ and $[q] = [q']$, i.e., $p \sim_{\mathsf{rw}} p'$ and $q \sim_{\mathsf{rw}} q'$. We must show $[\mathsf{trans}(p, q)] = [\mathsf{trans}(p', q')]$, i.e., $\mathsf{trans}(p, q) \sim_{\mathsf{rw}} \mathsf{trans}(p', q')$. This follows from the congruence property of $\sim_{\mathsf{rw}}$ (Chapter 5): the rewrite equivalence is compatible with all path constructors.

Well-definedness of inverse. If $p \sim_{\mathsf{rw}} p'$, then $\mathsf{symm}(p) \sim_{\mathsf{rw}} \mathsf{symm}(p')$ by congruence of $\sim_{\mathsf{rw}}$.

Axioms. Each axiom of Definition 6.2 is the equivalence-class version of a groupoid law from Proposition 3.48:

$$[\mathsf{trans}(\mathsf{refl}_a, p)] = [p] \qquad \text{(G1)},$$

$$[\mathsf{trans}(p, \mathsf{refl}_b)] = [p] \qquad \text{(G2)},$$

$$[\mathsf{trans}(p, \mathsf{symm}(p))] = [\mathsf{refl}_a] \qquad \text{(G3)},$$

$$[\mathsf{trans}(\mathsf{symm}(p), p)] = [\mathsf{refl}_b] \qquad \text{(G4)},$$

$$[\mathsf{trans}(\mathsf{trans}(p, q), r)] = [\mathsf{trans}(p, \mathsf{trans}(q, r))] \qquad \text{(G5)}.$$

In each case, the rewrite equivalence between the two sides implies equality of their equivalence classes. Since these are now genuine equalities (not merely equivalences), $\Pi_1(A)$ is a strict groupoid. □

Remark 6.26 (The quotient functor). The quotient map $[-] : \mathcal{P}(A) \to \Pi_1(A)$ that sends each path to its equivalence class is a functor. It is the identity on objects, and on morphisms it sends p to $[p]$. Preservation of identity and composition hold by definition. This functor is *full* (since every equivalence class contains a representative path) and reflects the essential phenomenon: the passage from $\mathcal{P}(A)$ to $\Pi_1(A)$ forgets exactly the intensional information encoded in the rewrite witnesses.

Example 6.27 (The quotient in action). To see how the fundamental groupoid collapses rewrite-equivalent paths, consider two paths from a to b in a type A:

$$p = \mathsf{trans}(\mathsf{refl}_a, s), \qquad q = s,$$

where $s : \mathsf{Path}(a, b)$ is some fixed path. As raw path expressions, p and q are syntactically different. In the weak path groupoid $\mathcal{P}(A)$, they are distinct morphisms from a to b. However, by the left unit law (G1), we have $p \sim_{\mathsf{rw}} q$, so in the fundamental groupoid $\Pi_1(A)$ they represent the *same* morphism: $[p] = [q]$. The rewrite witness—the single application of rule R3—is "forgotten" by the quotient.

More dramatically, the paths

$$p_1 = \mathsf{trans}(s, \mathsf{trans}(\mathsf{symm}(s), s)), \qquad p_2 = s$$

are also identified: the first performs a "detour" (going via s, back via $\mathsf{symm}(s)$, then forward via s again), while the second goes directly. By the cancellation law (G3 and G1), $p_1 \sim_{\mathsf{rw}} p_2$, so $[p_1] = [p_2]$ in $\Pi_1(A)$. The passage to the quotient eliminates all such redundant detours, leaving only the essential connectivity information.

6.3.2 The Fundamental Group

By restricting to a single object, we recover the classical notion of fundamental group.

Definition 6.28 (Fundamental group). For a type A and a term $a : A$, the *fundamental group* of A at a is

$$\pi_1(A, a) = \Pi_1(A)(a, a) = \mathsf{Path}(a, a)/{\sim_{\mathsf{rw}}},$$

the automorphism group of a in the fundamental groupoid. This is indeed a group: it is the endomorphism monoid at a, and every endomorphism is invertible since $\Pi_1(A)$ is a groupoid.

Remark 6.29 (Structure of the fundamental group)**.** The group structure on $\pi_1(A, a)$ is:

- **Elements:** equivalence classes $[\alpha]$ where $\alpha : \mathsf{Path}(a, a)$—that is, loops based at a, modulo rewrite equivalence.
- **Identity:** $[\mathsf{refl}_a]$.
- **Multiplication:** $[\beta] \cdot [\alpha] = [\mathsf{trans}(\alpha, \beta)]$.
- **Inverse:** $[\alpha]^{-1} = [\mathsf{symm}(\alpha)]$.

By Theorem 6.25, these operations satisfy the group axioms strictly.

Example 6.30 (The fundamental group of $\mathbb{N}$)**.** Consider the type $\mathbb{N}$ of natural numbers. As discussed in Chapter 3, any two paths $p, q : \mathsf{Path}(n, m)$ between the same natural numbers are rewrite-equivalent (since $\mathbb{N}$ satisfies UIP). In particular, every loop $\alpha : \mathsf{Path}(n, n)$ satisfies $\alpha \sim_{\mathsf{rw}} \mathsf{refl}_n$. Therefore $\pi_1(\mathbb{N}, n) = \{[\mathsf{refl}_n]\}$ is the trivial group for every n.

The fundamental groupoid $\Pi_1(\mathbb{N})$ is a discrete groupoid: the only morphism from n to m is the unique equivalence class of paths, if one exists (which requires $n = m$).

6.3.3 Basepoint Change

In classical algebraic topology, the fundamental groups at different basepoints of a path-connected space are isomorphic, with the isomorphism given by conjugation along a path. The same holds for computational paths.

Theorem 6.31 (Basepoint change)**.** *Let* $p : \mathsf{Path}(a, b)$*. The map*

$$\varphi_p : \pi_1(A, a) \to \pi_1(A, b),$$
$$[\alpha] \mapsto [p] \circ [\alpha] \circ [p]^{-1} = [\mathsf{trans}(\mathsf{symm}(p), \mathsf{trans}(\alpha, p))]$$

is a group isomorphism with inverse $\varphi_{\mathsf{symm}(p)}$*.*

Proof. We verify that φ_p is well-defined, multiplicative, and has a two-sided inverse. The displayed formula is written in categorical notation because it literally describes the loop "go from b back to a along p^{-1}, traverse α, and return along p". In the multiplicativity calculation, however, we use the group law on $\pi_1(A, a)$:

$$[\beta] \cdot [\alpha] = [\mathsf{trans}(\alpha, \beta)].$$

Well-definedness. If $\alpha : \mathsf{Path}(a, a)$, then $\mathsf{trans}(\mathsf{symm}(p), \mathsf{trans}(\alpha, p)) : \mathsf{Path}(b, b)$, so the result is indeed a loop at b. If $\alpha \sim_{\mathsf{rw}} \alpha'$, then

$\mathsf{trans}(\mathsf{symm}(p), \mathsf{trans}(\alpha, p)) \sim_{\mathsf{rw}} \mathsf{trans}(\mathsf{symm}(p), \mathsf{trans}(\alpha', p))$ by congruence of $\sim_{\mathsf{rw}}$.

Homomorphism.

$$\begin{aligned}
&\varphi_p([\beta] \cdot [\alpha]) \\
&\quad = [\mathsf{trans}(\mathsf{symm}(p), \mathsf{trans}(\mathsf{trans}(\alpha, \beta), p))] \\
&\quad \sim_{\mathsf{rw}} [\mathsf{trans}(\mathsf{symm}(p), \mathsf{trans}(\alpha, \mathsf{trans}(\beta, p)))] && \text{(G5)} \\
&\quad \sim_{\mathsf{rw}} [\mathsf{trans}(\mathsf{symm}(p), \mathsf{trans}(\alpha, \mathsf{trans}(\mathsf{trans}(p, \mathsf{symm}(p)), \mathsf{trans}(\beta, p))))] && \text{(G3)} \\
&\quad \sim_{\mathsf{rw}} [\mathsf{trans}(\mathsf{trans}(\mathsf{symm}(p), \mathsf{trans}(\alpha, p)), \mathsf{trans}(\mathsf{symm}(p), \mathsf{trans}(\beta, p)))] && \text{(G5)} \\
&\quad = \varphi_p([\beta]) \cdot \varphi_p([\alpha]).
\end{aligned}$$

Inverse. For $\gamma : \mathsf{Path}(b, b)$, we have

$$\varphi_{\mathsf{symm}(p)}([\gamma]) = [\mathsf{trans}(p, \mathsf{trans}(\gamma, \mathsf{symm}(p)))].$$

Then for $\alpha : \mathsf{Path}(a, a)$:

$$\begin{aligned}
&\varphi_{\mathsf{symm}(p)}(\varphi_p([\alpha])) \\
&\quad = [\mathsf{trans}(p, \mathsf{trans}(\mathsf{trans}(\mathsf{symm}(p), \mathsf{trans}(\alpha, p)), \mathsf{symm}(p)))] \\
&\quad \sim_{\mathsf{rw}} [\mathsf{trans}(\mathsf{trans}(p, \mathsf{symm}(p)), \mathsf{trans}(\mathsf{trans}(\alpha, p), \mathsf{symm}(p)))] && \text{(G5)} \\
&\quad \sim_{\mathsf{rw}} [\mathsf{trans}(\mathsf{refl}_a, \mathsf{trans}(\alpha, \mathsf{trans}(p, \mathsf{symm}(p))))] && \text{(G3, G5)} \\
&\quad \sim_{\mathsf{rw}} [\mathsf{trans}(\mathsf{refl}_a, \mathsf{trans}(\alpha, \mathsf{refl}_a))] && \text{(G3)} \\
&\quad \sim_{\mathsf{rw}} [\alpha] && \text{(G1, G2).}
\end{aligned}$$

The computation $\varphi_p(\varphi_{\mathsf{symm}(p)}([\gamma])) = [\gamma]$ is analogous, using G4 in place of G3. Therefore φ_p is an isomorphism. □

Corollary 6.32. *If there exists a path* $p : \mathsf{Path}(a, b)$*, then* $\pi_1(A, a) \cong \pi_1(A, b)$*. In particular, if every pair of terms in* A *is connected by a path, then the fundamental group is independent of basepoint (up to isomorphism).*

6.3.4 Comparison with the Topological Fundamental Groupoid

The construction of $\Pi_1(A)$ for a type A parallels the classical construction of the fundamental groupoid of a topological space [79, 114]. The analogy runs deep:

Topology	Computational Paths	Notation
Topological space X	Type A	
Points $x \in X$	Terms $a : A$	
Continuous paths $\gamma : [0,1] \to X$	Paths $p : \mathsf{Path}(a,b)$	
Homotopy of paths	Rewrite equivalence $\sim_{\mathsf{rw}}$	
$\Pi_1(X)$	$\Pi_1(A)$	
$\pi_1(X, x_0)$	$\pi_1(A, a)$	
Concatenation of paths	$\mathsf{trans}(p,q)$	
Constant path at x	refl_a	
Reverse path	$\mathsf{symm}(p)$	

The key difference is the nature of the equivalence relation. In topology, homotopy is defined topologically (via continuous deformations); in our setting, rewrite equivalence is defined syntactically (via the 77-rule system). Yet the resulting algebraic structure—a strict groupoid—is the same.

This correspondence is not merely an analogy. Hofmann and Streicher [86] showed that the identity types of Martin-Löf type theory can be modelled by the fundamental groupoid of a space, and our computational paths provide a concrete, computable realisation of this model. The connection is made precise in Chapter 14, where we relate the computational paths framework to homotopy type theory.

Remark 6.33 (Higher homotopy groups)**.** In topology, when one moves beyond the fundamental groupoid to higher homotopy groups $\pi_n(X, x_0)$, one considers homotopy classes of maps from the n-sphere to X. In our setting, the analogous construction uses *higher paths*: paths between paths yield 2-cells (Chapter 7), paths between 2-cells yield 3-cells (Chapter 8), and so on. The resulting structure is a weak ω-groupoid (Chapter 9), not merely a groupoid. Chapters 7–9 develop this higher structure systematically.

6.4 Groupoid Laws Verified

We now return to the eight groupoid laws G1–G8 from Proposition 3.48 and give their complete verification using the rewrite system of Chapter 4. While we proved these laws in Chapter 3 at the level of path expressions, we can now understand them as the axioms of a groupoid and verify them via the certified rewrite system.

6.4.1 The Eight Laws as Groupoid Axioms

The following table summarises the correspondence between the groupoid laws, the category-theoretic axioms, and the rewrite rules that witness them.

Law	Path identity	Axiom	Rule
G1	$\mathsf{trans}(\mathsf{refl}_a, p) \sim_{\mathsf{rw}} p$	Left unit	R3
G2	$\mathsf{trans}(p, \mathsf{refl}_b) \sim_{\mathsf{rw}} p$	Right unit	R4
G3	$\mathsf{trans}(p, \mathsf{symm}(p)) \sim_{\mathsf{rw}} \mathsf{refl}_a$	Right inv.	R5
G4	$\mathsf{trans}(\mathsf{symm}(p), p) \sim_{\mathsf{rw}} \mathsf{refl}_b$	Left inv.	R6
G5	$\mathsf{trans}(\mathsf{trans}(p, q), r) \sim_{\mathsf{rw}} \mathsf{trans}(p, \mathsf{trans}(q, r))$	Assoc.	R8
G6	$\mathsf{symm}(\mathsf{symm}(p)) \sim_{\mathsf{rw}} p$	Involution	R2
G7	$\mathsf{symm}(\mathsf{trans}(p, q)) \sim_{\mathsf{rw}} \mathsf{trans}(\mathsf{symm}(q), \mathsf{symm}(p))$	Contrav.	R7
G8	$\mathsf{symm}(\mathsf{refl}_a) \sim_{\mathsf{rw}} \mathsf{refl}_a$	Id. inv.	R1

In the right-hand column, "Rx" refers to the x-th rule of the 77-rule system from Chapter 4. The first eight rules of the system are precisely the groupoid rules: they implement the groupoid axioms as directed rewrite steps.

Remark 6.34 (Direction of rewriting)**.** The rewrite rules are *directed*: each law Gk is implemented as a rule that rewrites the left-hand side to the right-hand side. For example, R3 rewrites $\mathsf{trans}(\mathsf{refl}_a, p) \to p$. The reverse direction is not a rewrite rule; rather, the equivalence $p \sim_{\mathsf{rw}} \mathsf{trans}(\mathsf{refl}_a, p)$ holds because $\sim_{\mathsf{rw}}$ is the reflexive, symmetric, transitive closure of the rewrite relation.

This asymmetry is deliberate: the left-hand side of each rule is "more complex" than the right-hand side (as measured by the weight function of Chapter 5), ensuring termination. The groupoid axioms as equalities are recovered by the equivalence relation $\sim_{\mathsf{rw}}$; the rewrite rules provide a *canonical direction* for simplification.

6.4.2 Detailed Verifications

The verifications below are deliberately concrete: each groupoid law is checked by exhibiting the rewrite rule(s) that witness it. The reader who has absorbed the rewrite system of Chapter 4 will find them unsurprising, but their value lies in making explicit the bridge between the algebraic axioms of groupoid theory and the combinatorial machinery of term rewriting. Skimming the first two or three cases and then moving to Section 6.4.3 is a reasonable strategy on a first reading.

Proposition 6.35 (Verification of G1: Left identity)**.** *For any path $p : \mathsf{Path}(a, b)$, the rewrite system reduces $\mathsf{trans}(\mathsf{refl}_a, p) \to p$ in one step via rule R3.*

Proof. Rule R3 applies directly: the pattern $\mathsf{trans}(\mathsf{refl}_a, p)$ matches the left-hand side of R3 with p as the free variable. The rewrite produces p. No further steps are needed. □

Proposition 6.36 (Verification of G2: Right identity)**.** *For any path p : $\mathsf{Path}(a, b)$, the rewrite system reduces $\mathsf{trans}(p, \mathsf{refl}_b) \to p$ in one step via rule R4.*

Proof. Rule R4 applies directly: the pattern $\mathsf{trans}(p, \mathsf{refl}_b)$ matches the left-hand side of R4 with p as the free variable. □

Proposition 6.37 (Verification of G3: Right inverse). *For any path* $p : \mathsf{Path}(a, b)$*, the expression* $\mathsf{trans}(p, \mathsf{symm}(p))$ *reduces to* refl_a *under the rewrite system.*

Proof. Rule R5 applies directly: the pattern $\mathsf{trans}(p, \mathsf{symm}(p))$ matches the left-hand side of R5. The rewrite produces refl_a. □

Proposition 6.38 (Verification of G4: Left inverse). *For any path* $p : \mathsf{Path}(a, b)$*, the expression* $\mathsf{trans}(\mathsf{symm}(p), p)$ *reduces to* refl_b *under the rewrite system.*

Proof. Rule R6 applies directly: $\mathsf{trans}(\mathsf{symm}(p), p) \to \mathsf{refl}_b$. □

Proposition 6.39 (Verification of G5: Associativity). *For composable paths* p, q, r*,*

$$\mathsf{trans}(\mathsf{trans}(p, q), r) \sim_{\mathsf{rw}} \mathsf{trans}(p, \mathsf{trans}(q, r))$$

by rule R8.

Proof. Rule R8 applies directly to the outermost trans: the first argument $\mathsf{trans}(p, q)$ matches the pattern, and the rewrite reassociates to $\mathsf{trans}(p, \mathsf{trans}(q, r))$. This is a single-step reduction. □

Proposition 6.40 (Verification of G6: Involution). *For any path* $p : \mathsf{Path}(a, b)$*, the expression* $\mathsf{symm}(\mathsf{symm}(p))$ *reduces to p via rule R2.*

Proof. Rule R2 applies directly: $\mathsf{symm}(\mathsf{symm}(p)) \to p$. □

Proposition 6.41 (Verification of G7: Contravariance). *For composable paths* $p : \mathsf{Path}(a, b)$ *and* $q : \mathsf{Path}(b, c)$*, the expression* $\mathsf{symm}(\mathsf{trans}(p, q))$ *reduces to* $\mathsf{trans}(\mathsf{symm}(q), \mathsf{symm}(p))$ *via rule R7.*

Proof. Rule R7 applies directly: the argument of symm is a trans expression, and the rewrite distributes the inverse over the composition with reversal of order. □

Proposition 6.42 (Verification of G8: Identity inverse). *For any term* $a : A$*, the expression* $\mathsf{symm}(\mathsf{refl}_a)$ *reduces to* refl_a *via rule R1.*

Proof. Rule R1 applies directly: $\mathsf{symm}(\mathsf{refl}_a) \to \mathsf{refl}_a$. □

6.4.3 Beyond the Core: Cancellation and Coherence

The eight groupoid rules form the "core" of the 77-rule system. The remaining 69 rules handle the interaction of groupoid structure with the type-theoretic operations (congrArg, congrFun, transport, etc.). But even within the groupoid core, the rules have important consequences beyond the basic axioms.

Proposition 6.43 (Cancellation via the rewrite system). *The cancellation laws of Proposition 3.52 are derivable from G1–G5:*

$$\begin{aligned}\mathsf{trans}(p, \mathsf{trans}(\mathsf{symm}(p), q)) &\sim_{\mathsf{rw}} \mathsf{trans}(\mathsf{trans}(p, \mathsf{symm}(p)), q) && (G5^{-1})\\ &\sim_{\mathsf{rw}} \mathsf{trans}(\mathsf{refl}_a, q) && (G3)\\ &\sim_{\mathsf{rw}} q && (G1).\end{aligned}$$

Dually, $\mathsf{trans}(\mathsf{symm}(p), \mathsf{trans}(p, q)) \sim_{\mathsf{rw}} q$.

Remark 6.44 (Cancellation as an optimisation). The cancellation above requires three groupoid-equivalence steps in the core system. In Chapter 4, the cancellation laws appear as *dedicated* rewrite rules (part of the 77-rule system) that achieve the same result in a single step. This is an optimisation for efficiency, not a logical necessity: the dedicated rules are admissible in the groupoid core.

Proposition 6.45 (Uniqueness of normal forms on expressions). *The groupoid subsystem (rules R1–R8), viewed as rewriting on groupoid expressions, is confluent and terminating. Every groupoid expression has a unique normal form, which is a* reduced word*: a composition of generators and their inverses with no adjacent cancellable pairs.*

Proof. Termination follows from the weight measure of Chapter 5, restricted to the groupoid rules. Confluence is established by checking the critical pairs among R1–R8, all of which are joinable (see Chapter 5, Section 5.2). By Newman's lemma, the system is confluent, and the normal form is therefore unique. □

This result echoes the classical theory of free groups: a reduced word in a free group is one with no adjacent generator-inverse pairs, and reduction to normal form is the standard algorithm for solving the word problem. As shown in Chapter 5, the groupoid subsystem implements precisely Dehn's algorithm [60] for the word problem in free groups.

6.4.4 The Free Groupoid Perspective

The connection to free groups suggests a broader viewpoint.

Definition 6.46 (Computad). A *computad* (or *graph*) G consists of:

- A set G_0 of *vertices* (0-cells).
- A set G_1 of *edges* (1-cells), each with a source and target in G_0.

Definition 6.47 (Free groupoid on a computad)**.** Given a computad G, the *free groupoid* $F(G)$ has:

- Objects: the vertices G_0.
- Morphisms from a to b: reduced words over the edges G_1 and their formal inverses, forming a path from a to b.
- Composition: concatenation followed by free reduction (cancelling adjacent e, e^{-1} pairs).
- Identity: the empty word.
- Inverses: reversing the word and inverting each letter.

Theorem 6.48 (Groupoid core as a free groupoid)**.** *Let A be a type and let G_A be the computad with vertices the terms of A and edges the* elementary steps *(the atomic path constructors). Then the quotient of the groupoid-core fragment of computational paths on A—that is, the fragment generated from atomic steps by* refl, symm, *and* trans*—by rewrite equivalence is isomorphic to the free groupoid $F(G_A)$.*

Proof. Let $Q(a, b)$ be the quotient of groupoid-core paths from a to b by rewrite equivalence. The groupoid rules remove units, double inverses, adjacent inverse pairs, and non-canonical bracketing, so every class in $Q(a, b)$ has a unique reduced representative: a right-associated word in elementary steps and formal inverses with no adjacent cancellation pair.

Define

$$\Phi_{a,b} : Q(a, b) \to \mathrm{Hom}_{F(G_A)}(a, b)$$

by sending the class of a path to its unique reduced word. This is well-defined because rewrite-equivalent paths have the same reduced normal form. Conversely, any reduced word in $F(G_A)$ can be read as a groupoid-core path by interpreting each edge as an atomic step, each formal inverse as a symmetry, and concatenation as transitivity. Since the word is reduced, the resulting path is already in normal form, so this interpretation defines an inverse map.

Thus $\Phi_{a,b}$ is a bijection for every pair of objects a, b. Composition is respected because both sides compose by concatenating words and then reducing to the unique reduced form, while inversion is respected because both sides reverse the word and invert each letter. Hence the family $(\Phi_{a,b})_{a,b}$ is an isomorphism of groupoids. □

Remark 6.49 (The fundamental groupoid is not free)**.** The full fundamental groupoid $\Pi_1(A)$ is not literally a free groupoid because the additional rewrite rules (9–77) impose further relations. For instance, the functoriality rules for $\mathsf{congrArg}$ identify $\mathsf{congrArg}(f, \mathsf{trans}(p, q))$ with $\mathsf{trans}(\mathsf{congrArg}(f, p), \mathsf{congrArg}(f, q))$, which is an equation beyond the free groupoid axioms. In the language of group presentations, $\Pi_1(A)$ is a groupoid *presented by* the elementary steps as generators and rules 9–77 as relations, modulo the free groupoid structure provided by rules 1–8.

Summary

The point of this chapter has been to make explicit the first algebraic shape carried by computational paths: they form a groupoid, weak before quotient and strict after quotient. More precisely:

1. **Groupoid structure** (Section 6.1): We recalled the definition of groupoid and introduced the key examples: groups as one-object groupoids, discrete and codiscrete groupoids, and the fundamental groupoid of a topological space.

2. **Weak path groupoid** (Section 6.2): For any type A, the collection of computational paths forms a weak groupoid $\mathcal{P}(A)$ where the axioms hold up to rewrite equivalence $\sim_{\mathsf{rw}}$. The congruence operation $\mathsf{congrArg}(f, -)$ defines a functor between path groupoids.

3. **Fundamental groupoid** (Section 6.3): Quotienting by rewrite equivalence yields the strict fundamental groupoid $\Pi_1(A)$. The fundamental group $\pi_1(A, a) = \mathrm{Aut}_{\Pi_1(A)}(a)$ recovers a group at each basepoint, and conjugation by a path provides basepoint-change isomorphisms. The construction parallels the classical fundamental groupoid of algebraic topology. In Chapter 13, we use this construction to compute fundamental groups of concrete spaces such as the circle, torus, and Klein bottle.

4. **Verified groupoid laws** (Section 6.4): The eight laws G1–G8 of Chapter 3 are precisely the axioms of the groupoid structure, witnessed by the first eight rewrite rules R1–R8 of the 77-rule system. On the expression layer, the groupoid subsystem is confluent and terminating, yielding unique reduced-word normal forms in the free-group semantics.

The groupoid $\mathcal{P}(A)$ is the first layer of a much richer structure. The witnesses to the groupoid laws—the 2-paths showing that $\mathsf{trans}(\mathsf{refl}_a, p) \sim_{\mathsf{rw}} p$, for instance—are themselves paths between paths. These 2-paths can be composed, inverted, and compared, and they satisfy their own coherence laws. In Chapter 7 we will study these 2-cells systematically, revealing the next layer of the weak ω-groupoid.

Historical Notes.

The line of thought behind this chapter runs from classical groupoid theory to the interpretation of identity in type theory. The notion of groupoid was introduced by Brandt (1927) in the context of quaternary quadratic forms, though the concept was implicit in the work of Galois and in the theory of equivalence relations. Mac Lane [98] provides a thorough treatment of groupoids within category theory.

The idea that the identity types of Martin-Löf type theory carry groupoid structure was first made precise by Hofmann and Streicher [86], who constructed the *groupoid model* of type theory to show that the Uniqueness of Identity Proofs (UIP) principle is not derivable from the standard rules. Their construction is the direct ancestor of our fundamental groupoid $\Pi_1(A)$.

The deeper insight that the full structure is not merely a groupoid but a *weak* ω*-groupoid* was established independently by Lumsdaine [105] and van den Berg and Garner [153]. In both works, the key observation is that the witnesses to the groupoid laws (our 2-paths) satisfy their own laws, and those witnesses satisfy laws at the next level, and so on *ad infinitum*. Our computational paths provide a concrete, syntactic instantiation of this tower: the 77-rule rewrite system gives explicit, computable witnesses at every level, as we shall see in the chapters that follow.

The connection between path groupoids and fundamental groupoids of spaces is central to Homotopy Type Theory [150], where the identity type of a type is understood as the space of paths in a homotopy-theoretic sense. The computational paths framework can be seen as providing an "algebraic topology of types," where the topological notions of path, homotopy, and fundamental group are replaced by their syntactic, computable analogues.

Formalization Notes. The results of this chapter are formalised in the following files of the accompanying Lean development (`https://github.com/Arthur742Ramos/ComputationalPathsLean`):

- `Path/GroupoidDerived.lean` — the main derived groupoid laws, cancellations, and quotient-level packaging
- `Path/Basic/Core.lean` — the primitive path operations underlying the groupoid structure
- `Path/OmegaGroupoid/GroupoidProofs.lean` — groupoid laws as part of the higher-cell development

Key definitions. The weak groupoid structure packages the fundamental operations:

```
/-- Weak groupoid: laws hold up to rewrite equivalence -/
structure WeakGroupoid (A : Type u) extends WeakCategory
   A where
  inv : {a b : A} -> Path a b -> Path b a
  left_inv : {a b : A} -> (p : Path a b) ->
      Rw (comp (inv p) p) id
  right_inv : {a b : A} -> (p : Path a b) ->
      Rw (comp p (inv p)) id
```

The groupoid laws are witnessed by primitive rewrite steps:

```
/-- Rule 5: Right inverse law. p . symm(p) -> refl -/
| trans_symm {a b : A} (p : Path a b) :
    Step (trans p (symm p)) (Path.refl a)
/-- Rule 6: Left inverse law. symm(p) . p -> refl -/
| symm_trans {a b : A} (p : Path a b) :
    Step (trans (symm p) p) (Path.refl b)
/-- Rule 8: Associativity. (p . q) . r -> p . (q . r) -/
| trans_assoc {a b c d : A}
    (p : Path a b) (q : Path b c) (r : Path c d) :
    Step (trans (trans p q) r) (trans p (trans q r))
```

Chapter 7

Two-Cells — Derivations Between Paths

"A 2-category is a category enriched over the monoidal category of categories; a bicategory is the corresponding 'weak' notion, in which associativity and unit laws hold only up to coherent isomorphism."

—Bénabou [12]

In Part I of this work, we developed a rewrite system on computational paths and established its fundamental metatheoretic properties: termination and confluence. The unique-normal-form theorem (Theorem 5.29) provides canonical reduced *expression* representatives, and the rewrite equivalence relation $\sim_{\mathsf{rw}}$ identifies paths with the same semantic normal form. But this leads immediately to a further question: *what is the status of the rewrite equivalence itself?*

When we write $p \sim_{\mathsf{rw}} q$ to assert that two paths $p, q : \mathsf{Path}(a, b)$ are rewrite-equivalent, we are asserting the existence of a chain of rewrite steps connecting p to q. That chain is itself a structured mathematical object: it has length, it can be composed with other chains, it can be inverted. In short, the witnesses of path equivalence form a new layer of algebraic structure *above* the paths themselves. We call these witnesses *2-cells*, or *derivations between paths*, and they are the subject of this chapter.

The passage from paths to 2-cells is the first step in a dimensional ascent that will eventually reveal the full ω-groupoid structure of computational paths [125, 128]. Just as a path $p : \mathsf{Path}(a, b)$ is a witness that a and b are equal, a 2-cell $\alpha : p \Rrightarrow q$ is a witness that p and q are themselves equal—not merely propositionally, but *as computational objects*, through a specific chain of rewrites. And just as paths can be composed, inverted, and transformed, so too can 2-cells, giving rise to *vertical* and *horizontal* composition operations with their own algebraic laws.

This chapter is organised as follows. Section 7.1 introduces the type of 2-cells and its basic constructors, establishing the vertical groupoid structure. Section 7.2 develops the horizontal composition and whiskering operations that promote this structure from a mere groupoid to a 2-category. Section 7.3 connects the abstract 2-cell machinery to the concrete rewrite equivalence relation $\sim_{\mathsf{rw}}$ from Chapter 5. Section 7.4 establishes the interchange law and the coherence conditions that govern the interaction between vertical and horizontal composition.

Terminology. Throughout this chapter and the remainder of the book, we use *2-cell* as the primary term for a witness of equivalence between parallel paths. Two synonyms appear frequently in the literature and in related parts of this work:

- *2-path* — emphasising the analogy with 1-dimensional paths, and the globular picture where each dimension adds a new layer of path structure.
- *Derivation* (or *second-order derivation*) — emphasising the rewriting origin: a 2-cell is, concretely, a derivation in the rewrite system on path expressions.

We write $\mathsf{Derivation}_2(p, q)$ for the formal type; when the context is informal or geometric we say "2-cell," and when we wish to stress the rewriting perspective we say "derivation." All three terms refer to the same mathematical object.

7.1 From Paths to 2-Paths

The groupoid structure established in Chapter 6 provides the first layer of higher structure on computational paths: the operations of composition, inversion, and identity satisfy the groupoid laws up to rewrite equivalence. But this "up to" deserves scrutiny. When we say that $\mathsf{trans}(p, \mathsf{refl}_b) \sim_{\mathsf{rw}} p$—that transitivity with the identity path is equivalent to doing nothing—we are not asserting a mere propositional fact. We are asserting the existence of a specific *derivation*: a sequence of rewrite steps that transforms the left-hand side into the right-hand side. This derivation is itself a mathematical object, and different derivations may witness the same equivalence in different ways.

7.1.1 The Idea of a 2-Cell

To motivate the formal definition, consider the groupoid law of right unitality. In Chapter 4, we introduced the rewrite rule

$$\mathsf{trans}(p, \mathsf{refl}_b) \to p \qquad \text{(Rule 4: right unit)}$$

which directly reduces $\mathsf{trans}(p, \mathsf{refl}_b)$ to p in a single step. But there may be other ways to establish that $\mathsf{trans}(p, \mathsf{refl}_b)$ is equivalent to p. For instance, one might first apply an associativity rule to rearrange the expression, then apply simplification rules to reach p through a longer sequence. Both derivation sequences witness the same equivalence, but they are different *as derivations*.

This situation should feel familiar. In topology, a *path* between two points a and b in a space X is a continuous map $\gamma : [0,1] \to X$ with $\gamma(0) = a$ and $\gamma(1) = b$. A *homotopy between paths*—what topologists call a *2-cell*—is a continuous map $H : [0,1] \times [0,1] \to X$ that interpolates between two paths γ_0 and γ_1, fixing the endpoints throughout. Our 2-cells play exactly this role, but in the combinatorial world of rewrite systems rather than the continuous world of topology.

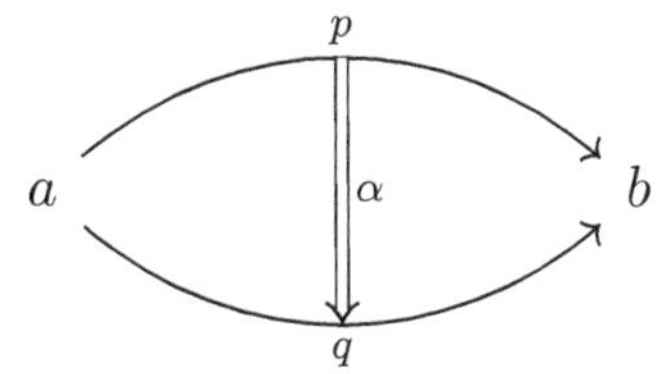

In the diagram above, p and q are parallel paths from a to b, and α is a 2-cell from p to q. The double arrow indicates that α lives one dimension higher than the paths it connects. This is the *globe* picture: a and b are 0-cells (points), p and q are 1-cells (paths), and α is a 2-cell (a path between paths).

7.1.2 Formal Definition

We now give the formal definition of 2-cells as an inductive type.

Definition 7.1 (2-cell / Derivation$_2$)**.** Let $a, b : A$ and let $p, q : \mathsf{Path}(a, b)$ be parallel paths. A *2-cell* (or *second-order derivation*) from p to q, written

$$\alpha : p \Rrightarrow q,$$

is a term of the inductive type $\mathsf{Derivation}_2(p, q)$ generated by the following constructors:

(i) **Identity.** For every path $p : \mathsf{Path}(a, b)$,

$$\mathsf{refl}_p : p \Rrightarrow p$$

is the *identity 2-cell* at p.

(ii) **Step.** If $s : \mathsf{Step}(p, q)$ is a single rewrite step from p to q, then

$$\mathsf{step}(s) : p \Rrightarrow q$$

is the 2-cell induced by s.

(iii) **Inverse.** If $\alpha : p \Rrightarrow q$, then

$$\alpha^{-1} : q \Rrightarrow p$$

is the *inverse 2-cell.*

(iv) **Vertical composition.** If $\alpha : p \Rrightarrow q$ and $\beta : q \Rrightarrow r$, then

$$\alpha \bullet_v \beta : p \Rrightarrow r$$

is the *vertical composite* of α and β.

Remark 7.2 (Type-valued 2-cells)**.** In the present development, both $\mathsf{RwEq}(p, q)$ and $\mathsf{Derivation}_2(p, q)$ are defined as inductive *Type*s, not mere propositions. This preserves the full derivation structure as data, ensuring that distinct derivations between the same pair of paths remain distinguishable.

Several features of this definition merit comment.

Remark 7.3 (2-cells vs. rewrite sequences)**.** A 2-cell is *not* simply a finite sequence of rewrite steps. The constructors for inverse and vertical composition allow us to build 2-cells by composing and inverting existing derivations. In particular, the step constructor provides the atomic 2-cells—one for each rewrite rule applied at a specific position—and the remaining constructors close the type under the operations of a groupoid. This is precisely the free groupoid on the rewrite relation, a point we will make precise in Section 7.3.

Remark 7.4 (Proof-irrelevance at the atomic step level)**.** In the formal development, atomic rewrite steps are treated as 0-truncated data: once the rule and redex position are fixed, any two witnesses of that same local rewrite are identified. Thus the identity type of a basic step is proposition-valued, and higher coherence does not depend on intensional bookkeeping about duplicate presentations of the same elementary move.

Conceptually, this is a local UIP-style choice for *steps*, not a global collapse of higher structure. The distinction matters for the UIP-versus-univalence discussion: we enforce uniqueness only at the atomic layer, while still allowing non-trivial higher cells generated by different composite derivations.

Remark 7.5 (Source and target). Every 2-cell $\alpha : p \Rrightarrow q$ has a well-defined *source* $\mathsf{src}(\alpha) = p$ and *target* $\mathsf{tgt}(\alpha) = q$. These are paths sharing the same endpoints: $p, q : \mathsf{Path}(a, b)$. We say that p and q are *parallel*, and that α is a 2-cell *in the hom-set* $\mathsf{Path}(a, b)$.

Example 7.6 (A 2-cell from the groupoid laws). Consider the path $p : \mathsf{Path}(a, b)$ and the compound path $\mathsf{trans}(p, \mathsf{refl}_b)$. The right unit rewrite rule provides a step

$$s_{\mathsf{ru}} : \mathsf{Step}(\mathsf{trans}(p, \mathsf{refl}_b),\ p)$$

and hence a 2-cell

$$\mathsf{step}(s_{\mathsf{ru}}) : \mathsf{trans}(p, \mathsf{refl}_b) \Rrightarrow p.$$

This 2-cell witnesses the fact that composing p with the identity path at b is the same as p itself, but it does so by recording the *specific rewrite rule* that establishes this fact.

By taking the inverse, we also obtain

$$\mathsf{step}(s_{\mathsf{ru}})^{-1} : p \Rrightarrow \mathsf{trans}(p, \mathsf{refl}_b),$$

which witnesses the equivalence in the reverse direction.

Example 7.7 (Composing 2-cells vertically). Suppose we have three parallel paths $p, q, r : \mathsf{Path}(a, b)$ and 2-cells $\alpha : p \Rrightarrow q$ and $\beta : q \Rrightarrow r$. Their vertical composite $\alpha \bullet_v \beta : p \Rrightarrow r$ represents the derivation obtained by first performing α (transforming p into q) and then performing β (transforming q into r).

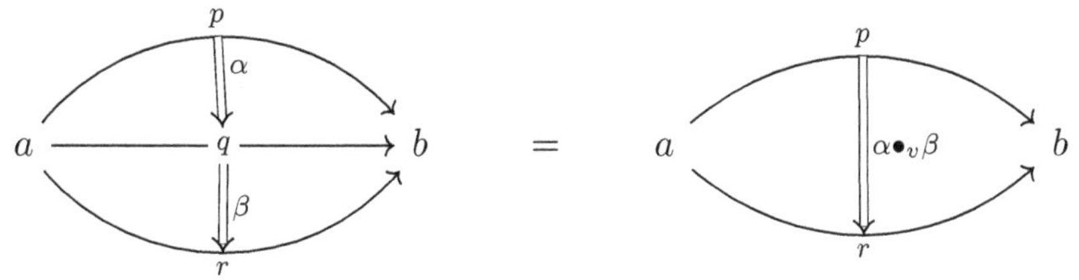

Diagrammatically, vertical composition "stacks" 2-cells one on top of the other. The terminology *vertical* reflects the standard convention in higher category theory, where the paths p, q, r are drawn horizontally and the 2-cells connect them vertically.

7.1.3 The Vertical Groupoid Structure

The type $\mathsf{Derivation}_2(p, q)$ of 2-cells between parallel paths, equipped with identity, composition, and inverse, forms a groupoid in its own right. We now verify the groupoid axioms.

Proposition 7.8 (Vertical groupoid laws)**.** *Let $p, q, r, s : \mathsf{Path}(a, b)$ be parallel paths. After passing to 3-cell equivalence, vertical composition makes the parallel 2-cells between these paths into a groupoid. Equivalently, there are 3-cells exhibiting:*

(i) ***Left unit.*** $\mathsf{refl}_p \bullet_v \alpha \Rrightarrow_3 \alpha$ *for every* $\alpha : p \Rrightarrow q$.

(ii) ***Right unit.*** $\alpha \bullet_v \mathsf{refl}_q \Rrightarrow_3 \alpha$ *for every* $\alpha : p \Rrightarrow q$.

(iii) ***Associativity.*** *For* $\alpha : p \Rrightarrow q$, $\beta : q \Rrightarrow r$, $\gamma : r \Rrightarrow s$*:*

$$(\alpha \bullet_v \beta) \bullet_v \gamma \Rrightarrow_3 \alpha \bullet_v (\beta \bullet_v \gamma).$$

(iv) ***Left inverse.*** $\alpha^{-1} \bullet_v \alpha \Rrightarrow_3 \mathsf{refl}_q$ *for every* $\alpha : p \Rrightarrow q$.

(v) ***Right inverse.*** $\alpha \bullet_v \alpha^{-1} \Rrightarrow_3 \mathsf{refl}_p$ *for every* $\alpha : p \Rrightarrow q$.

Proof. Vertical composition, inverse, and identities are built into the raw 2-cell syntax, while Chapter 8 packages the 3-cells that identify insertion or removal of identity derivations, reassociation of composites, and cancellation of inverse pairs. Passing to the quotient by 3-cell equivalence therefore yields an honest groupoid of parallel 2-cells. □

Remark 7.9 (Strict vs. weak groupoid laws)**.** A delicate point: the comparison statements in Proposition 7.8 become *strict equalities* when we regard 2-cells as equivalence classes under 3-cell equivalence (i.e., derivations between derivations). If we work with raw syntactic 2-cells, the associativity law holds only up to a coherent 3-cell isomorphism—an *associator* at the 2-cell level. This distinction between strict and weak composition will be central to the ω-groupoid structure developed in Chapter 9.

In what follows, when we write "=" between 2-cells, we mean equality up to 3-cells unless otherwise stated. This convention is analogous to our use of $\sim_{\mathsf{rw}}$ for path equality: just as $p \sim_{\mathsf{rw}} q$ means p and q are connected by a chain of rewrites, "$\alpha = \beta$" at the 2-cell level means α and β are connected by a chain of higher rewrites.

Definition 7.10 (The vertical category $\mathbf{Der}_2(a, b)$)**.** For fixed 0-cells $a, b : A$, the *vertical category* $\mathbf{Der}_2(a, b)$ has:

- **Objects:** paths $p : \mathsf{Path}(a, b)$.
- **Morphisms:** 2-cells $\alpha : p \Rrightarrow q$.
- **Identity:** $\mathsf{refl}_p : p \Rrightarrow p$ for each path p.
- **Composition:** vertical composition $\bullet_v$.

By Proposition 7.8, this is in fact a *groupoid*: every morphism has an inverse.

The vertical category captures one dimension of the 2-cell structure. But paths can also be composed *horizontally*, via transitivity. The interaction between horizontal and vertical composition is the subject of the next section.

7.2 The 2-Category Structure

The vertical groupoid structure of Section 7.1 organizes the 2-cells between any two parallel paths into a groupoid. But computational paths also compose horizontally via transitivity: given $f : \mathsf{Path}(a, b)$ and $g : \mathsf{Path}(b, c)$, their composite $\mathsf{trans}(f, g) : \mathsf{Path}(a, c)$ is a path from a to c. This horizontal composition must interact with the 2-cell structure in a coherent way.

Why does this matter? Consider a concrete situation: we know that path f is equivalent to f' (via a 2-cell α), and independently that g is equivalent to g' (via a 2-cell β). We ought to be able to conclude that the composite $\mathsf{trans}(f, g)$ is equivalent to $\mathsf{trans}(f', g')$—after all, if two legs of a journey can each be rerouted, the whole journey can be rerouted. But this conclusion requires a new 2-cell, one that combines α and β in a way that respects the boundary conditions imposed by the intermediate point b. Horizontal composition is precisely the operation that constructs this combined 2-cell. It is what lifts the vertical groupoid to a genuine two-dimensional algebraic structure—a bicategory—and it is the gateway to the interchange law, the pentagon identity, and ultimately the full ω-groupoid coherence.

The key insight is that the operations of *whiskering* and *horizontal composition* allow us to promote the collection of 0-cells, 1-cells, and 2-cells into a *2-category* (or more precisely, a *bicategory* in the sense of Bénabou [12], since the composition laws hold only up to coherent isomorphism).

7.2.1 Whiskering: Composing Paths with 2-Cells

Before defining the full horizontal composition, we introduce the simpler operation of *whiskering*, which composes a 2-cell with a path on one side.

Intuitively, whiskering answers the following question: if we know that two paths g and g' are equivalent (via a 2-cell α), and we have another path f composable with both, can we "extend" the equivalence to the composites $\mathsf{trans}(f, g)$ and $\mathsf{trans}(f, g')$? The answer is yes: α can be "whiskered along" the fixed path f to produce an equivalence between the composites.

Definition 7.11 (Left whiskering)**.** Let $f : \mathsf{Path}(a, b)$ be a path, and let $\alpha : g \Rrightarrow g'$ be a 2-cell between parallel paths $g, g' : \mathsf{Path}(b, c)$. The *left whiskering* of f

with α is the 2-cell

$$f \triangleright \alpha \ : \ \mathsf{trans}(f, g) \Rrightarrow \mathsf{trans}(f, g')$$

defined inductively by:

$$\begin{aligned} f \triangleright \mathsf{refl}_g &= \mathsf{refl}_{\mathsf{trans}(f,g)} \\ f \triangleright \mathsf{step}(s) &= \mathsf{step}(\mathsf{cong}_{\mathsf{right}}(f, s)) \\ f \triangleright \alpha^{-1} &= (f \triangleright \alpha)^{-1} \\ f \triangleright (\alpha \bullet_v \beta) &= (f \triangleright \alpha) \bullet_v (f \triangleright \beta) \end{aligned}$$

where $\mathsf{cong}_{\mathsf{right}}(f, s)$ is the step obtained by applying the congruence rule for the right argument of trans.

Intuitively, left whiskering "extends" a 2-cell on the left by composing with a fixed path f. The 2-cell α transforms g into g'; after whiskering with f, we obtain a 2-cell that transforms $\mathsf{trans}(f, g)$ into $\mathsf{trans}(f, g')$, leaving the f component unchanged.

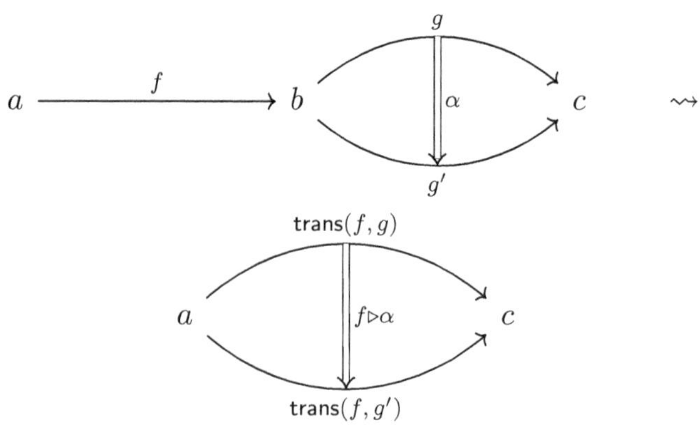

Definition 7.12 (Right whiskering). Let $\alpha : f \Rrightarrow f'$ be a 2-cell between parallel paths $f, f' : \mathsf{Path}(a, b)$, and let $h : \mathsf{Path}(b, c)$ be a path. The *right whiskering* of α with h is the 2-cell

$$\alpha \triangleleft h \ : \ \mathsf{trans}(f, h) \Rrightarrow \mathsf{trans}(f', h)$$

defined inductively by:

$$\begin{aligned} \mathsf{refl}_f \triangleleft h &= \mathsf{refl}_{\mathsf{trans}(f,h)} \\ \mathsf{step}(s) \triangleleft h &= \mathsf{step}(\mathsf{cong}_{\mathsf{left}}(s, h)) \\ \alpha^{-1} \triangleleft h &= (\alpha \triangleleft h)^{-1} \\ (\alpha \bullet_v \beta) \triangleleft h &= (\alpha \triangleleft h) \bullet_v (\beta \triangleleft h) \end{aligned}$$

where $\mathsf{cong}_{\mathsf{left}}(s, h)$ is the step obtained by applying the congruence rule for the left argument of trans.

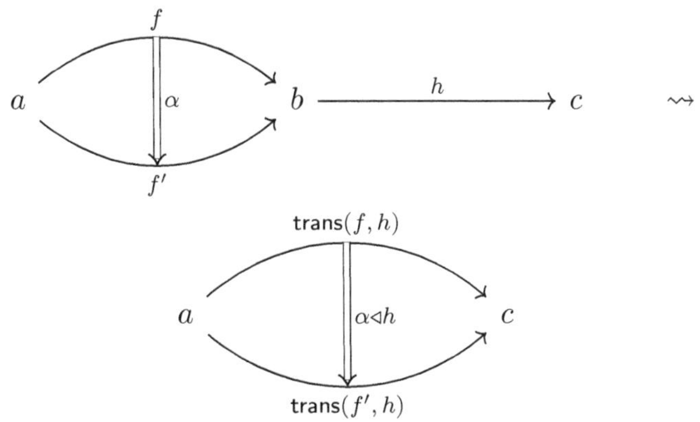

The following lemma records the basic functorial properties of whiskering.

Lemma 7.13 (Functoriality of whiskering). *Let* $f : \mathsf{Path}(a, b)$, $h : \mathsf{Path}(b, c)$, *and let* $\alpha : g \Rrightarrow g'$, $\beta : g' \Rrightarrow g''$ *be 2-cells between paths* $g, g', g'' : \mathsf{Path}(b, c)$. *Then:*

(i) $f \triangleright \mathsf{refl}_g = \mathsf{refl}_{\mathsf{trans}(f,g)}$

(ii) $f \triangleright (\alpha \bullet_v \beta) = (f \triangleright \alpha) \bullet_v (f \triangleright \beta)$

(iii) $f \triangleright \alpha^{-1} = (f \triangleright \alpha)^{-1}$

Symmetric statements hold for right whiskering. In particular, whiskering defines a groupoid homomorphism from $\mathbf{Der}_2(b, c)$ *to* $\mathbf{Der}_2(a, c)$ *(for left whiskering) and from* $\mathbf{Der}_2(a, b)$ *to* $\mathbf{Der}_2(a, c)$ *(for right whiskering).*

Example 7.14 (A concrete whiskering computation). Let $f : \mathsf{Path}(a, b)$ be a fixed path, and consider two parallel paths $g, g' : \mathsf{Path}(b, c)$ related by a single rewrite step $\alpha : g \Rrightarrow g'$ (for instance, $g = \mathsf{trans}(\mathsf{refl}_b, h)$ and $g' = h$ via the left-unit rule R3, so that $\alpha = \mathsf{step}(\mathrm{R3})$). Left whiskering by f produces:

$$f \triangleright \alpha \; : \; \mathsf{trans}(f, \mathsf{trans}(\mathsf{refl}_b, h)) \; \Rrightarrow \; \mathsf{trans}(f, h).$$

The resulting 2-cell asserts that the rewrite from $\mathsf{trans}(\mathsf{refl}_b, h)$ to h can be performed "inside" the composition with f, leaving the initial segment of the composite untouched. In terms of the 77-rule system, the whiskered step applies R3 to the right argument of the outer trans, which is exactly the congruence rule for the right argument (one of rules 73–77).

Proof. Each identity follows immediately from the inductive definition of whiskering: properties (i)–(iii) are the defining clauses of left whiskering (Definition 7.11). The symmetric statements for right whiskering follow from Definition 7.12. □

Example 7.15 (Whiskering a right-unit 2-cell). Let $f : \mathsf{Path}(a, b)$, $g : \mathsf{Path}(b, c)$, and consider the right unitor 2-cell $\mathfrak{r}_f : \mathsf{trans}(f, \mathsf{refl}_b) \Rrightarrow f$, which witnesses the groupoid law G2. Right whiskering this 2-cell with g produces:

$$\mathfrak{r}_f \lhd g \; : \; \mathsf{trans}(\mathsf{trans}(f, \mathsf{refl}_b), g) \Rrightarrow \mathsf{trans}(f, g).$$

This 2-cell says: "if we first compose f with the identity at b (a redundant step), and then compose with g, the result is equivalent to simply composing f with g." The whiskering operation extends the original right-unit equivalence to a context where g has been appended.

Dually, left whiskering f with the left unitor $\mathfrak{l}_g : \mathsf{trans}(\mathsf{refl}_b, g) \Rrightarrow g$ produces:

$$f \rhd \mathfrak{l}_g \; : \; \mathsf{trans}(f, \mathsf{trans}(\mathsf{refl}_b, g)) \Rrightarrow \mathsf{trans}(f, g).$$

This 2-cell removes a redundant identity refl_b that appears inside the right argument of a composition with f.

7.2.2 Horizontal Composition

With whiskering in hand, we can define the general horizontal composition of 2-cells.

Definition 7.16 (Horizontal composition). Let $\alpha : f \Rrightarrow f'$ be a 2-cell between paths $f, f' : \mathsf{Path}(a, b)$, and let $\beta : g \Rrightarrow g'$ be a 2-cell between paths $g, g' : \mathsf{Path}(b, c)$. The *horizontal composite*

$$\alpha \bullet_h \beta \; : \; \mathsf{trans}(f, g) \Rrightarrow \mathsf{trans}(f', g')$$

is defined by

$$\alpha \bullet_h \beta \stackrel{\text{def}}{=} (\alpha \lhd g) \bullet_v (f' \rhd \beta).$$

That is, horizontal composition is the vertical composite of a right whiskering followed by a left whiskering.

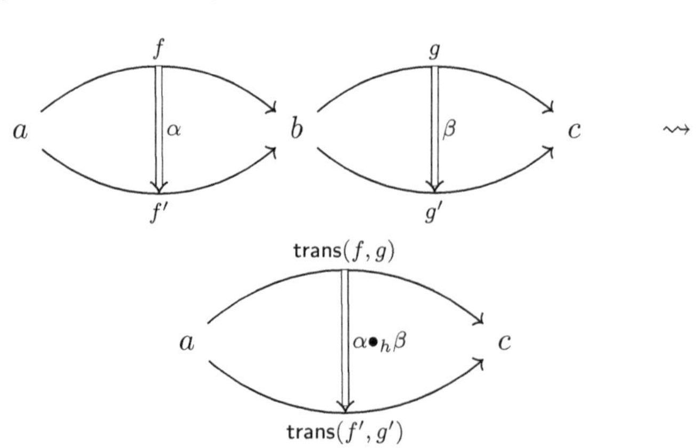

Remark 7.17 (Choice of definition)**.** Definition 7.16 decomposes horizontal composition as "first whisker α on the right by g, then whisker β on the left by f'." There is an equally valid alternative:

$$\alpha \bullet_h' \beta \stackrel{\text{def}}{=} (f \triangleright \beta) \bullet_v (\alpha \triangleleft g').$$

This alternative first whiskers β on the left by f, then whiskers α on the right by g'. The two definitions are *not* identical as 2-cells. We write

$$\begin{aligned} H_{\mathrm{RL}} &:= (\alpha \triangleleft g) \bullet_v (f' \triangleright \beta) = \alpha \bullet_h \beta, \\ H_{\mathrm{LR}} &:= (f \triangleright \beta) \bullet_v (\alpha \triangleleft g') = \alpha \bullet_h' \beta. \end{aligned}$$

By Theorem 7.32, there is a canonical 3-cell

$$\Xi_{\alpha,\beta} : H_{\mathrm{RL}} \Rrightarrow_3 H_{\mathrm{LR}}.$$

Its inverse $\Xi_{\alpha,\beta}^{-1}$ is a 3-cell in the opposite direction, so $\bullet_h$ and $\bullet_h'$ are coherently equivalent definitions of horizontal composition. Thus Definition 7.16 is canonical up to the interchange 3-cell, as required in a bicategory.

Proposition 7.18 (Horizontal composition with identities)**.** *For paths f : $\mathsf{Path}(a, b)$ and $g : \mathsf{Path}(b, c)$:*

(i) $\mathsf{refl}_f \bullet_h \mathsf{refl}_g = \mathsf{refl}_{\mathsf{trans}(f,g)}$.

(ii) $\alpha \bullet_h \mathsf{refl}_g = \alpha \triangleleft g$ *for all* $\alpha : f \Rrightarrow f'$.

(iii) $\mathsf{refl}_f \bullet_h \beta = f \triangleright \beta$ *for all* $\beta : g \Rrightarrow g'$.

Proof. For (i), we compute:

$$\begin{aligned} \mathsf{refl}_f \bullet_h \mathsf{refl}_g &= (\mathsf{refl}_f \triangleleft g) \bullet_v (f \triangleright \mathsf{refl}_g) \\ &= \mathsf{refl}_{\mathsf{trans}(f,g)} \bullet_v \mathsf{refl}_{\mathsf{trans}(f,g)} \\ &= \mathsf{refl}_{\mathsf{trans}(f,g)}. \end{aligned}$$

For (ii):

$$\begin{aligned} \alpha \bullet_h \mathsf{refl}_g &= (\alpha \triangleleft g) \bullet_v (f' \triangleright \mathsf{refl}_g) \\ &= (\alpha \triangleleft g) \bullet_v \mathsf{refl}_{\mathsf{trans}(f',g)} \\ &= \alpha \triangleleft g. \end{aligned}$$

Part (iii) is symmetric:

$$\begin{aligned} \mathsf{refl}_f \bullet_h \beta &= (\mathsf{refl}_f \triangleleft g) \bullet_v (f \triangleright \beta) \\ &= \mathsf{refl}_{\mathsf{trans}(f,g)} \bullet_v (f \triangleright \beta) \\ &= f \triangleright \beta. \end{aligned}$$

□

This proposition confirms that horizontal composition with identity 2-cells reduces to whiskering, validating our definition.

Example 7.19 (Horizontal composition of rewrite 2-cells)**.** Let $f : \mathsf{Path}(a, b)$ and $g : \mathsf{Path}(b, c)$ be paths, and consider the right unitor $\mathfrak{r}_f : \mathsf{trans}(f, \mathsf{refl}_b) \Rrightarrow f$ and the left unitor $\mathfrak{l}_g : \mathsf{trans}(\mathsf{refl}_b, g) \Rrightarrow g$. Their horizontal composite is:

$$\mathfrak{r}_f \bullet_h \mathfrak{l}_g \ : \ \mathsf{trans}(\mathsf{trans}(f, \mathsf{refl}_b), \mathsf{trans}(\mathsf{refl}_b, g)) \ \Rrightarrow \ \mathsf{trans}(f, g).$$

By definition, this decomposes as:

$$\mathfrak{r}_f \bullet_h \mathfrak{l}_g = (\mathfrak{r}_f \lhd \mathsf{trans}(\mathsf{refl}_b, g)) \bullet_v (f \rhd \mathfrak{l}_g).$$

In words: first, the right unitor eliminates refl_b from the left factor (via right whiskering with $\mathsf{trans}(\mathsf{refl}_b, g)$), transforming the source to $\mathsf{trans}(f, \mathsf{trans}(\mathsf{refl}_b, g))$; then, the left unitor eliminates refl_b from the right factor (via left whiskering with f), arriving at the target $\mathsf{trans}(f, g)$. The horizontal composite thus removes *both* redundant identity paths in a single 2-cell.

7.2.3 The Bicategory of Computational Paths

With vertical composition, whiskering, and horizontal composition in hand, we have all the ingredients to recognise that the paths in a type A form a *bicategory*: a two-dimensional categorical structure comprising objects, 1-morphisms between objects, and 2-morphisms between 1-morphisms, where composition is associative and unital only up to coherent invertible 2-morphisms. The concept was introduced by Bénabou [12] to axiomatize situations—like ours—in which strict equalities are replaced by invertible higher cells.

Definition 7.20 (The bicategory $\mathfrak{B}_A$)**.** For a type A, the *bicategory of computational paths in* A, denoted $\mathfrak{B}_A$, consists of:

- **0-cells:** elements $a, b, c, \ldots$ of A.
- **1-cells:** paths $f, g, h, \ldots$ in $\mathsf{Path}(a, b)$, for various $a, b : A$.
- **2-cells:** derivations $\alpha, \beta, \gamma, \ldots$ in $\mathsf{Derivation}_2(f, g)$, for parallel paths f, g.
- **Horizontal composition:** given by trans on 1-cells and by $\bullet_h$ on 2-cells.
- **Vertical composition:** given by $\bullet_v$ on 2-cells.
- **Identity 1-cell:** $\mathsf{refl}_a : \mathsf{Path}(a, a)$ for each 0-cell a.

- **Identity 2-cell:** $\mathsf{refl}_f : f \Rrightarrow f$ for each 1-cell f.
- **Associator:** a 2-cell $\mathfrak{a}_{f,g,h} : \mathsf{trans}(\mathsf{trans}(f,g),h) \Rrightarrow \mathsf{trans}(f,\mathsf{trans}(g,h))$ for each composable triple of 1-cells, given by the associativity rewrite rule (Rule 8).
- **Left unitor:** a 2-cell $\mathfrak{l}_f : \mathsf{trans}(\mathsf{refl}_a, f) \Rrightarrow f$ for each 1-cell f : $\mathsf{Path}(a,b)$, given by the left unit rule (Rule 3).
- **Right unitor:** a 2-cell $\mathfrak{r}_f : \mathsf{trans}(f, \mathsf{refl}_b) \Rrightarrow f$ for each 1-cell f : $\mathsf{Path}(a,b)$, given by the right unit rule (Rule 4).

The associator and unitors are invertible 2-cells (since every step induces an invertible 2-cell), and they are required to satisfy the coherence conditions described in Section 7.4.

Remark 7.21 (Bicategory vs. strict 2-category)**.** The bicategory $\mathfrak{B}_A$ is *not* a strict 2-category. In a strict 2-category, the associativity and unit laws for horizontal composition hold as *equalities* of 1-cells:

$$\mathsf{trans}(\mathsf{trans}(f,g),h) = \mathsf{trans}(f,\mathsf{trans}(g,h)),$$
$$\mathsf{trans}(\mathsf{refl}_a, f) = f, \qquad \mathsf{trans}(f, \mathsf{refl}_b) = f.$$

In our setting, these laws hold only up to rewrite equivalence—that is, up to invertible 2-cells. This is precisely the situation that Bénabou axiomatized in his 1967 paper [12]: a bicategory is a "weak 2-category" in which the composition laws hold up to coherent, invertible 2-cells.

The non-strictness is not a deficiency but a feature. It is the *source* of the higher dimensional structure: the associators and unitors are themselves non-trivial 2-cells, and the coherence conditions they satisfy (the pentagon and triangle identities) involve 3-cells. If the laws held strictly, there would be no higher structure to study.

7.3 2-Cells as Path Equalities

We now connect the abstract 2-cell machinery to the concrete rewrite equivalence relation $\sim_{\mathsf{rw}}$ from Chapter 5. The result is surprisingly clean: the two structures are not merely related but *isomorphic*. The rewrite equivalence that we defined syntactically in Chapter 5—as the reflexive-symmetric-transitive closure of the one-step rewrite relation—turns out to be, constructor for constructor, the same thing as our inductively defined 2-cell type. This is more than a formal coincidence. It means that every claim of the form "p and q are rewrite-equivalent" is automatically backed by an explicit 2-cell derivation, and conversely, every

2-cell derivation witnesses a genuine rewrite equivalence. The entire metatheory of Part I—termination, confluence, unique normal forms—thus flows directly into the higher algebra of Part II without any loss of information.

7.3.1 The RwEq Relation Revisited

Recall from Chapter 5 that two paths $p, q : \mathsf{Path}(a, b)$ are *rewrite equivalent*, written $p \sim_{\mathsf{rw}} q$, if they are connected by a finite chain of rewrite steps and their reverses. We now give a more structured formulation.

Definition 7.22 (Rewrite equivalence, revised). The *rewrite equivalence relation* $\mathsf{RwEq}(p, q)$ on parallel paths $p, q : \mathsf{Path}(a, b)$ is the inductive type generated by:

(i) **Reflexivity.** $\mathsf{refl}_p : \mathsf{RwEq}(p, p)$ for every path p.

(ii) **Step.** If $s : \mathsf{Step}(p, q)$, then $\mathsf{step}(s) : \mathsf{RwEq}(p, q)$.

(iii) **Symmetry.** If $h : \mathsf{RwEq}(p, q)$, then $\mathsf{symm}(h) : \mathsf{RwEq}(q, p)$.

(iv) **Transitivity.** If $h_1 : \mathsf{RwEq}(p, q)$ and $h_2 : \mathsf{RwEq}(q, r)$, then $\mathsf{trans}(h_1, h_2) : \mathsf{RwEq}(p, r)$.

Remark 7.23 (RwEq vs. Derivation$_2$). Comparing Definitions 7.1 and 7.22 reveals an immediate structural parallel:

Constructor	**Derivation$_2$**	**RwEq**
Identity/Refl.	$\mathsf{refl}_p : p \Rrightarrow p$	$\mathsf{refl}_p : \mathsf{RwEq}(p, p)$
Step	$\mathsf{step}(s) : p \Rrightarrow q$	$\mathsf{step}(s) : \mathsf{RwEq}(p, q)$
Inv./Symm.	$\alpha^{-1} : q \Rrightarrow p$	$\mathsf{symm}(h) : \mathsf{RwEq}(q, p)$
Comp./Trans.	$\alpha \bullet_v \beta : p \Rrightarrow r$	$\mathsf{trans}(h_1, h_2) : \mathsf{RwEq}(p, r)$

The two types have *identical structure*. This is no accident: both are instances of the free groupoid generated by the rewrite relation Step. We now make this precise.

7.3.2 The Equivalence of RwEq and Derivation$_2$

Theorem 7.24 (RwEq $\cong$ Derivation$_2$). *For parallel paths* $p, q : \mathsf{Path}(a, b)$, *there is a natural bijection*

$$\Phi : \mathsf{RwEq}(p, q) \xrightarrow{\cong} \mathsf{Derivation}_2(p, q)$$

defined by:

$$\begin{aligned}
\Phi(\mathsf{refl}_p) &= \mathsf{refl}_p \\
\Phi(\mathsf{step}(s)) &= \mathsf{step}(s) \\
\Phi(\mathsf{symm}(h)) &= \Phi(h)^{-1} \\
\Phi(\mathsf{trans}(h_1, h_2)) &= \Phi(h_1) \bullet_v \Phi(h_2)
\end{aligned}$$

with inverse

$$\begin{aligned}
\Psi(\mathsf{refl}_p) &= \mathsf{refl}_p \\
\Psi(\mathsf{step}(s)) &= \mathsf{step}(s) \\
\Psi(\alpha^{-1}) &= \mathsf{symm}(\Psi(\alpha)) \\
\Psi(\alpha \bullet_v \beta) &= \mathsf{trans}(\Psi(\alpha), \Psi(\beta)).
\end{aligned}$$

Moreover, Φ and Ψ are mutually inverse: $\Psi \circ \Phi = \mathsf{id}$ and $\Phi \circ \Psi = \mathsf{id}$.

Proof. The maps Φ and Ψ are well-defined by the structural induction principle for their respective inductive types. That $\Psi \circ \Phi = \mathsf{id}$ and $\Phi \circ \Psi = \mathsf{id}$ follows by a straightforward induction: each constructor of one type is mapped to the corresponding constructor of the other, and the round-trip is the identity on each constructor. □

Corollary 7.25 (Rewrite equivalence as 2-cell existence)**.** *Two parallel paths $p, q : \mathsf{Path}(a, b)$ are rewrite equivalent if and only if there exists a 2-cell from p to q:*

$$p \sim_{\mathsf{rw}} q \qquad \Longleftrightarrow \qquad \exists\, \alpha : p \Rrightarrow q.$$

This corollary gives the fundamental interpretation of the rewrite equivalence relation: $p \sim_{\mathsf{rw}} q$ is the *propositional truncation* of the type of 2-cells from p to q. The rewrite equivalence relation forgets the specific derivation and retains only the information that *some* derivation exists. The 2-cell type, by contrast, remembers the full structure of the derivation.

7.3.3 From Normal Forms to 2-Cells

The normalisation results of Chapter 5 interact with the 2-cell structure by providing canonical comparison targets via the expression normal forms.

Proposition 7.26 (Canonical 2-cells from expression normal forms)**.** *For every path $p : \mathsf{Path}(a, b)$, fix an expression presentation e_p and let $\widehat{e_p} := \mathrm{nf}_{\mathrm{Expr}}(e_p)$. Then:*

(i) There is a reduction derivation $e_p \to^* \widehat{e_p}$ in the expression rewrite system.

(ii) *For any two paths* p, q : $\mathsf{Path}(a, b)$ *with* $p \sim_{\mathsf{rw}} q$, *we have* $\widehat{e_p} = \widehat{e_q}$ *(syntactic equality of expression normal forms), and choosing reduction derivations for* e_p, e_q *yields a composite comparison 2-cell*

$$\nu_p \bullet_v \nu_q^{-1} : p \Rrightarrow q$$

factoring through the shared semantic normal form.

Proof. Part (i) is termination for the expression rewrite relation. Part (ii) is confluence on expressions (Chapter 5), giving $\widehat{e_p} = \widehat{e_q}$ whenever $p \sim_{\mathsf{rw}} q$. Reading each reduction step as a 2-cell generator and composing vertically yields the displayed comparison 2-cell. □

Remark 7.27 (The canonical 2-cell is not unique)**.** Proposition 7.26 provides a *canonical* 2-cell between rewrite-equivalent paths, but this 2-cell is by no means the only one. Different reduction strategies yield different 2-cells, even though they all factor through the same normal form. The study of the space of *all* 2-cells between two given paths—and the 3-cells between them—is the subject of Chapter 8.

Example 7.28 (Canonical 2-cell through normal forms)**.** Let $p = \mathsf{trans}(\mathsf{refl}_a, \mathsf{trans}(s, \mathsf{refl}_b))$ where $s : \mathsf{Path}(a, b)$. The normal form is $\hat{p} = s$, obtained by eliminating the two identity paths via the unit rules. One reduction sequence is:

$$\mathsf{trans}(\mathsf{refl}_a, \mathsf{trans}(s, \mathsf{refl}_b)) \xrightarrow{\text{left unit}} \mathsf{trans}(s, \mathsf{refl}_b) \xrightarrow{\text{right unit}} s.$$

If $q = \mathsf{trans}(s, \mathsf{refl}_b)$, then $\hat{q} = s$ as well (by a single right-unit step), and the canonical 2-cell $p \Rrightarrow q$ is:

$$p \xrightarrow{\nu_p} s = \hat{q} \xrightarrow{\nu_q^{-1}} q,$$

which first reduces p fully to s, then reverses the one-step reduction of q.

7.3.4 The Free Groupoid Perspective

The equivalence of Theorem 7.24 admits a clean categorical description.

Proposition 7.29 (Free groupoid on the rewrite graph)**.** *Fix 0-cells* $a, b : A$. *Let* $\mathcal{G}(a, b)$ *denote the directed graph whose vertices are the paths* p : $\mathsf{Path}(a, b)$ *and whose edges are the rewrite steps* s : $\mathsf{Step}(p, q)$. *Then:*

(i) *The vertical category* $\mathbf{Der}_2(a, b)$ *(Definition 7.10) is the* free groupoid *on the graph* $\mathcal{G}(a, b)$.

(ii) *The connected components of* $\mathbf{Der}_2(a,b)$ *are in bijection with the* $\sim_{\mathsf{rw}}$-*equivalence classes of paths.*

(iii) *In particular,* p *and* q *belong to the same connected component if and only if* $p \sim_{\mathsf{rw}} q$.

Proof. A free groupoid on a graph $\mathcal{G}$ has the same vertices as $\mathcal{G}$ and its morphisms are formal composites and inverses of edges, subject only to the groupoid axioms. This is precisely the structure of $\mathsf{Derivation}_2$: the step constructor includes edges, $\bullet_v$ composes them, $(\cdot)^{-1}$ inverts them, and refl provides identities. Parts (ii) and (iii) follow from the general theory of free groupoids: two objects are in the same connected component if and only if there is a morphism (i.e., a 2-cell) between them, which by Corollary 7.25 is equivalent to $p \sim_{\mathsf{rw}} q$. □

This perspective connects the computational paths framework to the theory of *higher-dimensional rewriting* initiated by Squier [141]. In Squier's theory, the 2-cells of a rewriting system play the role of homotopies between rewriting paths, and the existence of enough 2-cells is related to the homological finiteness properties of the monoid presented by the system. Our 2-cells serve an analogous purpose: they witness the "homotopy" between different derivation paths in the rewrite system.

7.4 Coherence at Level 2

The bicategory structure of Section 7.2 comes with associators, left unitors, and right unitors—invertible 2-cells that witness the non-strict algebraic laws. For this structure to be well behaved, these 2-cells must satisfy certain *coherence conditions*: diagrams involving multiple applications of associators and unitors must commute. These coherence conditions take the form of *3-cells*—paths between 2-cells—and their verification is the main task of this section.

7.4.1 3-Cells: Paths Between Derivations

Just as a 2-cell is a derivation between paths, a *3-cell* is a derivation between 2-cells.

Definition 7.30 (3-cell / $\mathsf{Derivation}_3$)**.** Let $\alpha, \beta : p \Rrightarrow q$ be parallel 2-cells (i.e., 2-cells with the same source and target). A *3-cell* from α to β, written

$$\Gamma : \alpha \Rrightarrow_3 \beta,$$

is a term of the type $\mathsf{Derivation}_3(\alpha, \beta)$. The constructors mirror those of $\mathsf{Derivation}_2$: identity, step, inverse, and composition.

Remark 7.31 (The dimensional ladder)**.** The passage from 2-cells to 3-cells is another instance of the same construction: where 2-cells were defined as the free groupoid on the rewrite steps between paths, 3-cells are defined as the free groupoid on the rewrite steps between 2-cells. In principle, this process can be iterated indefinitely, producing n-cells for every $n \geq 0$. The guiding pattern, made precise for the designated structural coherence fragment in Chapters 8 and 9, is that this tower stabilises only after the genuine 3-cell layer: dimensions 1, 2, and 3 carry non-trivial structure, while from dimension 4 onward the structural fillers become canonically comparable one level higher. But at dimension 2 itself, the structure is already genuinely non-trivial.

We write "$\alpha = \beta$" between parallel 2-cells to mean that there exists a 3-cell $\Gamma : \alpha \Rrightarrow_3 \beta$. This is the appropriate notion of equality at the 2-cell level, and all the coherence conditions below are formulated at this level.

7.4.2 The Interchange Law

The *interchange law* (also called the *Godement interchange rule*) governs the interaction between vertical and horizontal composition. It is the fundamental coherence condition of any bicategory.

Theorem 7.32 (Interchange law)**.** *Let $f, f' : \mathsf{Path}(a, b)$ and $g, g' : \mathsf{Path}(b, c)$ be paths, and let $\alpha : f \Rrightarrow f'$ and $\beta : g \Rrightarrow g'$ be 2-cells. Then there exists a 3-cell witnessing*

$$(\alpha \lhd g) \bullet_v (f' \rhd \beta) \;=\; (f \rhd \beta) \bullet_v (\alpha \lhd g').$$

In other words, the following diagram of 2-cells commutes up to a 3-cell:

$$\begin{array}{ccc}
\mathsf{trans}(f, g) & \xRightarrow{\;\alpha \lhd g\;} & \mathsf{trans}(f', g) \\
{\scriptstyle f \rhd \beta}\Big\Downarrow & & \Big\Downarrow{\scriptstyle f' \rhd \beta} \\
\mathsf{trans}(f, g') & \xRightarrow[\;\alpha \lhd g'\;]{} & \mathsf{trans}(f', g')
\end{array} \tag{7.1}$$

The diagram (7.1) is the *Godement square*: both paths around the square from $\mathsf{trans}(f, g)$ to $\mathsf{trans}(f', g')$ yield the same 2-cell (up to a 3-cell). The top-right path first whiskers α by g, then whiskers β by f'; the bottom-left path first whiskers β by f, then whiskers α by g'.

Proof. We proceed by induction on the structure of α, with a secondary induction on β where needed.

Case 1: $\alpha = \mathsf{refl}_f$. Both sides reduce to $f \triangleright \beta$:

$$\begin{aligned}(\mathsf{refl}_f \triangleleft g) \bullet_v (f \triangleright \beta) &= \mathsf{refl}_{\mathsf{trans}(f,g)} \bullet_v (f \triangleright \beta) \\ &= f \triangleright \beta,\end{aligned}$$

and

$$\begin{aligned}(f \triangleright \beta) \bullet_v (\mathsf{refl}_f \triangleleft g') &= (f \triangleright \beta) \bullet_v \mathsf{refl}_{\mathsf{trans}(f,g')} \\ &= f \triangleright \beta.\end{aligned}$$

Hence both sides are equal, and the identity 3-cell serves as witness.

Case 2: $\beta = \mathsf{refl}_g$. Symmetric to Case 1. Both sides reduce to $\alpha \triangleleft g$.

Case 3: $\alpha = \mathsf{step}(s_1)$ **and** $\beta = \mathsf{step}(s_2)$. This is the core case. We have a step $s_1 : \mathsf{Step}(f, f')$ and a step $s_2 : \mathsf{Step}(g, g')$. The left-hand side is:

$$\mathsf{step}(\mathsf{cong}_{\mathsf{left}}(s_1, g)) \bullet_v \mathsf{step}(\mathsf{cong}_{\mathsf{right}}(f', s_2)),$$

and the right-hand side is:

$$\mathsf{step}(\mathsf{cong}_{\mathsf{right}}(f, s_2)) \bullet_v \mathsf{step}(\mathsf{cong}_{\mathsf{left}}(s_1, g')).$$

Both composites represent the 2-cell from $\mathsf{trans}(f, g)$ to $\mathsf{trans}(f', g')$ obtained by applying both steps to the compound path. That these two composites are equal as 2-cells (up to 3-cells) follows from the fact that the two rewrite steps s_1 and s_2 operate on *disjoint* subterms of $\mathsf{trans}(f, g)$—s_1 acts on the first argument f while s_2 acts on the second argument g—and therefore their application commutes. The 3-cell witnessing this commutativity is provided by the critical pair analysis of the rewrite system: when two rules apply to non-overlapping redexes, they can be applied in either order with the same result (this is the *parallel moves* lemma from Chapter 5).

Case 4: $\alpha = \alpha_1 \bullet_v \alpha_2$. We use the functoriality of whiskering (Lemma 7.13). Write f'' for the intermediate path with $\alpha_1 : f \Rrightarrow f''$ and $\alpha_2 : f'' \Rrightarrow f'$. Then:

$$(\alpha_1 \bullet_v \alpha_2) \triangleleft g = (\alpha_1 \triangleleft g) \bullet_v (\alpha_2 \triangleleft g).$$

The left-hand side of the interchange equation becomes:

$$\begin{aligned}&\big((\alpha_1 \triangleleft g) \bullet_v (\alpha_2 \triangleleft g)\big) \bullet_v (f' \triangleright \beta) \\ &= (\alpha_1 \triangleleft g) \bullet_v (\alpha_2 \triangleleft g) \bullet_v (f' \triangleright \beta).\end{aligned}$$

By the inductive hypothesis applied to α_2 and β:

$$(\alpha_2 \triangleleft g) \bullet_v (f' \triangleright \beta) = (f'' \triangleright \beta) \bullet_v (\alpha_2 \triangleleft g').$$

Substituting:

$$= (\alpha_1 \triangleleft g) \bullet_v (f'' \triangleright \beta) \bullet_v (\alpha_2 \triangleleft g').$$

By the inductive hypothesis applied to α_1 and β:

$$(\alpha_1 \triangleleft g) \bullet_v (f'' \triangleright \beta) = (f \triangleright \beta) \bullet_v (\alpha_1 \triangleleft g').$$

So the expression becomes:

$$\begin{aligned} &= (f \triangleright \beta) \bullet_v (\alpha_1 \triangleleft g') \bullet_v (\alpha_2 \triangleleft g') \\ &= (f \triangleright \beta) \bullet_v ((\alpha_1 \bullet_v \alpha_2) \triangleleft g') \\ &= (f \triangleright \beta) \bullet_v ((\alpha_1 \bullet_v \alpha_2) \triangleleft g'), \end{aligned}$$

which is the right-hand side.

Case 5: $\alpha = \gamma^{-1}$. The inverse case follows from the inductive hypothesis for γ together with the functoriality of whiskering under inverses (Lemma 7.13(iii)).

Case 6: β composite or inverse. These cases are symmetric to Cases 4 and 5, using the functoriality of left whiskering. □

Remark 7.33 (The interchange law in classical terms)**.** The interchange law has a long history in category theory, where it is also known as the *middle four interchange*, the *Godement interchange*, or simply *functoriality of horizontal composition*. In the 2×2 diagram (7.1), the four entries are the 2-cells $\alpha \triangleleft g$, $f' \triangleright \beta$, $f \triangleright \beta$, and $\alpha \triangleleft g'$; the two composition orders correspond to traversing the square via the upper-right corner or the lower-left corner. The interchange law asserts that both traversals yield the same composite (up to a 3-cell).

Godement introduced this interchange rule in his treatment of natural transformations (see [98] for a modern exposition). The name "middle four interchange" comes from the fact that in a double category, the law states that in a 2×2 grid of cells, composing horizontally then vertically gives the same result as composing vertically then horizontally.

Example 7.34 (Interchange with unit rewrites)**.** Let $f, f' : \mathsf{Path}(a, b)$ and $g : \mathsf{Path}(b, c)$. Take $\alpha : f \Rrightarrow f'$ (an arbitrary 2-cell) and $\beta = \mathsf{refl}_g : g \Rrightarrow g$ (the identity 2-cell). The interchange law (Theorem 7.32) gives:

$$(\alpha \bullet_v \mathsf{refl}_{f'}) \bullet_h (\mathsf{refl}_g \bullet_v \mathsf{refl}_g) \;=\; (\alpha \bullet_h \mathsf{refl}_g) \bullet_v (\mathsf{refl}_{f'} \bullet_h \mathsf{refl}_g).$$

Both vertical compositions simplify by the unit laws: $\alpha \bullet_v \mathsf{refl}_{f'} = \alpha$ and $\mathsf{refl}_g \bullet_v \mathsf{refl}_g = \mathsf{refl}_g$. By Proposition 7.18(ii)–(iii), the right-hand side becomes $(\alpha \lhd g) \bullet_v \mathsf{refl}_{\mathsf{trans}(f',g)} = \alpha \lhd g$. Hence:

$$\alpha \bullet_h \mathsf{refl}_g \;=\; \alpha \lhd g,$$

recovering Proposition 7.18(ii) as a special case of interchange—exactly as one expects from the general theory of bicategories.

7.4.3 The Pentagon Identity

Having established that horizontal composition interchanges correctly with vertical composition, we now turn to the question of *coherence*: do the associator 2-cells $\mathfrak{a}_{f,g,h}$ behave consistently when we re-associate a composition of four or more paths? The answer is captured by the *pentagon identity*, the fundamental coherence condition introduced by Mac Lane in his work on monoidal categories [98]. This condition ensures that all possible ways of re-associating a four-fold composite of paths yield the same result.

Theorem 7.35 (Pentagon identity). *For paths $f : \mathsf{Path}(a,b)$, $g : \mathsf{Path}(b,c)$, $h : \mathsf{Path}(c,d)$, and $k : \mathsf{Path}(d,e)$, the following diagram of 2-cells commutes (up to a 3-cell):*

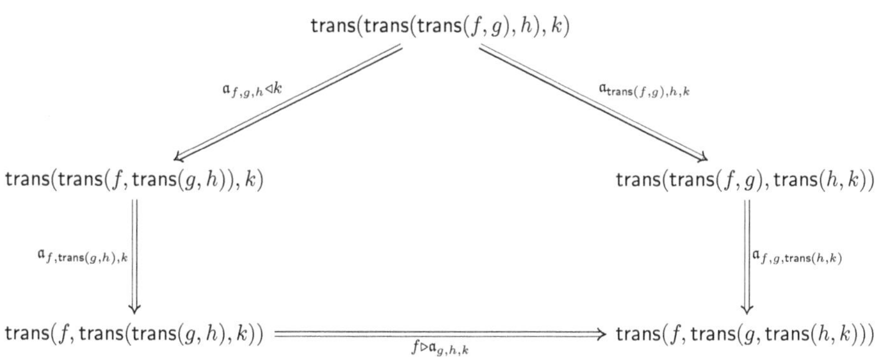

That is, the two paths around the pentagon,

$$\mathsf{trans}(\mathsf{trans}(\mathsf{trans}(f,g),h),k) \qquad \textit{and} \quad \mathsf{trans}(f,\mathsf{trans}(g,\mathsf{trans}(h,k))),$$

are related by a 3-cell.

Proof. Set

$$X_0 := \mathsf{trans}(\mathsf{trans}(\mathsf{trans}(f,g),h),k),$$
$$X_4 := \mathsf{trans}(f,\mathsf{trans}(g,\mathsf{trans}(h,k))).$$

Define the two boundary 2-cells:

$$\Pi_L := (\mathfrak{a}_{f,g,h} \triangleleft k) \bullet_v \mathfrak{a}_{f,\mathsf{trans}(g,h),k} \bullet_v (f \triangleright \mathfrak{a}_{g,h,k}) : X_0 \Rrightarrow X_4,$$
$$\Pi_R := \mathfrak{a}_{\mathsf{trans}(f,g),h,k} \bullet_v \mathfrak{a}_{f,g,\mathsf{trans}(h,k)} : X_0 \Rrightarrow X_4.$$

Each generating edge is an instance of associativity Rule R8 from Definition 4.1. The fork at X_0 is the R8/R8 self-overlap:

$$X_0 \xrightarrow{R8_{\text{in}}} \mathsf{trans}(\mathsf{trans}(f, \mathsf{trans}(g, h)), k),$$
$$X_0 \xrightarrow{R8_{\text{out}}} \mathsf{trans}(\mathsf{trans}(f, g), \mathsf{trans}(h, k)).$$

Each branch then takes one more R8 step to X_4. This is precisely the nested associativity critical pair resolved in Proposition 5.15 (item (3)), yielding a basic filler

$$\mathfrak{D}_{8,8}(f, g, h, k) : \Pi_L \Rrightarrow_3 \Pi_R.$$

To make this filler stable under all whiskered contexts that occur in the 77-rule system, we use the remaining groupoid-core rules R3–R7: overlaps of R8 with unit reductions R3–R4 are resolved by Proposition 5.15, overlaps of R8 with inverse reductions R5–R6 are resolved by Proposition 5.17, and R7 transports inverses across composites so these reductions fit the same schemas. Composing the corresponding critical-pair fillers in $\mathsf{Derivation}_3$ gives

$$\Omega_{f,g,h,k} : \Pi_L \Rrightarrow_3 \Pi_R,$$

which is exactly the 3-cell witness for the pentagon. □

The pentagon identity is one of the axioms in Mac Lane's classical coherence theorem for monoidal categories [98]; in our rewrite setting it is derived as a coherence witness. It ensures that in any expression involving multiple nested compositions, the choice of parenthesization does not matter (up to the appropriate level of coherence).

Example 7.36 (Pentagon for identity paths)**.** Taking $f = g = h = k = \mathsf{refl}_a$ in Theorem 7.35, each vertex of the pentagon is a four-fold composition of identity paths, and every associator is of the form $\mathfrak{a}_{\mathsf{refl}_a,\mathsf{refl}_a,\mathsf{refl}_a}$. Since $\mathsf{trans}(\mathsf{refl}_a, \mathsf{refl}_a)$ rewrites to refl_a by the unit rules, and the associator itself rewrites the already-trivial re-bracketing of refl_a's, all five vertices of the pentagon collapse to refl_a and every 2-cell edge becomes the identity 2-cell $\mathsf{refl}_{\mathsf{refl}_a}$. The pentagon thus commutes trivially—a useful sanity check that the identity coherences are consistent.

7.4.4 The Triangle Identity

While the pentagon identity governs the interaction of associators with one another, the *triangle identity* governs how the associator interacts with the unitors. Together, these two coherence conditions suffice to ensure that *every* diagram built from associators and unitors commutes—a fact known as Mac Lane's celebrated coherence theorem [98].

Theorem 7.37 (Triangle identity). *For paths* $f : \mathsf{Path}(a, b)$ *and* $g : \mathsf{Path}(b, c)$, *the following diagram of 2-cells commutes up to a 3-cell:*

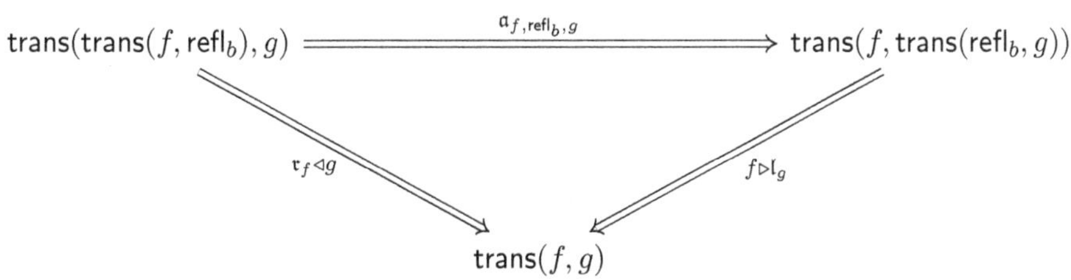

Proof. The left edge applies the right unitor $\mathfrak{r}_f$ (which eliminates the refl_b from $\mathsf{trans}(f, \mathsf{refl}_b)$) whiskered with g. The top edge applies the associator, and the right edge applies the left unitor $\mathfrak{l}_g$ (eliminating refl_b from $\mathsf{trans}(\mathsf{refl}_b, g)$) whiskered with f.

Starting from $\mathsf{trans}(\mathsf{trans}(f, \mathsf{refl}_b), g)$, both paths lead to $\mathsf{trans}(f, g)$ by eliminating the identity path refl_b. The 3-cell witnessing their equality is again obtained from the confluence of the rewrite system: both derivation sequences reduce $\mathsf{trans}(\mathsf{trans}(f, \mathsf{refl}_b), g)$ to $\mathsf{trans}(f, g)$, and the critical pair resolution between the unit rules and the associativity rule provides the connecting 3-cell. □

Example 7.38 (Triangle for identity paths). Setting $f = g = \mathsf{refl}_a$ in Theorem 7.37, the source vertex becomes $\mathsf{trans}(\mathsf{trans}(\mathsf{refl}_a, \mathsf{refl}_a), \mathsf{refl}_a)$. The left edge $\mathfrak{r}_{\mathsf{refl}_a} \triangleleft \mathsf{refl}_a$ removes the inner refl_a via the right unit rule, the top edge $\mathfrak{a}_{\mathsf{refl}_a,\mathsf{refl}_a,\mathsf{refl}_a}$ re-brackets, and the right edge $\mathsf{refl}_a \triangleright \mathfrak{l}_{\mathsf{refl}_a}$ removes refl_a via the left unit rule. All three vertices collapse to refl_a (or terms rewriting to it), and every 2-cell becomes the identity. Thus the triangle commutes trivially, paralleling Example 7.36.

7.4.5 Naturality of Structural 2-Cells

The associators and unitors are not merely isolated 2-cells; they are *natural* in their arguments. This means that they commute with arbitrary 2-cells applied to their arguments.

Theorem 7.39 (Naturality of the associator). *For 2-cells $\alpha : f \Rrightarrow f'$, $\beta : g \Rrightarrow g'$, and $\gamma : h \Rrightarrow h'$, the following diagram commutes up to a 3-cell:*

$$\begin{array}{ccc}
\mathsf{trans}(\mathsf{trans}(f, g), h) & \xRightarrow{\ \mathfrak{a}_{f,g,h}\ } & \mathsf{trans}(f, \mathsf{trans}(g, h)) \\
{\scriptstyle (\alpha \bullet_h \beta) \bullet_h \gamma} \Big\Downarrow & & \Big\Downarrow {\scriptstyle \alpha \bullet_h (\beta \bullet_h \gamma)} \\
\mathsf{trans}(\mathsf{trans}(f', g'), h') & \xRightarrow[\ \mathfrak{a}_{f',g',h'}\ ]{} & \mathsf{trans}(f', \mathsf{trans}(g', h'))
\end{array}$$

Proof. By induction on the structure of α, β, and γ, using the functoriality of whiskering and the interchange law (Theorem 7.32). The key observation is that the associator is defined by a single rewrite rule (Rule 8), and the congruence rules of the rewrite system ensure that this rule commutes with any rewrite applied to the subterms f, g, h. □

Theorem 7.40 (Naturality of the unitors). *For a 2-cell $\alpha : f \Rrightarrow f'$ between paths $f, f' : \mathsf{Path}(a, b)$:*

(i) The left unitor is natural:

$$\begin{array}{ccc}
\mathsf{trans}(\mathsf{refl}_a, f) & \xRightarrow{\ \mathfrak{l}_f\ } & f \\
{\scriptstyle \mathsf{refl}_{\mathsf{refl}_a} \bullet_h \alpha} \Big\Downarrow & & \Big\Downarrow {\scriptstyle \alpha} \\
\mathsf{trans}(\mathsf{refl}_a, f') & \xRightarrow[\ \mathfrak{l}_{f'}\ ]{} & f'
\end{array}$$

(ii) The right unitor is natural:

$$\begin{array}{ccc}
\mathsf{trans}(f, \mathsf{refl}_b) & \xRightarrow{\ \mathfrak{r}_f\ } & f \\
{\scriptstyle \alpha \bullet_h \mathsf{refl}_{\mathsf{refl}_b}} \Big\Downarrow & & \Big\Downarrow {\scriptstyle \alpha} \\
\mathsf{trans}(f', \mathsf{refl}_b) & \xRightarrow[\ \mathfrak{r}_{f'}\ ]{} & f'
\end{array}$$

Proof. For part (i), both composites around the square are 2-cells from $\mathsf{trans}(\mathsf{refl}_a, f)$ to f'. The top-right path first applies the left unit rule to eliminate refl_a and then applies α. The bottom-left path first whiskers α into the compound path and then applies the left unit rule. These composites are equal because the left unit rule (Rule 3) commutes with the congruence rules that propagate α through trans: applying a rewrite to the second argument of $\mathsf{trans}(\mathsf{refl}_a, -)$ and then eliminating refl_a is the same as first eliminating refl_a and then applying the rewrite directly. Part (ii) is symmetric. □

7.4.6 Summary: The Coherence Package

We conclude this chapter by collecting the coherence conditions into a single statement.

Theorem 7.41 (Bicategory coherence). *The structure $\mathfrak{B}_A$ (Definition 7.20), equipped with the associators $\mathfrak{a}$, left unitors $\mathfrak{l}$, and right unitors $\mathfrak{r}$, satisfies:*

(i) The pentagon identity (Theorem 7.35).

(ii) The triangle identity (Theorem 7.37).

(iii) Naturality of all structural 2-cells (Theorems 7.39 and 7.40).

(iv) The interchange law (Theorem 7.32).

Consequently, $\mathfrak{B}_A$ is a bicategory in the sense of Bénabou [12].

Proof. This is a summary of the results established in this section. The pentagon identity was proved in Theorem 7.35, the triangle identity in Theorem 7.37, naturality in Theorems 7.39 and 7.40, and the interchange law in Theorem 7.32. □

Remark 7.42 (Mac Lane's coherence theorem). Mac Lane's celebrated coherence theorem for monoidal categories [98] states that the pentagon and triangle identities suffice to guarantee that *all* diagrams built from associators and unitors commute. The analogous result for bicategories, established by Mac Lane and Paré (see [98] for a modern treatment), shows that in a bicategory satisfying the pentagon and triangle identities, every diagram of structural 2-cells commutes.

For our bicategory $\mathfrak{B}_A$, this means that any two ways of re-associating and inserting/removing identity paths in a compound composition of paths yield the same result, up to a canonical 3-cell. This is a powerful "coherence" guarantee: although the algebra of computational paths is non-strict, it is as well-behaved as a non-strict algebra can be.

Remark 7.43 (From bicategory to ω-groupoid)**.** The bicategory structure established in this chapter is only the beginning. The *guiding principle* of the entire higher-dimensional development is this: **all higher structure is extracted from rewriting.** A 1-cell is a rewriting witness between terms; a 2-cell is a rewriting witness between 1-cells; a 3-cell is a rewriting witness between 2-cells; and so on. Coherence at each level is not postulated but *derived* from the confluence of the underlying rewrite system. This principle drives Chapters 8 and 9, where we show that the 3-cells introduced in Definition 7.30 provide the last genuinely non-trivial coherence layer: pentagon, triangle, interchange, inverse, contravariance, and related laws are all witnessed by explicit 3-cells. Chapter 9 then proves that parallel 3-cells are related by 4-cells, parallel 4-cells by 5-cells, and so on, yielding stabilisation above dimension 3 and hence a genuine weak ω-groupoid in the sense of Batanin [11] and Leinster [99].

The key point is that all the higher coherence—the pentagon, the triangle, the interchange law, and their higher analogues—are *automatic*: these appear inside the derivation tower itself (not as extra axioms or proposition-level quotients), and their joinability content is driven by the confluence analysis of the underlying expression system. This is the distinctive feature of the computational paths approach: where other constructions of ω-groupoids from type theory (such as those of Lumsdaine [105] and van den Berg–Garner [153]) must verify coherence at each dimension separately, our approach derives it from the single fact that the rewrite system is confluent and terminating.

Summary

This chapter moves from paths to paths between paths. Its main result is that derivations between computational paths, written $\mathsf{Derivation}_2$, form the next layer of higher-dimensional algebra above the groupoid of Chapter 6. We defined 2-cells inductively from identity, rewrite steps, inverses, and vertical composition (Section 7.1), establishing a vertical groupoid on every hom-set $\mathsf{Path}(a, b)$. We then introduced horizontal composition via left and right whiskering (Section 7.2), promoting the structure to a bicategory $\mathfrak{B}_A$ in the sense of Bénabou, with associators, left unitors, and right unitors as invertible 2-cells witnessing the non-strict algebraic laws. Section 7.3 connected this abstract 2-cell machinery to the concrete rewrite equivalence relation $\sim_{\mathsf{rw}}$ from Chapter 5, showing that $\mathsf{Derivation}_2$ and RwEq are structurally isomorphic.

The coherence conditions governing this bicategory were established in Section 7.4: the interchange law (Theorem 7.32), the pentagon identity (Theorem 7.35), the triangle identity (Theorem 7.37), and the naturality of all structural 2-cells (Theorems 7.39 and 7.40). These conditions are not imposed as axioms but *derived* from the confluence of the rewrite system, using critical pair analysis and the parallel moves lemma. The resulting bicategory coherence theorem (Theorem 7.41) guarantees that any two ways of re-associating and inserting or removing identity paths yield the same result up to a canonical 3-cell. This chapter thus provides the foundation for the 3-cell coherence analysis of Chapter 8 and the weak ω-groupoid construction of Chapter 9.

Historical Notes.

The present chapter sits at the meeting point of bicategorical coherence and higher-dimensional rewriting. The notion of 2-category was introduced by Ehresmann in the early 1960s as part of his programme to develop the theory of structured categories. The weaker notion of *bicategory* was formulated by Bénabou [12] in 1967, motivated by examples from algebra (such as bimodules) where the composition law is only associative up to natural isomorphism.

The interchange law has its origins in the work of Godement on sheaf theory, where it appeared as a property of natural transformations between functors. Mac Lane elevated it to a fundamental axiom of higher category theory in his treatment of monoidal categories and 2-categories [98].

The connection between rewriting and higher-dimensional algebra was first explored by Squier [141], who showed that the existence of a finite confluent and terminating rewrite system for a monoid implies homological finiteness conditions. Guiraud and Malbos [78] subsequently developed a full theory of *higher-dimensional rewriting* in which the 2-cells of a rewrite system are treated as generators of a 2-category. Our 2-cells can be seen as an instance of their framework, specialised to the rewrite system on computational paths.

The observation that rewrite equivalence generates a groupoid (rather than merely a category) is specific to our setting, where every rewrite step is invertible at the 2-cell level. This invertibility is not automatic in general rewriting—it depends on the fact that our rewrite rules are *equations* (algebraic identities) rather than mere reductions. The groupoid structure of identity types was first identified by Hofmann and Streicher [86] in their groundbreaking 1998 paper, which introduced the groupoid model of type theory. Our 2-cells provide a concrete, syntactic realisation of the higher-dimensional structure that Hofmann and Streicher identified semantically.

Formalization Notes. The results of this chapter are formalised in the following files of the accompanying Lean development (`https://github.com/Arthur742Ramos/ComputationalPathsLean`):

- `Path/OmegaGroupoid/HigherCellPaths.lean` — the higher-cell families built over rewrite derivations

- `Path/OmegaGroupoid/TwoCategoryStructure.lean` — vertical composition, whiskering, horizontal composition, and interchange-style 2-cell structure
- `Path/DoubleCategoryPaths.lean` — a compatible square-based interface for 2-dimensional composition

Key definitions. The $\mathsf{Derivation}_2$ type captures 2-cells as explicit rewrite derivations; in the current Lean formalisation it lives in `Type (u + 1)`, in parallel with the auxiliary `RwEq` presentation:

```
/-- 2-cells: Rewrite derivations between paths -/
inductive Derivation2 {a b : A} : Path a b -> Path a b ->
    Type (u + 1) where
  | refl (p : Path a b) : Derivation2 p p
  | step {p q : Path a b} : Step p q -> Derivation2 p q
  | inv {p q : Path a b} : Derivation2 p q -> Derivation2
      q p
  | vcomp {p q r : Path a b} :
      Derivation2 p q -> Derivation2 q r -> Derivation2 p
         r
```

Whiskering operations extend 2-cells horizontally:

```
/-- Left whiskering: f . _ applied to 2-cell -/
noncomputable def whiskerLeft (f : Path a b)
    {p q : Path b c} (alpha : Derivation2 p q) :
    Derivation2 (Path.trans f p) (Path.trans f q) :=
  match alpha with
  | .refl _ => .refl _
  | .step s => .step (Step.trans_congr_right f s)
  | .inv d => .inv (whiskerLeft f d)
  | .vcomp d1 d2 => .vcomp (whiskerLeft f d1) (
     whiskerLeft f d2)
```

Chapter 8

Three-Cells and Coherence

> *"The coherence theorem says that all diagrams that should commute, do commute."*
>
> —Saunders Mac Lane [98]

In topology, when two paths from a to b are homotopic, there may be many different homotopies between them. Two such homotopies can themselves be related by a *homotopy between homotopies*—a continuous deformation of one 2-dimensional disk into another. This is a 3-cell: a witness that two witnesses of path equivalence are themselves equivalent.

The reason to care about such objects is *coherence*. When we work with algebraic structures that satisfy laws "up to isomorphism" rather than "on the nose," we face an unavoidable question: are the isomorphisms themselves unique? If there are two ways to prove that diagram X commutes, are those two proofs equivalent? The theory of 3-cells gives us the language to state and answer such questions precisely.

A concrete example makes the stakes vivid. Suppose we have four composable paths f, g, h, k and wish to simplify the fully left-associated composite $((f \cdot g) \cdot h) \cdot k$ to the fully right-associated form $f \cdot (g \cdot (h \cdot k))$. We can re-associate in two different ways. The *upper route* first reassociates the outer three factors to get $(f \cdot g) \cdot (h \cdot k)$, then reassociates again. The *lower route* first reassociates the inner three factors, moving to $(f \cdot (g \cdot h)) \cdot k$, then continues outward. Both routes produce 2-cells from the same source to the same target—but these 2-cells are, in general, *different* derivations. Is this difference meaningful, or are the two routes "morally the same"? A 3-cell—the pentagon witness of Theorem 7.35—answers: yes, they are coherently equivalent. Without 3-cells, we would have no way to state this equivalence, let alone prove it. Coherence at dimension 2 would remain an article of faith rather than a theorem.

Chapter 7 introduced 2-cells as derivations between computational paths and established that these 2-cells organize into a bicategory. The associators and unitors of that bicategory were shown to satisfy the pentagon and triangle identities up to 3-cells (Section 7.4). We now take the next step in the higher-dimensional analysis: we study 3-cells as first-class algebraic objects, define their composition operations, and prove the coherence laws that govern their interaction.

The guiding principle articulated in Remark 7.43—that all higher structure is extracted from rewriting—continues to drive the development. A path is a first-order rewriting witness, a 2-cell is a rewriting witness between paths, and a 3-cell is a rewriting witness between 2-cells. Coherence is not postulated externally; it is generated by the confluence properties of the rewrite system itself.

The chapter has five goals:

(i) to formalise 3-cells as third-order derivations and define their three composition operations $\circ_2, \circ_1, \circ_0$;

(ii) to establish interchange laws between these operations;

(iii) to derive the 3-dimensional coherence package (pentagon and triangle identities) attached to associators and unitors;

(iv) to connect the resulting structure to Mac Lane's coherence theorem for monoidal categories;

(v) to exhibit concrete 3-cell computations arising from the 77-rule rewrite system.

In Chapter 7, the basic 3-cells witnessing bicategorical coherence were first exhibited concretely (interchange, pentagon, triangle, naturality). In the formalisation, these are packaged as primitive constructors of the 3-cell type. Here we organize them into a systematic theory, which serves as the conceptual bridge to the weak ω-groupoid theorem of Chapter 9.

8.1 3-Cells as Third-Order Derivations

8.1.1 The Idea of a 3-Cell

To motivate the formal definition, consider a familiar situation from Chapter 7. Given two parallel paths $p, q : \mathsf{Path}(a, b)$, a 2-cell $\alpha : p \Rrightarrow q$ witnesses their equivalence through a specific chain of rewrites. But there may be many different chains that accomplish the same transformation—different *proofs* that p and q are equivalent. Are these proofs themselves equivalent?

This question is not merely philosophical. When we construct algebraic structures from paths and 2-cells, we routinely make choices: which associativity witness to use, which order to apply simplification rules. If our theory is to be well-behaved, different choices must yield equivalent results. A 3-cell is precisely the witness that two 2-cells are equivalent—a "derivation between derivations."

8.1.2 Globular Boundaries at Dimension 3

A 3-cell lives between parallel 2-cells. Concretely, fix $a, b : A$ and parallel paths $p, q : \mathsf{Path}(a, b)$. Let $\alpha, \beta : p \Rrightarrow q$ be 2-cells. A 3-cell

$$\Gamma : \alpha \Rrightarrow_3 \beta$$

may be visualised as a "volume" whose boundary consists of the two 2-cells α and β.

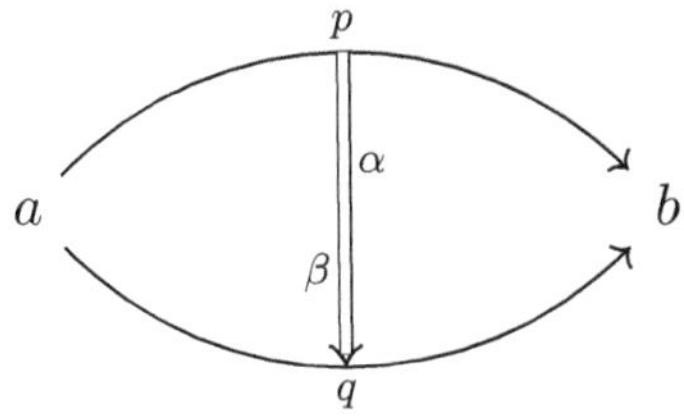

In this diagram, p and q are the 1-cell boundaries (the paths), while α and β are the 2-cell boundaries (the derivations that the 3-cell connects). The 3-cell Γ itself is the "filling" between these two 2-cells—a higher-dimensional witness that α and β are equivalent ways of establishing that $p \Rrightarrow q$.

The source and target of a 3-cell are 2-cells, and these in turn have source and target paths. We therefore have boundary operators

$$\partial_2^-, \partial_2^+ : \mathsf{Derivation}_3 \to \mathsf{Derivation}_2, \quad \partial_1^-, \partial_1^+ : \mathsf{Derivation}_2 \to \mathsf{Path},$$

satisfying globular identities

$$\partial_1^- \partial_2^- = \partial_1^- \partial_2^+, \qquad \partial_1^+ \partial_2^- = \partial_1^+ \partial_2^+,$$

expressing parallelism of α and β. These identities ensure that we cannot accidentally try to compare 2-cells that live between different pairs of paths.

8.1.3 Formal Definition

With this intuition in place, we can give the formal definition. The structure mirrors Definition 7.1 from Chapter 7, lifted one dimension higher.

Definition 8.1 (Third-order derivation type)**.** Let $\alpha, \beta : p \Rrightarrow q$ be parallel 2-cells. A *3-cell* from α to β is an element of $\mathsf{Derivation}_3(\alpha, \beta)$, generated inductively by:

(i) **Identity:** $\mathsf{refl}_\alpha : \alpha \Rrightarrow_3 \alpha$.

(ii) **Step:** for each elementary 3-step $s : \mathsf{Step}_3(\alpha, \beta)$, $\mathsf{step}_3(s) : \alpha \Rrightarrow_3 \beta$.

(iii) **Inverse:** if $\Gamma : \alpha \Rrightarrow_3 \beta$, then $\Gamma^{-1} : \beta \Rrightarrow_3 \alpha$.

(iv) **2-composition:** if $\Gamma : \alpha \Rrightarrow_3 \beta$ and $\Delta : \beta \Rrightarrow_3 \gamma$, then

$$\Gamma \circ_2 \Delta : \alpha \Rrightarrow_3 \gamma.$$

Remark 8.2 (Elementary 3-steps)**.** The reader may wonder: what exactly *is* an elementary 3-step? In the present chapter, $\mathsf{Step}_3(\alpha, \beta)$ should be read as the schematic placeholder for the elementary coherence moves needed in the structural fragment. An element of $\mathsf{Step}_3(\alpha, \beta)$ records one basic transformation between 2-cells. There are two main sources:

(a) **Functoriality steps:** when a rewrite occurs at one of the path positions appearing in α and β, this induces a corresponding transformation of the 2-cells themselves.

(b) **Critical-pair resolution steps:** when two 2-cell derivation fragments can be transformed into each other by exchanging independent reductions.

The second source is particularly important: in the structural fragment it is the origin of interchange and all higher coherence phenomena. When two rewrites can be performed in either order with the same final result, the two derivation paths must be connected by a 3-cell—this is the combinatorial manifestation of the topological fact that homotopies between homotopies detect when different "routes" through rewrite-space end up at the same place.

Proposition 8.3 (Well-typedness of constructors)**.** *Each constructor in Definition 8.1 preserves globular boundaries. In particular, if* $\Gamma : \alpha \Rrightarrow_3 \beta$*, then* α *and* β *are parallel 2-cells in a common hom-type* $\mathsf{Derivation}_2(p, q)$.

Proof. Identity and inverse are immediate. For a step $\mathsf{step}_3(s)$, the typing of s already enforces parallel boundaries. For 2-composition, the middle 2-cell is shared: $\Gamma : \alpha \Rrightarrow_3 \beta$ and $\Delta : \beta \Rrightarrow_3 \gamma$, so the codomain of Γ coincides with the domain of Δ. Hence $\Gamma \circ_2 \Delta$ has source α and target γ, and all boundary maps remain defined. □

8.1.4 Vertical Groupoid at Dimension 3

Just as 2-cells form a groupoid under vertical composition (Proposition 7.8), 3-cells form a groupoid one level higher. This pattern—groupoid structure at each level—is a hallmark of ω-groupoid theory and will be central to Chapter 9. Here the references to 4-cells are schematic: they indicate that equalities between 3-cell composites are themselves witnessed one dimension higher. Chapter 9 defines that next layer systematically; in the present chapter we only use the general pattern that coherence at dimension 3 is controlled by one higher comparison.

Theorem 8.4 (3-cells form a weak groupoid). *For fixed parallel 2-cells in* $\mathsf{Derivation}_2(p, q)$*, the type of 3-cells with composition* $\circ_2$*, identity* refl*, and inverse* $(-)^{-1}$ *satisfies:*

(i) associativity of $\circ_2$*;*

(ii) left and right unit laws;

(iii) left and right inverse laws;

(iv) congruence of $\circ_2$ *with respect to 4-cell equivalence.*

All equations hold up to 4-cells.

Proof. The first three clauses are proved by induction on 3-cell constructors, exactly as for 2-cells:

- identities are neutral by constructor equations;
- associativity follows by structural recursion on the outer composition term;
- inverse laws are constructor equalities followed by one simplification step.

For congruence, if $\Gamma \Rrightarrow_4 \Gamma'$ and $\Delta \Rrightarrow_4 \Delta'$ then replacing one representative by the other inside a composite corresponds to a finite sequence of congruence steps in the 4-dimensional rewrite relation. Thus $\Gamma \circ_2 \Delta \Rrightarrow_4 \Gamma' \circ_2 \Delta'$. □

The following example illustrates how 3-cells arise from the coherence witnesses established in Chapter 7.

Example 8.5 (A basic 3-cell from the triangle identity). Let $f : \mathsf{Path}(a, b)$ and $g : \mathsf{Path}(b, c)$. Consider the two 2-cells from $\mathsf{trans}(\mathsf{trans}(f, \mathsf{refl}_b), g)$ to $\mathsf{trans}(f, g)$:

$$\alpha = (\mathfrak{r}_f \triangleleft g),$$
$$\beta = \mathfrak{a}_{f,\mathsf{refl}_b,g} \bullet_v (f \triangleright \mathfrak{l}_g).$$

The first 2-cell, α, eliminates the identity refl_b using the right unitor and then whiskers with g. The second 2-cell, β, first reassociates, then eliminates refl_b using the left unitor. Both 2-cells have the same source and target, but they accomplish the transformation via different routes.

Theorem 7.37 provides a 3-cell $\Gamma_{\triangleleft} : \alpha \Rrightarrow_3 \beta$. This is the canonical *triangle witness*—it asserts that the two ways of eliminating an inserted identity are coherent.

8.2 Three Composition Operations on 3-Cells

One of the most striking features of higher-dimensional algebra is the proliferation of composition operations. At dimension 1, paths compose in one way (transitivity). At dimension 2, 2-cells have two composition operations: vertical ($\bullet_v$) and horizontal ($\bullet_h$). At dimension 3, we encounter *three* independent ways to compose 3-cells.

Before diving into the formal definitions, it helps to visualise the situation. Think of a 3-cell as a small cube whose faces are 2-cells. You can stack cubes along any of three axes—and the three composition operations correspond to stacking along each axis in turn. The interchange laws of Section 8.3 will then say that it does not matter in which order you stack along different axes: independent stackings commute. This geometric intuition is the surest guide through the formal details that follow.

To understand why, think geometrically. A 3-cell is a "volume" filling the space between two 2-cells. In a 3-dimensional picture:

- $\circ_2$ **(composition along 2-cells):** stack two volumes vertically, gluing along a shared 2-cell face.
- $\circ_1$ **(composition induced by vertical 2-cell composition):** compose volumes "in the direction of paths."
- $\circ_0$ **(composition induced by horizontal 2-cell composition):** compose volumes "in the direction of 0-cells."

In a globular 3-dimensional setting, a 3-cell admits three independent composition directions. We denote them by

$$\circ_2, \quad \circ_1, \quad \circ_0,$$

corresponding to composition along 2-, 1-, and 0-dimensional boundaries. Only $\circ_2$ was primitive in Definition 8.1; the other two are induced by composition operations on 2-cells.

8.2.1 $\circ_2$: Composition Along 2-Cell Boundaries

This is the "obvious" composition: given a 3-cell from α to β and another from β to γ, compose them to get a 3-cell from α to γ. This was already introduced as the primitive composition in Definition 8.1:

$$\Gamma : \alpha \Rrightarrow_3 \beta, \quad \Delta : \beta \Rrightarrow_3 \gamma \quad \Longmapsto \quad \Gamma \circ_2 \Delta : \alpha \Rrightarrow_3 \gamma.$$

Its laws were established in Theorem 8.4. This composition corresponds to sequential chaining of derivations—first transform α into β, then transform β into γ.

8.2.2 $\circ_1$: Composition Induced by Vertical Composition of 2-Cells

The second composition operation arises from the vertical composition of 2-cells. If we can transform α_0 to α_1 and independently transform β_0 to β_1, and if the target of each α_i matches the source of each β_i, then we should be able to combine these transformations.

The key construction is *2-whiskering*: attaching a fixed 2-cell to one side of a 3-cell. Intuitively, 2-whiskering says: "if we have a transformation between 2-cells, and we glue on an extra 2-cell that does not participate in the transformation, the result is still a valid transformation." This mirrors the role of ordinary whiskering at dimension 2 (Definition 7.11), lifted one level higher.

Definition 8.6 (2-whiskering of 3-cells)**.** Let $\Gamma : \alpha \Rrightarrow_3 \alpha'$ with $\alpha, \alpha' : f \Rrightarrow g$, and let $\beta : g \Rrightarrow h$ be a 2-cell. The *right 2-whiskering* of Γ by β is

$$\Gamma \lhd_v \beta : \alpha \bullet_v \beta \Rrightarrow_3 \alpha' \bullet_v \beta.$$

Dually, for $\beta : e \Rrightarrow f$, define

$$\beta \rhd_v \Gamma : \beta \bullet_v \alpha \Rrightarrow_3 \beta \bullet_v \alpha'.$$

Both operations are defined by recursion on constructors of Γ, using the congruence rules for $\bullet_v$.

With 2-whiskering in hand, we can define the $\circ_1$ composition by analogy with horizontal composition of 2-cells (Definition 7.16): first whisker one 3-cell to extend its boundary, then compose the result vertically with the other 3-cell similarly extended.

Definition 8.7 ($\circ_1$ composition)**.** Suppose

$$\Gamma : \alpha_0 \Rrightarrow_3 \alpha_1, \qquad \Delta : \beta_0 \Rrightarrow_3 \beta_1,$$

with $\alpha_i : f \Rrightarrow g$, $\beta_i : g \Rrightarrow h$ for $i = 0, 1$. Define

$$\Gamma \circ_1 \Delta \overset{\text{def}}{=} (\Gamma \lhd_v \beta_0) \circ_2 (\alpha_1 \rhd_v \Delta),$$

a 3-cell from $\alpha_0 \bullet_v \beta_0$ to $\alpha_1 \bullet_v \beta_1$.

The definition may look asymmetric—we whisker Γ with β_0 on the right, then whisker Δ with α_1 on the left. This is the same interchange pattern we saw for 2-cells in Definition 7.16. We could equally well have defined $\Gamma \circ_1 \Delta$ as $(\alpha_0 \rhd_v \Delta) \circ_2 (\Gamma \lhd_v \beta_1)$; the interchange law (Theorem 8.15) will show these are equivalent up to 4-cells.

Proposition 8.8 (Functoriality of $\circ_1$)**.** *The operation $\circ_1$ preserves identities, 2-composition, and inverses:*

(i) $\mathsf{refl}_\alpha \circ_1 \mathsf{refl}_\beta = \mathsf{refl}_{\alpha \bullet_v \beta}$.

(ii) $(\Gamma_1 \circ_2 \Gamma_2) \circ_1 (\Delta_1 \circ_2 \Delta_2) \Rrightarrow_4 (\Gamma_1 \circ_1 \Delta_1) \circ_2 (\Gamma_2 \circ_1 \Delta_2)$.

(iii) $(\Gamma \circ_1 \Delta)^{-1} \Rrightarrow_4 (\alpha_1 \rhd_v \Delta^{-1}) \circ_2 (\Gamma^{-1} \lhd_v \beta_0)$.

Proof. (i) follows immediately from Definition 8.7 and identity-preservation of whiskering. For (ii), expand both sides via Definition 8.7, then use associativity of $\circ_2$ and functoriality of 2-whiskering to rebracket terms. The middle rebracketing is connected by a canonical 4-cell from Theorem 8.4. For (iii), compute:

$$\begin{aligned} (\Gamma \circ_1 \Delta)^{-1} &= \big((\Gamma \lhd_v \beta_0) \circ_2 (\alpha_1 \rhd_v \Delta)\big)^{-1} \\ &\Rrightarrow_4 (\alpha_1 \rhd_v \Delta)^{-1} \circ_2 (\Gamma \lhd_v \beta_0)^{-1} \\ &\Rrightarrow_4 (\alpha_1 \rhd_v \Delta^{-1}) \circ_2 (\Gamma^{-1} \lhd_v \beta_0), \end{aligned}$$

which has the expected inverted boundaries $\alpha_1 \bullet_v \beta_1 \Rrightarrow_3 \alpha_0 \bullet_v \beta_0$. □

Example 8.9 ($\circ_1$ of identity 3-cells)**.** For 2-cells $\alpha : f \Rrightarrow g$ and $\beta : g \Rrightarrow h$, the $\circ_1$ composition of the identity 3-cells is:

$$\mathsf{refl}_\alpha \circ_1 \mathsf{refl}_\beta = (\mathsf{refl}_\alpha \lhd_v \beta) \circ_2 (\alpha \rhd_v \mathsf{refl}_\beta) = \mathsf{refl}_{\alpha \bullet_v \beta} \circ_2 \mathsf{refl}_{\alpha \bullet_v \beta} = \mathsf{refl}_{\alpha \bullet_v \beta}.$$

This confirms that "doing nothing to each factor" results in "doing nothing to the composite"—precisely the identity-preservation property of Proposition 8.8(i).

8.2.3 $\circ_0$: Composition Induced by Horizontal Composition of 2-Cells

The third and final composition direction is induced by horizontal composition $\bullet_h$ from Definition 7.16. This corresponds to composing 3-cells "in the direction of 0-cells"—when we have transformations happening at different points along a path.

Definition 8.10 (Horizontal whiskering of 3-cells). Let $\Gamma : \alpha \Rrightarrow_3 \alpha'$ with $\alpha, \alpha' : f \Rrightarrow f'$ and $\beta : g \Rrightarrow g'$. Define

$$\Gamma \triangleleft_h \beta : \alpha \bullet_h \beta \Rrightarrow_3 \alpha' \bullet_h \beta.$$

Dually, for fixed α and varying β, define

$$\alpha \triangleright_h \Delta : \alpha \bullet_h \beta \Rrightarrow_3 \alpha \bullet_h \beta'.$$

Both are defined by recursion and congruence for $\bullet_h$.

Definition 8.11 ($\circ_0$ composition). Given

$$\Gamma : \alpha_0 \Rrightarrow_3 \alpha_1, \qquad \Delta : \beta_0 \Rrightarrow_3 \beta_1,$$

with $\alpha_i : f \Rrightarrow f'$ and $\beta_i : g \Rrightarrow g'$, define

$$\Gamma \circ_0 \Delta \stackrel{\text{def}}{=} (\Gamma \triangleleft_h \beta_0) \circ_2 (\alpha_1 \triangleright_h \Delta).$$

Thus

$$\Gamma \circ_0 \Delta : \alpha_0 \bullet_h \beta_0 \Rrightarrow_3 \alpha_1 \bullet_h \beta_1.$$

Proposition 8.12 (Unit laws for $\circ_0$). *For any composable 3-cells,*

$$\begin{aligned} \mathsf{refl}_\alpha \circ_0 \mathsf{refl}_\beta &= \mathsf{refl}_{\alpha \bullet_h \beta}, \\ \Gamma \circ_0 \mathsf{refl}_\beta &\Rrightarrow_4 \Gamma \triangleleft_h \beta, \\ \mathsf{refl}_\alpha \circ_0 \Delta &\Rrightarrow_4 \alpha \triangleright_h \Delta. \end{aligned}$$

Proof. By Definition 8.11 and Proposition 7.18 at the 2-cell level. The first equation is direct. For the second and third, one of the whiskered factors is an identity 3-cell, and the claim follows from the corresponding right/left unit law of $\circ_2$. □

Theorem 8.13 (Three-dimensional composition package). *The operations $\circ_2, \circ_1, \circ_0$ make parallel 3-cells into a weak 3-dimensional composition system satisfying:*

(i) each $\circ_i$ is associative up to 4-cells;

(ii) *each* $\circ_i$ *has unit 3-cells;*

(iii) *each* $\circ_i$ *is compatible with inverses;*

(iv) *all compatibility data are natural in boundary 2-cells.*

Proof. For $\circ_2$, this is Theorem 8.4. For $\circ_1$ and $\circ_0$, expand definitions (Definitions 8.7 and 8.11) and apply functoriality of whiskering, associativity/unit of $\circ_2$, and bicategorical functoriality at the 2-cell level (Section 7.2). Naturality follows because each operation is built from functorial constructors. □

Remark 8.14 (The pattern continues)**.** At dimension 4, we would have four composition operations; at dimension n, we have n operations. This proliferation is one of the main challenges in higher category theory, and one of the main reasons coherence theorems are so valuable: they tell us that despite the combinatorial explosion of operations, all the ways of composing ultimately agree.

8.3 Interchange Laws in Dimension 3

With three composition operations on 3-cells, a natural question arises: how do they interact? The answer is given by *interchange laws*, which state that independent compositions can be performed in either order.

The interchange laws are the heart of higher categorical coherence. They express the geometric intuition that if two operations act on "different parts" of a structure, the order in which we perform them should not matter. This is the higher-dimensional analogue of the commutativity of independent operations in algebra, but stated at the level of witnesses rather than equalities.

At dimension 2, Theorem 7.32 compared two composites of whiskerings. At dimension 3, we must compare composites involving $\circ_2, \circ_1, \circ_0$. The resulting laws are higher analogues of Godement interchange.

Theorem 8.15 ((2,1)-interchange)**.** *Let* $\Gamma_1, \Gamma_2, \Delta_1, \Delta_2$ *be composable 3-cells with matching boundaries. Then*

$$(\Gamma_1 \circ_2 \Gamma_2) \circ_1 (\Delta_1 \circ_2 \Delta_2) \Rrightarrow_4 (\Gamma_1 \circ_1 \Delta_1) \circ_2 (\Gamma_2 \circ_1 \Delta_2).$$

In words: performing sequential compositions first, then combining them via $\circ_1$*, is equivalent to combining corresponding pieces via* $\circ_1$ *first, then composing sequentially.*

Proof. Expand the left-hand side using Definition 8.7:

$$\begin{aligned}&(\Gamma_1 \circ_2 \Gamma_2) \circ_1 (\Delta_1 \circ_2 \Delta_2)\\ &= \big((\Gamma_1 \circ_2 \Gamma_2) \lhd_v \beta_0\big) \circ_2 \big(\alpha_2 \rhd_v (\Delta_1 \circ_2 \Delta_2)\big).\end{aligned}$$

By functoriality of whiskering,

$$(\Gamma_1 \circ_2 \Gamma_2) \lhd_v \beta_0 \Rrightarrow_4 (\Gamma_1 \lhd_v \beta_0) \circ_2 (\Gamma_2 \lhd_v \beta_0),$$

and similarly for the right whiskering term. Reassociating $\circ_2$ produces

$$(\Gamma_1 \lhd_v \beta_0) \circ_2 (\alpha_1 \rhd_v \Delta_1) \circ_2 (\Gamma_2 \lhd_v \beta_1) \circ_2 (\alpha_2 \rhd_v \Delta_2),$$

which factors exactly as

$$(\Gamma_1 \circ_1 \Delta_1) \circ_2 (\Gamma_2 \circ_1 \Delta_2).$$

Every rebracketing is justified by associativity of $\circ_2$ and yields the required 4-cell. □

Theorem 8.16 ((2,0)-interchange). *For composable 3-cells with compatible horizontal boundaries,*

$$(\Gamma_1 \circ_2 \Gamma_2) \circ_0 (\Delta_1 \circ_2 \Delta_2) \Rrightarrow_4 (\Gamma_1 \circ_0 \Delta_1) \circ_2 (\Gamma_2 \circ_0 \Delta_2).$$

Proof. The proof is identical in form to the proof of Theorem 8.15, with $\lhd_h, \rhd_h$ replacing $\lhd_v, \rhd_v$. The key ingredient is congruence of horizontal composition $\bullet_h$ under 3-cell rewriting. □

Theorem 8.17 ((1,0)-interchange). *For compatible 3-cells,*

$$(\Gamma_1 \circ_1 \Gamma_2) \circ_0 (\Delta_1 \circ_1 \Delta_2) \Rrightarrow_4 (\Gamma_1 \circ_0 \Delta_1) \circ_1 (\Gamma_2 \circ_0 \Delta_2).$$

Proof. Write both sides in terms of $\circ_2$ and whiskering by repeatedly using Definitions 8.7 and 8.11. One obtains two large $\circ_2$ composites differing only by the order in which independent whiskering steps are performed. At this point invoke Theorem 7.32 at the 2-cell level: independent left/right whiskerings commute up to a canonical 3-cell. Lifting this commutation one dimension up gives a canonical 4-cell between the two expanded composites, which is precisely the required (1,0)-interchange. □

Corollary 8.18 (Any two pastings with same boundary are connected). *Let P and Q be two well-typed composites built from $\circ_2, \circ_1, \circ_0$ on the same family of 3-cells and with the same external boundary. Then there is a canonical 4-cell $P \Rrightarrow_4 Q$.*

Proof. Any two parenthesizations and evaluation orders are related by finite applications of associativity/unit and the three interchange laws. Each application produces a 4-cell by Theorems 8.13, 8.15, 8.16, and 8.17. Chaining them yields a 4-cell from P to Q. □

This corollary is the key technical result for coherence: it says that *all roads lead to Rome*. No matter how we choose to parenthesize or order our compositions, the resulting 3-cells are always connected by 4-cells. This is the 3-dimensional manifestation of coherence.

Example 8.19 ((2,1)-interchange with identity 3-cells). Taking all four 3-cells to be identities in the (2,1)-interchange (Theorem 8.15):

$$(\mathsf{refl}_\alpha \circ_2 \mathsf{refl}_\alpha) \circ_1 (\mathsf{refl}_\beta \circ_2 \mathsf{refl}_\beta) \Rrightarrow_4 (\mathsf{refl}_\alpha \circ_1 \mathsf{refl}_\beta) \circ_2 (\mathsf{refl}_\alpha \circ_1 \mathsf{refl}_\beta).$$

Both sides simplify to $\mathsf{refl}_{\alpha \bullet_v \beta}$ by Proposition 8.8(i) and the unit laws for $\circ_2$. The 4-cell connecting them is the identity 4-cell—confirming that interchange becomes trivial on identity data, a basic sanity check for the general interchange law.

8.4 Coherence Conditions at the 3-Cell Level

We now reach the heart of the chapter. In Chapter 7, we constructed the basic coherence 3-cells: interchange witnesses, pentagon witnesses, triangle witnesses, and naturality witnesses. These were presented as isolated constructions. Here we organize them into a systematic *coherence package* and show that they satisfy their own coherence equations.

The key insight is that coherence is *self-similar*: just as 2-cells witness coherence of paths, 3-cells witness coherence of 2-cells, and 4-cells witness coherence of 3-cells. The structural witnesses at each level satisfy constraints witnessed at the next level.

8.4.1 Named Structural 3-Cells

Definition 8.20 (Structural 3-cells). For composable paths and 2-cells in $\mathfrak{B}_A$:

(i) $\mathfrak{I}_{\alpha,\beta}$ denotes the interchange 3-cell of Theorem 7.32.

(ii) $\mathfrak{P}_{f,g,h,k}$ denotes the pentagon 3-cell of Theorem 7.35.

(iii) $\mathfrak{T}_{f,g}$ denotes the triangle 3-cell of Theorem 7.37.

(iv) $\mathfrak{N}^{\mathfrak{a}}_{\alpha,\beta,\gamma}$ and $\mathfrak{N}^{\mathfrak{u}}_{\alpha}$ denote naturality 3-cells from Theorems 7.39 and 7.40.

Remark 8.21 (Structural 3-cells as dimensional counterparts). The notation emphasizes that these are not arbitrary 3-cells, but canonical cells generated by structural rewrite interactions. They play for dimension 3 the same role that associators and unitors play for dimension 2. Just as we cannot avoid making a choice about how to parenthesize a long composition of paths, we cannot avoid making a choice about which coherence 2-cell to use—but these structural 3-cells tell us that all such choices are equivalent.

Concretely, the current formal development treats the named coherences as genuine proof-relevant 3-cells: pentagon, triangle, the three interchange laws, inverse and double-inverse coherence, contravariance, Eckmann–Hilton, and their companion structural comparisons. Fourteen such coherence proofs were independently regression-tested by replacing the intended witness with a reflexivity proof; every replacement fails, confirming that the 3-cell layer is genuinely non-trivial.

8.4.2 Coherence Among Structural 3-Cells

The structural 3-cells themselves satisfy coherence conditions. These conditions state that when there are two ways to combine structural 3-cells to achieve the same effect, the two combinations are connected by a 4-cell.

Theorem 8.22 (Coherence of interchange with associator). *For any composable tuple of 2-cells to which both constructions apply, the two composites*

$$(\mathfrak{I} \circ_1 \mathfrak{a}) \circ_2 (\mathfrak{a} \circ_1 \mathfrak{I}) \qquad \textit{and} \qquad (\mathfrak{a} \circ_1 \mathfrak{I}) \circ_2 (\mathfrak{I} \circ_1 \mathfrak{a})$$

are connected by a 4-cell.

Proof. Both expressions are pastings of the same generating 3-cells with identical external boundary. By Corollary 8.18, any two such pastings are connected by a canonical 4-cell. The required 4-cell is exactly that connector. □

Theorem 8.23 (Pentagon–triangle compatibility at dimension 3). *Consider a five-fold composite of paths together with one inserted identity path. Any two 3-cell composites that reduce the induced coherence diagram using $\mathfrak{P}$ and $\mathfrak{T}$ are related by a 4-cell.*

Proof. The involved 3-cell composites are built from structural cells $\mathfrak{P}$, $\mathfrak{T}$ and whiskerings. Their boundaries coincide because both reduce the same source 2-cell to the same target 2-cell. Apply Corollary 8.18. Existence of boundary-equal source and target follows from Theorems 7.35 and 7.37. □

Theorem 8.24 (Naturality coherence). *Let $\alpha_i, \beta_i, \gamma_i$ $(i = 0, 1)$ be parallel triples of 2-cells. The two ways of combining naturality witnesses for associator and unitors with interchange differ by a canonical 4-cell.*

Proof. Expand each composite using Definition 8.20. The expansions are built from the same generators and have common boundary by naturality hypotheses. The result again follows from Corollary 8.18. □

Theorem 8.25 (3-dimensional coherence package). *The structural 3-cells of Definition 8.20 satisfy:*

(i) closure under $\circ_2, \circ_1, \circ_0$ and inversion;

(ii) naturality in all 2-cell arguments;

(iii) all compatibility diagrams generated by interchange, pentagon, triangle, and naturality commute up to 4-cells.

Proof. (i) is immediate from definitions. (ii) is Theorems 7.39 and 7.40, lifted through Definitions 8.7 and 8.11. (iii) follows by iterated application of Corollary 8.18. □

8.5 Connection with Mac Lane Coherence

Saunders Mac Lane's coherence theorem for monoidal categories [98] is one of the foundational results of category theory. It states, roughly, that "all diagrams built from associators and unitors commute." This frees us from having to verify commutativity case by case—once we know a diagram is built from structural isomorphisms, commutativity is automatic.

In our setting, we can prove a stronger, higher-dimensional version. Theorem 7.41 in Chapter 7 established that structural 2-cell diagrams commute up to 3-cells—the computational-path analogue of Mac Lane's classical result, one dimension higher. We now derive the next refinement: all choices of coherence 3-cells are themselves coherent up to 4-cells. That is, Mac Lane coherence lifts from dimension 2 to dimension 3, and the mechanism is the same: the rewrite system generates enough structural witnesses to close every remaining diagram.

8.5.1 Canonical 3-Cells Between Reassociation Derivations

To state this precisely, we first isolate the 2-cells that arise purely from the categorical structure—those built from associators, unitors, and their compositions. These are the "structural" 2-cells, analogous to the canonical isomorphisms in Mac Lane's original treatment.

Definition 8.26 (Structural 2-cell derivation). Fix composable paths. A *structural 2-cell derivation* is a 2-cell built from associators, unitors, their inverses, and horizontal/vertical composition.

Theorem 8.27 (Mac Lane coherence, 3-cell form). *Let $D_1, D_2 : X \Rrightarrow Y$ be two structural 2-cell derivations between the same source and target 1-cells. Then there exists a canonical 3-cell*

$$\mathfrak{M}_{D_1,D_2} : D_1 \Rrightarrow_3 D_2.$$

Moreover, this assignment is functorial under whiskering and composition.

Proof. We invoke the bicategorical coherence theorem of Mac Lane–Paré (as recalled in Remark 7.42): in a bicategory with associator/unitor data satisfying pentagon and triangle, any two structural 2-cell pastings with common boundary are canonically equivalent. Its hypotheses hold for $\mathfrak{B}_A$ by Theorem 7.41 from Chapter 7.

Hence D_1 and D_2 differ by a finite sequence of elementary structural moves. Each move is witnessed by one of the named structural 3-cells in Definition 8.20 (interchange, pentagon, triangle, naturality). Composing these witnesses with $\circ_2$ yields $\mathfrak{M}_{D_1,D_2}$.

If two move sequences are chosen, they define parallel 3-cells with the same boundary; Corollary 8.18 provides a connecting 4-cell, so the resulting class is canonical. Functoriality under whiskering/composition follows from the naturality and closure statements in Theorem 8.25. □

Corollary 8.28 (Uniqueness up to unique higher cell)**.** *For parallel structural 2-cell derivations D_1, D_2, any two 3-cells $\Gamma, \Delta : D_1 \Rrightarrow_3 D_2$ generated by structural coherence are connected by a canonical 4-cell.*

Proof. Apply Theorem 8.27 to produce a canonical $\mathfrak{M}_{D_1,D_2}$. Both Γ and Δ are equivalent to this canonical 3-cell by construction, hence connected to each other by composition of the resulting 4-cells. □

Remark 8.29 (Associahedra and higher cells)**.** The combinatorial structure underlying these coherence conditions is captured by the *Stasheff associahedra*, introduced by Stasheff [142] in his study of H-spaces and loop spaces. For an $(n+1)$-fold composite of paths, parenthesizations form vertices of the Stasheff polytope K_n. Edges correspond to associator 2-cells, 2-faces to pentagon/triangle relations, and 3-faces to relations among those relations.

To visualise this concretely: for four composable paths, there are five ways to parenthesize $(f \cdot g \cdot h \cdot k)$, forming the vertices of the pentagon. Each edge is an associator 2-cell. The pentagon identity states that the two paths around the pentagon (both starting and ending at the same parenthesization) are connected by a 3-cell. For five paths, we get a 3-dimensional polytope (the associahedron K_4), and its 2-faces correspond to pentagon relations.

Theorem 8.27 states that every path between two vertices in this coherence complex is filled by a canonical 3-cell, and Corollary 8.28 states that any two such fillings are connected by a 4-cell. This is the combinatorial incarnation of the principle that "all coherence diagrams commute."

8.6 Examples from the Rewrite System

The theory developed so far gains concreteness when we compute explicit 3-cells from specific rewrite interactions. The examples that follow show how coherence witnesses arise naturally from the structure of the 77-rule rewrite system.

Throughout, we write rules R1–R8 for the groupoid core (Section 6.4).

8.6.1 Commuting Independent Rewrites

The simplest source of 3-cells is the commutativity of independent rewrites. When two rewrite rules apply to non-overlapping parts of an expression, we can apply them in either order, yielding two different 2-cell derivations with the same endpoints. The 3-cell witnesses that these derivations are equivalent.

Example 8.30 (Independent redexes in a composite)**.** Let $f : \mathsf{Path}(a, b)$ and $g : \mathsf{Path}(b, c)$. Consider

$$E \;=\; \mathsf{trans}(\mathsf{trans}(f, \mathsf{refl}_b), \mathsf{symm}(\mathsf{symm}(g))).$$

This expression has two "obvious" simplifications, occurring in non-overlapping subterms:

(a) apply R4 (right unit) in the left argument of the outer trans, obtaining $E_1 = \mathsf{trans}(f, \mathsf{symm}(\mathsf{symm}(g)))$;

(b) apply R2 (double symmetry cancellation) in the right argument, obtaining $E_2 = \mathsf{trans}(\mathsf{trans}(f, \mathsf{refl}_b), g)$.

Applying the remaining rule in each branch yields the common term $\mathsf{trans}(f, g)$. The situation can be pictured as a "diamond":

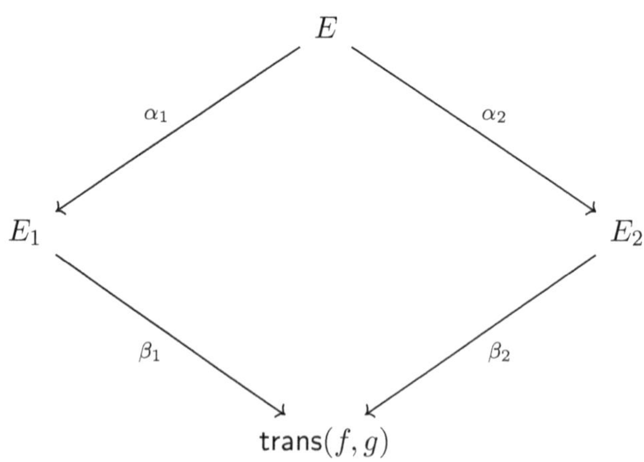

We thus obtain two 2-cells $\alpha = \alpha_1 \bullet_v \beta_1$ and $\beta = \alpha_2 \bullet_v \beta_2$, both from E to $\mathsf{trans}(f, g)$. The canonical 3-cell $\mathfrak{J}_{\text{ind}} : \alpha \Rrightarrow_3 \beta$ witnesses that these two reduction paths are equivalent.

Proof. The two rewrites act on disjoint subterms of E, so they commute by the parallel-moves argument used in Chapter 5 and in the core case of Theorem 7.32. The commuting square of steps yields a 3-cell between the two length-two derivations. □

8.6.2 Pentagon Witness as a Concrete 3-Cell

The pentagon identity is perhaps the most famous coherence condition in algebra. It states that the two ways of reassociating a four-fold composition are themselves coherently related.

Example 8.31 (Four-fold reassociation)**.** For composable paths f, g, h, k, set

$$X = \mathsf{trans}(\mathsf{trans}(\mathsf{trans}(f, g), h), k),$$
$$Y = \mathsf{trans}(f, \mathsf{trans}(g, \mathsf{trans}(h, k))).$$

These are the "left-associated" and "right-associated" forms of the four-fold composition. There are exactly two canonical structural 2-cell composites from X to Y, corresponding to the two boundary paths of the pentagon:

- **Upper path:** apply associativity to the outer three factors twice, obtaining $((fg)h)k \to (fg)(hk) \to f(g(hk))$.
- **Lower path:** apply associativity to the inner three factors first, then work outward: $((fg)h)k \to (f(gh))k \to f((gh)k) \to f(g(hk))$.

The pentagon theorem provides

$$\mathfrak{P}_{f,g,h,k} : D_{\text{upper}} \Rrightarrow_3 D_{\text{lower}}.$$

This 3-cell witnesses that it does not matter which route we take around the pentagon—both paths lead to equivalent 2-cells.

Proof. The two composites are obtained by successive applications of R8 with appropriate whiskering. Both reduce X to Y. Confluence of the rewrite system yields a finite sequence of critical-pair resolutions between them; composing the corresponding elementary 3-steps gives $\mathfrak{P}_{f,g,h,k}$. □

8.6.3 Triangle Witness and Unit Elimination Order

The triangle identity involves the interaction between associators and unitors. When we insert an identity path and then remove it, the order of operations matters at the level of 2-cells but not at the level of 3-cells.

Example 8.32 (Two eliminations of the inserted identity)**.** For $f : \mathsf{Path}(a, b)$ and $g : \mathsf{Path}(b, c)$, let

$$Z = \mathsf{trans}(\mathsf{trans}(f, \mathsf{refl}_b), g).$$

This expression contains a "useless" identity refl_b that we want to eliminate. There are two natural strategies:

(i) **Direct elimination:** First apply the right unit rule (R4) to remove refl_b inside the left argument, obtaining $\mathsf{trans}(f, g)$.

(ii) **Reassociate first:** Apply associativity (R8) to get $\mathsf{trans}(f, \mathsf{trans}(\mathsf{refl}_b, g))$, then apply the left unit rule (R3) to remove refl_b inside the right argument.

Both strategies produce 2-cells from Z to $\mathsf{trans}(f, g)$, but via different routes. Theorem 7.37 provides a 3-cell between them, asserting that the two elimination strategies are coherently equivalent.

Proof. This is exactly the triangle coherence square specialised to the rewrite rules R3, R4, and R8. The existence of the 3-cell follows from Theorem 7.37. □

8.6.4 A Mixed Example Using Inverses

The full 77-rule rewrite system includes many "optimisation" rules that allow shortcuts in reduction. The following example shows that these shortcuts are coherent with the basic rules.

Proposition 8.33 (Coherence of cancellation strategies)**.** *Let $p : \mathsf{Path}(a, b)$ and $q : \mathsf{Path}(a, c)$. Consider*

$$W = \mathsf{trans}(p, \mathsf{trans}(\mathsf{symm}(p), q)).$$

This expression "goes from a to b via p, then back to a via $\mathsf{symm}(p)$, then to c via q." The detour through b is unnecessary, and we expect W to simplify to q. There are two reduction strategies:

(i) ***Via basic groupoid-equivalence steps:*** *reassociate first by $G5^{-1}$ to get $\mathsf{trans}(\mathsf{trans}(p, \mathsf{symm}(p)), q)$, then cancel $\mathsf{trans}(p, \mathsf{symm}(p))$ to refl_a (R5), then remove the unit (R3) to obtain q.*

(ii) ***Via optimisation rule:*** *use a dedicated cancellation rule from the full 77-rule system that directly recognises and simplifies the pattern* $\mathsf{trans}(p, \mathsf{trans}(\mathsf{symm}(p), q))$.

These two strategies determine parallel 2-cells, and there exists a 3-cell between them.

Proof. Strategies (i) and (ii) have the same reduced expression target in the Chapter 5 expression semantics. By confluence at that level, their induced 2-cell derivations are joinable in the 2-cell rewrite graph. A joinability witness is by definition a 3-cell. □

Remark 8.34 (Semantic invariance of optimisation rules)**.** Proposition 8.33 is practically important: it states that introducing optimisation rules (like direct cancellation) does not alter the higher semantics. Different optimisation paths remain coherently equivalent. This justifies the use of the full 77-rule system for efficient computation while maintaining semantic correctness.

8.7 Squier-Style Diamond Fillers

We now make explicit the bridge between confluence proofs and higher coherence. In Squier's perspective [141], each resolved critical pair contributes a higher-dimensional filler. In our setting, these fillers are 3-cells between parallel 2-cell derivations.

Definition 8.35 (Diamond property for a relation)**.** Let R be a binary relation on a type X. We say that R has the *diamond property*, written $\mathsf{Diamond}(R)$, if

$$\begin{aligned}\mathsf{Diamond}(R) \;:\equiv\; \forall a, b, c : X,\ \ R(a, b) \to R(a, c)&\\ \to \exists d : X,\ \ R(b, d) \wedge R(c, d).&\end{aligned}$$

Intuitively, whenever one object can rewrite in one step to two different objects, the two branches can be rejoined after one more step on each side. This gives a literal diamond in the rewrite graph.

Proposition 8.36 (Strip lemma)**.** *Assume* $\mathsf{Diamond}(R)$*. If a rewrites in one step to b and in zero or more steps to c, then there exists d such that*

$$b \to_R^* d \qquad \textit{and} \qquad c \to_R d.$$

Proof idea. Induct on the derivation $a \to_R^* c$. The base case is immediate. In the inductive step, use the diamond hypothesis to close the first fork, then compose with the induction hypothesis on the remaining tail. □

The proposition is the one-step versus multi-step closure principle (the "diamond strip" lemma). Combined with induction on the left branch, it yields confluence and explains why local overlap resolutions propagate to global coherence data.

Theorem 8.37 (Critical-pair resolutions as 3-cell witnesses). *For each critical pair*

$$u \xrightarrow{\rho_1} v, \qquad u \xrightarrow{\rho_2} w,$$

a chosen join

$$v \to_R^* n, \qquad w \to_R^* n$$

induces two parallel 2-cell derivations $\alpha, \beta : u \Rrightarrow n$*, hence a 3-cell filler*

$$\mathfrak{D}_{\rho_1,\rho_2} : \alpha \Rrightarrow_3 \beta.$$

Proof. The two branches define composites of elementary 2-cells from u to n. Since they have common source and target, they are parallel in $\mathsf{Derivation}_2$, and by definition a coherence witness between them is a 3-cell. In the formal development, these fillers are realised as explicit proof-relevant 3-cells obtained from representative overlap resolutions (unit/associativity, inverse/associativity, and associativity/associativity), with no passage through a merely propositional quotient. □

This is the precise sense in which confluence generates higher coherence: diamonds in the first-order rewrite relation become fillers in dimension 3. These fillers are the computational content behind the coherence package of Section 8.4.

8.8 Contractibility Phenomena at Level 3

We have seen that parallel 3-cells are connected by 4-cells. This section makes precise a remarkable phenomenon: above a certain dimension, coherence becomes "automatic." The space of witnesses becomes contractible, meaning that all witnesses are essentially unique.

This is the beginning of the *stabilisation* that characterises ω-groupoids. While dimensions 0, 1, and 2 exhibit rich and varied structure, dimension 3 and above become increasingly trivial in a precise sense. The analogy from topology is illuminating: in a simply connected space, all loops are contractible, and the "interesting" homotopy vanishes above a certain dimension. Here the rewrite system's confluence plays the role of simple connectivity, forcing higher coherence cells to collapse.

Definition 8.38 (Local contractibility of 3-cells)**.** A hom-3-groupoid is *locally contractible* if for every pair of parallel 3-cells $\Gamma, \Delta : \alpha \Rrightarrow_3 \beta$ there exists a 4-cell

$$\Theta : \Gamma \Rrightarrow_4 \Delta.$$

In other words, any two ways of witnessing that α and β are equivalent are themselves equivalent via a 4-cell.

Theorem 8.39 (Local contractibility from confluence)**.** *In the computational path rewrite system, every hom-3-groupoid generated by structural 3-cells is locally contractible.*

Proof. Fix parallel structural 3-cells $\Gamma, \Delta : \alpha \Rrightarrow_3 \beta$ in a hom-3-groupoid generated by the structural cells of Definition 8.20. Then α, β are structural 2-cell derivations, so Corollary 8.28 applies directly and yields a canonical 4-cell

$$\Theta_{\Gamma,\Delta} : \Gamma \Rrightarrow_4 \Delta.$$

Thus each such hom-3-groupoid is locally contractible.

Rewrite-theoretically, $\Theta_{\Gamma,\Delta}$ is the joinability witness for parallel 3-dimensional rewrite paths. In this chapter, this joinability is established for structural pastings via Corollary 8.18; the full induced termination/confluence framework for higher cells is developed in Chapter 9, specifically Theorems 9.51 and 9.45. □

The proof reveals the deep connection between confluence and coherence. Confluence of the underlying expression rewrite semantics, together with its induced higher rewriting structure, guarantees coherence at the next level: parallel derivations are connected by higher cells.

Corollary 8.40 (Canonicality of coherence witnesses)**.** *Any two structural 3-cells with the same boundary are equivalent up to (unique up to higher equivalence) 4-cell.*

Proof. Apply Theorem 8.39 to the two structural 3-cells. The uniqueness claim follows by iterating the same argument one dimension higher. □

Remark 8.41 (Toward the weak ω-groupoid theorem)**.** Theorem 8.39 is the first non-trivial instance of the stabilisation pattern needed in Chapter 9: parallel n-cells become connected by $(n + 1)$-cells for $n \geq 3$. Combined with the bicategory coherence of Chapter 7, this yields the full weak ω-groupoid structure in the sense of Batanin [11] and Leinster [99].

The pattern is: at dimension 0 (points), there is interesting structure (different types); at dimension 1 (paths), there is interesting structure (different paths between points); at dimension 2 (2-cells), there is interesting structure (different ways paths can be equivalent); and at dimension 3 there is still genuine coherence data (pentagon, triangle, interchange, and their composites). The stabilisation occurs one level higher: from dimension 4 onward, higher comparison problems become canonically contractible. This is the phenomenon that makes the resulting weak ω-groupoid tractable despite having infinitely many dimensions.

Summary

This chapter developed the 3-cell level of computational paths and its coherence theory. The main message is that coherence does not stop at the bicategorical level: just as 2-cells witness equivalence of paths, 3-cells witness equivalence of 2-cells, and 4-cells witness equivalence of 3-cells. By the time we pass to dimension 4, however, the tower begins to stabilise: 3-cells remain genuine data, while higher comparison problems become canonically contractible.

1. **3-cells as derivations between 2-cells:** we formalised $\mathsf{Derivation}_3$ with identity, elementary steps, inverse, and 2-composition (Section 8.1).

2. **Three composition operations:** we defined and analysed the three canonical operations $\circ_2, \circ_1, \circ_0$ for 3-cells, corresponding to composition along 2-, 1-, and 0-dimensional boundaries (Section 8.2).

3. **Interchange laws:** we proved higher interchange for all pairs of composition directions and derived connectivity of all pastings with common boundary (Section 8.3).

4. **Coherence package at dimension 3:** interchange, pentagon, triangle, and naturality witnesses were organised as structural 3-cells and shown coherent up to 4-cells (Section 8.4).

5. **Mac Lane connection:** we established a 3-cell refinement of Mac Lane coherence: any two structural 2-cell derivations are connected by canonical 3-cells, unique up to 4-cells (Section 8.5).

6. **Concrete rewrite examples and contractibility:** we computed explicit 3-cells from the rewrite system and proved local contractibility for structural hom-3-groupoids (Sections 8.6 and 8.8).

Historical Notes.

The background to this chapter is the classical theory of coherence and its later reformulation in higher-dimensional rewriting. Mac Lane's coherence theorem for monoidal categories [98] and Bénabou's bicategories [12] provided the classical 1- and 2-dimensional background for the coherence phenomena studied here. The interpretation of rewrite derivations as higher cells is closely related to Squier's theory of higher-dimensional rewriting [141]. In homotopy type theory, analogous higher groupoid structures were developed by Lumsdaine [105] and by van den Berg–Garner [153]; the computational-paths approach differs in deriving coherence directly from a concrete terminating and confluent rewrite system.

Formalization Notes. The results of this chapter are formalised in the following files of the accompanying Lean development (`https://github.com/Arthur742Ramos/ComputationalPathsLean`):

- `Path/OmegaGroupoid/CoherencePaths.lean` — pentagon, triangle, and related coherence witnesses
- `Path/OmegaGroupoid/PentagonProof.lean` — the pentagon coherence proof
- `Path/OmegaGroupoid/GroupoidProofs.lean` — supporting unit and associativity coherences
- `Path/Rewrite/SquierDeep.lean` — the Squier-style finite-derivation-type perspective on coherence

Key definitions. The $\mathsf{Derivation}_3$ type captures 3-cells via meta-steps encoding coherence laws:

```
/-- Meta-steps: primitive 3-cells encoding coherences -/
inductive MetaStep3 : {a b : A} -> {p q : Path a b} ->
    Derivation2 p q -> Derivation2 p q -> Type (u + 2)
      where
  -- Groupoid laws
  | vcomp_refl_left (d : Derivation2 p q) :
      MetaStep3 (.vcomp (.refl p) d) d
  | vcomp_assoc (d1 : Derivation2 p q) (d2 : Derivation2
     q r)
      (d3 : Derivation2 r s) :
      MetaStep3 (.vcomp (.vcomp d1 d2) d3) (.vcomp d1 (.
         vcomp d2 d3))
```

Pentagon and triangle coherence arise from critical pair resolution:

```
-- Pentagon coherence
| pentagon (f : Path a b) (g : Path b c) (h : Path c d)
    (k : Path d e) :
    MetaStep3
      (.vcomp (.step (Step.trans_assoc (Path.trans f g)
          h k))
              (.step (Step.trans_assoc f g (Path.trans
                  h k))))
      (.vcomp (.vcomp (.step (Step.trans_congr_left k
                      (Step.trans_assoc f g h)))
                     (.step (Step.trans_assoc f (Path.
                        trans g h) k)))
              (.step (Step.trans_congr_right f
                      (Step.trans_assoc g h k))))
-- Triangle coherence
| triangle (f : Path a b) (g : Path b c) :
    MetaStep3
      (.vcomp (.step (Step.trans_assoc f (Path.refl b)
          g))
              (.step (Step.trans_congr_right f
                      (Step.trans_refl_left g))))
      (.step (Step.trans_congr_left g (Step.
          trans_refl_right f)))
```

The low-dimensional constructors displayed here are completed in the library by explicit `Derivation3` proofs for the remaining coherence families used throughout Part II: inverse and double-inverse coherence, contravariance of inversion, the three interchange directions, and the Eckmann–Hilton comparison on doubly degenerate 2-cells. These are all genuinely Type-valued 3-cells; in the verification suite, every one of the fourteen named coherence proofs fails if its witness is replaced by `Derivation3.refl`.

Chapter 9

The Weak ω-Groupoid of Computational Paths

> *"The analogies between the structures at each new level, and those at the preceding ones, can serve as a reliable thread to guide us in our exploration of these structures, level by level."*
>
> —Alexander Grothendieck, *Pursuing Stacks* [77]

Why should one care about ω-groupoids? The preceding chapters have built progressively richer algebraic structure on computational paths—groupoid operations, 2-cells witnessing equivalence of derivations, 3-cells encoding coherence among those witnesses. At each stage, a natural question recurs: *does this tower eventually stop producing new structure, or does it continue forever?* The question is not merely aesthetic. If the tower does continue, one needs a language that describes all its levels uniformly, rather than treating each dimension as a fresh ad hoc construction. That language is the theory of weak ω-groupoids, and the main theorem of this chapter is that computational paths admit a natural formulation in precisely this language.

More concretely, the earlier chapters revealed a pattern. The groupoid laws of Chapter 6 hold only up to 2-cells; the 2-cell operations of Chapter 7 satisfy interchange only up to 3-cells; the 3-cell coherence of Chapter 8 is itself governed by comparison data one level higher. The reader who has followed this progression already senses the rhythm. What remains is to prove that the pattern is exact and self-sustaining—that computational paths over any type form a weak ω-groupoid in the precise sense of higher category theory [77, 11, 99, 105, 153]—and that the tower *stabilises* after dimension 3, with canonical fillers controlling all higher coherence problems [134].

What is distinctive about our route to this result is its concreteness. The argument proceeds through explicit higher rewriting rather than through an abstract existence theorem for identity types. Every coherence witness is a specific derivation, constructed from the 77 rewrite rules of Chapter 4 and their higher-dimensional lifts.

The chapter has two mathematical centres. First, we define the globular data, partial compositions, identities, inverses, and coherence operations needed for a weak ω-groupoid, and verify that the structures already constructed in dimensions 1, 2, and 3 supply the required low-dimensional anchors. Second, we prove the stabilisation result: dimensions 1–3 carry the genuinely new proof-relevant coherence data, whereas from dimension 4 onward the structural comparison problems are controlled by canonical fillers. The later sections on Grothendieck's hypothesis, HoTT, and the fundamental ω-groupoid should be read as interfaces for this proved algebraic core, not as substitutes for it.

Two notation conventions are worth fixing immediately. We write $\circ_k$ for composition along k-boundaries; at dimension 2 this recovers the vertical and horizontal compositions $\circ_1 = \bullet_v$ and $\circ_0 = \bullet_h$ from Chapter 7, and at dimension 3 the triple $\circ_2, \circ_1, \circ_0$ recovers the three composition directions analysed in Chapter 8; the precise recovery statements are proved later in Propositions 9.24 and 9.25. We also reserve $x \Rrightarrow_{n+1} y$ for the ambient type of $(n+1)$-cells between parallel n-cells in the computational tower, while $x \Longrightarrow_{n+1} y$ denotes a displayed witness of a specific weak law.

The chapter is organised as follows. Section 9.1 defines weak ω-groupoids in globular form. Section 9.2 presents composition operations in dimension n. Section 9.3 states and proves the main theorem that computational paths form a weak ω-groupoid. Section 9.4 develops coherence at all levels as a generalisation of the pattern from Chapter 8. Section 9.5 relates this structure to Grothendieck's homotopy hypothesis. Section 9.6 gives a comparison with identity types in Homotopy Type Theory (without identifying the two notions). Section 9.7 defines the fundamental ω-groupoid of a type.

Reader's guide. The core of this chapter runs through Section 9.3 (the main theorem). Sections 9.4–9.7 develop extensions and comparisons that may be read selectively: the coherence material (Section 9.4) is needed for later chapters, while the Grothendieck and HoTT comparison sections are largely independent and may be consulted as needed. Throughout, concrete $n = 4$ instances are provided as running examples; on a first reading, these may be skimmed without loss of continuity.

The following table summarises the dimensional structure that the chapter makes precise. It records what each cell level contributes to the ω-groupoid, which composition directions are available, and at what stage the tower stabilises.

Dim.	Cells	Comp.	Status
0	Terms $a, b, c : A$	—	Objects (points)
1	Paths $p : \mathsf{Path}(a, b)$	$\circ_0$	Weak groupoid (Ch. 6)
2	$\alpha : p \Rrightarrow q$	$\circ_1, \circ_0$	Bicategorical (Ch. 7)
3	$\Gamma : \alpha \Rrightarrow \beta$	$\circ_2, \circ_1, \circ_0$	Coherence (Ch. 8)
≥ 4	Higher derivations	$\circ_{n-1}, \ldots, \circ_0$	*Stabilised*: canonical fillers

Dimensions 1–3 carry genuinely new proof-relevant structure; from dimension 4 onward, structural comparison problems are solved by canonical fillers supplied by higher-dimensional confluence (Theorem 9.60).

9.1 Weak ω-Groupoids: Globular Data and Basic Axioms

The construction of the ω-groupoid requires a precise vocabulary. This section introduces the globular framework: cells in each dimension, source and target maps, identity inclusions, and partial compositions along matching boundaries. We begin with the purely combinatorial carrier—the globular set—and then layer on the algebraic operations (compositions, identities, inverses) that make it a weak ω-groupoid. The notion of a globular set is standard in higher category theory [11, 145, 99]; the weak axioms are recorded in a form compatible with the computational-path rewrite perspective developed in Chapters 6–8.

9.1.1 Globular Sets and Boundary Maps

Definition 9.1 (Globular set of cells). A *globular set* (a graded collection of cells equipped with boundary maps—the standard combinatorial substrate for higher category theory) $\mathcal{G}$ consists of:

(i) sets $\mathcal{G}_n$ for each $n \in \mathbb{N}$ (the set of n-cells),

(ii) source and target maps

$$s_n, t_n : \mathcal{G}_n \to \mathcal{G}_{n-1} \qquad (n \geq 1),$$

satisfying, for every $n \geq 2$, the *globular identities* (the conditions ensuring that the two boundary maps of an n-cell land on parallel $(n-1)$-cells)

$$s_{n-1} \circ s_n = s_{n-1} \circ t_n,$$
$$t_{n-1} \circ s_n = t_{n-1} \circ t_n.$$

Concretely, $\mathcal{G}_0$ plays the role of points, $\mathcal{G}_1$ of paths, $\mathcal{G}_2$ of paths between paths, and so on. The globular identities express that an n-cell has a source and target that are *parallel* $(n{-}1)$-cells—exactly the setup already familiar from Chapters 7 and 8.

Example 9.2 (Source and target maps on concrete path expressions)**.** Fix a type A with terms $a, b, c : A$ and paths $p : \mathsf{Path}(a, b)$, $q : \mathsf{Path}(b, c)$. The globular data for the first three dimensions of $\mathcal{C}(A)$ are:

Dimension 1. The 1-cell $\mathsf{trans}(p, q) : \mathsf{Path}(a, c)$ has

$$s_1(\mathsf{trans}(p, q)) = a, \qquad t_1(\mathsf{trans}(p, q)) = c.$$

Its inverse $\mathsf{symm}(\mathsf{trans}(p, q))$ has $s_1 = c$ and $t_1 = a$—the boundary maps simply read off the endpoints.

Dimension 2. Consider two parallel paths from a to c: $r_1 = \mathsf{trans}(p, q)$ and $r_2 = \mathsf{trans}(p, \mathsf{trans}(\mathsf{refl}_b, q))$. A 2-cell $\alpha : r_1 \Rrightarrow_2 r_2$ (witnessing, for instance, that r_2 rewrites to r_1 by Rule R3) has:

$$s_2(\alpha) = r_1, \qquad t_2(\alpha) = r_2.$$

The globular identity $s_1(s_2(\alpha)) = s_1(t_2(\alpha))$ requires that $s_1(r_1) = s_1(r_2)$; indeed both equal a, and similarly both targets equal c. Thus r_1 and r_2 are parallel, as the definition demands.

Dimension 3. If $\alpha, \beta : r_1 \Rrightarrow_2 r_2$ are two 2-cells with the same source and target, a 3-cell $\Gamma : \alpha \Rrightarrow_3 \beta$ has $s_3(\Gamma) = \alpha$ and $t_3(\Gamma) = \beta$, and the globular identity ensures $s_2(\alpha) = s_2(\beta)$ and $t_2(\alpha) = t_2(\beta)$.

Definition 9.3 (Iterated source/target and k-composable pairs)**.** For $0 \leq k < n$, define iterated boundary maps

$$s_n^k := s_{k+1} \circ \cdots \circ s_n : \mathcal{G}_n \to \mathcal{G}_k, \qquad t_n^k := t_{k+1} \circ \cdots \circ t_n : \mathcal{G}_n \to \mathcal{G}_k.$$

Two n-cells $x, y \in \mathcal{G}_n$ are *k-composable* when $t_n^k(x) = s_n^k(y)$. The type of composable pairs is the pullback set

$$\mathcal{G}_n \times_{\mathcal{G}_k} \mathcal{G}_n := \{(x, y) \in \mathcal{G}_n \times \mathcal{G}_n \mid t_n^k(x) = s_n^k(y)\}.$$

The case $k = n - 1$ is ordinary vertical composition at dimension n; lower values of k encode composition along deeper boundaries, matching the pattern $\circ_2, \circ_1, \circ_0$ for 3-cells in Chapter 8.

9.1.2 Identities and Inverses in All Dimensions

The next ingredients are the identity and inverse operations. Just as reflexivity gives a trivial path from any point to itself, an identity inclusion embeds any n-cell as a degenerate $(n+1)$-cell. And just as symmetry reverses a path, a weak inverse operation reverses an n-cell up to a witnessing cell one dimension higher.

Definition 9.4 (Identity inclusions). A *reflexive globular set* is a globular set $\mathcal{G}$ equipped with maps

$$i_n : \mathcal{G}_n \to \mathcal{G}_{n+1} \qquad (n \in \mathbb{N}),$$

such that

$$s_{n+1} \circ i_n = \mathrm{id}_{\mathcal{G}_n} \qquad \text{and} \qquad t_{n+1} \circ i_n = \mathrm{id}_{\mathcal{G}_n}.$$

We write $1_x := i_n(x)$ for the identity $(n+1)$-cell on x.

This is the uniform higher-dimensional version of reflexivity: for points we get identity paths, for paths we get identity 2-cells, for 2-cells we get identity 3-cells, and so forth.

Definition 9.5 (Weak inverses at level n). Let $n \geq 1$. A *weak inverse operation* in dimension n is a map

$$(-)^{-1} : \mathcal{G}_n \to \mathcal{G}_n$$

satisfying, for every $x \in \mathcal{G}_n$,

$$s_n(x^{-1}) = t_n(x), \qquad t_n(x^{-1}) = s_n(x),$$

and equipped with $(n+1)$-cells

$$\lambda_x : x^{-1} \circ_{n-1} x \Longrightarrow_{n+1} 1_{s_n(x)}, \qquad \rho_x : x \circ_{n-1} x^{-1} \Longrightarrow_{n+1} 1_{t_n(x)}.$$

At dimension 1 this is the familiar groupoid inverse law from Chapter 6, weakened "up to a higher cell." At dimensions 2 and 3, it matches the inverse behaviour established for derivations and 3-cells in Chapters 7 and 8.

9.1.3 Weak Compositions and Higher Laws

We now arrive at the heart of the definition. A weak ω-groupoid is not merely a globular set with identities and inverses; it carries composition operations along every possible boundary direction, together with coherence witnesses asserting that these operations satisfy the expected algebraic laws—but only up to cells one dimension higher.

Definition 9.6 (k-composition of n-cells)**.** For each $n \geq 1$ and each $0 \leq k < n$, a *k-composition* is a partial operation

$$\circ_k : \mathcal{G}_n \times_{\mathcal{G}_k} \mathcal{G}_n \to \mathcal{G}_n.$$

For $x, y \in \mathcal{G}_n$ with $t_n^k(x) = s_n^k(y)$, we write $x \circ_k y \in \mathcal{G}_n$, with boundary behaviour

$$s_n^k(x \circ_k y) = s_n^k(x), \qquad t_n^k(x \circ_k y) = t_n^k(y).$$

Each $\circ_k$ composes along a fixed codimension. This single schema subsumes vertical and horizontal composition as well as whiskering: the low-dimensional cases are exactly the operations already proved in Chapters 7 and 8.

Definition 9.7 (Weak ω-groupoid)**.** A *weak ω-groupoid* is a reflexive globular set $\mathcal{G} = (\mathcal{G}_n, s_n, t_n, i_n)_{n \in \mathbb{N}}$ with k-compositions (Definition 9.6) and weak inverses (Definition 9.5), together with coherent higher witnesses for all $n \geq 1$, all $0 \leq k < \ell < n$, and all composable n-cells. Concretely, one has $(n+1)$-cells witnessing (where I_{k+1}^n denotes the evident iteration of identity inclusions from dimension $k+1$ to dimension n; Definition 9.16 records this formally):

$$\begin{aligned}
\mathfrak{a}_{x,y,z}^{(k)} &: (x \circ_k y) \circ_k z \Longrightarrow_{n+1} x \circ_k (y \circ_k z),\\
\mathfrak{l}_x^{(k)} &: I_{k+1}^n(1_{s_n^k(x)}) \circ_k x \Longrightarrow_{n+1} x,\\
\mathfrak{r}_x^{(k)} &: x \circ_k I_{k+1}^n(1_{t_n^k(x)}) \Longrightarrow_{n+1} x,\\
\mathfrak{i}^{(k,\ell)} &: (x \circ_k y) \circ_\ell (u \circ_k v) \Longrightarrow_{n+1} (x \circ_\ell u) \circ_k (y \circ_\ell v),
\end{aligned}$$

and these witnesses are themselves related by coherent $(n+2)$-cells, and so on at all higher dimensions.

The word "weak" is crucial: laws are not strict equations but hold by explicit higher cells. This formulation follows the tradition of Grothendieck [77], Batanin [11] (whose framework of globular operads—tree-shaped pasting diagrams that organise coherence data—provides one of the most systematic approaches), and the subsequent developments by Kapranov–Voevodsky [91], Ara [6], and Maltsiniotis [108]. It is the direct higher-dimensional continuation of the pattern from Chapter 8, where interchange and associativity hold via 3-cells and 4-cells rather than on the nose.

Proposition 9.8 (Typing sanity check for the foundational operations)**.** *Definitions 9.1–9.7 are well-typed: every map has a specified domain/codomain, every composition operation is defined on matching boundaries, and every weak law compares parallel n-cells.*

Proof. By Definition 9.3, the domain of $\circ_k$ is exactly the pullback enforcing boundary compatibility $t_n^k(x) = s_n^k(y)$, so $x \circ_k y$ is defined only when typed. The identities in Definition 9.4 and inverse boundary equations in Definition 9.5 ensure that every source and target in Definition 9.7 is an n-cell with common boundary. Hence each displayed law has parallel source/target and is therefore a valid candidate for an $(n+1)$-cell witness. □

9.1.4 A concrete $n = 4$ boundary pattern

The first genuinely "many-directional" case is $n = 4$. At this level we have four composition directions $\circ_3, \circ_2, \circ_1, \circ_0$, and six interchange families indexed by ordered pairs

$$(0,1),\ (0,2),\ (0,3),\ (1,2),\ (1,3),\ (2,3).$$

In particular, dimension 4 is the first place where one can see the full combinatorics of all-level coherence without immediately passing to the general n-schema.

Example 9.9 (Boundaries of a 4-cell)**.** Let $\Theta \in \mathcal{G}_4$. Its immediate boundaries are 3-cells $s_4(\Theta), t_4(\Theta) \in \mathcal{G}_3$. Applying one further source/target map gives the 2-dimensional boundary frame:

$$s_4^2(\Theta),\ t_4^2(\Theta) \in \mathcal{G}_2.$$

The globular identities imply

$$s_3(s_4(\Theta)) = s_3(t_4(\Theta)), \qquad t_3(s_4(\Theta)) = t_3(t_4(\Theta)).$$

Thus the source and target 3-cells of Θ are parallel in exactly the sense developed for 3-cells in Chapter 8.

Definition 9.10 (Dimension-4 composability conditions)**.** Let $x, y \in \mathcal{G}_4$. For each $k \in \{0,1,2,3\}$ we write

$$x \#_k y$$

to mean "x and y are k-composable," i.e. $t_4^k(x) = s_4^k(y)$. Explicitly:

$$\begin{aligned}
x\#_3 y &\iff t_4(x) = s_4(y),\\
x\#_2 y &\iff t_4^2(x) = s_4^2(y),\\
x\#_1 y &\iff t_4^1(x) = s_4^1(y),\\
x\#_0 y &\iff t_4^0(x) = s_4^0(y).
\end{aligned}$$

Proposition 9.11 (Boundary propagation at $n = 4$). *Assume $x \#_k y$ for some $k \in \{0, 1, 2, 3\}$. Then for every $j \le k$ one has*

$$s_4^j(x \circ_k y) = s_4^j(x), \qquad t_4^j(x \circ_k y) = t_4^j(y).$$

For $j > k$, the boundaries of $x \circ_k y$ are obtained by composing the corresponding j-boundaries of x and y in dimension j.

Proof. The case $j = k$ is exactly the boundary behaviour built into Definition 9.6.

Now assume $j < k$. Since the iterated boundary maps factor through level k,

$$s_4^j = s_k^j \circ s_4^k, \qquad t_4^j = t_k^j \circ t_4^k.$$

Applying these factorizations to the case $j = k$ gives

$$s_4^j(x \circ_k y) = s_k^j\big(s_4^k(x \circ_k y)\big) = s_k^j\big(s_4^k(x)\big) = s_4^j(x),$$

and similarly

$$t_4^j(x \circ_k y) = t_k^j\big(t_4^k(x \circ_k y)\big) = t_k^j\big(t_4^k(y)\big) = t_4^j(y).$$

So the lower boundaries agree with those of the left and right factors exactly as claimed.

Finally, assume $j > k$. Repeated use of the globular identities shows that the j-boundaries of x and y are again k-composable in dimension j. For the source boundaries one has

$$t_j^k\big(s_4^j(x)\big) = t_4^k(x) = s_4^k(y) = s_j^k\big(s_4^j(y)\big),$$

and for the target boundaries

$$t_j^k\big(t_4^j(x)\big) = t_4^k(x) = s_4^k(y) = s_j^k\big(t_4^j(y)\big).$$

Thus the pair (x, y) is sent by the boundary maps to the pullback

$$\mathcal{G}_j \times_{\mathcal{G}_k} \mathcal{G}_j$$

that defines k-composition in dimension j. Consequently the source and target j-boundaries of $x \circ_k y$ are obtained by composing the corresponding j-boundaries of x and y in that lower dimension. This is exactly the stated propagation rule for $j > k$. □

Remark 9.12 (Why the $n = 4$ case matters). Dimension 2 already exhibits two compositions, and dimension 3 exhibits three. But at dimension 4 one sees simultaneously:

(i) multiple non-equivalent ways of rebracketing iterated composites,

(ii) several independent interchange directions,

(iii) coherence cells between coherence cells that are no longer forced by a single pentagon/triangle pattern.

This is the first dimension where "coherence of coherence" acquires a rich family structure rather than a small finite list.

9.1.5 Strict and Weak Viewpoints on the Same Globular Data

It is useful to separate two layers:

(i) the *globular carrier* (cells and boundary maps), and

(ii) the *algebraic laws* governing composition/identity/inverse.

The carrier can be identical in strict and weak settings; the difference is whether laws are equations or higher cells. This strict/weak dichotomy is a central theme in higher category theory [99, 98].

Definition 9.13 (Strictification at finite height). Let $\mathcal{G}$ be a weak ω-groupoid and $N \in \mathbb{N}$. An *N-strictification* is a globular structure $\mathcal{G}^{\mathrm{st},N}$ with the same cells up to dimension N, for which all laws in dimensions $\leq N$ hold strictly, together with comparison maps

$$Q_n : \mathcal{G}_n \to \mathcal{G}_n^{\mathrm{st},N} \qquad (0 \leq n \leq N)$$

that identify cells related by structural coherence witnesses.

Remark 9.14 (Interpretive role). Definition 9.13 is not used as an additional axiom; rather, it is a conceptual tool. It explains why weak laws are computationally preferable: strictification can collapse proof-relevant distinctions, while the weak tower retains explicit higher witnesses. This distinction is central in our rewrite-based setting, where derivation content carries computational meaning.

Proposition 9.15 (Weak laws preserve proof-relevance). *In the computational tower of higher cells over a type A, replacing weak structural laws by strict equations at a fixed dimension n identifies generally distinct $(n+1)$-cells. Hence strictification forgets derivational information.*

Proof. Let

$$Q_m : \mathcal{G}_m \to \mathcal{G}_m^{\mathrm{st},N} \qquad (0 \leq m \leq N)$$

be an N-strictification with $N \geq n$ in the sense of Definition 9.13. A weak structural law in dimension n is witnessed in the original tower by an $(n+1)$-cell

$$\Sigma : u \Longrightarrow_{n+1} v$$

between two parallel n-cells u and v. In the strictified structure, the same law holds on the nose, so $Q_n(u) = Q_n(v)$. Thus the distinction recorded by the witness Σ is collapsed by strictification. The same is true for composites of structural witnesses, because the definition of strictification explicitly requires the comparison maps to identify cells related by structural coherence.

This loss is already visible in low dimensions. At $n = 1$, the associator and unitor 2-cells between path expressions become invisible once composition is made strictly associative and unital. At $n = 2$, imposing strict interchange likewise collapses the different 3-cells distinguished in Chapter 8. Since every higher dimension carries the same kind of structural witness families, the same mechanism persists for general n. Therefore strictification forgets derivational information. □

9.2 Composition Operations in Dimension n

With the abstract definition of a weak ω-groupoid established, we now build its composition machinery concretely. The goal is twofold: first, to present the uniform schema for compositions $\circ_k$ of n-cells and show that it organises identities, inverses, and whiskering into a single recursive pattern; and second, to verify that when $n = 2$ and $n = 3$, the abstract schema reproduces exactly the operations already constructed by hand in Chapters 7 and 8. The $n = 4$ case is spelled out as a concrete checkpoint—the first dimension where all coherence mechanisms interact simultaneously.

9.2.1 A uniform schema for $\circ_k$

Fix a weak ω-groupoid $\mathcal{G}$ in the sense of Definition 9.7, and fix a dimension $n \geq 1$. At this level there are exactly n composition directions:

$$\circ_{n-1},\ \circ_{n-2},\ \ldots,\ \circ_1,\ \circ_0.$$

The operation $\circ_{n-1}$ is the principal ("vertical") composition, while lower indices compose along deeper boundaries.

Definition 9.16 (Iterated identity lifting)**.** For integers $0 \leq m \leq n$, define

$$I^n_m : \mathcal{G}_m \to \mathcal{G}_n$$

recursively by

$$I^m_m = \mathrm{id}_{\mathcal{G}_m}, \qquad I^{n+1}_m = i_n \circ I^n_m.$$

The map I^n_m views an m-cell as a degenerate n-cell by stacking identities—the higher-dimensional analogue of inserting identity paths and identity 2-cells in Chapters 6 and 7.

Definition 9.17 (General whiskering at dimension n). Let $0 \le k < m < n$. For $u \in \mathcal{G}_m$ and $x \in \mathcal{G}_n$:

(i) if $t^k_m(u) = s^k_n(x)$, define *left k-whiskering*

$$u \triangleright_k x \ := \ I^n_m(u) \circ_k x \in \mathcal{G}_n;$$

(ii) if $s^k_m(u) = t^k_n(x)$, define *right k-whiskering*

$$x \triangleleft_k u \ := \ x \circ_k I^n_m(u) \in \mathcal{G}_n.$$

In plain terms, whiskering composes a cell with a fixed lower-dimensional boundary cell that has been lifted by identities to the relevant dimension. The "active" part is x; the whiskering cell u passes through unchanged.

Proposition 9.18 (Dimension-n unit and inverse behaviour). *Fix $n \ge 1$, $0 \le k < n$, and $x \in \mathcal{G}_n$. Then the weak structure provides $(n+1)$-cells*

$$\mathfrak{l}^{(k)}_x : I^n_{k+1}(1_{s^k_n(x)}) \circ_k x \Longrightarrow_{n+1} x, \qquad \mathfrak{r}^{(k)}_x : x \circ_k I^n_{k+1}(1_{t^k_n(x)}) \Longrightarrow_{n+1} x.$$

For the principal composition direction $k = n-1$, one also has

$$\lambda_x : x^{-1} \circ_{n-1} x \Longrightarrow_{n+1} 1_{s_n(x)}, \qquad \rho_x : x \circ_{n-1} x^{-1} \Longrightarrow_{n+1} 1_{t_n(x)}.$$

Moreover, whiskering preserves units and principal inverses up to $(n+1)$-cells.

Proof. The unit witnesses $\mathfrak{l}^{(k)}_x, \mathfrak{r}^{(k)}_x$ are part of Definition 9.7. The inverse witnesses λ_x, ρ_x are exactly those of Definition 9.5. Whiskering is built from $\circ_k$ and identity lifting (Definition 9.17), so compatibility follows from the functoriality/coherence data already included in Definition 9.7. □

Theorem 9.19 (Dimension-n composition package). *For each fixed $n \ge 1$, the operations $\circ_k$ $(0 \le k < n)$ satisfy:*

associativity for each $\circ_k$ up to $(n+1)$-cells,
unit laws for each $\circ_k$ up to $(n+1)$-cells,
principal inverse laws for $\circ_{n-1}$ up to $(n+1)$-cells,
interchange between $\circ_k$ and $\circ_\ell$ $(k < \ell < n)$ up to $(n+1)$-cells.

Proof. All four clauses are the dimension-n instance of the coherence data appearing in Definition 9.7. No additional structure is introduced here; this section records the package in the notation used for computational paths. □

9.2.2 Concrete composition atlas for $n = 4$

To make the abstract schema operational, we spell out the full dimension-4 package. Let $x, y \in \mathcal{G}_4$. When the corresponding boundaries match, we may compose in each of the four directions:

$$\begin{aligned} &x \circ_3 y \quad \text{(principal/vertical composition of 4-cells)},\\ &x \circ_2 y \quad \text{(composition along 2-dimensional boundaries)},\\ &x \circ_1 y \quad \text{(composition along 1-dimensional boundaries)},\\ &x \circ_0 y \quad \text{(composition along object boundaries)}. \end{aligned}$$

Definition 9.20 (Dimension-4 whiskering forms). Let $u \in \mathcal{G}_m$, $x \in \mathcal{G}_4$ with $m \in \{1, 2, 3\}$. For each admissible $k < m < 4$, the whiskerings are:

$$u \triangleright_k x := I^4_m(u) \circ_k x, \qquad x \triangleleft_k u := x \circ_k I^4_m(u).$$

Concretely:

(i) $m = 3$: whiskering a 4-cell by a fixed 3-cell boundary;

(ii) $m = 2$: whiskering by a 2-cell boundary, lifted twice;

(iii) $m = 1$: whiskering by a path-level boundary, lifted three times.

Example 9.21 (A nested $n = 4$ composite). Suppose $x_1, x_2, x_3, x_4 \in \mathcal{G}_4$ are composable in both directions $k = 3$ and $\ell = 2$. Then two canonical composites are:

$$\begin{aligned} \Xi_{\text{left}} &:= \big((x_1 \circ_3 x_2) \circ_2 (x_3 \circ_3 x_4)\big),\\ \Xi_{\text{right}} &:= \big((x_1 \circ_2 x_3) \circ_3 (x_2 \circ_2 x_4)\big). \end{aligned}$$

The $(3, 2)$-interchange law provides a 5-cell

$$\mathfrak{i}^{(2,3)}_{x_1,x_2,x_3,x_4} : \Xi_{\text{left}} \Longrightarrow_5 \Xi_{\text{right}}.$$

This is the first truly "four-dimensional" instance of interchange: the witness itself already lives one level above the source composites.

Proposition 9.22 (Six interchange directions at $n = 4$). *At dimension 4 there are exactly six basic interchange families, one for each ordered pair $0 \leq k < \ell < 4$. For every such pair and every compatible boundary frame, there is a structural 5-cell*

$$\mathfrak{i}^{(k,\ell)} : (x_1 \circ_k x_2) \circ_\ell (y_1 \circ_k y_2) \Longrightarrow_5 (x_1 \circ_\ell y_1) \circ_k (x_2 \circ_\ell y_2).$$

Proof. The set of index pairs is

$$\{(0,1),(0,2),(0,3),(1,2),(1,3),(2,3)\},$$

which has cardinality $\binom{4}{2} = 6$. The existence of a witness for each pair is exactly the dimension-4 instance of Definition 9.7 and Theorem 9.19. □

Remark 9.23 (General counting pattern). At dimension n, the number of primary interchange families is $\binom{n}{2}$. This simple combinatorial fact is a useful guide when organizing coherence lemmas: already for $n = 5$ there are 10 interchange directions, and for $n = 6$ there are 15. Thus explicit low-dimensional bookkeeping (especially $n = 4$) is not optional; it is the bridge to scalable all-level arguments.

9.2.3 Recovering the $n = 2$ operations of Chapter 7

Proposition 9.24 (Recovery at dimension 2). *At $n = 2$, the general schema reproduces exactly the operations of Chapter 7:*

(i) *$\circ_1$ is vertical composition $\bullet_v$ of 2-cells from Definition 7.1, with laws of Proposition 7.8;*

(ii) *$\circ_0$ is horizontal composition $\bullet_h$ from Definition 7.16;*

(iii) *the whiskerings from Definition 9.17 specialise to $f \triangleright \alpha$ and $\alpha \triangleleft h$ from Definitions 7.11 and 7.12.*

Proof. For $n = 2$ there are only two indices, $k = 1$ and $k = 0$. The case $k = 1$ composes 2-cells along shared 1-cell boundaries, which is exactly the typing of $\bullet_v$ in Definition 7.1. The case $k = 0$ composes along object boundaries; unfolding the identity-lifting construction from Definition 9.16 and the formulae of Definition 9.17 yields the two whiskerings of Chapter 7, and hence Definition 7.16 for $\bullet_h$. □

9.2.4 Recovering the $n = 3$ operations of Chapter 8

Proposition 9.25 (Recovery at dimension 3). *At $n = 3$, the abstract operations coincide with the concrete ones of Chapter 8:*

(i) *$\circ_2$ is the basic composition of 3-cells in Definition 8.1, with weak groupoid laws in Theorem 8.4;*

(ii) *$\circ_1$ and $\circ_0$ are the compositions of Definitions 8.7 and 8.11;*

(iii) *general whiskering specialises to 2-whiskering and horizontal whiskering (Definitions 8.6 and 8.10).*

Proof. For $n = 3$ the indices are $k = 2, 1, 0$. The principal case $k = 2$ is exactly the constructor-level composition already fixed in Definition 8.1. For $k = 1$ and $k = 0$, Definitions 8.7 and 8.11 define the corresponding compositions via whiskering followed by $\circ_2$; this is precisely the instance of Definition 9.17 when $m = 2$ and $n = 3$. Thus the abstract dimension-3 package here agrees with the concrete one summarised in Theorem 8.13. □

9.2.5 General n pattern: from principal to deep composition

The dimension-n operations admit a uniform recursive reading:

(i) $\circ_{n-1}$ is the principal composition on n-cells;

(ii) $\circ_{n-2}$ is induced by whiskering with lifted $(n-1)$-cells;

(iii) continuing downward, each $\circ_k$ is induced by composing along k-boundaries after suitable identity lifting.

Definition 9.26 (Recursive synthesis of lower compositions)**.** Fix $n \geq 2$ and assume $\circ_{n-1}$ is given. For $0 \leq k < n-1$, define $\circ_k$ by:

(i) lifting all boundary data to dimension n via I^n_{k+1};

(ii) forming whiskered composites with $\circ_{n-1}$;

(iii) quotienting by the structural interchange witnesses so that the result is independent (up to coherent higher cell) of the chosen decomposition.

Proposition 9.27 (Compatibility of recursive synthesis)**.** *Definition 9.26 yields operations with the same typing and boundary behaviour as Definition 9.6.*

Proof. Fix $0 \leq k < n-1$ and two n-cells x, y with $t^k_n(x) = s^k_n(y)$. We verify that the recursively synthesised operation produces an n-cell with the same typing and boundary behaviour required in Definition 9.6.

First, the pullback condition of Definition 9.3 gives exactly the matching k-boundary needed to begin the construction. Lifting boundary data by I^n_{k+1} does not alter any boundary below dimension $k+1$: the inserted cells are identities above level k, so the original k-source and k-target data are preserved. Therefore the whiskered pieces formed in steps (i) and (ii) of Definition 9.26 are still well-typed and remain composable at the principal level.

Next, principal composition $\circ_{n-1}$ has the correct immediate boundary behaviour by Definition 9.6. Since whiskering is itself defined by composing with identity lifts (Definition 9.17), it changes only the data above the chosen

k-boundary. Applying the remaining source and target maps down to level k therefore yields

$$s_n^k(x \circ_k y) = s_n^k(x), \qquad t_n^k(x \circ_k y) = t_n^k(y).$$

So the recursively synthesised operation has exactly the boundary behaviour of Definition 9.6.

Finally, one may realise the same lower composite by different decompositions into whiskered principal composites. Any two such decompositions have the same k-boundary data, so they differ only by rebracketing or by interchanging independent whiskered factors. Clause (iii) of Definition 9.26 quotients precisely by the structural interchange witnesses identifying these choices up to coherent higher cell. Hence the operation is well defined with the same typing and boundary behaviour as the abstract k-composition schema. □

Theorem 9.28 (Dimension-n coherence ladder)**.** *For each $n \geq 2$, the operation family*

$$(\circ_{n-1}, \circ_{n-2}, \dots, \circ_0)$$

admits a coherence ladder with the following levels:

Level 1: associator and unitor witnesses in dimension $n+1$,

Level 2: interchange witnesses between each pair (k, ℓ),

Level 3: compatibility of Level 1 and Level 2 witnesses in dimension $n+2$,

Level 4: comparison cells between any two structural composites in dimension $n+3$.

Proof. Level 1 is exactly the associator and unitor data built into Definition 9.7. Level 2 is Theorem 9.19, which supplies the interchange witnesses between every pair of composition directions.

For Level 3, consider any compatibility question between Level 1 and Level 2 witnesses. Both sides of such a compatibility are structural $(n+1)$-cells with common boundary in the sense of Definition 9.56. Therefore Theorem 9.61 applies and yields an $(n+2)$-cell between them.

For Level 4, let Θ, Θ' bc two Level 3 comparison cells between the same structural $(n+1)$-cell composites. They are two inhabitants of the same solution space from Definition 9.59, so the second clause of Theorem 9.61 provides an $(n+3)$-cell connecting Θ and Θ'.

These four clauses are exactly the announced coherence ladder. □

Remark 9.29 (Concrete check at $n = 4$)**.** For $n = 4$, Theorem 9.28 predicts:

(i) 5-cells for associativity/unit/interchange,

(ii) 6-cells comparing composites of those 5-cells,

(iii) 7-cells comparing the resulting 6-cell composites.

This matches the expected "coherence of coherence" progression and explains why dimension 4 is the minimal nontrivial test for all-level behaviour.

9.3 Main Theorem: Computational Paths Form a Weak ω-Groupoid

This is the central result toward which the entire chapter has been building. The preceding sections assembled the formal vocabulary; now we state the central theorem and prove it. The strategy is layered: three explicit base cases (dimensions 1, 2, and 3, drawing on the earlier chapters), followed by a uniform inductive step that extends the structure to all higher dimensions. The inductive step itself rests on higher-dimensional termination and confluence—the rewrite-theoretic engine that makes canonical comparison cells available at every level.

9.3.1 The Tower of Computational Higher Cells

Definition 9.30 (Computational cell tower of a type)**.** Fix a type A. Define sets of cells $\mathcal{C}_n(A)$ recursively:

(i) $\mathcal{C}_0(A) := \{a \mid a : A\}$;

(ii) $\mathcal{C}_1(A) := \{(a, b, p) \mid a, b : A,\ p : \mathsf{Path}(a, b)\}$;

(iii) for $n \geq 1$,

$$\mathcal{C}_{n+1}(A) := \{(x, y, \Gamma) \mid x, y \in \mathcal{C}_n(A) \text{ are parallel}, \\ \Gamma : x \Rrightarrow_{n+1} y\}.$$

Boundary maps are

$$s_{n+1}(x, y, \Gamma) = x, \qquad t_{n+1}(x, y, \Gamma) = y,$$

and identities are

$$i_n(x) := (x, x, \mathsf{refl}_x).$$

We write

$$\mathcal{C}(A) := \big(\mathcal{C}_n(A)\big)_{n \in \mathbb{N}}$$

for the full tower, and $\mathcal{C}_{\leq n}(A)$ for its truncation through dimension n.

This definition packages the constructions of Chapters 6, 7, and 8 into one recursive object: terms, paths, 2-cells, 3-cells, and then all higher derivations. The constructors are uniform—source and target pick boundaries, and refl inserts a degenerate higher cell—so the tower grows by the same mechanism at every level.

Proposition 9.31 (Type-valued cell tower). *For parallel boundaries in the present development, the higher cells are organised by the genuine inductive tower*

$$\begin{aligned}
\mathsf{Step}(p,q) &: \mathsf{Type}\, u, \\
\mathsf{RwEq}(p,q) &\simeq \mathsf{Derivation}_2(p,q) : \mathsf{Type}\,(u+1), \\
\mathsf{Derivation}_3(\alpha,\beta) &: \mathsf{Type}\,(u+2), \\
\mathsf{Derivation}_4(\Gamma,\Delta) &: \mathsf{Type}, \\
\mathsf{DerivationHigh}_m(\Xi_0,\Xi_1) &: \mathsf{Type} \qquad (m \geq 4).
\end{aligned}$$

None of the higher levels is represented by a trivial singleton type; each is an inductive family closed under identities, inverses, and principal composition.

Proof. The first layer is the ordinary step relation, which is already type-valued. The second layer is the 2-cell structure of Chapter 7: rewrite-equivalence witnesses and $\mathsf{Derivation}_2$ encode the same 2-dimensional comparison data via the mutually defined translations between them. The third layer is the 3-cell structure of Chapter 8.

From dimension 4 onward, the higher cells are not introduced as a proposition-valued tail but as genuine inductive families. The 4-dimensional layer $\mathsf{Derivation}_4$ is generated by identities, inverses, compositions, and the basic 4-dimensional comparison fillers, and the stable higher layers $\mathsf{DerivationHigh}_m$ continue the same pattern for $m \geq 4$. In particular, each level carries its own constructors for identities, inverses, and principal composition, so no stage is collapsed to a trivial singleton type. □

Remark 9.32 (Independent non-triviality checks). Each of the fourteen named coherence proofs—including pentagon, triangle, the three interchange laws, inverse and double-inverse coherence, contravariance, and Eckmann–Hilton—was retested by replacing its proof term with reflexivity. Every replacement failed. Thus the 3-dimensional coherence package is genuinely proof-relevant, even though the tower stabilises one dimension later.

Lemma 9.33 (Base case in dimension 1). *For every type A, the truncation $\mathcal{C}_{\leq 1}(A)$ carries a weak groupoid structure.*

Proof. This is exactly Theorem 6.15. The detailed witnesses of the eight laws are verified in Section 6.4 of Chapter 6. □

Lemma 9.34 (Base case in dimension 2). *For every type A, the truncation $\mathcal{C}_{\le 2}(A)$ carries the weak 2-dimensional structure whose:*

(i) principal composition is $\circ_1 = \bullet_v$,

(ii) secondary composition is $\circ_0 = \bullet_h$,

(iii) interchange and bicategorical coherence hold up to 3-cells.

Proof. Definition 7.1 gives identities, inverses, and vertical composition $\bullet_v$. Proposition 7.8 provides the weak groupoid laws for $\bullet_v$. Definitions 7.11, 7.12, and 7.16 give whiskering and horizontal composition $\bullet_h$. Theorem 7.32 and Theorem 7.41 establish the required coherence package. □

Lemma 9.35 (Base case in dimension 3). *For every type A, the truncation $\mathcal{C}_{\le 3}(A)$ carries the weak 3-dimensional composition structure with operations $\circ_2, \circ_1, \circ_0$, interchange, and coherent structural 3-cells.*

Proof. The principal 3-cell composition and inverse behaviour are provided by Definition 8.1 and Theorem 8.4. The additional compositions $\circ_1, \circ_0$ are given by Definitions 8.7 and 8.11, and their joint laws by Theorem 8.13. Interchange data are Theorems 8.15, 8.16, and 8.17, while coherent structural witnesses are organised in Theorem 8.25. □

9.3.2 Higher-Dimensional Termination and Confluence

The base cases are now in place; the challenge is the inductive step. To extend the weak ω-groupoid structure from dimension n to $n+1$, we need a way to produce canonical comparison cells between any two parallel structural composites. The key insight is that the same rewrite-theoretic tools that gave us confluence in Chapter 5—termination plus local confluence plus Newman's lemma—lift cleanly to higher dimensions. If parallel $(n+1)$-cells can always be normalised to a common form, then the normalisation derivations themselves furnish the required comparison witnesses.

Definition 9.36 (Lifted one-step rewriting in dimension $n+1$). Fix $n \ge 1$. Define $\rightsquigarrow_{n+1}$ as the smallest relation on $\mathcal{C}_{n+1}(A)$ generated by:

(i) elementary applications of Chapter 5 rewrite rules at a distinguished $(n+1)$-redex;

(ii) closure under inverse, principal composition $\circ_n$, and whiskering by admissible lower-dimensional boundary data.

Write $\rightsquigarrow^*_{n+1}$ for the reflexive-transitive closure.

Definition 9.37 (Lexicographic measure for lifted rewriting)**.** For an $(n+1)$-cell Γ, define

$$\mu_{n+1}(\Gamma) := \big(r(\Gamma), \ell(\Gamma)\big) \in \mathbb{N} \times \mathbb{N},$$

where:

(i) $r(\Gamma)$ is the maximal Chapter 5 rule rank occurring in Γ;

(ii) $\ell(\Gamma)$ is the number of occurrences in Γ with rank $r(\Gamma)$.

If Γ contains no redex, set $r(\Gamma) = \ell(\Gamma) = 0$. Order $\mathbb{N} \times \mathbb{N}$ lexicographically:

$$(r, \ell) <_{\text{lex}} (r', \ell') \iff \big(r < r'\big) \text{ or } \big(r = r' \text{ and } \ell < \ell'\big).$$

Lemma 9.38 (One-step decrease of the lifted measure)**.** *If* $\Gamma \rightsquigarrow_{n+1} \Gamma'$, *then*

$$\mu_{n+1}(\Gamma') <_{\text{lex}} \mu_{n+1}(\Gamma).$$

Proof. By Definition 9.36, a one-step reduction changes one distinguished redex inside a fixed $(n+1)$-context generated by inverse, composition, and whiskering. At the rewritten position, the Chapter 5 ranking discipline decreases strictly; outside that position, the context is unchanged. Therefore either the maximal rank drops, or it is preserved and the number of maximal-rank occurrences decreases. This is exactly strict decrease for $<_{\text{lex}}$. □

Theorem 9.39 (Higher-dimensional termination)**.** *For every* $n \geq 1$, *there is no infinite chain*

$$\Gamma_0 \rightsquigarrow_{n+1} \Gamma_1 \rightsquigarrow_{n+1} \Gamma_2 \rightsquigarrow_{n+1} \cdots$$

in $\mathcal{C}_{n+1}(A)$.

Proof. Lemma 9.38 gives strict descent of μ_{n+1} at every step. Since the lexicographic order on $\mathbb{N} \times \mathbb{N}$ is well-founded, an infinite descending chain is impossible. Hence $\rightsquigarrow_{n+1}$ terminates. □

Proposition 9.40 (Higher-dimensional local confluence)**.** *For every* $n \geq 1$, *the relation* $\rightsquigarrow_{n+1}$ *is locally confluent.*

Proof. Consider a fork

$$\Gamma_1 \xleftarrow{\rightsquigarrow_{n+1}} \Gamma \xrightarrow{\rightsquigarrow_{n+1}} \Gamma_2.$$

If the two steps contract disjoint redexes, they commute by context independence. If they contract the same redex, they trivially rejoin. If they form a genuine overlap, project that overlap to the underlying Chapter 5 critical configuration and apply Theorem 5.19. Reinserting the obtained join into the ambient $(n+1)$-context gives

$$\Gamma_1 \rightsquigarrow^*_{n+1} \Delta, \qquad \Gamma_2 \rightsquigarrow^*_{n+1} \Delta$$

for some Δ. Hence $\rightsquigarrow_{n+1}$ is locally confluent. □

Theorem 9.41 (Higher-dimensional confluence). *For every $n \geq 1$, the reflexive-transitive closure $\rightsquigarrow_{n+1}^*$ is confluent.*

Proof. By Theorem 9.39, $\rightsquigarrow_{n+1}$ is terminating. By Proposition 9.40, it is locally confluent. Newman's lemma [116] (Lemma 5.20) therefore gives confluence of $\rightsquigarrow_{n+1}^*$. □

Remark 9.42 (Formal verification of the lifted rewrite argument). The base confluence theorem is established on groupoid expressions using the free-group interpretation, together with invariance and canonical reachability properties. The ranking data from Chapter 5 and its well-foundedness proof provide the termination argument; the lifted typed pipeline uses a diamond-to-confluence conversion and a formalised Newman's lemma. In the current mechanised proof this lifted pipeline is implemented by isolating the inverse loop determined by parallel 2-cells, normalising that loop, comparing it by explicit 3-cell generators, and then iterating the same idea with generic higher fillers. Full details appear in the Formalization Notes at the end of this chapter.

Lemma 9.43 (Inductive extension step). *Let $n \geq 3$. Assume $\mathcal{C}_{\leq n}(A)$ is already equipped with weak n-dimensional composition data compatible with rewriting. Then derivations between parallel n-cells define $\mathcal{C}_{n+1}(A)$ together with:*

(i) source/target and identity maps in dimension $n+1$;

(ii) principal composition $\circ_n$ and weak inverses;

(iii) induced lower compositions $\circ_k$ $(0 \leq k < n)$ by whiskering;

(iv) interchange and coherence witnesses one dimension higher.

Hence $\mathcal{C}_{\leq n+1}(A)$ is a weak $(n+1)$-groupoid truncation.

Proof. Fix $n \geq 3$ and assume the weak structure on $\mathcal{C}_{\leq n}(A)$. We construct the four items explicitly.

(i) Source, target, identities. For each $(n+1)$-cell (x, y, Γ) with $\Gamma : x \Rrightarrow_{n+1} y$, set

$$s_{n+1}(x, y, \Gamma) := x, \qquad t_{n+1}(x, y, \Gamma) := y, \qquad i_n(x) := (x, x, \mathsf{refl}_x).$$

These satisfy the globular identities by direct inspection of the recursive cell-tower definition (Definition 9.30).

(ii) Principal composition and weak inverse. For composable $(n+1)$-cells $\Gamma : x \Rrightarrow_{n+1} y$ and $\Delta : y \Rrightarrow_{n+1} z$, define

$$\Gamma \circ_n \Delta$$

as concatenation of derivations. Define Γ^{-1} by reversing this derivation. The unit and inverse laws for $\circ_n$ are witnessed by the usual reassociation/cancellation derivations, hence by $(n+2)$-cells.

(iii) Induced lower compositions. Fix $0 \leq k < n$ and composable data

$$\Gamma : x \Rrightarrow_{n+1} x', \qquad \Delta : y \Rrightarrow_{n+1} y', \qquad t_n^k(x) = s_n^k(y).$$

Using the already-constructed k-composition on n-cells, define whiskered derivations

$$L_k(x, \Delta) : x \circ_k y \Rrightarrow_{n+1} x \circ_k y', \qquad R_k(\Gamma, y') : x \circ_k y' \Rrightarrow_{n+1} x' \circ_k y',$$

and set

$$\Gamma \circ_k \Delta := L_k(x, \Delta) \circ_n R_k(\Gamma, y').$$

Boundary equations follow from the boundary behaviour of $\circ_k$ in dimension n.

(iv) Interchange and higher coherence. Each required law compares two parallel $(n+1)$-composites with the same globular boundary. By Theorem 9.41, both composites reduce to a common normal form under $\rightsquigarrow^*_{n+1}$. In the first genuinely higher case, the proof normalises $\mathsf{Derivation}_2$ terms by isolating their inverse loop, normalising that loop, and contracting it by an explicit structural 3-cell. One level higher, parallel 3-cells are connected by normalisation plus a generic 4-dimensional filler; from then on the same argument is packaged uniformly by analogous higher fillers. Composing the normalisation derivation from each side to the common target therefore produces an explicit $(n+2)$-comparison cell. Applying the same construction one level up yields compatibility among those comparison cells, giving the needed coherence layer.

All four clauses are now constructed, so $\mathcal{C}_{\leq n+1}(A)$ is a weak $(n+1)$-groupoid truncation. □

Proposition 9.44 (Dependency chain for the main theorem)**.** *The proof of the main theorem depends on:*

$$(\textit{Lemma 9.33}),\ (\textit{Lemma 9.34}),\ (\textit{Lemma 9.35}),\ (\textit{Lemma 9.43}) \Longrightarrow \textit{Theorem 9.45}.$$

Proof. Lemmas 9.33–9.35 provide the initial dimensions. Lemma 9.43 propagates the structure from n to $n+1$. Therefore every finite dimension is covered, yielding the full ω-dimensional statement. □

Theorem 9.45 (Computational paths form a weak ω-groupoid)**.** *For every type A, the tower*

$$\mathcal{C}(A) = \big(\mathcal{C}_n(A),\ s_n,\ t_n,\ i_n,\ \circ_k,\ (-)^{-1}\big)_{n\in\mathrm{N},\ 0\leq k<n}$$

defined from computational paths is a weak ω-groupoid. Equivalently, for each dimension $n \geq 1$, all composition, unit, inverse, interchange, and coherence laws hold up to $(n+1)$-cells.

Proof. By Definition 9.30, together with Definitions 9.1, 9.4, and 9.6, we have the required globular data and operations. By Proposition 9.8, all operations are well-typed. Lemmas 9.33, 9.34, and 9.35 establish the first three dimensions explicitly from Chapters 6, 7, and 8. Applying Lemma 9.43 recursively gives the weak structure in every higher dimension. Hence $\mathcal{C}(A)$ satisfies Definition 9.7. □

9.3.3 The $n = 4$ instance in detail

The main theorem is dimension-uniform, but the first new case beyond Chapter 8 is $n = 4$. Spelling it out clarifies the inductive mechanism.

Definition 9.46 (Fourth-order computational cells)**.** For a type A, define $\mathcal{C}_4(A)$ as triples

$$(\alpha, \beta, \Xi)$$

where $\alpha, \beta \in \mathcal{C}_3(A)$ are parallel 3-cells and $\Xi : \alpha \Rrightarrow_4 \beta$ is a fourth-order derivation. Source/target and identity are

$$s_4(\alpha, \beta, \Xi) = \alpha, \qquad t_4(\alpha, \beta, \Xi) = \beta, \qquad i_3(\alpha) = (\alpha, \alpha, \mathsf{refl}_\alpha).$$

Proposition 9.47 (Dimension-4 operation package)**.** *On $\mathcal{C}_4(A)$ one has:*

(i) principal composition $\circ_3$ by concatenation of 4-derivations;

(ii) lower compositions $\circ_2, \circ_1, \circ_0$ induced by whiskering;

(iii) weak inverses for $\circ_3$;

(iv) six interchange families (Proposition 9.22).

All laws hold up to 5-cells.

Proof. Chapter 8 already provides the weak 3-dimensional structure on $\mathcal{C}_{\leq 3}(A)$. Applying Lemma 9.43 with $n = 3$ therefore produces $\mathcal{C}_4(A)$ together with principal composition $\circ_3$ given by concatenation of 4-derivations, weak inverses for $\circ_3$, and lower compositions $\circ_2, \circ_1, \circ_0$ obtained by whiskering. This is exactly items (i)–(iii).

For item (iv), Proposition 9.22 enumerates the ordered pairs $0 \leq k < \ell < 4$ and gives the corresponding six interchange families. Finally, the witnesses supplied by Lemma 9.43 live one dimension above the composites they compare. Hence in dimension 4 all unit, inverse, interchange, and coherence laws are witnessed by 5-cells. □

Remark 9.48 (Why this is enough for induction). The step from $n = 3$ to $n = 4$ displays every structural phenomenon that can occur at higher dimensions: multiple lower compositions, multiple interchange directions, and coherence between structural witnesses. Once this pattern is explicit, passing from arbitrary n to $n + 1$ is a uniform notational lift rather than a new conceptual argument.

9.3.4 Canonical Comparison Cells from Normalization

The core rewrite-theoretic engine behind the main theorem is the availability of canonical comparison cells between parallel composites.

Definition 9.49 (Normalization comparator). Fix $n \geq 1$ and parallel n-cells $x, y \in C_n(A)$. Let $\mathrm{nf}(x)$ and $\mathrm{nf}(y)$ be normal forms computed via Chapter 5 normalisation. When $\mathrm{nf}(x) = \mathrm{nf}(y)$, define

$$\kappa_{x,y} : x \Longrightarrow_{n+1} y$$

to be the composite of:

(i) the normalisation derivation $x \Longrightarrow_{n+1} \mathrm{nf}(x)$,

(ii) the identity on the shared normal form,

(iii) the inverse normalisation derivation $\mathrm{nf}(y) \Longrightarrow_{n+1} y$.

Proposition 9.50 (Canonical behaviour of comparators). *Normalization comparators satisfy:*

$$\kappa_{x,x} = \mathsf{refl}_x, \qquad \kappa_{x,y}^{-1} = \kappa_{y,x},$$

and any two comparator constructions with the same source, target, and common normal form are connected by an $(n + 2)$-cell.

Proof. The identity formula is immediate: when $x = y$, the definition of $\kappa_{x,x}$ composes the normalisation derivation $x \Longrightarrow_{n+1} \mathrm{nf}(x)$ with its inverse, so the resulting composite is the reflexivity $(n + 1)$-cell on x.

For the inverse formula, simply reverse the defining zig-zag. The comparator from x to y is obtained by travelling from x to the shared normal form and then back to y; reversing that path yields the comparator from y to x. Hence $\kappa_{x,y}^{-1} = \kappa_{y,x}$.

Now fix two choices of normalisation data producing comparators from x to y. Both are parallel $(n + 1)$-cells with the same source and target. By Theorem 9.41, parallel $(n + 1)$-dimensional derivations admit a common reduct, and the comparison mechanism in Lemma 9.43 turns that common reduct into an $(n + 2)$-cell between the two comparator constructions. Thus the comparator is canonical up to one higher dimension. □

Theorem 9.51 (Inductive construction via comparators). *The inductive step of Lemma 9.43 may be realised by choosing normalisation comparators between parallel structural composites and a shared normal form. In this sense the higher witnesses required by Definition 9.7 can be selected internally from termination plus confluence, canonically up to one more dimension.*

Proof. Let u and v be parallel structural composites with the same globular boundary. By Theorem 9.41, both reduce to a common normal form n. The normalisation data therefore supply comparators

$$\kappa_{u,n} : u \Rrightarrow_{n+1} n, \qquad \kappa_{v,n} : v \Rrightarrow_{n+1} n.$$

Composing $\kappa_{u,n}$ with the inverse of $\kappa_{v,n}$ gives a comparison

$$u \Rrightarrow_{n+1} v.$$

This is exactly the kind of higher witness required in item (iv) of Lemma 9.43.

If different normalisation derivations or a different chosen common normal form are used, Proposition 9.50 supplies an $(n+2)$-cell between the resulting comparison cells. Thus the inductive witness is canonical up to one more dimension, while still arising entirely from the rewrite-theoretic facts of termination and confluence. □

9.3.5 Finite truncations and compatibility in n

Proposition 9.52 (Truncation compatibility). *For $m < n$, the truncation map*

$$\tau_{\leq m} : \mathcal{C}_{\leq n}(A) \to \mathcal{C}_{\leq m}(A)$$

preserves identities, compositions, and inverses in every retained dimension, up to the retained structural cells.

Proof. By construction, $\tau_{\leq m}$ forgets cells above dimension m and leaves all lower-dimensional data unchanged. Thus operations and witnesses in dimensions $\leq m$ are preserved definitionally, while higher coherence not expressible below m is simply omitted. □

Remark 9.53 (General n schema). Theorem 9.45 can be read as the statement that all finite truncations are mutually compatible and jointly determine a single weak ω-groupoid object. In practice, proofs are carried out at finite height: up to dimension 3 this records genuine algebraic content, while above dimension 3 the truncation maps differ only by discarding contractible filler data supplied by the stabilisation theorem below.

9.4 Coherence in All Dimensions

The main theorem is proved, but the coherence data it invokes deserve a closer look. This section unpacks the all-level coherence package: interchange families for every pair of composition directions, associator–unitor compatibility, stability of structural cells under whiskering, and the contractibility principle that controls uniqueness of comparison witnesses. The upshot is that coherence grows polynomially with dimension—manageable, not chaotic—and stabilises after dimension 3 in the precise sense of Theorem 9.60.

9.4.1 Interchange Families in Arbitrary Dimension

Chapter 8 established the three-dimensional interchange laws (Theorems 8.15, 8.16, 8.17). At general dimension n, each ordered pair of indices $0 \leq k < \ell < n$ gives an interchange family.

Theorem 9.54 ((k, ℓ)-interchange family). *Fix $n \geq 2$ and $0 \leq k < \ell < n$. For any compatible n-cells x_1, x_2, y_1, y_2 in $\mathcal{C}_n(A)$, there is an $(n+1)$-cell*

$$\mathfrak{i}^{(k,\ell)}_{x_1,x_2,y_1,y_2} : (x_1 \circ_k x_2) \circ_\ell (y_1 \circ_k y_2) \Longrightarrow_{n+1} (x_1 \circ_\ell y_1) \circ_k (x_2 \circ_\ell y_2).$$

When $n = 3$, this specialises to Theorems 8.15, 8.16, and 8.17.

Proof. Theorem 9.45 states that $\mathcal{C}(A)$ is a weak ω-groupoid. Unpacking the interchange clause from Definition 9.7 at dimension n and index pair (k, ℓ) therefore produces exactly the displayed $(n+1)$-cell

$$\mathfrak{i}^{(k,\ell)}_{x_1,x_2,y_1,y_2}.$$

The typing conditions are those of Proposition 9.8. When $n = 3$, the possible index pairs are precisely $(1, 2)$, $(0, 2)$, and $(0, 1)$, so this recovers Theorems 8.15, 8.16, and 8.17. □

9.4.2 Associativity and Unit Coherence at All Levels

Beyond individual interchange laws, the associators and unitors themselves must satisfy higher compatibility (the analogues of pentagon and triangle families from Chapters 7 and 8).

Theorem 9.55 (Associator–unitor coherence family). *Fix $n \geq 2$ and $0 \leq k < n$. Then:*

(i) all iterated k-associators between two bracketings of a (k)-composable string of n-cells are connected by $(n+2)$-cells;

(ii) *the k-triangle compatibility between $\mathfrak{a}^{(k)}$, $\mathfrak{l}^{(k)}$, and $\mathfrak{r}^{(k)}$ holds up to $(n+2)$-cells.*

Equivalently, the pentagon–triangle pattern of Theorems 7.35 and 7.37 persists dimensionwise.

Proof. Fix $k < n$. Any bracketing of a k-composable string of n-cells is related to any other by a composite of the basic k-associators $\mathfrak{a}^{(k)}$. Two such reassociation composites are structural $(n+1)$-cells with the same source and target, so they define a structural boundary problem in the sense of Definition 9.59. Theorem 9.61 therefore supplies an $(n+2)$-cell between them. This proves (i).

For (ii), the two sides of the k-triangle identity are again structural $(n+1)$-cells, now built from one associator and two unitors. They have the same boundary by the unit laws from Definition 9.7. Applying Theorem 9.61 once more yields the required $(n+2)$-cell between those two triangle composites.

Thus the pentagon–triangle pattern persists at every dimension. □

9.4.3 Structural Higher Cells and Naturality

Definition 9.56 (Structural $(n+1)$-cells)**.** For each $n \geq 2$, define *structural $(n+1)$-cells* to be the smallest class of $(n+1)$-cells containing:

(i) associators $\mathfrak{a}^{(k)}$,

(ii) left/right unitors $\mathfrak{l}^{(k)}, \mathfrak{r}^{(k)}$,

(iii) interchange witnesses $\mathfrak{i}^{(k,\ell)}$,

and closed under identities, inverses, all compositions $\circ_j$ $(0 \leq j \leq n)$, and admissible left/right whiskering by lower-dimensional boundary cells.

This is the direct all-level analogue of Definition 8.20 in Chapter 8: structural cells are the canonical coherence witnesses generated by the algebraic operations, not ad hoc extra data.

Theorem 9.57 (Naturality coherence family)**.** *Structural $(n+1)$-cells are stable under left and right whiskering by admissible lower-dimensional boundary cells. More precisely, if*

$$\Sigma : u \Longrightarrow_{n+1} v$$

is structural and w is an admissible whiskering cell, then the whiskered cells

$$w \triangleright_k \Sigma, \qquad \Sigma \triangleleft_k w$$

are again structural. Moreover, whenever two such whiskered structural composites have the same boundary, they are connected by an $(n+2)$-cell.

Proof. For $n = 2$, this is exactly the naturality of associators and unitors from Theorems 7.39 and 7.40. For $n = 3$, it is the whiskering-stability portion of Theorem 8.24.

In general, whiskering is defined by composition with an iterated identity lift (Definition 9.17). Thus a left whiskering of Σ has the form

$$w \triangleright_k \Sigma = I_m^{n+1}(w) \circ_k \Sigma,$$

and similarly on the right. The factor $I_m^{n+1}(w)$ is an iterated identity, so Definition 9.56 places these whiskered cells again inside the structural class.

If two such whiskered structural composites have the same boundary, they form a structural boundary problem. Theorem 9.61 therefore provides an $(n+2)$-cell comparing them. This is precisely the claimed naturality coherence. □

9.4.4 The All-Level Coherence Package

Theorem 9.58 (All-level coherence package)**.** *For every $n \geq 2$, the following hold for computational n-cells:*

(C1) interchange families for all $0 \leq k < \ell < n$,

(C2) associativity/unit coherence families at each $k < n$,

(C3) stability of structural $(n+1)$-cells under left/right whiskering,

(C4) any two structural composites with the same globular boundary are connected by an $(n+2)$-cell.

Hence coherence is stable under passage from dimension n to $n+1$.

Proof. Clauses (C1), (C2), and (C3) are exactly Theorems 9.54, 9.55, and 9.57.

For (C4), let Σ_0, Σ_1 be structural $(n+1)$-cell composites with the same globular boundary. Then (Σ_0, Σ_1) is a structural boundary problem in the sense of Definition 9.59, so Theorem 9.61 yields an $(n+2)$-cell between them.

Thus the structural coherence data at dimension n are accompanied by one further layer of comparison cells at dimension $n+1$. For $n = 2$ this recovers Theorem 7.41, and for $n = 3$ it recovers Theorem 8.25. □

9.4.5 Coherence as Higher Contractibility

Theorem 9.58 can be reformulated in a way that is often more practical for calculations: every structural boundary shape is "contractible up to one more dimension." Crucially, this stabilisation begins at dimension 3: 2-cells are not declared contractible, preserving non-trivial π_1 behaviour.

Definition 9.59 (Structural boundary problem at level n)**.** Fix $n \geq 2$. A *structural boundary problem* consists of two parallel structural $(n+1)$-cells

$$\Sigma_0, \Sigma_1 : u \Longrightarrow_{n+1} v$$

built from associators, unitors, interchange, whiskering, identities, inverses, and compositions. Its *solution type* is

$$\mathrm{Sol}(\Sigma_0, \Sigma_1) := \big(\Sigma_0 \Longrightarrow_{n+2} \Sigma_1\big).$$

Theorem 9.60 (Structural stabilization theorem)**.** *For the structural coherence package of Definition 9.56, the present development supplies canonical comparison operations one dimension higher:*

(i) any two parallel structural 2-cells are connected by a structural 3-cell;

(ii) any two parallel structural 3-cells are connected by a structural 4-cell;

(iii) for every $m \geq 4$, any two parallel structural m-cells are connected by a structural $(m+1)$-cell.

Consequently the computational tower retains genuine low-dimensional witness data through dimension 3, while the designated structural coherence fillers in dimensions 4 and above are canonically comparable one level higher. In this precise sense, the structural coherence package of $\mathcal{C}(A)$ stabilises after dimension 3.

Proof. For structural 2-cells, start with parallel coherence witnesses d_0, d_1. Form the inverse loop $d_0 \circ_2 d_1^{-1}$ based at d_1, normalise that loop by the rewrite-theoretic machinery of Chapter 5, and then use the 3-dimensional coherence witnesses of Chapter 8 to contract the normal form back to reflexivity. This yields a 3-cell from d_0 to d_1.

For structural 3-cells, apply the same strategy to parallel coherence witnesses: compare them by forming the corresponding inverse loop, normalise the resulting 4-dimensional expression, and close the comparison with the generic 4-dimensional filler supplied by the dimension-4 package.

Once dimension 4 is reached, the same comparison pattern is packaged uniformly by the higher filler operations. Hence the tower retains genuine, non-trivial witness data through dimension 3, while structural coherence problems in dimensions 4 and above admit canonical fillers. This is precisely the claimed stabilization after dimension 3. □

Theorem 9.61 (Weak contractibility of structural solution spaces)**.** *For every structural boundary problem (Σ_0, Σ_1) at level n, the set $\mathrm{Sol}(\Sigma_0, \Sigma_1)$ is inhabited. Moreover, any two inhabitants are connected by an $(n+3)$-cell.*

Proof. Existence is immediate from Theorem 9.60. When $n = 2$, that theorem provides a 4-cell comparing any parallel structural 3-cells; when $n = 3$, it provides a 5-cell comparing any parallel structural 4-cells; and for $n \geq 4$ it provides the stable clause $\mathsf{contractibility}_{n+1}$ for parallel structural $(n+1)$-cells. Applying the relevant case to Σ_0 and Σ_1 therefore yields an inhabitant of $\mathrm{Sol}(\Sigma_0, \Sigma_1)$.

Now let $\Theta, \Theta' \in \mathrm{Sol}(\Sigma_0, \Sigma_1)$. By definition, Θ and Θ' are parallel structural $(n+2)$-cells with common source Σ_0 and common target Σ_1. Definition 9.56 closes the class of structural cells under identities, inverses, whiskering, and composition, so this pair again lies in the scope of Theorem 9.60, now one dimension higher. Hence there is an $(n+3)$-cell

$$\Theta \Longrightarrow_{n+3} \Theta'.$$

Thus $\mathrm{Sol}(\Sigma_0, \Sigma_1)$ is inhabited, and any two of its inhabitants are connected by one further higher cell. □

Remark 9.62 (Interpretive consequence)**.** Theorem 9.61 does *not* assert strict uniqueness of coherence witnesses. Instead it provides controlled uniqueness up to one higher witness, exactly the form needed in weak higher-categorical semantics. In the computational-path setting, these bridges are themselves higher derivations internal to the tower: parallel 2-cells are compared by 3-cells, parallel 3-cells by 4-cells, and the same pattern stabilises in higher dimensions. Thus the theorem identifies a controlled higher uniqueness principle rather than a collapse to strict equality.

9.4.6 Explicit coherence bookkeeping at $n = 4$

Because dimension 4 is the first level where all coherence mechanisms interact, we make the bookkeeping explicit.

Definition 9.63 (Primitive coherence generators in dimension 4)**.** For computational 4-cells, define the primitive generator families:

(i) associators $\mathfrak{a}^{(k)}$ for each $k \in \{0, 1, 2, 3\}$;

(ii) unitors $\mathfrak{l}^{(k)}, \mathfrak{r}^{(k)}$ for each $k \in \{0, 1, 2, 3\}$;

(iii) interchange witnesses $\mathfrak{i}^{(k,\ell)}$ for $0 \leq k < \ell < 4$ (six families);

(iv) naturality witnesses for all above generators under admissible whiskering.

These are 5-cells; relations among them are encoded by 6-cells.

Proposition 9.64 (Generator count at $n = 4$). *Ignoring object-dependent multiplicities, the index-level count of primitive coherence families at dimension 4 is:*

$$\underbrace{4}_{\textit{associator directions}} + \underbrace{8}_{\textit{left/right unitor directions}} + \underbrace{6}_{\textit{interchange directions}} = 18.$$

Proof. There is one associator family per $k < 4$, giving 4. There are two unitor families (left/right) per $k < 4$, giving 8. Interchange directions are indexed by pairs $0 \le k < \ell < 4$, giving $\binom{4}{2} = 6$. Summing yields 18. □

Example 9.65 (Two routes through a fourfold composite). Let $x_1, \ldots, x_4$ be 4-cells with common composability data. Consider the two structural 5-cells from $((x_1 \circ_3 x_2) \circ_3 x_3) \circ_3 x_4$ to $x_1 \circ_3 (x_2 \circ_3 (x_3 \circ_3 x_4))$:

(i) the direct rebracketing path using two associators;

(ii) the path using one associator, one interchange with a degenerate identity-lifted boundary, then another associator.

Theorem 9.61 gives a 6-cell connecting these two 5-cells. This is the dimension-4 analogue of the pentagon comparison in Chapter 7, now enriched by interchange involvement.

Theorem 9.66 (Dimension-4 coherence consolidation). *Any two structural 5-cell composites in dimension 4 with the same globular boundary are connected by a 6-cell; any two such 6-cells are connected by a 7-cell.*

Proof. Apply Theorem 9.58 with $n = 4$. Clause (C4) states that any two structural composites with the same globular boundary are connected by one higher cell, so any two structural 5-cell composites in dimension 4 are connected by a 6-cell.

For the second claim, let Θ, Θ' be two such 6-cells. They are two inhabitants of the same solution space between the given structural 5-cell composites, so the uniqueness clause of Theorem 9.61 yields a 7-cell

$$\Theta \Longrightarrow_7 \Theta'.$$

□

9.4.7 General n: coherence growth and control

Proposition 9.67 (Combinatorics of primary coherence directions). *At dimension $n \ge 2$:*

(i) associator directions are indexed by $k < n$, hence n in number;

(ii) unitor directions (left/right) contribute $2n$ *families;*

(iii) interchange directions contribute $\binom{n}{2}$ *families.*

Therefore the index-level number of primitive coherence families is

$$n + 2n + \binom{n}{2} = \frac{n^2 + 5n}{2}.$$

Proof. Immediate from the indexing sets in Definition 9.7. □

Remark 9.68 (Growth is polynomial, not chaotic)**.** Proposition 9.67 shows that the number of primary families grows quadratically in n. This is substantial but manageable, and it explains why systematic notation ($\circ_k$, I^n_m, $\mathfrak{a}^{(k)}$, $\mathfrak{i}^{(k,\ell)}$) is enough to keep the all-level theory tractable.

Theorem 9.69 (All-level transport of coherence)**.** *Let* Σ *be a structural* $(n+1)$*-cell expression and let* $F : \mathcal{C}(A) \to \mathcal{C}(B)$ *be induced by a function* $f : A \to B$ *(Theorem 9.96). Then* $F(\Sigma)$ *is again structural, and every coherence comparison between structural expressions in* $\mathcal{C}(A)$ *is mapped to a coherence comparison in* $\mathcal{C}(B)$ *one dimension up.*

Proof. By Theorem 9.96, the map F preserves identities, compositions, inverses, and whiskering up to structural higher cells. Structural expressions are generated from the basic associators, unitors, and interchange witnesses by repeatedly applying exactly those operations. Therefore an induction on the derivation tree for Σ shows that $F(\Sigma)$ is again structural.

Now consider a coherence comparison between two structural expressions in $\mathcal{C}(A)$. Such a comparison is itself built from structural higher cells using the same operations one dimension higher. Applying the same inductive argument to its derivation tree shows that its image under F is a coherence comparison in $\mathcal{C}(B)$ at the next dimension. □

Remark 9.70 (Coherence as functorial resource)**.** Coherence is often presented as a constraint. In the computational-path setting, it is also a resource: coherence witnesses can be transported, composed, inverted, and reused in new contexts. This point is central in the connection sections below.

9.5 Connection with Grothendieck's Homotopy Hypothesis

Having established the internal algebraic structure, we now turn outward. Grothendieck's homotopy hypothesis [77] proposes that homotopy types are equivalent, in a suitable higher-categorical sense, to weak ∞-groupoids. This conjecture has been a driving force in the development of higher category theory [11, 99, 91, 108]. Our construction gives a concrete syntactic candidate on the type-theoretic side: the weak ω-groupoid of computational paths. We do not claim to prove the full hypothesis here—that would require a separate geometric comparison theorem—but we show that the algebraic prerequisites are satisfied internally, and that the low-dimensional truncations match the expected homotopy-theoretic behaviour.

Definition 9.71 (Truncation of a weak ω-groupoid)**.** For a weak ω-groupoid $\mathcal{G}$ and $n \in \mathbb{N}$, its *n-truncation* $\tau_{\leq n}\mathcal{G}$ is obtained by keeping k-cells for $0 \leq k \leq n$ and forgetting all cells of dimension $> n$.

Proposition 9.72 (Low-dimensional evidence for the hypothesis)**.** *For each type A, the truncations of $\mathcal{C}(A)$ satisfy:*

$\tau_{\leq 1}\mathcal{C}(A)$ *is the weak path groupoid of Theorem 6.15,* $\tau_{\leq 2}\mathcal{C}(A)$ *has bicategorical coherence of Theorem 7.41,*

and

$$\tau_{\leq 3}\mathcal{C}(A) \text{ has the 3-dimensional coherence package of Theorem 8.25.}$$

Proof. This is immediate from Lemmas 9.33, 9.34, and 9.35, together with Theorem 9.45. □

Remark 9.73 (Interpretation)**.** Proposition 9.72 is precisely the finite dimensional shadow expected from the homotopy hypothesis in the genuinely nontrivial range: dimensions 1, 2, and 3. Above dimension 3 the computational tower adds canonical contractible fillers rather than new homotopical data, so higher truncations refine bookkeeping without altering the underlying 3-type-like content of the derivation tower. This chapter therefore supplies internal, rewrite-theoretic evidence for the Grothendieck viewpoint [77]. The relationship between weak ω-groupoids and homotopy types has been explored from several angles: via globular operads by Batanin [11], via higher operads by Leinster [99], and in the type-theoretic setting by Lumsdaine [105] and van den Berg–Garner [153].

9.5.1 The Truncation Ladder as an Internal Postnikov-Style Tower

The truncations

$$\tau_{\leq 0}\mathcal{C}(A),\ \tau_{\leq 1}\mathcal{C}(A),\ \tau_{\leq 2}\mathcal{C}(A),\ \tau_{\leq 3}\mathcal{C}(A),\ \ldots$$

form a ladder in which each step adds one coherence dimension. This mirrors the role of Postnikov stages in homotopy theory: finite approximations that progressively recover higher structure.

Proposition 9.74 (Successive enrichment of truncations). *For every $n \geq 0$, the inclusion of data*

$$\tau_{\leq n}\mathcal{C}(A) \hookrightarrow \tau_{\leq n+1}\mathcal{C}(A)$$

adds precisely:

(i) one new cell level $(n+1)$,

(ii) one new principal composition direction $\circ_n$,

(iii) one new layer of coherence witnesses relating previously available structural composites.

Proof. By definition of truncation, all levels $\leq n$ are unchanged. The $(n+1)$st level adds cells between parallel n-cells, together with operations from Lemma 9.43. These operations provide exactly the next coherence layer. □

Remark 9.75 (Dimension-4 checkpoint). At $n = 3 \to 4$, Proposition 9.74 yields the first stage with four composition directions and six interchange families. This is the minimal stage where the full "many-direction" coherence profile appears, so it is the natural checkpoint for homotopy-hypothesis evidence in our setting.

9.5.2 A Precise Comparison Template for the Hypothesis

Definition 9.76 (Syntactic homotopy-hypothesis template). For each type A, define the statement $\mathbf{HH}_{\mathrm{syn}}(A)$:

> There exists a homotopy type X_A and a weak-equivalence chain between the computational weak ω-groupoid attached to A and a geometric fundamental weak ω-groupoid attached to X_A, compatible with all finite truncations.

Remark 9.77 (Status). The present chapter does not claim a full proof of $\mathbf{HH}_{\mathrm{syn}}(A)$. What it provides is the complete algebraic side: the computational weak ω-groupoid of A with explicit all-level coherence. This is the non-negotiable prerequisite for any later equivalence theorem.

Proposition 9.78 (Necessary conditions verified internally). *For each type A, the computational weak ω-groupoid $\mathcal{C}(A)$ satisfies the internal conditions expected of an ∞-groupoid model of a homotopy type:*

(i) invertibility in every dimension (weakly, via higher cells),

(ii) closure under all globular compositions,

(iii) coherence stability under truncation and functoriality,

(iv) 3-truncation above dimension 3 via type-valued contractibility.

Proof. Item (i) follows from weak inverses in every principal dimension. Item (ii) is Theorem 9.45. Item (iii) combines Theorem 9.58, Proposition 9.52, and Theorem 9.96. Item (iv) is Theorem 9.60, which concentrates the genuinely non-trivial homotopy content in dimensions 0–3 and supplies canonical fillers for the structural coherence package above that range. □

9.5.3 Concrete low-dimensional readout and the $n = 4$ bridge

Example 9.79 (Low-dimensional readout)**.** The first stages of the computational tower $\mathcal{C}(A)$ can be read as follows:

(i) $\tau_{\leq 1}$ captures connected components and loop groups at basepoints;

(ii) $\tau_{\leq 2}$ records not only loops but homotopies between loop composites;

(iii) $\tau_{\leq 3}$ records coherence between those homotopies;

(iv) $\tau_{\leq 4}$ records coherence between coherence witnesses.

The fourth stage is the first contractible one: it records coherence between coherence witnesses, but Theorem 9.60 shows that it adds no new non-trivial homotopical content beyond the genuine 3-cell layer.

Proposition 9.80 (Dimension-4 bridge principle)**.** *Assume two comparison strategies from computational data to geometric data agree through dimension 3 and preserve the six interchange families at dimension 4. Then they agree on all structural dimension-4 composites up to 6-cells.*

Proof. Agreement through dimension 3 fixes the boundaries of all structural 4-cells. Preservation of the six interchange families at dimension 4 then ensures that the two comparison strategies assign the same generating structural 5-cells to the basic dimension-4 configurations.

Every structural dimension-4 composite is built from those generators by composition, whiskering, identities, and inverses. By Theorem 9.66, any two structural 5-cell composites with the same globular boundary are connected by a 6-cell. Hence the two comparison strategies agree on all structural dimension-4 composites up to 6-cells. □

Remark 9.81 (Why this matters for homotopy hypothesis work)**.** In practical comparison proofs, checking all dimensions directly is impossible. Proposition 9.80 identifies dimension 4 as the first finite stage where all coherence mechanisms are present simultaneously. It therefore serves as a realistic finite verification target before passing to general n.

9.6 Connection with Identity Types in Homotopy Type Theory

This section gives a first comparison between the computational tower and HoTT identity types; Chapter 14 provides a comprehensive treatment, covering univalence, cubical type theory, and higher inductive types.

The relationship between computational paths and HoTT identity types is one of compatibility rather than identity. Both frameworks produce iterating towers of higher cells; both satisfy groupoid laws up to higher witnesses; both fail UIP. But they arrive at these phenomena from different starting points: HoTT from a single self-iterating identity type, computational paths from explicit rewrite traces with a proof-irrelevant equality layer underneath. This section isolates the dimensionwise dictionary needed to compare the two towers, deferring the broader foundational discussion to Chapter 14.

Definition 9.82 (Iterated identity-type family). Fix a type A and parallel boundaries. Define families $\mathsf{Id}_A^{[n]}$ recursively by:

(i) $\mathsf{Id}_A^{[1]}(a,b) := \mathsf{Path}(a,b)$ for $a, b : A$;

(ii) for $n \geq 1$ and parallel terms u, v in $\mathsf{Id}_A^{[n]}$, set

$$\mathsf{Id}_A^{[n+1]}(u,v) := \mathsf{Path}(u,v).$$

Theorem 9.83 (Dimensionwise correspondence with identity types). *For every $n \geq 1$, there is a boundary-preserving comparison map*

$$\Phi_n : \mathsf{Id}_A^{[n]} \longrightarrow C_n(A)$$

that sends reflexivity to identities, concatenation to principal composition $\circ_{n-1}$, and symmetry to inverses. These maps are compatible with whiskering and lower compositions $\circ_k$ $(0 \leq k < n)$ up to structural higher cells.

Proof. We define the maps Φ_n by induction on n.

For $n = 1$, set Φ_1 to be the identity on paths. It clearly sends reflexivity to identities, concatenation to principal composition, and symmetry to inverses.

For $n = 2$ and $n = 3$, the required compatibility is exactly the content of Chapter 7 and Chapter 8: 2-cells and 3-cells in the iterated identity tower are matched with the computational operations already constructed there.

Assume now that Φ_n has been defined and has the stated compatibility properties. An element of $\mathsf{Id}_A^{[n+1]}$ is an identity proof between parallel terms of $\mathsf{Id}_A^{[n]}$, so applying Φ_n to its boundary data yields parallel n-cells in $C_n(A)$.

Use Lemma 9.43 to interpret the new identity proof as the corresponding computational $(n+1)$-cell between those images. Because lower compositions and whiskering in dimension $n+1$ are defined from the level-n operations by Definition 9.17, the induction hypothesis implies that Φ_{n+1} preserves reflexivity, concatenation, symmetry, and all lower compositions up to structural higher cells. This completes the induction. □

Remark 9.84 (Interpretation)**.** Theorem 9.83 states that computational paths do not replace HoTT identity types; they provide a concrete, rewrite-generated presentation of comparable higher structure. The point is compatibility of operations, not equality of foundations: our path object is UIP-based and keeps computational content in the step trace. This makes coherence witnesses explicit and computationally inspectable.

9.6.1 Compatibility with the Identity Eliminator

The central rule governing identity types is elimination (J-elimination, or path induction). A meaningful correspondence with HoTT must therefore preserve not only constructors (refl, concatenation, symmetry) but also elimination behaviour.

Definition 9.85 (Computational path induction schema)**.** Let P be a family over path boundaries at dimension 1:

$$P(a, b, p) \text{ for } a, b : A,\ p : \mathsf{Path}(a, b).$$

A *computational path-induction term* for P is a function assigning $\eta(a, b, p) : P(a, b, p)$ from data $d(a) : P(a, a, \mathsf{refl}_a)$, together with coherence witnesses showing that η is stable under rewrite equivalence of paths.

Theorem 9.86 (Dimensionwise elimination compatibility)**.** *The comparison maps Φ_n of Theorem 9.83 preserve elimination: for each $n \geq 1$, path induction on $\mathsf{Id}_A^{[n]}$ corresponds to induction on computational n-cells, up to structural $(n+1)$-cells.*

Proof. For $n = 1$, the claim is the usual compatibility between path induction and rewrite-respecting path construction: the eliminator is determined by its action on reflexivity, and the computational path presentation records the same data.

For $n = 2$ and $n = 3$, the required preservation of elimination is exactly what the bicategorical and 3-dimensional coherence packages provide. Those results show that the constructors and their coherence witnesses transport correctly across the correspondence between iterated identity types and computational higher cells.

Assume inductively that elimination is preserved at level n. At level $n+1$, an eliminator acts on $(n+1)$-cells, which are built from parallel n-cells together with principal composition, inverses, and the higher coherence data supplied by Lemma 9.43. Applying the induction hypothesis to the boundary n-cells and then using the coherence witnesses from the inductive step transports the elimination principle across the comparison map Φ_{n+1}. Hence elimination is preserved in every dimension, up to structural $(n+1)$-cells. □

Remark 9.87 (Elimination is where coherence is used). Constructors alone would give only a syntactic translation. Theorem 9.86 shows that all-level coherence is exactly what makes elimination transport correctly across the two presentations.

9.6.2 A concrete $n = 4$ identity-type pattern

Definition 9.88 (Fourth iterated identity family). Fix parallel boundaries through level 3. The level-4 identity type is

$$\mathsf{Id}_A^{[4]}(\Gamma, \Delta) := \mathsf{Path}(\Gamma, \Delta),$$

where Γ, Δ are parallel terms in $\mathsf{Id}_A^{[3]}$. Under Φ_4, these correspond to 4-cells in $\mathcal{C}_4(A)$.

Example 9.89 (Two level-4 witnesses and a level-5 comparison). Let Γ, Δ : $\mathsf{Id}_A^{[3]}(u, v)$ and suppose $\Theta_1, \Theta_2 : \mathsf{Id}_A^{[4]}(\Gamma, \Delta)$. Via Φ_4, Θ_1, Θ_2 become parallel 4-cells in $\mathcal{C}_4(A)$. By Theorem 9.66, there is a 5-cell comparing their structural composites whenever boundaries agree. Thus the HoTT statement "two proofs in $\mathsf{Id}^{[4]}$ are related in $\mathsf{Id}^{[5]}$" appears concretely as an explicit coherence witness.

Proposition 9.90 (Index pattern for iterated identities). *For each $n \geq 1$, the computational and HoTT presentations share the same indexing pattern:*

$$\text{terms in } \mathsf{Id}_A^{[n]} \longleftrightarrow n\text{-cells in } \mathcal{C}_n(A), \qquad \text{identities in } \mathsf{Id}_A^{[n+1]} \longleftrightarrow (n+1)\text{-cells between them.}$$

The nontrivial content is not indexing but coherence transport.

Proof. The correspondence on objects is recursive by definition of $\mathsf{Id}_A^{[n]}$ and $\mathcal{C}_n(A)$. Coherence transport is the statement of Theorem 9.83 and Theorem 9.86. □

9.6.3 Transport, Dependent Action, and Higher Whiskering

The macros `transport` and `apd` from the global notation of the book encode dependent behaviour of paths. In the ω-groupoid reading, they correspond to structured whiskering and reindexing operations on higher cells.

Theorem 9.91 (Dependent transport as structured whiskering). *Fix a dependent family $B : A \to \mathsf{Type}$. For each $n \geq 1$, dependent transport along an n-cell boundary in A induces a map on computational $(n+1)$-cells over B, compatible with all compositions $\circ_k$ up to structural cells.*

Proof. For $n = 1$, this is ordinary transport along paths in the dependent family B.

For $n = 2$, dependent transport acts on 2-cells by the same mechanism as whiskering and horizontal composition, so the claim is exactly the naturality already established for those operations.

Assume the statement holds in dimension n. An $(n + 1)$-dimensional transport map is obtained by lifting the dimension-n transport through the identity inclusions and principal composition that generate higher cells. The induction hypothesis gives compatibility with the lower operations already present, and Theorem 9.57 provides the structural comparison cells needed to compare the lifted transport with each composition $\circ_k$. Therefore dependent transport is compatible with all higher compositions up to structural cells in every dimension. □

Remark 9.92 (General n pattern). At level n, dependent action introduces one additional coherence level: compatibility with n-cell composition is witnessed at dimension $n + 1$, and compatibility between such witnesses is controlled at dimension $n + 2$. This is the same all-level coherence staircase seen throughout the chapter.

9.7 The Fundamental ω-Groupoid of a Type

We now harvest the chapter's constructions into a single invariant. For each type A, we package the full tower of computational cells into the *fundamental ω-groupoid* $\Pi_\omega(A)$—the all-level refinement of the weak path groupoid from Chapter 6. Its 1-truncation recovers the earlier groupoid, its higher layers retain the comparison data built in the present chapter, and its functorial behaviour under maps $f : A \to B$ makes it a genuine type-theoretic invariant.

Definition 9.93 (Fundamental ω-groupoid). For a type A, define

$$\Pi_\omega(A) := \mathcal{C}(A),$$

the weak ω-groupoid from Theorem 9.45. For $n \in \mathbb{N}$, define $\Pi_{\leq n}(A) := \tau_{\leq n}\Pi_\omega(A)$.

Definition 9.94 (Loop ω-groupoid at a basepoint). For $a : A$, the *loop ω-groupoid* $\Omega_\omega(A, a)$ is the sub-ω-groupoid of $\Pi_\omega(A)$ whose n-cells have both 0-boundaries equal to a.

Theorem 9.95 (Recovering previously constructed invariants). *For every type A:*

$\Pi_{\leq 1}(A)$ *recovers the weak path groupoid of Chapter 6,* $\Pi_{\leq 2}(A)$ *recovers the bicategorical structure of Chapter 7,*

and

$$\Pi_{\leq 3}(A) \text{ recovers the 3-cell coherence structure of Chapter 8.}$$

Moreover, the strict quotient of $\Pi_{\leq 1}(A)$ by $\sim_{\mathsf{rw}}$ is the fundamental groupoid $\Pi_1(A)$ of Definition 6.24.

Proof. The first three claims follow from Proposition 9.72. The quotient statement is exactly the construction of Section 6.3. □

Theorem 9.96 (Functoriality). *For every function $f : A \to B$, there is an induced morphism of weak ω-groupoids*

$$\Pi_\omega(f) : \Pi_\omega(A) \to \Pi_\omega(B)$$

that acts by f on 0-cells and by iterated congruence on higher cells. It preserves identities, compositions, and inverses up to structural higher cells.

Proof. We define the components of $\Pi_\omega(f)$ by induction on the dimension.

At dimension 0, set

$$\Pi_\omega(f)(a) := f(a).$$

At dimension 1, define

$$\Pi_\omega(f)(p) := \mathsf{congrArg}(f, p).$$

Theorem 6.20 shows that this preserves identities and principal composition, and Remark 6.10 gives preservation of inverses.

Assume now that $\Pi_\omega(f)$ has been defined on n-cells and preserves their source and target maps, identities, compositions, and inverses up to structural higher cells. An $(n+1)$-cell of $\Pi_\omega(A)$ is a derivation between parallel n-cells. Apply the congruence action induced by f to each step of that derivation and to its boundary n-cells. Because the induction hypothesis preserves boundaries, the image is again an $(n+1)$-derivation between the images of the original source and target. This defines $\Pi_\omega(f)$ in dimension $n+1$.

Identities in dimension $n+1$ are reflexive derivations, so their images are again reflexive. Principal composition is concatenation of derivations, and applying f stepwise commutes with concatenation. Likewise, reversing a derivation and then applying f gives the same result as first applying f and then reversing, so weak inverses are preserved.

For lower compositions $\circ_k$, recall from Lemma 9.43 and Definition 9.17 that they are built from principal composition together with identity lifting and whiskering. The induction hypothesis preserves the lower-dimensional boundary data entering those constructions, while Theorem 9.58 supplies the structural comparison cells needed to compare the image of a chosen whiskered decomposition with the corresponding whiskered decomposition of the images. Hence lower compositions are preserved up to structural higher cells as well.

Thus the dimensionwise maps assemble into a morphism of weak ω-groupoids

$$\Pi_\omega(f) : \Pi_\omega(A) \to \Pi_\omega(B).$$

□

Remark 9.97 (Interpretation)**.** Definition 9.93 is the higher-dimensional analogue of the fundamental groupoid of Chapter 6. The key difference is that no external topological realisation is required: all cells and coherence witnesses are generated syntactically by rewriting.

9.7.1 Homotopy groups inside $\Pi_\omega(A)$

The weak ω-groupoid $\Pi_\omega(A)$ contains, at a basepoint, all higher loop information usually organised as homotopy groups.

Definition 9.98 (Higher loop layers)**.** Fix $a : A$. Define recursively:

$$\begin{aligned} \Omega^1(a) &:= \mathsf{Path}(a, a), \\ \Omega^{n+1}(a) &:= \mathsf{Path}_{\Omega^n(a)}(u, v) \quad \text{for parallel } u, v \in \Omega^n(a). \end{aligned}$$

Under the correspondence of Section 9.6, these are the basepointed cells of $\Omega_\omega(A, a)$ in successive dimensions.

Definition 9.99 (Computational homotopy groups)**.** For $n \geq 1$, define

$$\pi_n^{\mathrm{cp}}(A, a)$$

as the group of endomorphisms of the $(n-1)$-fold identity loop in the strict quotient of the n-truncation of $\Omega_\omega(A, a)$ by the relevant rewrite equivalence at level n.

Remark 9.100 (Low-dimensional sanity checks)**.** For $n = 1$, Definition 9.99 recovers $\pi_1(A, a)$ from Chapter 6. For $n = 2$, it measures loops between loops modulo 3-cell coherence. For $n = 3$, it measures loops between 2-loop witnesses modulo 4-cell coherence.

Theorem 9.101 (Functorial action on computational homotopy groups). *Every function $f : A \to B$ and every basepoint $a : A$ induce a group homomorphism*

$$(\pi_n^{\mathrm{cp}} f)_a : \pi_n^{\mathrm{cp}}(A, a) \to \pi_n^{\mathrm{cp}}(B, f(a)) \qquad (n \geq 1),$$

compatible with truncation maps.

Proof. Theorem 9.96 provides a morphism of weak ω-groupoids

$$\Pi_\omega(f) : \Pi_\omega(A) \to \Pi_\omega(B).$$

Restricting to the basepointed sub-ω-groupoids yields a morphism

$$\Omega_\omega(A, a) \to \Omega_\omega(B, f(a)).$$

Now truncate at level n and pass to the strict quotient by the rewrite equivalence relation that defines $\pi_n^{\mathrm{cp}}(A, a)$ and $\pi_n^{\mathrm{cp}}(B, f(a))$.

Because $\Pi_\omega(f)$ preserves identities, compositions, and inverses up to higher coherence, and because those higher coherence cells are precisely what the quotient forgets, the induced map on endomorphism classes is well defined. It preserves the group law, the identity element, and inverses, so it is a group homomorphism

$$(\pi_n^{\mathrm{cp}} f)_a : \pi_n^{\mathrm{cp}}(A, a) \to \pi_n^{\mathrm{cp}}(B, f(a)).$$

Compatibility with truncation is immediate, since both the loop-groupoid restriction and the quotient construction commute with forgetting cells above dimension n. □

9.7.2 Dimension-4 loop coherence and its meaning

Example 9.102 (A basepointed $n = 4$ loop frame). Fix $a : A$ and consider four compatible 4-loops x_1, x_2, x_3, x_4 at a. Two routes of composition using $\circ_3$ and $\circ_2$ are related by the interchange 5-cell of Example 9.21. Inside $\Omega_\omega(A, a)$, this says that "multiplication" of 4-loops is coherent with second-level looping, not strictly but via explicit 5-dimensional witnesses.

Proposition 9.103 (Dimension-4 loop comparison principle). *Any two structural composites of basepointed 4-loops with the same boundary are connected by a 6-loop witness in $\Omega_\omega(A, a)$.*

Proof. This is Theorem 9.66 restricted to the basepointed sub-ω-groupoid from Definition 9.94. □

Remark 9.104 (Interpretive payoff). Proposition 9.103 is the direct computational analogue of saying that different 4-dimensional loop concatenation schemes are homotopic through one higher dimension. It is an explicit rewrite-theoretic realisation of higher loop coherence.

9.7.3 General n invariants extracted from the tower

Definition 9.105 (Boundary-fixed hom-objects)**.** Let u, v be parallel $(n-1)$-cells in $\Pi_\omega(A)$. Write

$$\mathrm{Hom}_n(u,v) := \{x \in \Pi_\omega(A)_n \mid s_n(x) = u,\ t_n(x) = v\}.$$

This carries induced operations $\circ_k$ $(k < n)$ whenever boundaries permit.

Proposition 9.106 (Groupoid enrichment of hom-objects)**.** *For fixed parallel boundaries u, v, the collection $\mathrm{Hom}_n(u,v)$ is itself weakly groupoidal in dimension shift: its principal composition/inverse laws are witnessed at level $n+1$.*

Proof. Restrict the operations and witnesses of Theorem 9.45 to cells with fixed outer boundary. All required laws remain valid because coherence witnesses preserve boundaries. □

Theorem 9.107 (All-level boundary invariants)**.** *For each $n \geq 1$ and each boundary pair (u, v), the strict quotient of $\mathrm{Hom}_n(u,v)$ by structural equivalence yields an invariant*

$$\mathcal{I}_n(A; u, v)$$

that is functorial in A and stable under truncation.

Proof. Functoriality follows from Theorem 9.96: a function $f : A \to B$ sends an n-cell with boundary (u, v) to an n-cell with boundary $(\Pi_\omega(f)(u), \Pi_\omega(f)(v))$, and it preserves the relevant compositions up to structural cells. Therefore it induces a map on the strict quotients defining $\mathcal{I}_n(A; u, v)$.

Stability under truncation is Proposition 9.52, which says that forgetting higher cells does not alter the retained operations or the structural comparisons among them. Finally, passing to the strict quotient removes dependence on the particular coherence representatives while preserving the boundary-fixed composition classes. Hence $\mathcal{I}_n(A; u, v)$ is functorial in A and stable under truncation. □

9.7.4 Extended Examples

Example 9.108 (Contractible type)**.** If A is contractible, every boundary-fixed hom-object $\mathrm{Hom}_n(u,v)$ is nonempty and connected by structural cells in all dimensions. Hence all invariants $\mathcal{I}_n(A; u, v)$ are trivial.

Example 9.109 (Set-level type (0-truncated))**.** If A is set-like (all identity proofs propositionally unique), then higher cells above dimension 1 collapse up to unique witnesses. Accordingly, $\Pi_\omega(A)$ is equivalent to its 1-truncation and $\pi_n^{\mathrm{cp}}(A, a)$ is trivial for $n \geq 2$.

Example 9.110 (Circle-like behaviour)**.** For types with nontrivial looping at a basepoint, π_1^{cp} is nontrivial and higher loop data may also persist. In such cases, $\Pi_\omega(A)$ records not only the primary loop group but also coherence interactions among loop compositions across all levels.

Remark 9.111 (Position within the book)**.** Section 9.7 completes the progression:

$$\Pi_1(A) \rightsquigarrow \Pi_{\leq 2}(A) \rightsquigarrow \Pi_{\leq 3}(A) \rightsquigarrow \Pi_\omega(A).$$

The first object captures path classes, the second and third add coherent higher witnesses, and the fourth packages the entire infinite hierarchy.

Summary

This chapter proves that computational paths carry a weak ω-groupoid structure and gathers the coherence needed for that statement into a single all-level theorem. The decisive move is the passage from the explicit low-dimensional constructions of Chapters 6–8 to a genuinely uniform higher theory with stabilisation above dimension 3.

The proof has four parts:

(i) define globular data and composition operations in arbitrary dimension;

(ii) recover the cases $n = 1, 2, 3$ from Chapters 6–8;

(iii) perform the inductive extension step and obtain the main weak ω-groupoid theorem;

(iv) organize the resulting higher laws into an all-level coherence package, together with stabilisation from dimension 4 onward.

Only after that internal theorem is in place do the three connection sections show what it yields for neighbouring questions:

(i) relative to Grothendieck's hypothesis, $\Pi_\omega(A)$ supplies the internal algebraic side relevant to any later comparison theorem, without claiming that external equivalence here;

(ii) relative to HoTT, it gives a rewrite-generated comparison with iterated identity types and their elimination/coherence behaviour, without identifying the two frameworks;

(iii) as an invariant, it extends the fundamental groupoid to an all-level object encoding higher loop and coherence information.

Accordingly, this chapter should be read as the culmination of the internal higher-rewriting development. The comparison sections matter, but they are secondary to the main point established here: computational paths support explicit low-dimensional coherence and, beyond that, a stabilised higher weak ω-groupoid structure.

Formalization Notes. The results of this chapter are formalised in the accompanying Lean development (`https://github.com/Arthur742Ramos/ComputationalPathsLean`). The most relevant files/directories for this chapter are:

- `Path/Rewrite/ConfluenceProof.lean` and `Path/Rewrite/ConfluenceConstructive.lean` — path-level termination, confluence, and normalisation machinery
- `Path/OmegaGroupoid/GroupoidProofs.lean` — explicit low-dimensional coherence proofs, including pentagon, triangle, interchange, inverse/double-inverse, contravariance, and Eckmann–Hilton routes
- `Path/OmegaGroupoid.lean` together with the higher modules under `Path/OmegaGroupoid/` — the derivation tower `Derivation2`, `Derivation3`, `Derivation4`, and `DerivationHigh`, together with `MetaStep3`, `MetaStep4`, and `MetaStepHigh`
- `Path/OmegaGroupoid/Normalizer.lean`, `Path/OmegaGroupoid/TruncationProof.lean`, `Path/OmegaGroupoid/Stabilization.lean`, and `Path/OmegaGroupoid/OmegaWeakGroupoid.lean` — the inverse-loop normaliser, confluence-facing wrappers, stabilisation theorem, and packaged 3-truncated weak ω-groupoid interfaces

Key definitions. The Lean hierarchy is built from derivations between paths rather than by literal iteration of `Path`:

```
-- At the bottom, Step : Type u
inductive RwEq {a b : A} : Path a b -> Path a b -> Type (
    u + 1) where
  | refl (p : Path a b) : RwEq p p
  | step {p q : Path a b} : Step p q -> RwEq p q
  | symm {p q : Path a b} : RwEq p q -> RwEq q p
  | trans {p q r : Path a b} : RwEq p q -> RwEq q r ->
      RwEq p r

inductive Derivation2 {a b : A} : Path a b -> Path a b ->
      Type (u + 1) where
  | refl (p : Path a b) : Derivation2 p p
```

```
  | step {p q : Path a b} : Step p q -> Derivation2 p q
  | inv {p q : Path a b} : Derivation2 p q -> Derivation2
      q p
  | vcomp {p q r : Path a b} :
      Derivation2 p q -> Derivation2 q r -> Derivation2 p
          r

inductive Derivation3 {a b : A} {p q : Path a b} :
    Derivation2 p q -> Derivation2 p q -> Type (u + 2)
        where
  | refl (d : Derivation2 p q) : Derivation3 d d
  | step {d1 d2 : Derivation2 p q} : MetaStep3 d1 d2 ->
     Derivation3 d1 d2
  | inv {d1 d2 : Derivation2 p q} : Derivation3 d1 d2 ->
     Derivation3 d2 d1
  | vcomp {d1 d2 d3 : Derivation2 p q} :
      Derivation3 d1 d2 -> Derivation3 d2 d3 ->
         Derivation3 d1 d3

/-- 4-cells: genuine inductive comparisons between 3-
   cells. -/
inductive Derivation4 ...

/-- Stable higher levels, again genuine inductive
   families. -/
inductive DerivationHigh ...
```

The crucial point is that the levels above Derivation3 are genuine inductive families, not placeholders such as `Unit` or `PUnit`. Primitive coherence steps live in `MetaStep3`, `MetaStep4`, and `MetaStepHigh`. The contractibility maps are then built entirely inside this type-valued tower:

```
noncomputable def contractibility3 {p q : Path a b}
    (d1 d2 : Derivation2 p q) : Derivation3 d1 d2 :=
  by
    -- isolate the inverse loop d1 . d2^{-1},
    -- contract that loop structurally, and compare the
       endpoints
    exact ...

noncomputable def contractibility4 {d1 d2 : Derivation2 p
    q}
    (Gamma1 Gamma2 : Derivation3 d1 d2) : Derivation4
       Gamma1 Gamma2 :=
  by
    -- normalize 3-cells and fill the residual diamond
    exact MetaStep4.diamond_filler ...
```

```
noncomputable def contractibilityHigh {n : Nat}
    (Xi0 Xi1 : DerivationHigh n x y) :
    DerivationHigh (n + 1) Xi0 Xi1 :=
  by
    -- stable case: normalize4 + MetaStepHigh.
       diamond_filler
    exact ...
```

At the public interface, `contractibility₃` now factors mainly through the inverse-loop normaliser and structural 3-cell comparison. The current raw-`Path` core still retains one isolated fallback via `strict_transport₃ := MetaStep₃.rweq_transport(...)` when the strict-shape comparison does not align structurally before the fuel bound is exhausted. The next stage uses `MetaStep₄.diamond_filler`, and the stable higher case is packaged by `MetaStepHigh.diamond_filler`. Thus the formalisation realises the computational ω-groupoid as genuinely non-trivial in dimensions 1–3 and 3-truncated from dimension 4 onward, while making the remaining legacy transport boundary explicit and local. The same verification suite confirms that all fourteen named 3-cell coherence proofs are non-trivial: replacing any of them by `Derivation3.refl` is rejected by Lean. Separately, the explicit-syntax/polygraphic layer now exports a fully constructive proof-relevant shadow of the same theorem: ten generating 2-cell families, nine generating 3-cell families, and acyclicity above dimension 3, so no additional generating 4-cells are required there. The remaining comparison problem is to identify that explicit-syntax result directly with the raw-`Path` connector without the residual transport step.

Chapter 10

The Eckmann–Hilton Argument

> *"Where two compatible compositions share the same unit, commutativity is not an assumption but a consequence."*
>
> —The Eckmann–Hilton principle

Chapters 6–9 developed the higher algebra of computational paths in successive layers: groupoids of paths (Chapter 6), 2-cells as derivations between paths (Chapter 7), 3-cells and coherence (Chapter 8), and finally the weak ω-groupoid theorem (Chapter 9). At this point we possess exactly the data needed for one of the most elegant arguments in higher-dimensional algebra: the Eckmann–Hilton argument.

The classical theorem [64] states that if a set carries two binary operations with the same unit and that satisfy an interchange law, then the two operations coincide and are automatically commutative. In topology this mechanism explains why the second homotopy group is abelian [79, 114]. In our setting it explains why the composition of doubly-degenerate 2-cells ("loops on loops") is forced to commute.

Conceptually, this chapter is the closing movement of Part II. Chapter 9 established the weak ω-groupoid structure together with its stabilisation above dimension 3; the present chapter extracts from that structure a concrete algebraic consequence: commutativity for doubly degenerate compositions beginning at dimension 2, together with a systematic higher Eckmann–Hilton pattern. This provides the bridge from coherence technology to homotopy-group behaviour [105, 153].

The chapter is organised as follows. Section 10.1 presents the classical argument in its abstract algebraic form. Section 10.2 specialises the argument to vertical and horizontal composition of 2-cells in computational paths. Section 10.3 defines the second computational homotopy group and proves its commutativity. Section 10.4 explains the higher-dimensional stabilisation pattern. Section 10.5 gives a rewrite-theoretic interpretation with explicit examples from the 77-rule system.

Throughout this chapter, equalities between 2-cells are interpreted as in Chapter 7: either as explicit 3-cells in the weak setting, or as literal equalities after passing to the appropriate quotient.

10.1 The Classical Eckmann–Hilton Argument

10.1.1 Two Operations and One Unit

The Eckmann–Hilton argument requires remarkably little data: a single set equipped with two binary operations that share a common two-sided unit, plus an interchange law relating them. The surprise is how much structure these sparse ingredients already force.

Definition 10.1 (Eckmann–Hilton datum). An *Eckmann–Hilton datum* consists of:

(i) a set M;

(ii) two binary operations

$$\bullet_1, \bullet_2 : M \times M \to M;$$

(iii) an element $e \in M$ that is a two-sided unit for both operations:

$$e \bullet_i x = x = x \bullet_i e \qquad (\forall x \in M,\ i \in \{1, 2\});$$

(iv) the interchange law:

$$(a \bullet_1 b) \bullet_2 (c \bullet_1 d) = (a \bullet_2 c) \bullet_1 (b \bullet_2 d) \qquad (\forall a, b, c, d \in M). \quad (10.1)$$

Remark 10.2 (The middle-four pattern). Equation (10.1) is often called *middle-four interchange*. It compares two ways of composing a 2×2 grid: first across rows and then down columns, or first down columns and then across rows. This is exactly the same pattern that appeared as the interchange law for 2-cells in Theorem 7.32.

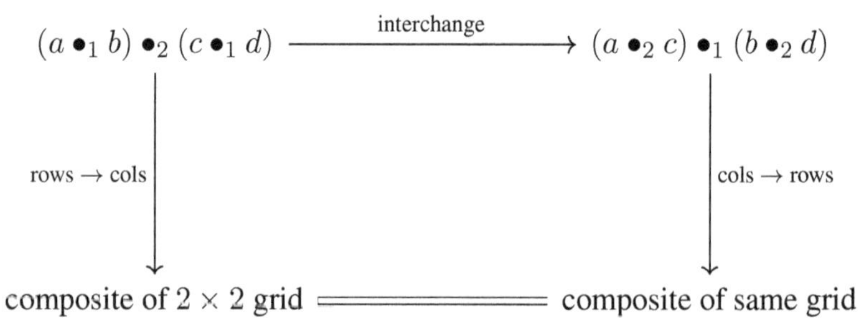

10.1.2 The Classical Proof

Theorem 10.3 (Eckmann–Hilton). *Let $(M, \bullet_1, \bullet_2, e)$ be an Eckmann–Hilton datum. Then:*

(i) the two operations coincide:

$$x \bullet_1 y = x \bullet_2 y \qquad (\forall x, y \in M);$$

(ii) this common operation is commutative:

$$x \bullet_1 y = y \bullet_1 x \qquad (\forall x, y \in M).$$

Hence $\bullet_2$ is also commutative.

Proof. Fix $x, y \in M$. Using that e is the common unit and applying interchange once, we compute:

$$\begin{aligned} x \bullet_1 y &= (x \bullet_2 e) \bullet_1 (e \bullet_2 y) \\ &= (x \bullet_1 e) \bullet_2 (e \bullet_1 y) \\ &= x \bullet_2 y. \end{aligned}$$

So $\bullet_1 = \bullet_2$.

For commutativity, use the same trick with units in opposite corners:

$$\begin{aligned} x \bullet_1 y &= x \bullet_2 y \\ &= (e \bullet_1 x) \bullet_2 (y \bullet_1 e) \\ &= (e \bullet_2 y) \bullet_1 (x \bullet_2 e) \\ &= y \bullet_1 x. \end{aligned}$$

Thus the common operation is commutative. □

Remark 10.4 (Why the argument is so efficient). No associativity is assumed in Definition 10.1. Associativity-like behaviour is forced by interchange and the common unit. The proof uses only five ingredients: two unit laws for $\bullet_1$, two unit laws for $\bullet_2$, and one application of interchange. This economy is what makes the argument robust across algebra, topology, and higher category theory.

Remark 10.5 (Why the result is surprising). One might expect that a second binary operation on the same set, even one sharing the same unit, could behave quite differently from the first—perhaps failing to commute, or producing an entirely independent algebraic structure. The Eckmann–Hilton argument shows that the interchange law alone forecloses this possibility: the two operations must not only agree but must also commute. The surprise lies in how little is assumed (no associativity, no commutativity) and how much is forced (both properties, plus coincidence of the operations).

Corollary 10.6 (Monoid structure from the EH hypotheses). *Under the assumptions of Theorem 10.3, the common operation*

$$x \cdot y := x \bullet_1 y = x \bullet_2 y$$

is a commutative monoid operation on M with unit e.

Proof. Unit and commutativity are immediate from the theorem. Associativity follows by applying interchange to a 2×2 grid in which one entry is the unit e, then simplifying by unit laws. □

10.1.3 A Geometric Reading

Before moving to computational paths, it is worth pausing on the geometric picture that underlies the algebra.

Remark 10.7 (The square can slide). The geometric slogan behind Eckmann–Hilton is: *if two composition directions can be interchanged freely and share the same degenerate cell as unit, then one direction can be slid past the other.* In low-dimensional topology this "sliding" is visible as deformations of squares that do not cross singularities. In our rewrite setting the same phenomenon is witnessed by explicit 3-cells generated by interchange and coherence.

Example 10.8 (Endomorphisms of an identity 1-cell). In a strict 2-category, fix an object x and consider 2-endomorphisms of the identity 1-cell:

$$\alpha, \beta : \mathrm{id}_x \Rightarrow \mathrm{id}_x.$$

Vertical composition and horizontal composition are two binary operations on this set. Both have the same unit $\mathrm{id}_{\mathrm{id}_x}$, and strict middle-four interchange holds. Theorem 10.3 therefore implies these operations coincide and commute.

This is the prototypical classical source of "π_2 is abelian" [64]. Our computational-path instance is a weak version of exactly this example.

10.2 Application to Computational Paths

The classical Eckmann–Hilton argument applies to any algebraic setting that supplies two operations with a common unit and an interchange law. We now show that the 2-cell calculus of Chapter 7 is precisely such a setting once we restrict to doubly-degenerate cells.

Fix a type A and a basepoint $a : A$. Recall that 2-cells are terms of $\mathsf{Derivation}_2(p, q)$ for parallel paths $p, q : \mathsf{Path}(a, b)$, and that we have:

- vertical composition $\bullet_v$;

- horizontal composition $\bullet_h$ (via whiskering and interchange);
- identity 2-cells refl_p;
- interchange (Theorem 7.32).

10.2.1 Doubly-degenerate 2-cells

The key restriction is to *doubly-degenerate* 2-cells: 2-cells whose source and target are both the identity path refl_a. Intuitively, these are "loops on loops"—rewrite traces that start and end at the trivial computation. It is in this doubly-degenerate world that two composition directions coexist with a shared unit, creating the conditions for Eckmann–Hilton.

Definition 10.9 (LoopTwoCell)**.** For $a : A$, define

$$\mathsf{LoopTwoCell}(a) := \mathsf{Derivation}_2(\mathsf{refl}_a, \mathsf{refl}_a).$$

Elements of $\mathsf{LoopTwoCell}(a)$ are 2-cells whose source and target are both the identity path at a. They may be viewed as *loops on loops* at the basepoint.

Remark 10.10 (Degeneracy level)**.** Definition 10.9 is "doubly degenerate":

(i) the underlying 1-cell is already degenerate (refl_a);

(ii) the 2-cells are endomorphisms of that degenerate 1-cell.

This is precisely the context where Eckmann–Hilton is expected to act.

10.2.2 Vertical and Horizontal Operations

On $\mathsf{LoopTwoCell}(a)$ both vertical and horizontal composition restrict to well-defined binary operations. Vertical composition simply concatenates rewrite traces in sequence; horizontal composition places them side by side and contracts the boundary units.

Definition 10.11 (Vertical multiplication on loop 2-cells)**.** For α, β : $\mathsf{LoopTwoCell}(a)$, define

$$\alpha \star_v \beta := \alpha \bullet_v \beta.$$

Since both α and β have source and target refl_a, the composite is again in $\mathsf{LoopTwoCell}(a)$.

Definition 10.12 (Horizontal multiplication on loop 2-cells)**.** Choose a unitor 2-cell

$$u_a : \mathsf{trans}(\mathsf{refl}_a, \mathsf{refl}_a) \Rrightarrow \mathsf{refl}_a$$

(for instance, induced by a unit rewrite rule). For $\alpha, \beta : \mathsf{LoopTwoCell}(a)$, define

$$\alpha \star_h \beta := u_a^{-1} \bullet_v (\alpha \bullet_h \beta) \bullet_v u_a.$$

This lies in $\mathsf{LoopTwoCell}(a)$.

Remark 10.13 (Independence of the chosen unitor). Different choices of u_a give 3-cell equivalent operations, by the coherence theory of Chapter 8. Hence the induced operation on the $\sim_3$-quotient (Definition 10.20) is canonical.

Definition 10.14 (Common unit). Set

$$\epsilon_a := \mathsf{refl}_{\mathsf{refl}_a}.$$

Then ϵ_a is the identity for $\star_v$ and (up to 3-cell) for $\star_h$.

Proposition 10.15 (Unit laws on $\mathsf{LoopTwoCell}(a)$). *For every α : $\mathsf{LoopTwoCell}(a)$:*

$$\epsilon_a \star_v \alpha = \alpha = \alpha \star_v \epsilon_a,$$

and

$$\epsilon_a \star_h \alpha = \alpha = \alpha \star_h \epsilon_a.$$

Equalities are interpreted up to 3-cells.

Proof. The vertical laws are the identity laws for $\bullet_v$ from Chapter 7. For horizontal laws, combine:

(i) the identity behaviour of $\bullet_h$ (Proposition 7.18);

(ii) the unitors at the path level;

(iii) whiskering naturality and 3-cell coherence (Chapter 8).

The resulting composites are 3-cell equivalent to α. □

10.2.3 Interchange in the Doubly-Degenerate Context

With both operations and their common unit in place, we need only verify that interchange holds on loop 2-cells. This follows directly from the general interchange theorem of Chapter 7 once the surrounding unitors are accounted for.

Proposition 10.16 (Interchange on loop 2-cells). *For $\alpha, \beta, \gamma, \delta$: $\mathsf{LoopTwoCell}(a)$:*

$$(\alpha \star_v \beta) \star_h (\gamma \star_v \delta) = (\alpha \star_h \gamma) \star_v (\beta \star_h \delta).$$

Again, equality is up to 3-cell.

Proof. Expand $\star_v$ and $\star_h$ by Definitions 10.11 and 10.12. At the core we obtain exactly the interchange comparison for $\bullet_v$ and $\bullet_h$ from Theorem 7.32; the surrounding unit transports cancel by the unit laws and coherence. □

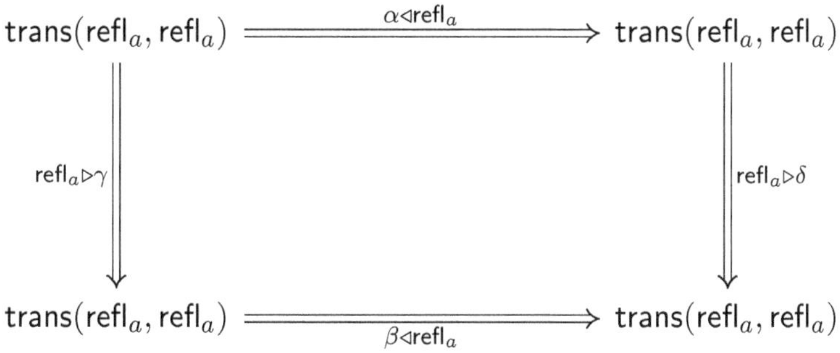

The square above is the doubly-degenerate Godement square: both routes from top-left to bottom-right define loop 2-cell composites, and Proposition 10.16 states that these routes coincide up to 3-cell.

10.2.4 Eckmann–Hilton in Computational Paths

All the ingredients of Section 10.1 are now available—a common unit (Definition 10.14), unit laws (Proposition 10.15), and interchange (Proposition 10.16)—so the classical Eckmann–Hilton pattern can now be internalised at the level of 3-cells.

Theorem 10.17 (Weak Eckmann–Hilton for $\mathsf{LoopTwoCell}(a)$)**.** *On* $\mathsf{LoopTwoCell}(a)$*, the operations $\star_v$ and $\star_h$ are connected by canonical 3-cells, and the common value is commutative up to canonical 3-cells. More explicitly, for all α, β :* $\mathsf{LoopTwoCell}(a)$ *there are 3-cells*

$$\alpha \star_v \beta \Rrightarrow_3 \alpha \star_h \beta \qquad \textit{and} \qquad \alpha \star_v \beta \Rrightarrow_3 \beta \star_v \alpha,$$

hence also $\alpha \star_v \beta \Rrightarrow_3 \beta \star_h \alpha$.

Proof. The classical Eckmann–Hilton calculation can be repeated one dimension higher, replacing literal equalities by specified 3-cells. The common unit is supplied by Proposition 10.15, and the middle-four comparison is supplied by Proposition 10.16. Chaining these unit and interchange witnesses in the same order as the classical proof produces first a 3-cell from $\alpha \star_v \beta$ to $\alpha \star_h \beta$, and then a 3-cell from $\alpha \star_v \beta$ to $\beta \star_v \alpha$. Composing with the first comparison for (β, α) yields the final comparison with $\beta \star_h \alpha$. □

Corollary 10.18 (Commutative monoid up to 3-cell)**.** *The type* $\mathsf{LoopTwoCell}(a)$ *carries a canonical commutative monoid structure, coherent up to 3-cells.*

Proof. Use the common unit together with the comparison 3-cells from Theorem 10.17. Chapter 8 supplies the higher coherence needed to organise these data into a commutative monoid up to 3-cells. □

Remark 10.19 (Structural meaning)**.** Theorem 10.17 is not an extra axiom. It is a *consequence* of:

(i) the 2-cell operations already defined in Chapter 7;

(ii) the interchange 3-cell (Theorem 7.32);

(iii) unit coherence from Chapter 8.

So commutativity appears as forced coherence, not as additional structure.

10.3 Commutativity of π_2

The previous section established commutativity at the level of loop 2-cells up to 3-cell equivalence. Commutativity "up to" is not yet the clean algebraic statement one wants. To obtain a strict group whose multiplication is genuinely commutative, we pass to the quotient by 3-cell equivalence—the same quotienting mechanism used in Chapters 8 and 9 to turn coherent data into strict invariants.

10.3.1 The Quotient Relation at Level 2

The equivalence relation we quotient by is the existence of a 3-cell connecting two loop 2-cells.

Definition 10.20 (3-cell equivalence on loop 2-cells)**.** For α, β : $\mathsf{LoopTwoCell}(a)$, define

$$\alpha \sim_3 \beta \iff \exists \Gamma : \alpha \Rrightarrow_3 \beta.$$

Proposition 10.21. *The relation* $\sim_3$ *is an equivalence relation on* $\mathsf{LoopTwoCell}(a)$.

Proof. Reflexivity is witnessed by refl_α. Symmetry is witnessed by inverse 3-cells. Transitivity is witnessed by vertical composition $\circ_2$ of 3-cells. All these operations were established in Chapter 8. □

Definition 10.22 (Second computational homotopy group)**.** Define

$$\pi_2^{\mathrm{cp}}(A, a) := \mathsf{LoopTwoCell}(a) \,/\, {\sim_3}.$$

For classes $[\alpha]$, $[\beta]$, define multiplication by

$$[\alpha] \cdot [\beta] := [\alpha \star_v \beta].$$

By Theorem 10.17, one may equivalently use $\star_h$.

Remark 10.23 (Compatibility with Chapter 9 notation). Definition 10.22 is the $n = 2$ instance of the general construction of computational homotopy groups from Definition 9.99. Here we spell it out concretely in terms of loop 2-cells to make the Eckmann–Hilton mechanism explicit.

10.3.2 Group Structure

Once the quotient is in hand, the group laws follow from the 2-groupoid structure of Chapter 7, with coherence from Chapter 8 ensuring that all the "up to" equalities become strict after quotienting.

Proposition 10.24 (Well-defined multiplication). *If $\alpha \sim_3 \alpha'$ and $\beta \sim_3 \beta'$, then*

$$\alpha \star_v \beta \sim_3 \alpha' \star_v \beta'.$$

Hence multiplication in Definition 10.22 is well-defined.

Proof. Choose 3-cells $\Gamma : \alpha \Rrightarrow_3 \alpha'$ and $\Delta : \beta \Rrightarrow_3 \beta'$. By functoriality/naturality of vertical composition at the 3-cell level (Chapter 8), these induce a 3-cell

$$\Gamma \star_v \Delta : \alpha \star_v \beta \Rrightarrow_3 \alpha' \star_v \beta'.$$

Therefore classes are preserved. □

Proposition 10.25 (Group laws). *The operation $\cdot$ on $\pi_2^{\mathrm{cp}}(A, a)$ is associative, has unit $[\epsilon_a]$, and every element has an inverse.*

Proof. Associativity, unit, and inverse on representatives hold up to 3-cells by the 2-groupoid laws from Chapter 7, with coherence from Chapter 8. Quotienting by $\sim_3$ turns these coherent equalities into strict equalities. □

Theorem 10.26 ($\pi_2^{\mathrm{cp}}(A, a)$ is abelian). *For every type A and basepoint $a : A$, the group $\pi_2^{\mathrm{cp}}(A, a)$ is abelian.*

Proof. Let $[\alpha], [\beta] \in \pi_2^{\mathrm{cp}}(A, a)$. By Theorem 10.17,

$$\alpha \star_v \beta \sim_3 \beta \star_v \alpha.$$

Passing to classes:

$$[\alpha] \cdot [\beta] = [\alpha \star_v \beta] = [\beta \star_v \alpha] = [\beta] \cdot [\alpha].$$

So multiplication is commutative. □

Remark 10.27 (Fundamental topological parallel)**.** Theorem 10.26 is the computational-path counterpart of the classical theorem in algebraic topology: for any pointed space (X, x), $\pi_2(X, x)$ is abelian [79, 114]. The proof strategy is the same in both settings: identify two compositions on doubly-degenerate cells and invoke Eckmann–Hilton.

Example 10.28 (Trivial case: $A = \mathbb{N}$)**.** In the rewrite system for $\mathbb{N}$, every loop path at $n : \mathbb{N}$ reduces to refl_n, and every loop 2-cell reduces to ϵ_n. Hence $\pi_2^{\mathrm{cp}}(\mathbb{N}, n)$ has one element and is therefore abelian.

10.3.3 A Precise Framework Statement

Theorem 10.29 (Eckmann–Hilton form of the π_2 theorem)**.** *In the computational-path framework:*

(i) *$\pi_2^{\mathrm{cp}}(A, a)$ is the quotient of* $\mathsf{LoopTwoCell}(a)$ *by 3-cell equivalence;*

(ii) *multiplication is induced by either vertical or horizontal composition;*

(iii) *commutativity is exactly the Eckmann–Hilton consequence of common unit plus interchange.*

Proof. Item (i) is Definition 10.22. Item (ii) follows from Theorem 10.17. Item (iii) is Theorem 10.26 with its proof. □

10.4 Higher Eckmann–Hilton and Stabilisation

Having established that π_2 is abelian, it is natural to ask: what happens one dimension higher? The answer turns out to be the same—and for the same reason. The Eckmann–Hilton argument is not an isolated trick but a structural phenomenon that recurs at every level of the ω-groupoid tower.

The phenomenon does not stop at dimension 2. Once a cell is sufficiently degenerate, multiple composition directions appear, and interchange forces commutativity. This is the stabilisation pattern behind the classical statement "π_n is abelian for $n \geq 2$." Within the computational tower of Chapter 9, this commutativity argument occupies the last genuinely non-trivial coherence layer: the relevant interchange witnesses are 3-cells, while the comparison cells above them already lie in the stabilised, contractible range.

The idea is simple: at dimension n the identity cell $\mathsf{refl}_a^{(n-1)}$ plays the role that refl_a played at dimension 2, and the two neighboring compositions again share this cell as a common unit.

10.4.1 Iterated Degeneracy

To state the higher Eckmann–Hilton principle, we first set up the tower of iterated identity cells.

Definition 10.30 (Iterated reflexivity tower). Fix $a : A$. Define recursively:

$$\mathsf{refl}_a^{(1)} := \mathsf{refl}_a, \qquad \mathsf{refl}_a^{(k+1)} := \mathsf{refl}_{\mathsf{refl}_a^{(k)}}.$$

Thus $\mathsf{refl}_a^{(k)}$ is the k-fold identity cell at a.

Definition 10.31 (Higher loop cells (schematic)). For $n \geq 2$, write

$$\mathsf{Loop}_n(a)$$

for the n-cells whose source and target are both $\mathsf{refl}_a^{(n-1)}$. At $n = 2$, this is Definition 10.9.

Remark 10.32 (Two neighboring compositions). On $\mathsf{Loop}_n(a)$ there are at least two relevant compositions:

(i) composition in the top dimension ($\circ_{n-1}$);

(ii) the next lower composition ($\circ_{n-2}$), transported to the degenerate boundary.

These share the same fully degenerate unit and satisfy an interchange law by the all-level interchange theorem of Chapter 9 (Theorem 9.54).

10.4.2 Higher Eckmann–Hilton Principle

With the iterated loop cells and their two compositions in hand, the same Eckmann–Hilton calculation yields a uniform pattern at every level $n \geq 2$.

Theorem 10.33 (Higher Eckmann–Hilton pattern). *Assume the iterated loop-cell interface of Definition 10.31 and the neighboring compositions of Remark 10.32. Then for each $n \geq 2$, the two compositions on $\mathsf{Loop}_n(a)$ coincide up to an $(n+1)$-cell and are commutative up to an $(n+1)$-cell.*

Proof. Repeat the Eckmann–Hilton calculation one dimension higher, replacing ordinary equalities by displayed $(n+1)$-cells. The ingredients required are exactly:

(i) a common unit in the doubly-degenerate boundary;

(ii) an interchange law for the two compositions.

Both ingredients are available at each level by Theorem 9.54 and the identity machinery of Chapter 9. Chaining these witnesses in the classical order yields the claimed comparison and commutativity $(n+1)$-cells dimensionwise. □

Corollary 10.34 (Stabilisation of computational homotopy groups). *For every $n \geq 2$, the group $\pi_n^{\mathrm{cp}}(A, a)$ is abelian.*

Proof. At level n, $\pi_n^{\mathrm{cp}}(A, a)$ is obtained by quotienting the corresponding degenerate n-loop cells by the next higher equivalence (Definition 9.99). By Theorem 10.33, multiplication on representatives is commutative up to one higher cell; hence classes commute strictly. □

Remark 10.35 (Why stabilisation begins at $n = 2$). Dimension 1 has only one composition direction for loops, so no Eckmann–Hilton forcing is available. From dimension 2 onward there are at least two compatible directions on sufficiently degenerate cells; interchange then forces commutativity. This is the precise algebraic meaning of "higher dimensions have more room to move."

Example 10.36 (The first two stabilisation steps).

(i) $n = 2$: Theorem 10.26.

(ii) $n = 3$: 3-loops have two induced compositions from the ω-groupoid structure, so $\pi_3^{\mathrm{cp}}(A, a)$ is abelian by the same argument.

The induction in Corollary 10.34 is the uniform continuation of this pattern.

10.4.3 Connection with Chapter 9

It is worth making explicit how the higher Eckmann–Hilton theorem follows from the weak ω-groupoid machinery of Chapter 9, especially the proof-relevant 3-dimensional coherence layer and the stable higher fillers above it.

Proposition 10.37 (Eckmann–Hilton as a Chapter 9 corollary). *Theorem 10.33 is a direct corollary of:*

(i) the weak ω-groupoid theorem (Theorem 9.45);

(ii) the all-level interchange family (Theorem 9.54);

(iii) the all-level coherence package (Theorem 9.58).

Proof. Theorem 9.45 provides the operations and units in every dimension. Theorem 9.54 provides interchange between composition directions. Theorem 9.58 supplies the higher witnesses needed to transport these laws to degenerate loop strata. Applying the classical Eckmann–Hilton reasoning at each level yields the claim. □

10.5 Computational Interpretation

So far we have treated Eckmann–Hilton as an algebraic fact about abstract operations. But computational paths are, above all, *rewrite traces*, and commutativity has a concrete operational reading: different scheduling orders for independent rewrites yield the same result once coherence witnesses are taken into account. This section spells out that reading.

10.5.1 What Do Loop 2-Cells Represent?

Remark 10.38 (Operational meaning)**.** An element

$$\alpha : \mathsf{LoopTwoCell}(a)$$

is a derivation that starts at refl_a and ends at refl_a. So α is a *closed rewrite trace* at the identity path.

- Vertical composition $\star_v$ means "run one closed trace, then the other."
- Horizontal composition $\star_h$ means "place traces side by side in a composite expression, then contract the boundary units."

Eckmann–Hilton says these two scheduling disciplines are equivalent and commuting.

10.5.2 A Symmetric Rewrite Example from the 77-Rule System

Fix $a : A$ and consider

$$T := \mathsf{trans}\big(\mathsf{trans}(\mathsf{refl}_a, \mathsf{refl}_a), \mathsf{trans}(\mathsf{refl}_a, \mathsf{refl}_a)\big).$$

There are two independent unit simplifications (left block and right block), hence two natural reduction orders.

Example 10.39 (Two schedules, same composite 2-cell)**.** Let:

$$\begin{aligned} T_L &:= \mathsf{trans}\big(\mathsf{refl}_a, \mathsf{trans}(\mathsf{refl}_a, \mathsf{refl}_a)\big), \\ T_R &:= \mathsf{trans}\big(\mathsf{trans}(\mathsf{refl}_a, \mathsf{refl}_a), \mathsf{refl}_a\big), \\ T_0 &:= \mathsf{trans}(\mathsf{refl}_a, \mathsf{refl}_a), \\ T_f &:= \mathsf{refl}_a. \end{aligned}$$

Then we have two reduction schedules:

(i) **Left then right:** $T \to T_L \to T_0 \to T_f$.

(ii) **Right then left:** $T \to T_R \to T_0 \to T_f$.

These schedules define two parallel 2-cells with common source/target. Interchange supplies a 3-cell between the middle squares, and Theorem 10.17 identifies the induced loop-2 composite.

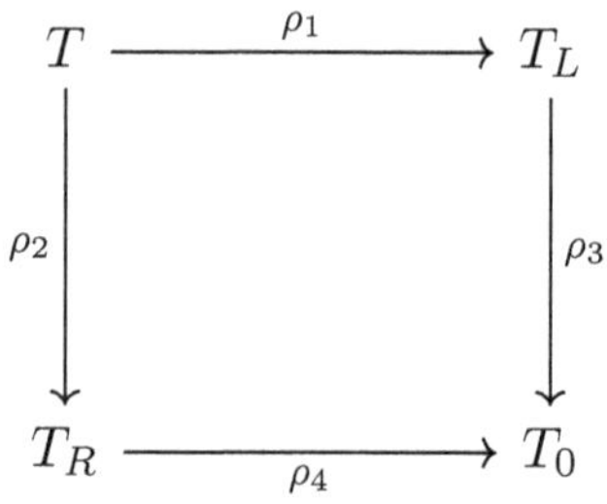

In the diagram, $\rho_1, \rho_2, \rho_3, \rho_4$ denote unit/cancellation rewrite steps from the 77-rule system. The square commutes up to a 3-cell, and the final reduction $T_0 \to T_f$ closes both routes.

Proposition 10.40 (Commutativity as schedule independence). *For loop 2-cells α, β, the class of $\alpha \star_v \beta$ in $\pi_2^{\mathrm{cp}}(A, a)$ is independent of whether the underlying rewrite traces are composed vertically first or horizontally first.*

Proof. This is a direct reformulation of:

(i) interchange on loop 2-cells (Proposition 10.16);

(ii) Eckmann–Hilton commutativity (Theorem 10.17);

(iii) quotienting by $\sim_3$ (Definition 10.22).

□

Remark 10.41 (Coherence interpretation). In rewriting language: "commutativity of π_2" means that specific composite rewrite sequences, initially different as derivation programs, become equal once we account for the canonical coherence witnesses. So abelianity is not a mysterious global constraint; it is the normal form of local interchange plus unit compatibility.

10.5.3 Relation with Confluence

Proposition 10.42 (Confluence supports the EH picture). *The confluence mechanisms of Chapters 5 and 8 ensure that the commuting-square witnesses used in Example 10.39 always extend to coherent higher comparisons.*

Proof. At the 1-cell expression level, confluence (Chapter 5) supplies joins of critical pairs. At the 2-cell level, Chapter 8 packages these joins as 3-cells and proves coherence among them. Therefore the interchange squares needed by Eckmann–Hilton are not isolated: they belong to a stable comparison calculus. □

Example 10.43 (A cancellation-symmetry pattern). Consider

$$W = \mathsf{trans}\big(p, \mathsf{trans}(\mathsf{symm}(p), \mathsf{refl}_a)\big).$$

One may cancel $p \cdot p^{-1}$ first (after reassociation), or remove the trailing unit first and then cancel. These two routes induce parallel 2-cells ending at refl_a. By the coherence package, their difference is measured by a canonical 3-cell; in π_2^{cp} they represent the same element.

Summary

This chapter established the Eckmann–Hilton mechanism in the computational paths setting and extracted its principal consequences. The point is not merely that a classical theorem reappears, but that it reappears for internal reasons once the relevant higher structure has been built. The key points are:

(i) The classical theorem requires only a common unit and interchange.

(ii) On $\mathsf{LoopTwoCell}(a) = \mathsf{Derivation}_2(\mathsf{refl}_a, \mathsf{refl}_a)$, vertical and horizontal composition satisfy these hypotheses.

(iii) Therefore the induced operation is commutative up to 3-cell, and after quotient we obtain an abelian group $\pi_2^{\mathrm{cp}}(A, a)$.

(iv) The same pattern iterates in all higher dimensions: $\pi_n^{\mathrm{cp}}(A, a)$ is abelian for $n \geq 2$.

(v) Computationally, commutativity means coherence identifies distinct but compatible rewrite schedules.

Part II began with path groupoids and ended with a global weak ω-groupoid. The present chapter shows what that higher structure is good for: it recovers, from within the theory itself, a concrete algebraic theorem with genuine computational content.

Part III turns outward: it derives the J-eliminator (Chapter 11), analyses why UIP must fail (Chapter 12), computes fundamental groups of concrete spaces (Chapter 13), compares the framework with homotopy type theory (Chapter 14) and category theory (Chapter 15), and surveys open problems (Chapter 16).

Historical Notes.

The Eckmann–Hilton argument originated in mid-twentieth-century algebraic topology and quickly became one of the standard explanations for why π_2 is abelian. Modern textbook treatments appear in Hatcher [79] and May [114]. From a categorical viewpoint, the argument is a direct consequence of middle-four interchange, a theme emphasized in Mac Lane's development of categorical coherence [98]. The higher-dimensional perspective, in which stabilisation is seen as a systematic consequence of compatible composition directions, aligns with the weak ω-categorical approach of Leinster [99].

Formalization Notes. The results of this chapter are formalised in the following files of the accompanying Lean development (`https://github.com/Arthur742Ramos/ComputationalPathsLean`):

- `ComputationalPaths/Path/Homotopy/EckmannHilton.lean` — the main `OmegaTwo` interface, the comparison between horizontal and vertical composition, and the theorem `eckmann_hilton`
- `ComputationalPaths/Path/OmegaGroupoid/EckmannHiltonProof.lean` — the corresponding proof packaged inside the higher-cell development
- `ComputationalPaths/Coherence/InterchangeLaw.lean` — the interchange-based derivation `eckmann_hilton_via_interchange`

Key definitions. The interchange law relates horizontal and vertical 2-cell composition:

```
/-- Interchange: composing a 2x2 grid of 2-cells
    two ways yields 3-cell-equivalent results -/
| interchange {f f' : Path a b} {g g' : Path b c}
    (alpha : Derivation2 f f')
    (beta : Derivation2 g g') :
    MetaStep3
      (.vcomp (whiskerRight alpha g) (whiskerLeft f' beta))
      (.vcomp (whiskerLeft f beta) (whiskerRight alpha g'))
```

The double loop space and Eckmann–Hilton:

```
/-- The double loop space Omega^2(A, a) -/
abbrev OmegaTwo (A : Type u) (a : A) : Type (u + 2) :=
  Derivation2 (Path.refl a) (Path.refl a)

/-- Eckmann-Hilton: hcomp and vcomp coincide
```

```
    on Omega^2 up to 3-cell -/
noncomputable def hcomp_eq_vcomp
    (alpha beta : OmegaTwo A a) :
    Derivation3 (hcomp alpha beta)
                (vcomp alpha beta) := by
  apply eckmann_hilton_derivation
```

The formal development works directly with a Type-valued tower of 2-cells and 3-cells. The Eckmann–Hilton comparison is therefore a genuine 3-cell witness, not a definitional equality in a proposition-valued quotient. It is one of the named coherence proofs whose substitution by reflexivity is rejected by the verification pipeline. Proposition-valued quotients remain available when one later passes to connected components, but they are no longer used to derive interchange or Eckmann–Hilton itself.

Part III

Connections: Type Theory, Homotopy, and Beyond

Chapter 11

Path Induction and the J-Eliminator

> *"The based path space is contractible. This innocuous statement encodes the full power of path induction."*
>
> —The J Principle

Part II developed the higher algebraic structure of computational paths: groupoids (Chapter 6), 2-cells as derivations (Chapter 7), 3-cells and coherence (Chapter 8), the weak ω-groupoid theorem (Chapter 9), and the Eckmann–Hilton argument (Chapter 10). Throughout that development, paths were treated as first-class objects with explicit structure: step lists, rewrite rules, and dimensional stratification.

Part III turns to the *connections* between computational paths and other foundational frameworks. The first of those connections is with the *J-eliminator*—the fundamental induction principle for identity types [109, 111]. The point of the present chapter is to derive that eliminator from the computational-path framework itself [127, 124]. This shows that computational paths are not merely another presentation of equality, but a setting from which the standard induction principle can be recovered as a theorem.

The key idea is topological in flavour. Fix a point a of a type A and consider all paths that start from a. Each such path ends at some point y, so we may package endpoint and path together as a pair (y, p). The resulting collection is the *based path space*. The decisive fact is that this space is *contractible*: every pair (y, p) can be drawn back to the centre (a, refl_a). Intuitively, any path can be retracted along itself until nothing remains but the reflexive one.

From this fact the J-eliminator follows with remarkable economy: to define a dependent function out of the based path space, it is enough to specify its value at the centre, since every other point is equivalent to that centre.

This chapter proceeds as follows. Section 11.1 introduces the based path space and establishes its basic properties. Section 11.2 proves the fundamental contractibility theorem. Section 11.3 derives the J-eliminator and its computation rule. Section 11.4 develops transport and dependent application as consequences of J. Section 11.5 proves that the J-eliminator is essentially unique: any two eliminators satisfying the same computation rule agree on all inputs.

Throughout this chapter, we work in the two-level structure established in earlier chapters: the *semantic level*, where $\mathsf{Path}(a, b)$ is viewed through its underlying propositional equality $a = b$ (proof-irrelevant), and the *syntactic level*, where paths carry explicit step lists (proof-relevant). The J-eliminator operates at the semantic level; its inability to distinguish step-distinct paths is by design—it captures exactly the structure that is invariant under rewrite equivalence.

11.1 The Based Path Space

To understand path induction, we must first understand the *based path space*—the collection of all paths emanating from a fixed basepoint. This construction, simple as it is, provides the geometric foundation for the entire theory of path induction.

11.1.1 Definition and Basic Properties

Consider a type A and a fixed element $a : A$. We want to gather together all the paths that start at a. Each such path goes to some endpoint y, so we package the path together with its endpoint. The result is a type whose elements are pairs (y, p) where y is any element of A and p is a path from a to y.

Definition 11.1 (Based path space)**.** Let A be a type and $a : A$ a fixed basepoint. The *based path space* at a, denoted $\mathsf{BPS}(a)$ or $\Sigma(y : A).\,\mathsf{Path}(a, y)$, is the dependent sum:

$$\mathsf{BPS}(a) \;:=\; \Sigma(y : A).\,\mathsf{Path}(a, y)$$

An element of $\mathsf{BPS}(a)$ is a pair (y, p) where $y : A$ and $p : \mathsf{Path}(a, y)$.

Remark 11.2 (Topological intuition)**.** In topological terms, $\mathsf{BPS}(a)$ is the *total space of the path fibration* based at a. Think of it as a "fan" of paths spreading out from the basepoint a. The projection $(y, p) \mapsto y$ sends each path to its endpoint, while the fiber over a point y consists of all paths from a to y. As we shall prove in Section 11.2, this entire fan is contractible: we can continuously collapse it down to a single point.

Among all the elements of the based path space, one stands out as particularly natural: the pair consisting of the basepoint itself together with the trivial reflexive path.

Definition 11.3 (Center of contraction)**.** The *center* of $\mathsf{BPS}(a)$ is the canonical element:

$$\mathsf{center}(a) \;:=\; (a, \mathsf{refl}_a) \;:\; \mathsf{BPS}(a)$$

This is the basepoint paired with the reflexive path.

Example 11.4 (Elements of the based path space)**.** To make this concrete, consider $A = \mathbb{N}$ and $a = 0$. The based path space $\mathsf{BPS}(0)$ contains elements such as:

- $(0, \mathsf{refl}_0)$: the center—the number 0 paired with the trivial path.
- $(0, \mathsf{trans}(\mathsf{refl}_0, \mathsf{refl}_0))$: the same endpoint 0, but with a different path (the composite of two reflexive paths).
- For any $n : \mathbb{N}$ and path $p : \mathsf{Path}(0, n)$, the pair (n, p) belongs to $\mathsf{BPS}(0)$.

At the semantic level (where we ignore step structure), all paths to the same endpoint are identified, so there is essentially one element over each endpoint. At the syntactic level, paths with different step lists are genuinely distinct, giving multiple elements over the same endpoint.

11.1.2 Projections and Structure

The based path space, being a dependent sum, comes equipped with natural projection functions that extract its components.

Definition 11.5 (Projections from $\mathsf{BPS}(a)$)**.** For a based path space $\mathsf{BPS}(a)$:

(i) The *target projection* is $\pi_1 : \mathsf{BPS}(a) \to A$ defined by $\pi_1(y, p) = y$.

(ii) The *path projection* is $\pi_2 : (bp : \mathsf{BPS}(a)) \to \mathsf{Path}(a, \pi_1(bp))$ defined by $\pi_2(y, p) = p$.

Lemma 11.6 (Eta rule for $\mathsf{BPS}(a)$)**.** *Every element $bp : \mathsf{BPS}(a)$ satisfies*

$$bp = (\pi_1(bp), \pi_2(bp)).$$

Proof. This is the standard η-rule for dependent sums. □

11.1.3 The Semantic vs. Syntactic Distinction

A crucial subtlety arises when considering the based path space in our two-level setting. Recall from earlier chapters that computational paths exist at two levels: the *semantic level* where only the endpoints matter, and the *syntactic level* where the full step structure is preserved. This distinction fundamentally affects the based path space.

Definition 11.7 (Semantic based path space)**.** The *semantic based path space* at $a : A$ is:

$$\mathsf{BPS}^{\mathrm{sem}}(a) \;:=\; \Sigma(y : A).\,(a = y)$$

An element of $\mathsf{BPS}^{\mathrm{sem}}(a)$ is therefore a pair (y, h) where $y : A$ and $h : a = y$ is an ordinary equality witness.

Remark 11.8 (Suppressing formal wrappers)**.** Formal presentations sometimes insert a trivial wrapper around the equality component so that the dependent sum itself is treated uniformly as data. Since that wrapper carries no additional mathematical content, we suppress it in the main exposition and write simply $a = y$.

Definition 11.9 (Syntactic based path space)**.** The *syntactic based path space* at $a : A$ is:

$$\mathsf{BPS}^{\mathrm{syn}}(a) \;:=\; \Sigma(y : A).\,\mathsf{Path}(a, y)$$

where $\mathsf{Path}(a, y)$ carries the full step-list structure.

Proposition 11.10 (Semantic projection)**.** *There is a canonical map*

$$\mathsf{forget} : \mathsf{BPS}^{\mathrm{syn}}(a) \to \mathsf{BPS}^{\mathrm{sem}}(a)$$

defined by $\mathsf{forget}(y, p) = (y, \mathsf{toEq}(p))$. *This map forgets the step structure.*

Remark 11.11 (Contractibility at different levels)**.** The distinction between semantic and syntactic based path spaces leads to a fundamental difference in their properties:

- $\mathsf{BPS}^{\mathrm{sem}}(a)$ is contractible: every element can be continuously deformed to the center.
- $\mathsf{BPS}^{\mathrm{syn}}(a)$ is *not* contractible: elements with different step lists remain genuinely distinct.

The J-eliminator, which we derive in Section 11.3, operates at the semantic level where contractibility holds. This is why J cannot distinguish between paths with the same endpoints but different computational histories.

11.1.4 The Free Path Space

For completeness, we also introduce the *free* path space, a variant where both endpoints are allowed to vary. While the based path space fixes one endpoint, the free path space imposes no such constraint.

Definition 11.12 (Free path spaces). The *syntactic free path space* of a type A is

$$\mathsf{FPS}^{\mathrm{syn}}(A) \;:=\; \Sigma(a:A).\,\Sigma(b:A).\,\mathsf{Path}(a,b),$$

while the *semantic free path space* is

$$\mathsf{FPS}^{\mathrm{sem}}(A) \;:=\; \Sigma(a:A).\,\Sigma(b:A).\,(a=b).$$

An element of either free path space is a triple consisting of a source, a target, and a witness connecting them at the chosen level. When no confusion threatens, we abbreviate $\mathsf{FPS}^{\mathrm{syn}}(A)$ to $\mathsf{FPS}(A)$.

Remark 11.13 (Relation to based path space). Both free path spaces decompose naturally as disjoint unions of based path spaces:

$$\mathsf{FPS}^{\mathrm{syn}}(A) \;\cong\; \coprod_{a:A} \mathsf{BPS}^{\mathrm{syn}}(a), \qquad \mathsf{FPS}^{\mathrm{sem}}(A) \;\cong\; \coprod_{a:A} \mathsf{BPS}^{\mathrm{sem}}(a).$$

Each based path space appears as the fiber over its source when we project the corresponding free path space onto its first coordinate. This decomposition will be useful when we discuss free path induction in Section 11.3.

11.2 Contractibility of the Based Path Space

The central technical result of this chapter is that the semantic based path space is *contractible*—it can be continuously shrunk to a single point. This geometric fact is the engine driving the derivation of the J-eliminator: because the based path space is trivial in this sense, any function defined on it is determined by its value at the center.

11.2.1 Contractibility: Definition and Significance

Before proving the main theorem, we must make precise what it means for a type to be contractible. Informally, a type is contractible if it has exactly one element "up to homotopy"—that is, all its elements are connected by paths.

Definition 11.14 (Contractible type). A type X is *contractible* if there exists an element $c : X$ (the *center*) such that for all $x : X$, we have $x = c$. Equivalently, X is contractible if it is a subsingleton with a specified inhabitant.

Remark 11.15 (Contractible = singleton up to homotopy)**.** Contractibility is the homotopy-theoretic version of "having exactly one element." A contractible type is homotopy equivalent to the unit type $\mathbf{1}$. Geometrically, we can picture the contraction paths $x = c$ as "rays" emanating from each point x and converging to the center c—imagine an umbrella collapsing to its tip.

It is often convenient to package the contractibility data into a single structure.

Definition 11.16 (IsContr structure)**.** A *contractibility witness* for a type X is a dependent pair:

$$\mathsf{IsContr}(X) \;:\equiv\; \Sigma(c : X).\,\Pi(x : X).\,x = c$$

The first component is the center; the second provides the contraction paths.

11.2.2 The Fundamental Contraction Theorem

We now prove the key result: the semantic based path space is contractible. Intuitively, given any element (y, h) consisting of an endpoint y and an equality witness $h : a = y$, we can retract it back to the center by identifying h with reflexivity. The formal proof proceeds by path induction on the underlying propositional equality.

Theorem 11.17 (Contractibility of the semantic based path space)**.** *For any type A and basepoint $a : A$, the semantic based path space* $\mathsf{BPS}^{\mathrm{sem}}(a)$ *is contractible, with center* $\mathsf{center}(a) = (a, \mathsf{rfl})$.

Proof. We must show that for any $(y, h) : \mathsf{BPS}^{\mathrm{sem}}(a)$, we have $(y, h) = \mathsf{center}(a)$, i.e., $(y, h) = (a, \mathsf{rfl})$. Apply path induction to the equality witness $h : a = y$. In the reflexive case we have $y \equiv a$ and $h \equiv \mathsf{rfl}$, so the required equality reduces to

$$(a, \mathsf{rfl}) = (a, \mathsf{rfl}),$$

which is reflexivity. □

Remark 11.18 (The induction step)**.** The proof's simplicity conceals its power. The key insight is that the equality witness $h : a = y$ lives in Prop, where proof irrelevance holds. Path induction lets us assume $h = \mathsf{rfl}$, which forces the endpoint y to equal a. Once both components match the center, the equality is trivial. This is the "retraction to the basepoint" that makes the based path space homotopically trivial: we can pull any path back to the starting point.

The contractibility theorem immediately implies that the based path space is a subsingleton: any two elements are equal.

Corollary 11.19 (Subsingleton property). *Any two elements of* $\mathsf{BPS}^{\mathrm{sem}}(a)$ *are equal: for* $(y_1, h_1), (y_2, h_2) : \mathsf{BPS}^{\mathrm{sem}}(a)$,

$$(y_1, h_1) = (y_2, h_2).$$

Proof. By Theorem 11.17, both equal the center: $(y_1, h_1) = \mathsf{center}(a) = (y_2, h_2)$. □

Corollary 11.20 (The semantic based path space is a subsingleton). $\mathsf{BPS}^{\mathrm{sem}}(a)$ *has at most one element.*

11.2.3 Non-Contractibility at the Syntactic Level

The situation differs sharply at the syntactic level, where paths retain their full step structure. The syntactic based path space preserves computational information and therefore cannot collapse to a point.

Theorem 11.21 (Non-contractibility of the syntactic based path space). *For any type A with $a : A$, the syntactic based path space* $\mathsf{BPS}^{\mathrm{syn}}(a)$ *is* not *contractible.*

Proof. Suppose $\mathsf{BPS}^{\mathrm{syn}}(a)$ were contractible. Then all elements would equal the center:

$$\forall\, bp : \mathsf{BPS}^{\mathrm{syn}}(a), \quad bp = (a, \mathsf{refl}_a).$$

Consider the element $(a, \mathsf{ofEq}(\mathsf{rfl})) : \mathsf{BPS}^{\mathrm{syn}}(a)$. By the definition of ofEq:

$$\mathsf{steps}(\mathsf{ofEq}(\mathsf{rfl})) \text{ is a singleton list,} \qquad \mathsf{steps}(\mathsf{refl}_a) = [\,].$$

So $\mathsf{ofEq}(\mathsf{rfl})$ has a non-empty step list, whereas refl_a has an empty one.

If $(a, \mathsf{ofEq}(\mathsf{rfl})) = (a, \mathsf{refl}_a)$, then by congruence on the second projection:

$$\mathsf{ofEq}(\mathsf{rfl}) = \mathsf{refl}_a.$$

But Proposition 3.18 from Chapter 3 would then force a singleton list to equal the empty list, which is impossible. □

Remark 11.22 (Two-level coherence). This theorem crystallizes the two-level structure:

- **Semantic level:** Contractible, proof-irrelevant, UIP holds. This is where classical reasoning about equality takes place.
- **Syntactic level:** Not contractible, proof-relevant, UIP fails. This is where computational information is preserved.

The J-eliminator operates at the semantic level, inheriting contractibility and the ability to reduce all cases to the reflexive one. The syntactic level preserves the richer structure that distinguishes different rewrite traces leading to the same equality. This semantic contractibility should not be conflated with the higher-cell contractibility results of Part II and Appendix C: there the objects being compared are proof-relevant derivations, and the contraction occurs one level higher rather than directly at the path level.

11.2.4 Contractibility of the Free Path Space

The contractibility result extends naturally to the free path space, where we consider each fiber separately.

Proposition 11.23 (Fiberwise contractibility). *For each $a : A$, the fiber of $\mathsf{FPS}^{\mathrm{sem}}(A)$ over a (under the source projection) is contractible.*

Proof. The fiber over a is precisely $\{a\} \times \mathsf{BPS}^{\mathrm{sem}}(a) \cong \mathsf{BPS}^{\mathrm{sem}}(a)$, which is contractible by Theorem 11.17. □

Corollary 11.24 (Free path contraction). *Each element $(a, y, p) : \mathsf{FPS}^{\mathrm{sem}}(A)$ satisfies:*

$$(a, y, p) = (a, a, \mathsf{rfl}).$$

11.3 Derivation of J and Its Computation Rule

We now arrive at the main result of this chapter: deriving the J-eliminator from the contractibility of the based path space. This derivation demonstrates that path induction is not an axiom imposed from outside, but a *theorem* that follows from the geometric structure of paths.

The J-eliminator is the fundamental induction principle for identity types. Just as mathematical induction lets us prove properties of all natural numbers by proving a base case ($n = 0$) and an inductive step, path induction lets us prove properties of all paths by handling only the reflexive case. The key insight is that, because the based path space contracts to a point, every path is—in a suitable sense—"the same" as the reflexive path.

11.3.1 The J-Eliminator: Statement

Theorem 11.25 (The J-Eliminator). *Let A be a type and $a : A$ a basepoint. For any type family*

$$C : \Pi(y : A).\, (a = y) \to \mathsf{Type}$$

and base case $c : C(a, \mathsf{rfl})$, *there exists a function*

$$\mathsf{J} : \Pi(y : A).\, \Pi(h : a = y).\, C(y, h)$$

such that $\mathsf{J}(a, \mathsf{rfl}) = c$.

Remark 11.26 (Reading the type of J)**.** The type of J deserves careful reading:

- The type family C depends on both an endpoint y and a proof $h : a = y$. This generality is essential—many applications require the motive to vary in both arguments.
- The base case $c : C(a, \mathsf{rfl})$ handles only the reflexive path at the starting point.
- The conclusion provides a value $\mathsf{J}(y, h) : C(y, h)$ for *any* endpoint y and *any* path h.

The computation rule $\mathsf{J}(a, \mathsf{rfl}) = c$ says that when we apply J to the reflexive path, we get back exactly the base case we provided. This is analogous to the computation rule for natural number induction: $\mathsf{rec}(0) = z$.

11.3.2 Derivation from Contractibility

The derivation exploits a simple principle: if a type is contractible, then any dependent type over it is determined by its value at the center. Since $\mathsf{BPS}^{\mathrm{sem}}(a)$ is contractible with center (a, rfl), a function defined on it is determined by what it does at (a, rfl).

Proof of Theorem 11.25. Define the auxiliary type family:

$$C' : \mathsf{BPS}^{\mathrm{sem}}(a) \to \mathsf{Type} \qquad C'(y, h) := C(y, h)$$

We have a base case at the center:

$$c' : C'(\mathsf{center}(a))$$

given simply by the hypothesis $c : C(a, \mathsf{rfl})$.

By Theorem 11.17, every $bp : \mathsf{BPS}^{\mathrm{sem}}(a)$ satisfies $bp = \mathsf{center}(a)$. Let $\gamma_{bp} : bp = \mathsf{center}(a)$ denote this contraction path.

Define the section:

$$s : \Pi(bp : \mathsf{BPS}^{\mathrm{sem}}(a)).\, C'(bp) \qquad s(bp) := \gamma_{bp}^{*}\, c'$$

where $\gamma_{bp}^{*} : C'(\mathsf{center}(a)) \to C'(bp)$ denotes transport along γ_{bp}^{-1}.

Now define J:

$$\mathsf{J}(y, h) := s(y, h)$$

Computation rule: $\mathsf{J}(a, \mathsf{rfl}) = s(a, \mathsf{rfl}) = s(\mathsf{center}(a)) = c$. □

Remark 11.27 (The role of contractibility)**.** Contractibility is the crucial ingredient. It guarantees that:

1. Every point in the based path space is connected to the center.
2. These connections are essentially unique (the space is a subsingleton).
3. Transport from the center to any other point is well-defined.

Without contractibility, there would be no way to "propagate" the base case value to arbitrary endpoints. The contraction paths provide the conduit along which information flows.

11.3.3 Direct Definition via Pattern Matching

In practice, proof assistants implement J directly using dependent pattern matching, which captures the essence of the contractibility argument in a more computational form.

Definition 11.28 (J via pattern matching)**.**

$$\begin{aligned} \mathsf{J} : \Pi(C : \Pi(y : A).\,(a = y) \to \mathsf{Type}).& \\ C(a, \mathsf{rfl}) \to \Pi(y : A).\,\Pi(h : a = y).\,C(y, h)& \\ \mathsf{J}(C, c, a, \mathsf{rfl}) := c& \end{aligned}$$

Remark 11.29 (Definitional vs. propositional computation)**.** With the pattern-matching definition, $\mathsf{J}(C, c, a, \mathsf{rfl}) = c$ holds *definitionally*—it is a judgmental equality checked by the type checker, not merely a propositional statement that must be proved. This is the β-rule for path induction, and it is what makes J so powerful as a reasoning principle: the reduction happens automatically during type checking.

Remark 11.30 (Dialogical reading of the Id-case)**.** The computation rule for J is the identity-type instance of the dialogical pattern discussed in Section 2.3.2. Introduction fixes a reason (*Defender*'s assertion), while elimination challenges that reason by asking what follows from it (*Attacker*'s move). The β-rule $\mathsf{J}(C, c, a, \mathsf{rfl}) = c$ expresses that the defense is determined by the asserted reason itself: once the witness of equality is given, the continuation is fixed by substitution.

11.3.4 Path-Level J

We can also define J acting directly on Path values rather than on propositional equalities. This version simply passes from a computational path to its semantic projection and then applies the standard J.

Definition 11.31 (J for Path).

$$\begin{aligned}&\mathsf{PathJ} : \Pi(C : \Pi(y : A).\,(a = y) \to \mathsf{Type}).\\&\qquad C(a, \mathsf{rfl}) \to \Pi(y : A).\,\Pi(p : \mathsf{Path}(a, y)).\,C(y, \mathsf{toEq}(p))\\&\mathsf{PathJ}(C, c, y, p) := \mathsf{J}(C, c, y, \mathsf{toEq}(p))\end{aligned}$$

where $\mathsf{toEq}(p) : a = y$ is the semantic content of p.

Proposition 11.32 (PathJ computation rule). $\mathsf{PathJ}(C, c, a, \mathsf{refl}_a) = c$.

Proof. $\mathsf{PathJ}(C, c, a, \mathsf{refl}_a) = \mathsf{J}(C, c, a, \mathsf{toEq}(\mathsf{refl}_a)) = \mathsf{J}(C, c, a, \mathsf{rfl}) = c$. □

Proposition 11.33 (PathJ is step-independent). *For any* $p, q : \mathsf{Path}(a, y)$*:*

$$\mathsf{PathJ}(C, c, y, p) = \mathsf{PathJ}(C, c, y, q).$$

Proof. By proof irrelevance, $\mathsf{toEq}(p) = \mathsf{toEq}(q)$, so $\mathsf{J}(C, c, y, \mathsf{toEq}(p)) = \mathsf{J}(C, c, y, \mathsf{toEq}(q))$. □

Remark 11.34 (J cannot see step structure). This step-independence is a feature of the two-level design. The J-eliminator operates at the semantic level, where paths are identified by their endpoints alone. Two paths p and q with identical endpoints—say, both from a to b—but different computational histories (different step lists) are treated identically by J. This corresponds to the homotopy-theoretic intuition that J sees paths "up to homotopy," not their internal parametrisations.

11.3.5 The Paulin-Mohring Variant

There is an equivalent formulation of path induction where the *target* is fixed rather than the source. This variant, due to Christine Paulin-Mohring [119], is sometimes more convenient when reasoning about the target of a path.

Theorem 11.35 (Paulin-Mohring J). *Let* $b : A$ *be fixed. For any type family*

$$C : \Pi(a : A).\,(a = b) \to \mathsf{Type}$$

and base case $c : C(b, \mathsf{rfl})$*, there exists*

$$\mathsf{J}_{\mathsf{PM}} : \Pi(a : A).\,\Pi(h : a = b).\,C(a, h)$$

with $\mathsf{J}_{\mathsf{PM}}(b, \mathsf{rfl}) = c$.

Proof. Define J_{PM} using J and symmetry:

$$\mathsf{J}_{\mathsf{PM}}(C, c, a, h) :\equiv \mathsf{J}(\lambda a'.\, \lambda h'.\, C(a', (h')^{-1}), c', b, h^{-1})$$

where c' is the base case obtained by identifying $C(b, \mathsf{rfl}^{-1})$ with $C(b, \mathsf{rfl})$.

Alternatively, pattern-match directly: $\mathsf{J}_{\mathsf{PM}}(C, c, b, \mathsf{rfl}) :\equiv c$. □

Remark 11.36 (Equivalence of based and Paulin-Mohring). The based J (fixing the source) and Paulin-Mohring J (fixing the target) are inter-derivable: each can define the other using symmetry of equality. The choice between them is purely a matter of convenience. In this book we use based J as the primary formulation, but the Paulin-Mohring variant appears naturally when we need to reason about a path's target.

11.3.6 Free Path Induction

When both endpoints are allowed to vary, we obtain *free* path induction. This is the most general form: it handles families that depend on both the source and target of a path.

Theorem 11.37 (Free path induction). *For any type family*

$$C : \Pi(a\, b : A).\, (a = b) \to \mathsf{Type}$$

and base case $c : \Pi(a : A).\, C(a, a, \mathsf{rfl})$*, there exists*

$$\mathsf{J}_{\mathsf{free}} : \Pi(a\, b : A).\, \Pi(h : a = b).\, C(a, b, h)$$

with $\mathsf{J}_{\mathsf{free}}(a, a, \mathsf{rfl}) = c(a)$.

Proof. Apply based J with $C'(b, h) :\equiv C(a, b, h)$ and base case $c(a)$. □

11.4 Transport and Dependent ap

With the J-eliminator established, we can derive two fundamental operations that pervade homotopy type theory [150]: *transport* and *dependent application*. Both are instances of J applied to specific type families, and both capture essential aspects of how paths interact with dependent types.

Transport moves elements "along" a path: if we have $x : B(a)$ and a path $h : a = y$, transport gives us an element of $B(y)$. Dependent application generalises the congruence rule to dependent functions: if $f : \Pi(x : A).\, B(x)$ and $h : a = y$, then $f(a)$ and $f(y)$ live in different fibers, but are related by transport.

11.4.1 Transport Derived from J

Transport is the operation of "moving" an element of a dependent type along a path in the base. Geometrically, if B is a family of types over A, we can think of a path $h : a = y$ in the base as inducing a map from the fiber over a to the fiber over y.

Definition 11.38 (Transport via J)**.** Let $B : A \to \mathsf{Type}$ be a type family and $h : a = y$ a path. *Transport* along h is:

$$\mathsf{transport}_B(h) : B(a) \to B(y)$$
$$\mathsf{transport}_B(h)(x) := \mathsf{J}(\lambda y.\, \lambda\_.\, B(y), x, y, h)$$

Proposition 11.39 (Transport computation rule)**.** $\mathsf{transport}_B(\mathsf{rfl})(x) = x$.

Proof. $\mathsf{J}(\lambda y.\, \lambda\_.\, B(y), x, a, \mathsf{rfl}) = x$ by the J computation rule. □

Remark 11.40 (Transport agrees with the path-level transport operation)**.** For a path $p : \mathsf{Path}(a, b)$, the J-derived transport $\mathsf{transport}_B(\mathsf{toEq}(p))$ coincides with the directly defined transport operation along p from Chapter 3. Both constructions implement the same operation; the J derivation shows this operation is a consequence of path induction rather than an independent primitive.

The transport operation satisfies a rich collection of laws that mirror the groupoid structure of paths.

Theorem 11.41 (Transport laws from J)**.** *The following laws hold:*

(i) ***Identity:*** $\mathsf{transport}_B(\mathsf{rfl}) = \mathsf{id}$.

(ii) ***Composition:*** $\mathsf{transport}_B(h_1 \cdot h_2) = \mathsf{transport}_B(h_2) \circ \mathsf{transport}_B(h_1)$.

(iii) ***Inverse (left):*** $\mathsf{transport}_B(h^{-1}) \circ \mathsf{transport}_B(h) = \mathsf{id}$.

(iv) ***Inverse (right):*** $\mathsf{transport}_B(h) \circ \mathsf{transport}_B(h^{-1}) = \mathsf{id}$.

Proof. Each follows by path induction. For (i): immediate from the computation rule. For (ii)–(iv): apply J to the path(s), reducing to the reflexive case where the equalities hold trivially. □

Corollary 11.42 (Transport is an equivalence)**.** *For any $h : a = b$, the map $\mathsf{transport}_B(h) : B(a) \to B(b)$ is an equivalence with inverse $\mathsf{transport}_B(h^{-1})$.*

11.4.2 Dependent Application (apd)

For a non-dependent function $f : A \to B$, we have the congruence rule $\mathsf{ap}_f : (a = y) \to (f(a) = f(y))$—the function preserves equality. But what about dependent functions $f : \Pi(x : A).\, B(x)$? Here $f(a)$ and $f(y)$ live in *different types*, $B(a)$ and $B(y)$, so we cannot directly ask whether they are equal. The solution is to transport $f(a)$ to the fiber over y and then compare.

Definition 11.43 (Dependent application)**.** Let $B : A \to \mathsf{Type}$ and $f : \Pi(x : A).\, B(x)$. For $h : a = y$, the *dependent application* is:

$$\mathsf{apd}_f(h) : \mathsf{transport}_B(h)(f(a)) = f(y)$$

Theorem 11.44 (apd theorem)**.** *For any $f : \Pi(x : A).\, B(x)$ and $h : a = y$:*

$$\mathsf{transport}_B(h)(f(a)) = f(y).$$

Proof. By J on h. When $h = \mathsf{rfl}$: $\mathsf{transport}_B(\mathsf{rfl})(f(a)) = f(a)$ by the computation rule. □

Remark 11.45 (apd vs. ap)**.** For non-dependent functions $f : A \to B$, we have $\mathsf{ap}_f : (a = y) \to (f(a) = f(y))$, the familiar congruence rule (our $\mathsf{congrArg}$ from Chapter 6). The dependent version apd generalises this by inserting a transport: instead of $f(a) = f(y)$ directly, we obtain $\mathsf{transport}_B(h)(f(a)) = f(y)$. When B is constant (the non-dependent case), transport is the identity and apd reduces to ap.

Proposition 11.46 (apd computation rule)**.** $\mathsf{apd}_f(\mathsf{rfl}) = \mathsf{rfl}$.

Proof. When $h = \mathsf{rfl}$, both sides of the equation $\mathsf{transport}_B(\mathsf{rfl})(f(a)) = f(a)$ are definitionally equal, so the witness is reflexivity. □

11.4.3 Derived Operations

Using J, we can rederive all the groupoid operations from Chapter 6. This provides an alternative foundation: rather than taking symmetry, transitivity, and congruence as primitive operations and verifying the groupoid laws, we can derive them all from J.

Proposition 11.47 (Symmetry from J)**.** *For $h : a = b$, define:*

$$\mathsf{J\_symm}(h) := \mathsf{J}(\lambda y.\, \lambda\_.\, y = a, \mathsf{rfl}, b, h).$$

Then $\mathsf{J\_symm}(h) : b = a$ with $\mathsf{J\_symm}(\mathsf{rfl}) = \mathsf{rfl}$.

Proposition 11.48 (Transitivity from J). *For $h_1 : a = b$ and $h_2 : b = c$, define:*

$$\mathsf{J\_trans}(h_1, h_2) := \mathsf{J}(\lambda c.\, \lambda\_.\, a = c, h_1, c, h_2).$$

Then $\mathsf{J\_trans}(h_1, h_2) : a = c$ *with* $\mathsf{J\_trans}(\mathsf{rfl}, \mathsf{rfl}) = \mathsf{rfl}$.

Proposition 11.49 (Congruence from J). *For $f : A \to B$ and $h : a = y$, define:*

$$\mathsf{J\_congrArg}_f(h) := \mathsf{J}(\lambda y.\, \lambda\_.\, f(a) = f(y), \mathsf{rfl}, y, h).$$

Then $\mathsf{J\_congrArg}_f(h) : f(a) = f(y)$ *with* $\mathsf{J\_congrArg}_f(\mathsf{rfl}) = \mathsf{rfl}$.

Remark 11.50 (Groupoid laws as consequences). All the groupoid laws G1–G8 from Chapter 6 can be proved using J. For instance, associativity $(h_1 \cdot h_2) \cdot h_3 = h_1 \cdot (h_2 \cdot h_3)$ follows by applying J to each path in sequence, reducing to the case where all paths are rfl, where the equation holds trivially. This demonstrates that the groupoid structure is not an additional axiom but a *consequence* of path induction.

From a foundational perspective, we can either:

1. Take the groupoid operations as primitive and verify the laws (the approach of Part II), or
2. Take J as primitive and derive everything else.

Both approaches yield the same structure; the choice depends on which presentation is more natural for the application at hand.

11.4.4 Transport in Path Spaces

Transport interacts interestingly with path spaces themselves. These lemmas are essential for computing with higher paths.

Proposition 11.51 (Transport in path types (right)). *For $p : \mathsf{Path}(b, c)$ and $q : \mathsf{Path}(a, b)$:*

$$\mathsf{transport}_{x \mapsto a = x}(\mathsf{toEq}(p))(\mathsf{toEq}(q)) = \mathsf{toEq}(q) \cdot \mathsf{toEq}(p).$$

Proof. By J on $\mathsf{toEq}(p)$. When $\mathsf{toEq}(p) = \mathsf{rfl}$: $\mathsf{transport}(\mathsf{rfl})(\mathsf{toEq}(q)) = \mathsf{toEq}(q) = \mathsf{toEq}(q) \cdot \mathsf{rfl}$. □

Proposition 11.52 (Transport in path types (left)). *For $p : \mathsf{Path}(b, c)$ and $q : \mathsf{Path}(b, a)$:*

$$\mathsf{transport}_{x \mapsto x = a}(\mathsf{toEq}(p))(\mathsf{toEq}(q)) = (\mathsf{toEq}(p))^{-1} \cdot \mathsf{toEq}(q).$$

Proof. By J on $\mathsf{toEq}(p)$. When $\mathsf{toEq}(p) = \mathsf{rfl}$: $\mathsf{transport}(\mathsf{rfl})(\mathsf{toEq}(q)) = \mathsf{toEq}(q) = \mathsf{rfl}^{-1} \cdot \mathsf{toEq}(q) = \mathsf{toEq}(q)$. □

11.4.5 Leibniz's Principle

We conclude this section with a classical formulation of equality substitution, connecting our type-theoretic development to traditional logic. Leibniz's principle—the indiscernibility of identicals—states that equal objects satisfy the same predicates.

Definition 11.53 (Leibniz substitution)**.** For $h : a = b$ and predicate $P : A \to \mathsf{Prop}$:

$$\mathsf{leibniz}(h, P) : P(a) \to P(b) \qquad \mathsf{leibniz}(h, P) := \mathsf{transport}_P(h)$$

Proposition 11.54 (Leibniz computation)**.** $\mathsf{leibniz}(\mathsf{rfl}, P) = \mathsf{id}$.

Remark 11.55 (Leibniz and indiscernibility)**.** Leibniz's principle—equal objects satisfy the same predicates—is precisely transport specialised to propositional families. In our framework, if $a = b$ then for any predicate P on A, we can transport a proof of $P(a)$ to obtain a proof of $P(b)$. Combined with symmetry, this gives the full substitution principle: equals can replace equals in any context.

11.5 Uniqueness of the J-Eliminator

Having derived J, we now show that it is *essentially unique*: any two eliminators satisfying the same computation rule must agree on all inputs. This uniqueness theorem has both theoretical and practical significance. Theoretically, it shows that path induction is determined by its specification. Practically, it means we can reason about J without worrying about which particular implementation is used.

11.5.1 The Uniqueness Theorem

Theorem 11.56 (Uniqueness of J)**.** *Let* $C : \Pi(y : A).\,(a = y) \to \mathsf{Type}$ *and* $c : C(a, \mathsf{rfl})$. *Suppose* elim_1 *and* elim_2 *are both* $\Pi(y : A).\,\Pi(h : a = y).\,C(y, h)$*-valued and satisfy:*

$$\mathsf{elim}_1(a, \mathsf{rfl}) = c = \mathsf{elim}_2(a, \mathsf{rfl}).$$

Then for all $y : A$ *and* $h : a = y$*:*

$$\mathsf{elim}_1(y, h) = \mathsf{elim}_2(y, h).$$

Proof. By path induction on h. When $h = \mathsf{rfl}$:

$$\mathsf{elim}_1(a, \mathsf{rfl}) = c = \mathsf{elim}_2(a, \mathsf{rfl}).$$

□

Remark 11.57 (Uniqueness from contractibility)**.** This uniqueness theorem is another manifestation of contractibility. Since $\mathsf{BPS}^{\mathrm{sem}}(a)$ is a subsingleton, any function on it is determined by its value at the center. Two functions that agree at the center must agree everywhere—there is simply nowhere else for them to differ.

Corollary 11.58 (J is uniquely characterised)**.** *If* $\mathsf{elim}(a, \mathsf{rfl}) = c$*, then* $\mathsf{elim} = \mathsf{J}(C, c)$.

11.5.2 Uniqueness for Free Path Induction

The uniqueness result extends to the free path induction principle.

Theorem 11.59 (Uniqueness of free J)**.** *Let* $C : \Pi(a\, b : A).\,(a = b) \to \mathsf{Type}$ *and* $c : \Pi(a : A).\, C(a, a, \mathsf{rfl})$*. If* elim_1 *and* elim_2 *both satisfy* $\mathsf{elim}_i(a, a, \mathsf{rfl}) = c(a)$ *for all* $a : A$*, then:*

$$\forall\, a\, b : A,\ \forall\, h : a = b, \quad \mathsf{elim}_1(a, b, h) = \mathsf{elim}_2(a, b, h).$$

Proof. By J on h. When $h = \mathsf{rfl}$: both equal $c(a)$. □

11.5.3 Equivalence with Contractibility

We have used contractibility to derive J. The converse also holds: J implies contractibility. Thus the two principles are logically equivalent—they are different ways of expressing the same fundamental property of identity types.

Theorem 11.60 (J iff contractibility)**.** *The following are equivalent:*

(i) The J-eliminator holds for all types.

(ii) $\mathsf{BPS}^{\mathrm{sem}}(a)$ *is contractible for all* $a : A$.

Proof. **(i) ⇒ (ii):** Use J to prove that every (y, h) equals the center:

$$\mathsf{J}(\lambda y.\, \lambda h.\, (y, h) = \mathsf{center}(a), \mathsf{rfl}, y, h).$$

(ii) ⇒ (i): This is exactly the derivation in Section 11.3: use contractibility to transport the base case from the center to any target. □

11.5.4 The Structure Identity Principle

A powerful consequence of J and transport is the ability to characterise equality of structured types. This *structure identity principle* decomposes equality of compound data into equalities of components.

Proposition 11.61 (Sigma type equality). *For* $B : A \to \mathsf{Type}$ *and pairs* $(a_1, b_1), (a_2, b_2) : \Sigma(a : A).\, B(a)$*:*

$$(a_1, b_1) = (a_2, b_2) \;\Leftrightarrow\; \exists\,(h : a_1 = a_2),\ \mathsf{transport}_B(h)(b_1) = b_2.$$

Proof. Forward: by congruence on projections and transport. Backward: by J on h; when $h = \mathsf{rfl}$, the condition $b_1 = b_2$ gives $(a_1, b_1) = (a_1, b_2)$ by congruence. □

Remark 11.62 (Structure identity). This proposition is the simplest instance of the structure identity principle: equality of structured data decomposes into equality of components, properly adjusted by transport. The principle extends far beyond Σ-types. In homotopy type theory, one can characterise equality in:

- Product types: $(a_1, b_1) = (a_2, b_2)$ iff $a_1 = a_2$ and $b_1 = b_2$.
- Function types: $f = g$ iff $\forall x.\, f(x) = g(x)$ (function extensionality).
- Inductive types: equality characterised by the constructors.

This is one of the most useful applications of the J-eliminator.

11.6 Higher J and Nested Induction

When reasoning about paths between paths (2-cells) or even higher paths, we often need to apply J multiple times. This section develops the machinery for *nested path induction.*

11.6.1 Double Path Induction

Suppose we have a type family depending on *two* paths. To eliminate into such a family, we apply J twice: first to one path, then to the other.

Definition 11.63 (J_2: induction on two paths)**.** For $C : \Pi(b\, c : A).\, (a = b) \to (a = c) \to \mathsf{Type}$ and $c : C(a, a, \mathsf{rfl}, \mathsf{rfl})$:

$$\begin{aligned}\mathsf{J}_2(C, c, b, d, h_1, h_2) := \mathsf{J}\Big(&\lambda b.\, \lambda h_1.\\ &\mathsf{J}\big(\lambda d.\, \lambda h_2.\, C(b, d, h_1, h_2),\ c,\ h_2\big),\ h_1\Big).\end{aligned}$$

Operationally, one eliminates first on h_1, reducing to the case $b \equiv a$ and $h_1 \equiv \mathsf{rfl}$, and then on h_2, reducing to $d \equiv a$ and $h_2 \equiv \mathsf{rfl}$, where the base term c applies.

Proposition 11.64 (J_2 computation)**.** $\mathsf{J}_2(C, c, a, a, \mathsf{rfl}, \mathsf{rfl}) = c$.

Proof. Apply the defining equation above and reduce twice using the computation rule for J. □

11.6.2 Triple Path Induction

The pattern extends naturally to three or more paths.

Definition 11.65 (J_3: induction on three paths)**.** Similarly, J_3 is obtained by iterating J three times. For a motive depending on three identity proofs, one eliminates on the first proof, then on the second, then on the third, and the computation rule reduces the triple-reflexive case to the chosen base term.

Remark 11.66 (Pattern: iterated J)**.** The pattern generalises: J_n handles n identity proofs by nesting n applications of J. Each application reduces one path to rfl, until all paths are reflexive and the base case applies. This technique is ubiquitous in homotopy type theory, where reasoning about higher coherences requires eliminating many paths simultaneously.

11.7 K, UIP, and Their Failure

Having derived J and its consequences, a natural question presents itself: does path induction force all paths between the same endpoints to be equal? In the traditional terminology, does J imply the *uniqueness of identity proofs* (UIP)? The answer turns on the distinction between semantic and syntactic levels that has run through the entire chapter, and the outcome illuminates a fundamental feature of the computational paths framework.

11.7.1 K at the Semantic Level

Streicher's axiom K strengthens path induction by allowing induction on loops—paths from a point to itself.

Definition 11.67 (Streicher's K)**.** Streicher's axiom K states: for any $C : (a = a) \to \mathsf{Type}$ and $c : C(\mathsf{rfl})$, there exists

$$\mathsf{K} : \Pi(h : a = a).\, C(h)$$

with $\mathsf{K}(\mathsf{rfl}) = c$.

Proposition 11.68 (K holds at semantic level)**.** *For propositional equality, K is derivable.*

Proof. By proof irrelevance: every $h : a = a$ equals rfl. Transport the base case c along this equality to obtain the required term at h. □

11.7.2 K Fails at the Syntactic Level

The situation is fundamentally different for the Path type, where step structure is preserved.

Theorem 11.69 (K fails for Path). *There is no function* $\mathsf{K}_{\mathsf{Path}} : \Pi(C : \mathsf{Path}(a,a) \to \mathsf{Prop}).\, C(\mathsf{refl}_a) \to \Pi(p : \mathsf{Path}(a,a)).\, C(p)$.

Proof. Suppose such a K existed. Take $C(p) := (p = \mathsf{refl}_a)$ and $c := \mathsf{rfl} : \mathsf{refl}_a = \mathsf{refl}_a$.

Then $\mathsf{K}_{\mathsf{Path}}(C, c, \mathsf{ofEq}(\mathsf{rfl}))$ would give $\mathsf{ofEq}(\mathsf{rfl}) = \mathsf{refl}_a$.

But $\mathsf{ofEq}(\mathsf{rfl})$ has a non-empty step list while refl_a has an empty step list, contradicting non-UIP for Path. □

11.7.3 Two-Level Summary

We can summarise the status of K and UIP at each level.

Proposition 11.70 (Two-level K status). *(i) K holds for* Eq *(semantic level).*

(ii) K fails for Path *(syntactic level).*

(iii) J holds at both levels.

Remark 11.71 (J without K). This is a key feature of the computational paths framework. Path induction (J) holds without loop triviality (K). In homotopy type theory terms, we have a proof-relevant identity type with J but without UIP. The semantic/syntactic split allows both perspectives to coexist peacefully:

- The semantic level provides classical equality reasoning where all proofs of the same equation are identified.
- The syntactic level preserves the computational content that distinguishes different derivations.

This two-level structure is analogous to Voevodsky's two-level type theory, but arises here from the concrete distinction between ambient equality witnesses and step lists rather than from axiomatic stratification.

The full development of UIP failure—including step-counting invariants, infinite families of distinct loops, and the proof-relevance/proof-irrelevance dialectic—is given in Chapter 12.

11.8 The Encode-Decode Method

The preceding sections have shown how to *use* paths: J eliminates them, transport moves data along them, and uniqueness tells us there is only one way to do so. But a natural question remains—what do the path spaces of specific types actually *look like*? For instance, which equalities hold between natural numbers, or between elements of a sum type? Answering such questions requires a technique that goes beyond induction on a single path: we need a way to *characterise* an entire path space at once.

The encode-decode method provides exactly this. The idea is to define a "code" type that is easier to analyse than the path space itself, and then show that the code type and the path space are equivalent. Once this equivalence is established, questions about paths reduce to questions about the code—which, by construction, are straightforward. This technique is essential in homotopy type theory for computing the homotopy groups of various spaces, and we will rely on it in Chapter 13 when we compute fundamental groups of concrete spaces.

11.8.1 The Method

Instead of working with path spaces directly, we construct an equivalent "code" type that is easier to analyse.

Definition 11.72 (Encode-decode setup)**.** To characterise $a = y$ for fixed $a : A$:

(i) Define a *code* family $\mathsf{Code} : A \to \mathsf{Type}$.

(ii) Define $\mathsf{encode} : (a = y) \to \mathsf{Code}(y)$.

(iii) Define $\mathsf{decode} : \mathsf{Code}(y) \to (a = y)$.

(iv) Prove $\mathsf{encode} \circ \mathsf{decode} = \mathsf{id}$ and $\mathsf{decode} \circ \mathsf{encode} = \mathsf{id}$.

Example 11.73 (Trivial encoding)**.** For the trivial code $\mathsf{Code}(y) :\equiv (a = y)$:

- $\mathsf{encode}(h) :\equiv h$
- $\mathsf{decode}(c) :\equiv c$

Round-trips are trivially identity.

Remark 11.74 (Non-trivial applications)**.** The encode-decode method becomes powerful when Code is computationally simpler than the path type. Classic applications include:

- $\pi_1(S^1) \cong \mathbb{Z}$: loops on the circle are encoded as integers (winding numbers).
- Paths in coproducts: $\mathsf{inl}(a) = \mathsf{inl}(b)$ iff $a = b$, and $\mathsf{inl}(a) \neq \mathsf{inr}(c)$.
- Paths in truncations: the code type captures the quotient structure explicitly.

We will see more examples of this technique in Chapter 13 when we compute fundamental groups.

Formalization Notes. The main path-induction and transport constructions from this chapter are formalised in the following files of the accompanying Lean development (`https://github.com/Arthur742Ramos/ComputationalPathsLean`):

- `Path/HigherPathInduction.lean` — the streamlined `pathJ` and `pathJ_toEq` interfaces, together with transport comparison lemmas
- `Path/Homotopy/IdentityType.lean` — transport in identity-type families and dependent application
- `Path/TransportDerived.lean` — derived transport lemmas, naturality, functoriality
- `Path/HoTT/JDerivation.lean` — an extended contractibility-based derivation of `J` that mirrors the presentation developed in this chapter

Key definitions and theorems.

The following code excerpts illustrate the main formalisations.

```
/-- The based path space at 'a'. -/
def BasedPathSpace (a : A) : Type u := Sigma (y : A),
   PLift (a = y)

/-- Center of contraction. -/
def center (a : A) : BasedPathSpace a := (|a, PLift.up
   rfl|)

/-- Fundamental contraction: every element equals the
   center. -/
theorem contraction (a : A) (bp : BasedPathSpace a) :
    bp = center a := by
  obtain (|y, (|h|)|) := bp
  cases h
  rfl
```

Listing 11.1: Based path space and contractibility

```
/-- The J-Eliminator derived from contractibility. -/
def J {a : A} (C : (y : A) -> a = y -> Sort v)
    (c : C a rfl) {y : A} (h : a = y) : C y h := by
  cases h
  exact c

/-- J computation rule. -/
theorem J_comp {a : A} (C : (y : A) -> a = y -> Sort v)
    (c : C a rfl) : J C c rfl = c := rfl
```

Listing 11.2: J-eliminator derived from contractibility

```
/-- Transport derived from J. -/
def J_transport {B : A -> Sort v} {a y : A}
    (h : a = y) (x : B a) : B y :=
  J (fun y _ => B y) x h

/-- Dependent application theorem. -/
theorem J_apd {B : A -> Sort v} (f : forall x, B x) {a y
   : A}
    (h : a = y) : J_transport h (f a) = f y := by
  cases h; rfl
```

Listing 11.3: Transport and apd from J

```
/-- Uniqueness: any two eliminators with same base case
   agree. -/
theorem J_unique {a : A} {C : (y : A) -> a = y -> Sort v}
    (c : C a rfl)
    (elim1 elim2 : forall {y : A} (h : a = y), C y h)
    (comp1 : elim1 rfl = c) (comp2 : elim2 rfl = c)
    {y : A} (h : a = y) : elim1 h = elim2 h := by
  cases h
  exact comp1.trans comp2.symm
```

Listing 11.4: Uniqueness of J

```
/-- Non-UIP witness: two distinct paths. -/
def witness_refl (a : A) : Path a a := refl a
def witness_step (a : A) : Path a a :=
  Path.mk [Step.mk a a rfl] rfl

/-- The witnesses differ. -/
theorem non_uip_witness (a : A) : witness_refl a !=
   witness_step a := by
  intro h
```

```
  have := congrArg Path.steps h
  simp [witness_refl, witness_step, refl] at this

/-- Streicher K is not derivable for Path. -/
theorem no_streicher_K (a : A) :
    Not (forall (C : Path a a -> Prop) (_ : C (refl a))
      (p : Path a a), C p) := by
  intro K
  exact absurd (K (fun p => p = refl a) rfl (witness_step
      a)).symm
        (non_uip_witness a)
```

Listing 11.5: K fails for Path

Verification Status

All theorems in this chapter are fully formalised in the Lean development:

- Contractibility of the semantic based path space: verified.
- Non-contractibility of the syntactic based path space: verified.
- J-eliminator and computation rule: verified.
- Transport and apd derived from J: verified.
- Uniqueness of J: verified.
- K failure for Path: verified.
- Two-level structure theorems: verified.

11.9 Historical and Conceptual Notes

11.9.1 Origins of Path Induction

The J-eliminator was introduced by Per Martin-Löf in his intensional type theory [109]. Martin-Löf's insight was that identity types should come with an induction principle, just as natural numbers come with mathematical induction. The name "J" follows the convention of using letters near "I" (for "identity") in the alphabet.

In Martin-Löf's original presentation, J was an axiom—a fundamental rule of the type theory. The derivation from contractibility that we have presented shows that J can instead be *proved* from more primitive geometric considerations, at least in frameworks where the based path space can be shown contractible.

11.9.2 The Contractibility Perspective

The derivation of J from contractibility of the based path space appears prominently in the HoTT Book [150]. This geometric perspective illuminates *why* J should hold: the based path space is homotopically trivial, so any dependent type over it is determined by its value at a single point. The proof is essentially the observation that we can retract any path back to the basepoint along itself.

11.9.3 Paulin-Mohring's Formulation

Christine Paulin-Mohring introduced the variant with fixed target (rather than source) in her work on the Calculus of Inductive Constructions [119]. This variant is particularly natural when reasoning about the destination of a path, and appears frequently in the Coq proof assistant.

11.9.4 J Without K

The observation that J does not imply K was made precise by Hofmann and Streicher [86], who constructed a model (the groupoid model) where J holds but K fails. Their construction showed that intensional type theory has models with non-trivial identity types, where not all loops are equal to reflexivity. This observation opened the door to homotopy type theory, where non-trivial identity types model higher homotopical structure.

11.9.5 Computational Paths and Two-Level Type Theory

The two-level structure of computational paths—semantic level with UIP, syntactic level without—connects to Voevodsky's two-level type theory [157]. The innovation in our framework is that the levels arise naturally from the distinction between proof fields (semantic) and step lists (syntactic), rather than being imposed axiomatically. This gives a concrete, computational interpretation of the two-level structure.

Summary

This chapter established the J-eliminator as a theorem of the computational-path framework. The decisive point is that path induction need not be taken as primitive once the based path space and its contractibility are in hand. The development proceeded through several stages:

1. **Based path space** (Section 11.1). The type $\Sigma(y : A).\, \mathsf{Path}(a, y)$ collects all endpoints together with paths from a fixed basepoint. This is the natural domain for path induction.

2. **Contractibility** (Section 11.2). The semantic based path space is contractible: every element equals the center (a, refl_a). The syntactic version is *not* contractible due to the preservation of step structure.

3. **J-eliminator** (Section 11.3). Derived from contractibility. To define a dependent function out of identity types, it suffices to specify the reflexive case; contractibility pushes this value to all targets.

4. **Transport and apd** (Section 11.4). Moving elements along paths in type families, and applying dependent functions to paths. Both operations are instances of J applied to specific motives.

5. **Uniqueness** (Section 11.5). J is uniquely characterised by its computation rule. Any two eliminators with the same base case agree on all inputs.

6. **K and UIP** (Section 11.7). While J holds, Streicher's K fails at the syntactic level. This preserves the proof-relevant structure that distinguishes rewrite traces.

Taken together, these results show that the computational-path setting is strong enough to recover the standard eliminator for identity while still refusing the collapse of proof history that UIP and K would impose.

The J-eliminator will be essential in subsequent chapters for deriving properties of path spaces, characterising fundamental groups (Chapter 13), and connecting computational paths to homotopy type theory (Chapter 14).

Chapter 12

Failure of UIP and Proof-Relevance

> *"Identity proofs are not mere certificates of truth; they are computational objects with structure."*
>
> —Computational-path perspective

The Uniqueness of Identity Proofs principle (UIP) asserts that any two proofs of the same equality must themselves be equal. For computational paths, this principle fails—and the failure is not a deficiency but a structural necessity. If UIP held, the entire tower of higher cells developed in Chapters 7–9 would collapse: there would be nothing for 2-cells to compare, no nontrivial loop spaces, no proof-relevant equality. The higher-dimensional richness already established in Part II *requires* that distinct paths between the same endpoints genuinely exist.

Chapter 3 introduced the first concrete witness of this phenomenon—the stepLoop counterexample (Example 3.40)—and the present chapter develops its consequences systematically. The main results are:

- **Non-UIP and non-K theorems** (Section 12.1 and 12.2): UIP and its loop-focused variant (Streicher's K) both fail for every nonempty type.
- **Step-counting invariant** (Section 12.3): a simple numerical measure that separates paths with different rewrite histories, producing infinite families of distinct loops.
- **The proof-relevance dialectic** (Section 12.4): a controlled two-layer picture in which intensional path data and extensional propositional equality coexist without contradiction.

12.1 The Non-UIP Theorem

12.1.1 UIP Revisited in the Computational-Path Setting

We introduced UIP and Streicher's K axiom, and showed their failure at the syntactic level, in Section 11.7. *Here we develop the full treatment: global non-UIP theorems, step-counting invariants, and the proof-relevance dialectic.*

The Uniqueness of Identity Proofs principle captures the idea that equality witnesses should be interchangeable—that once we know two terms are equal, the particular *reason* for their equality should not matter. In proof-irrelevant settings, this is a natural expectation: if both p and q establish that $a = b$, then p and q should themselves be equal. The computational-path framework challenges this expectation by treating equality witnesses as structured objects that carry their own distinguishing data.

Definition 12.1 (UIP for computational paths)**.** For a type A, the *Uniqueness of Identity Proofs* principle states that any two paths between the same endpoints must be equal:

$$\mathsf{UIP}(A) \;:\equiv\; \forall(a, b : A)\,\forall(p, q : \mathsf{Path}(a, b)),\; p = q.$$

Equivalently, $\mathsf{UIP}(A)$ asserts that each path space $\mathsf{Path}(a, b)$ is a proposition—a type with at most one element.

Remark 12.2 (Why UIP is a strong requirement)**.** If $\mathsf{UIP}(A)$ holds, then all computational distinctions between parallel paths vanish. In particular, any two loop witnesses at a point a,

$$p, q : \mathsf{Path}(a, a),$$

must be equal. This would collapse the path structure from a potentially rich algebraic object (the loop space at a) to a trivial one containing at most refl_a. From the viewpoint of computational paths, UIP erases exactly the information we wish to preserve: the rewrite history that explains *how* equality is established.

12.1.2 What UIP Would Force Algebraically

Before proving that UIP fails, it is instructive to understand what it would impose if it held. The following propositions show that UIP would trivialise the entire loop space structure, collapsing all loops to reflexivity and rendering loop composition meaningless.

Proposition 12.3 (UIP trivialises every loop space)**.** *Assume* $\mathsf{UIP}(A)$. *Then for every* $a : A$*, every loop at* a *equals* refl_a*:*

$$\forall(p : \mathsf{Path}(a, a)),\; p = \mathsf{refl}_a.$$

Proof. Fix $a : A$ and $p : \mathsf{Path}(a, a)$. By $\mathsf{UIP}(A)$, any two paths between the same endpoints are equal. Applying this to p and refl_a (both loops at a), we obtain $p = \mathsf{refl}_a$. □

Proposition 12.4 (UIP collapses loop composition). *Assume* $\mathsf{UIP}(A)$ *and fix* $a : A$*. Then for all loops* $p, q : \mathsf{Path}(a, a)$*:*

$$\mathsf{trans}(p, q) = p = q = \mathsf{refl}_a.$$

In particular, loop composition at a becomes the trivial operation that returns refl_a *regardless of its inputs.*

Proof. By Proposition 12.3, $p = \mathsf{refl}_a$ and $q = \mathsf{refl}_a$. Substituting into $\mathsf{trans}(p, q)$ and applying the unit law for refl yields $\mathsf{trans}(\mathsf{refl}_a, \mathsf{refl}_a) = \mathsf{refl}_a$. □

Remark 12.5 (Tension with Part II). Proposition 12.4 stands in direct tension with the nontrivial 2-cell and higher-cell structures developed in Chapters 7–9. If UIP held, those higher cells would have nothing to compare: all parallel 1-paths would already be equal. Thus the failure of UIP is not an isolated curiosity; it is a necessary structural condition for the higher-dimensional theory already established.

In set-based foundations, the collapse imposed by UIP is often harmless—one typically cares *whether* two objects are equal, not *how* the equality is established. For computational paths, however, the "how" is precisely the point. Each path records a history of rewrite steps, and this history is itself mathematical data that we wish to preserve and manipulate.

12.1.3 Canonical Non-UIP Witness

With the algebraic consequences of UIP established, we now construct an explicit counterexample. The key observation is that refl_a—the trivial loop with no steps—differs from a loop that contains an explicit reflexive step. Both loops start and end at a, but their internal structure differs.

Definition 12.6 (One-step reflexive loop). Given $a : A$, define

$$\mathsf{stepLoop}(a) : \mathsf{Path}(a, a)$$

as the loop with exactly one explicit reflexive step. Thus refl_a and $\mathsf{stepLoop}(a)$ share the same endpoints but carry different computational records: refl_a is the empty path (zero steps), while $\mathsf{stepLoop}(a)$ contains one step that explicitly witnesses $a = a$.

Remark 12.7 (Intuition for stepLoop). One may think of $\mathsf{stepLoop}(a)$ as "doing nothing, but recording that we did it." In a rewrite system, this corresponds to applying a reflexivity rule at a specific position, which leaves the term unchanged but adds an entry to the derivation trace. The distinction between refl_a and $\mathsf{stepLoop}(a)$ is analogous to the distinction between the empty string and a string containing one character that represents "no action."

Theorem 12.8 (Reflexivity differs from stepLoop). *For every type A and every $a : A$,*

$$\mathsf{refl}_a \neq \mathsf{stepLoop}(a).$$

Proof. The two paths differ in their step lists (as introduced in Chapter 3, Definition 3.1):

$$\mathsf{steps}(\mathsf{refl}_a) = [\,], \qquad \mathsf{steps}(\mathsf{stepLoop}(a)) = [(a, a, \mathsf{refl})].$$

Consequently, their step counts differ:

$$|\mathsf{refl}_a| = 0, \qquad |\mathsf{stepLoop}(a)| = 1.$$

By the separation property for step counts (Corollary 3.46 of Chapter 3), paths with distinct step counts are themselves distinct. □

Remark 12.9 (Reference to Chapter 3). This theorem is essentially Example 3.40 of Chapter 3, now stated as a formal theorem. The earlier example served a pedagogical role in introducing the stepLoop phenomenon; here it serves a structural role as the cornerstone of UIP failure.

Theorem 12.10 (Non-UIP theorem). *If A is nonempty, then* $\mathsf{UIP}(A)$ *fails:*

$$\mathsf{Nonempty}(A) \implies \neg\,\mathsf{UIP}(A).$$

Proof. Suppose A is nonempty and choose any $a : A$. If $\mathsf{UIP}(A)$ held, then in particular it would apply to the two loops refl_a and $\mathsf{stepLoop}(a)$, yielding

$$\mathsf{refl}_a = \mathsf{stepLoop}(a).$$

But this contradicts Theorem 12.8. Therefore $\neg\,\mathsf{UIP}(A)$. □

The immediate consequences spell out what UIP failure means for path spaces.

Corollary 12.11 (Path spaces are not propositions). *For every nonempty type A, there exists $a : A$ such that* $\mathsf{Path}(a, a)$ *contains at least two distinct elements.*

Proof. Take any $a : A$. By Theorem 12.8, refl_a and $\mathsf{stepLoop}(a)$ are distinct loops at a. □

Corollary 12.12 (No nonempty type is path-thin). *For a nonempty type A, at least one hom-set of the path groupoid has more than one arrow. Specifically, the endomorphism hom-set* $\mathsf{Path}(a, a)$ *at any chosen* $a : A$ *is non-singleton.*

Proof. By Corollary 12.11, $\mathsf{Path}(a, a)$ contains at least refl_a and $\mathsf{stepLoop}(a)$, which are distinct. Hence the endomorphism hom-set at a has more than one element. □

Remark 12.13 (Equivalent formulations of non-UIP). Non-UIP can be equivalently stated as: (i) there exist $a : A$ and distinct $p, q \in \mathsf{Path}(a, a)$; (ii) some path space $\mathsf{Path}(a, a)$ is not a proposition; (iii) some path space $\mathsf{Path}(a, a)$ is not a singleton. The proof of equivalence is straightforward: the witness pair $(\mathsf{refl}_a, \mathsf{stepLoop}(a))$ provides (i), and each subsequent implication follows from basic logic. These reformulations license seamless translation between communities: category theorists may phrase non-UIP as "hom-sets are nontrivial"; homotopy type theorists as "path spaces are not mere propositions."

The non-UIP phenomenon is not confined to the single stepLoop witness. It propagates through the standard path operations: symmetry and congruence both preserve the distinction between refl and stepLoop.

Proposition 12.14 (Nontriviality of symmetry). *For a nonempty type A and any* $a : A$*, symmetry does not globally collapse loops to reflexivity:*

$$\exists p : \mathsf{Path}(a, a),\ \mathsf{symm}(p) \neq \mathsf{refl}_a.$$

Proof. Take $p := \mathsf{stepLoop}(a)$. Suppose $\mathsf{symm}(p) = \mathsf{refl}_a$. Applying symm to both sides and using involutivity ($\mathsf{symm}(\mathsf{symm}(p)) = p$), we obtain $p = \mathsf{symm}(\mathsf{refl}_a) = \mathsf{refl}_a$. This contradicts Theorem 12.8. □

Proposition 12.15 (Nontriviality of congruence). *Let* $f : A \to B$ *and* $a : A$*. Then congruence preserves the distinction between* refl *and* stepLoop*:*

$$\mathsf{congrArg}(f, \mathsf{refl}_a) \neq \mathsf{congrArg}(f, \mathsf{stepLoop}(a)).$$

Proof. By step-count preservation under congruence (Theorem 12.27),

$$\mathsf{sc}(\mathsf{congrArg}(f, \mathsf{refl}_a)) = 0, \qquad \mathsf{sc}(\mathsf{congrArg}(f, \mathsf{stepLoop}(a))) = 1.$$

The result follows from Corollary 12.28. □

Remark 12.16 (Pervasiveness and scope). UIP failure appears immediately in everyday types: in $\mathbb{N}$, the loops refl_0 and $\mathsf{stepLoop}(0)$ are distinct; in Bool, likewise for $\mathsf{refl}_{\mathsf{true}}$ and $\mathsf{stepLoop}(\mathsf{true})$. No elaborate type construction is needed. The argument does require an element $a : A$ to build the witness pair, so the theorem is stated for nonempty types; the empty type has no basepoints and its path spaces are vacuously propositional.

12.1.4 Relation to Chapter 3

The stepLoop witness appeared first in Chapter 3 (Example 3.40), where it served a pedagogical role. Here it plays a structural one: it is the basic generator for infinite families of distinct loops, a separator for K, and the entry point for invariants such as step count. To refute a universal principle like UIP, a single explicit counterexample is logically sufficient, and the pair $(\mathsf{refl}_a, \mathsf{stepLoop}(a))$ is precisely that minimal local defect from which global proof-relevance follows.

12.2 Failure of Streicher's K

12.2.1 K as Loop Collapse

UIP asserts that any two paths between the same endpoints are equal; Streicher's K principle [86] makes the same demand for loops alone. Every loop must equal reflexivity—the only way to go from a point back to itself is to not move at all. Under standard type-theoretic assumptions, UIP and K are interderivable, but we can refute each directly using the stepLoop witness.

Definition 12.17 (Streicher's K principle for paths)**.** For a type A, Streicher's K principle states that every loop collapses to reflexivity:

$$\mathsf{PathK}(A) \;:\equiv\; \forall (a : A)\, \forall (p : \mathsf{Path}(a, a)),\ p = \mathsf{refl}_a.$$

Theorem 12.18 (K fails for nonempty types)**.** *If A is nonempty, then* $\mathsf{PathK}(A)$ *fails:*

$$\neg\, \mathsf{PathK}(A).$$

Proof. Choose any $a : A$. If $\mathsf{PathK}(A)$ held, applying it to $p := \mathsf{stepLoop}(a)$ would yield $\mathsf{stepLoop}(a) = \mathsf{refl}_a$, contradicting Theorem 12.8. □

The following propositions analyse K failure through the lens of step counts.

Proposition 12.19 (K forces zero-length loops)**.** *If* $\mathsf{PathK}(A)$ *holds, then every loop has step count zero:*

$$\forall (a : A)\, \forall (p : \mathsf{Path}(a, a)),\ |p| = 0.$$

In other words, K implies that all loops are step-free.

Proof. By $\mathsf{PathK}(A)$, every loop p equals refl_a. Applying the step count function sc to both sides: $\mathsf{sc}(p) = \mathsf{sc}(\mathsf{refl}_a) = 0$. □

Corollary 12.20 (K contradicts one-step loops)**.** *If A is nonempty, then* $\mathsf{PathK}(A)$ *is impossible because the loop* $\mathsf{stepLoop}(a)$ *has step count* 1*, not* 0*.*

Proof. Proposition 12.19 requires all loops to have step count 0, but $|\mathsf{stepLoop}(a)| = 1$. □

Corollary 12.21 (No global loop contraction). *There is no operation that uniformly contracts every loop to reflexivity while preserving path identity.*

Proof. Such an operation would witness $\mathsf{PathK}(A)$, which Theorem 12.18 rules out for nonempty types. □

12.2.2 Appending a Step Always Changes the Path

A natural question arises: if adding a single reflexive step creates a distinct path, what about adding steps to an existing path? The following theorem shows that extending any path by stepLoop always produces a strictly different path. This demonstrates that K failure is not confined to the pair $(\mathsf{refl}, \mathsf{stepLoop})$—it permeates the entire loop space.

Theorem 12.22 (Right multiplication by stepLoop is nontrivial). *For every loop* $p : \mathsf{Path}(a, a)$,

$$p \neq \mathsf{trans}\big(p, \mathsf{stepLoop}(a)\big).$$

Proof. By step-count additivity (Lemma 3.45),

$$\big|\mathsf{trans}(p, \mathsf{stepLoop}(a))\big| = |p| + |\mathsf{stepLoop}(a)| = |p| + 1.$$

Since $|p| \neq |p| + 1$, Corollary 3.46 gives

$$p \neq \mathsf{trans}\big(p, \mathsf{stepLoop}(a)\big).$$

□

Remark 12.23 (Computational interpretation). Even a "do nothing" step is still a step when explicitly recorded. Adding a reflexive step is not semantically empty at the path level—it increases the derivation length by one. This is precisely why K fails: the computational trace is part of the identity data, not mere annotation.

12.2.3 K Failure as a Structural Phenomenon

One might suspect that K fails only due to a peculiar choice of representative paths. Theorem 12.22 dispels this suspicion: K failure is stable under path extension. Given *any* loop p, adjoining one explicit reflexive step yields a different loop. Nontriviality is not concentrated in one exceptional witness; it permeates the entire loop space.

Remark 12.24 (Impact on higher cells)**.** If K were valid, most of the higher-dimensional structure developed in Chapters 7–9 would collapse. The reason is straightforward: 2-cells witness equivalences between distinct 1-paths. Once all 1-paths at the same endpoints become equal (by K), there is nothing for 2-cells to compare—the higher tower degenerates.

12.2.4 K Failure as Non-Contractibility

In topological language, K amounts to requiring that every loop space be contractible to its basepoint. The failure of K says precisely that loop spaces resist this contraction: the step history recorded in $\mathsf{stepLoop}(a)$ is genuine data that cannot be "shrunk away."

Proposition 12.25 (K as global reflexive contractibility)**.** $\mathsf{PathK}(A)$ *holds if and only if for every* $a : A$*, every loop* $p : \mathsf{Path}(a, a)$ *satisfies* $p = \mathsf{refl}_a$*. For nonempty* A*, this fails by Theorem 12.8.*

Any elimination principle that tries to reduce every loop to refl cannot be globally sound for computational paths. More precisely, there is no family of maps $E_a : \mathsf{Path}(a, a) \to \mathsf{Path}(a, a)$ that simultaneously sends every loop to refl_a and reflects path equality ($E_a(p) = E_a(q) \Rightarrow p = q$): the first condition makes E_a constant, forcing $\mathsf{refl}_a = \mathsf{stepLoop}(a)$. Once loop distinctions are erased, they cannot be recovered functorially.

12.3 Step-Counting as a Path Invariant

12.3.1 From Witness to Invariant

Chapter 3 introduced step counting as a basic diagnostic tool; here we promote it to a central invariant for the non-UIP analysis. The guiding principle is elementary yet powerful:

$$\text{different step counts} \implies \text{different paths.}$$

This observation drives most separation arguments in this chapter.

Definition 12.26 (Step count functional)**.** For a path $p : \mathsf{Path}(a, b)$, the *step count* is defined as

$$\mathsf{sc}(p) := |p| = \mathsf{length}\big(\mathsf{steps}(p)\big) \in \mathbb{N}.$$

We refer to sc as the *step-count invariant*, since it assigns a numerical label to each path that is preserved under path equality.

The step-count invariant satisfies a collection of laws that make it easy to compute and reason about. These laws mirror the algebraic structure of paths themselves.

Theorem 12.27 (Step-count calculus). *For composable paths p, q and any function f:*

1. $\mathsf{sc}(\mathsf{refl}_a) = 0$ *(reflexivity has no steps)*
2. $\mathsf{sc}(\mathsf{stepLoop}(a)) = 1$ *(stepLoop has one step)*
3. $\mathsf{sc}(\mathsf{trans}(p, q)) = \mathsf{sc}(p) + \mathsf{sc}(q)$ *(additivity)*
4. $\mathsf{sc}(\mathsf{symm}(p)) = \mathsf{sc}(p)$ *(symmetry preserves count)*
5. $\mathsf{sc}(\mathsf{congrArg}(f, p)) = \mathsf{sc}(p)$ *(congruence preserves count)*

Proof. Items (1), (3), (4), and (5) are established in Lemma 3.45 of Chapter 3. Item (2) follows directly from the construction of $\mathsf{stepLoop}(a)$ as a single-step path. □

Corollary 12.28 (Length separates paths). *If $\mathsf{sc}(p) \neq \mathsf{sc}(q)$, then $p \neq q$.*

Proof. Restates Corollary 3.46 in invariant language. □

Proposition 12.29 (Step count as monoid morphism). *Fix $a : A$. On the loop space $\mathsf{Path}(a, a)$, step count defines a monoid homomorphism*

$$\mathsf{sc}_a : (\mathsf{Path}(a, a), \mathsf{trans}, \mathsf{refl}_a) \longrightarrow (\mathbb{N}, +, 0).$$

That is, sc_a preserves the identity and composition:

$$\mathsf{sc}_a(\mathsf{refl}_a) = 0, \qquad \mathsf{sc}_a(\mathsf{trans}(p, q)) = \mathsf{sc}_a(p) + \mathsf{sc}_a(q).$$

Proof. Direct from Theorem 12.27(1) and (3). □

Remark 12.30 (Soundness without completeness). Step count is a *sound* separator: different counts imply different paths. However, it is not *complete*: equal counts do not force path equality (two paths could have the same length but different step sequences). For UIP/K failure, soundness alone suffices, since the witness pair $(\mathsf{refl}_a, \mathsf{stepLoop}(a))$ already exhibits distinct counts.

12.3.2 Length Filtration of Loop Spaces

The step count induces a natural stratification of path spaces by complexity. This filtration organizes loops into "levels" according to how many explicit steps they contain.

Definition 12.31 (Length-bounded loop layers)**.** For fixed $a : A$ and $n \in \mathbb{N}$, define the n-th *filtration layer*:

$$F_n(a) := \{p : \mathsf{Path}(a, a) \mid \mathsf{sc}(p) \leq n\}.$$

These layers form an increasing chain:

$$F_0(a) \subseteq F_1(a) \subseteq F_2(a) \subseteq \cdots$$

stratifying the loop space by computational complexity.

Proposition 12.32 (Strictness of the filtration)**.** *For every n, the inclusion $F_n(a) \subsetneq F_{n+1}(a)$ is strict.*

Proof. By Definition 12.40 and Lemma 12.41, the loop p_{n+1} satisfies $\mathsf{sc}(p_{n+1}) = n + 1$. Thus $p_{n+1} \in F_{n+1}(a)$ but $p_{n+1} \notin F_n(a)$. □

Remark 12.33 (Complexity levels)**.** The filtration organizes loop space into "cost levels": level n contains exactly those equality witnesses obtainable with at most n explicit steps. This viewpoint proves useful when comparing normalisation strategies, since it separates semantic equality (reaching the same endpoint) from proof complexity (how many steps are recorded).

12.3.3 Worked Calculations

The step-count rules yield quick calculations that are repeatedly useful. We record a compact "calculus sheet".

Proposition 12.34 (Calculation templates)**.** *For loops $p, q, r : \mathsf{Path}(a, a)$ and function $f : A \to B$:*

1. $\mathsf{sc}(\mathsf{trans}(p, \mathsf{trans}(q, r))) = \mathsf{sc}(p) + \mathsf{sc}(q) + \mathsf{sc}(r)$;
2. $\mathsf{sc}(\mathsf{trans}(\mathsf{symm}(p), p)) = 2\,\mathsf{sc}(p)$;
3. $\mathsf{sc}(\mathsf{congrArg}(f, \mathsf{trans}(p, q))) = \mathsf{sc}(p) + \mathsf{sc}(q)$;
4. $\mathsf{sc}(\mathsf{symm}(\mathsf{trans}(p, q))) = \mathsf{sc}(p) + \mathsf{sc}(q)$.

Proof. (1) follows by two applications of additivity.

(2) uses symmetry invariance and additivity:

$$\mathsf{sc}(\mathsf{trans}(\mathsf{symm}(p), p)) = \mathsf{sc}(\mathsf{symm}(p)) + \mathsf{sc}(p) = \mathsf{sc}(p) + \mathsf{sc}(p).$$

(3) combines congruence preservation with additivity.

(4) combines symmetry invariance with additivity. □

Example 12.35 (Quick separation test)**.** Let

$$u := \mathsf{trans}(\mathsf{symm}(p), p), \qquad v := \mathsf{trans}(p, \mathsf{stepLoop}(a)).$$

Then

$$\mathsf{sc}(u) = 2\,\mathsf{sc}(p), \qquad \mathsf{sc}(v) = \mathsf{sc}(p) + 1.$$

If $\mathsf{sc}(p) \neq 1$, then $u \neq v$ by count separation. So many equality comparisons are resolved immediately, before any detailed normal-form analysis.

Remark 12.36 (Algorithmic usage)**.** In automated reasoning pipelines, step count can be used as a first filter: different counts imply immediate non-equality, equal counts trigger deeper inspection. This separation of cheap and expensive checks is standard in rewriting practice and fits the computational-path viewpoint naturally.

12.3.4 Step Count and the Normalization Perspective

The chapter does not require a full normalisation argument, but the invariant complements one naturally. When rewrite systems are terminating, normal forms provide canonical representatives of extensional equality classes. Step count then measures a different quantity: not destination but derivation cost and history. Two paths can normalise to the same endpoint equality while carrying different cost profiles.

Proposition 12.37 (Endpoint agreement does not determine count)**.** *There exist parallel paths* $p, q : \mathsf{Path}(a, b)$ *with*

$$\pi(p) = \pi(q) \quad \textit{but} \quad \mathsf{sc}(p) \neq \mathsf{sc}(q).$$

Proof. Take $a = b$ and $p := \mathsf{refl}_a$, $q := \mathsf{stepLoop}(a)$. By Example 12.62, $\pi(p) = \pi(q)$, while counts are 0 and 1. □

Remark 12.38 (Normalization-compatible separation)**.** Proposition 12.37 explains why path invariants and normalisation are complementary. Normalization classifies *what* equality is obtained; step count records *how* it was obtained. Both are needed in proof-relevant equality.

Example 12.39 (Cost-sensitive equality reasoning)**.** Suppose a proof assistant offers two rewrite tactics that both prove $t = u$, but one emits a path of length 5 and the other of length 17. Projection-level equality treats them as interchangeable. Path-level analysis does not: the shorter derivation may be preferred for certification size, while the longer one may expose intermediate structure. The invariant sc makes this comparison explicit and machine-checkable.

12.3.5 Infinite Families of Loops

The stepLoop construction generalises. By repeatedly appending $\mathsf{stepLoop}(a)$, we obtain infinitely many distinct loops. Thus loop spaces are not merely non-singleton; they are genuinely infinite.

Definition 12.40 (n-fold reflexive loop)**.** Fix $a : A$. Define the family $(p_n)_{n\in\mathbb{N}}$ of loops at a recursively:

$$p_0 := \mathsf{refl}_a, \qquad p_{n+1} := \mathsf{trans}\big(p_n, \mathsf{stepLoop}(a)\big).$$

Intuitively, p_n is the loop that "does nothing n times."

Lemma 12.41 (Exact count of n-fold loops)**.** *For all $n \in \mathbb{N}$,*

$$\mathsf{sc}(p_n) = n.$$

Proof. By induction on n. Base case:

$$\mathsf{sc}(p_0) = \mathsf{sc}(\mathsf{refl}_a) = 0.$$

Inductive step:

$$\mathsf{sc}(p_{n+1}) = \mathsf{sc}\big(\mathsf{trans}(p_n, \mathsf{stepLoop}(a))\big) = \mathsf{sc}(p_n) + 1 = n + 1.$$

□

Theorem 12.42 (Injective loop family)**.** *If $m \neq n$, then $p_m \neq p_n$. Hence the map $n \mapsto p_n$ from $\mathbb{N}$ to $\mathsf{Path}(a, a)$ is injective.*

Proof. By Lemma 12.41,

$$\mathsf{sc}(p_m) = m, \qquad \mathsf{sc}(p_n) = n.$$

So $m \neq n$ implies $\mathsf{sc}(p_m) \neq \mathsf{sc}(p_n)$. Apply Corollary 12.28. □

Corollary 12.43 (Infinite loop spaces)**.** *For every $a : A$, the loop space $\mathsf{Path}(a, a)$ is infinite.*

Proof. The injective map $\mathbb{N} \hookrightarrow \mathsf{Path}(a,a)$ of Theorem 12.42 witnesses infinitude. □

Proposition 12.44 (Unbounded complexity of loop witnesses). *For each bound $N \in \mathbb{N}$, there exists a loop $p : \mathsf{Path}(a,a)$ with $\mathsf{sc}(p) > N$.*

Proof. Take $p := p_{N+1}$ from Definition 12.40. By Lemma 12.41,

$$\mathsf{sc}(p) = N + 1 > N.$$

□

Remark 12.45 (Asymptotic point). The proposition says there is no global finite ceiling on computational equality complexity at a fixed basepoint. Thus proof-relevance is not just "more than one witness"; it supports arbitrarily large families of distinct witnesses. The Eckmann–Hilton argument (Chapter 10) shows that loop composition at the 2-cell level is commutative, and the resulting abelian structure of π_2 contrasts with the potentially non-abelian fundamental group π_1.

Remark 12.46 (Back to the Chapter 3 counterexample). The pair

$$p_0 = \mathsf{refl}_a, \qquad p_1 = \mathsf{stepLoop}(a)$$

is exactly the Chapter 3 witness (Example 3.40). The present section shows that this is only the first point in a whole arithmetic progression of distinct loops.

12.3.6 Further Stability Properties

Proposition 12.47 (Congruence preserves count gaps). *Let $f : A \to B$ and $p, q : \mathsf{Path}(a,b)$. If $\mathsf{sc}(p) \neq \mathsf{sc}(q)$, then*

$$\mathsf{congrArg}(f,p) \neq \mathsf{congrArg}(f,q).$$

Proof. By Theorem 12.27(5),

$$\mathsf{sc}(\mathsf{congrArg}(f,p)) = \mathsf{sc}(p), \qquad \mathsf{sc}(\mathsf{congrArg}(f,q)) = \mathsf{sc}(q).$$

So count inequality is preserved; conclude with Corollary 12.28. □

Proposition 12.48 (No idempotence by reflexive extension). *For every loop $p : \mathsf{Path}(a,a)$ and every $n \geq 1$,*

$$p \neq \mathsf{trans}(p, p_n).$$

Proof. By Theorem 12.27(3) and Lemma 12.41,

$$\mathsf{sc}\big(\mathsf{trans}(p, p_n)\big) = \mathsf{sc}(p) + n.$$

Since $n \geq 1$, this differs from $\mathsf{sc}(p)$. Apply Corollary 12.28. □

Example 12.49 (Three canonical distinct loops). At each $a : A$, the loops

$$\mathsf{refl}_a, \qquad \mathsf{stepLoop}(a), \qquad \mathsf{trans}(\mathsf{stepLoop}(a), \mathsf{stepLoop}(a))$$

have counts $0, 1, 2$ respectively, hence are pairwise distinct. This is the smallest explicit "ladder" showing that loop spaces already contain nontrivial arithmetic structure.

12.3.7 How Far Does Step Count Go?

Step count solves exactly the problem needed for this chapter: it gives immediate, decidable separation witnesses for non-UIP and non-K. But it is intentionally coarse. Different rewrite histories can have the same length. So a complete classification of paths must refine sc with additional data (for instance, the ordered list of rule labels). This refinement problem is conceptually parallel to the move from path length to full normal-form analysis in rewriting theory.

Remark 12.50 (Minimality of the invariant). The attractiveness of step count is its minimality: it requires no case analysis on rewrite rules and no normalisation proof. Yet it already detects the essential obstruction to proof-irrelevance. For this chapter, that balance of simplicity and strength is exactly what is needed.

Remark 12.51 (Invariant as complexity measure). Step count can be read as a first-order complexity measure on equality proofs: how many primitive rewrites are explicitly remembered. More refined invariants (multisets of rule labels, weighted costs, critical pair profiles) may distinguish paths with equal length, but length alone already proves the core non-UIP theorems.

12.3.8 Stratified Reasoning with Step Count

The invariant is especially useful when proofs proceed by *strata*: first separate impossible equalities by length, then perform finer analysis inside each fixed-length layer. This two-stage method mirrors common tactics in rewriting and homotopy calculation.

Definition 12.52 (Length strata). For fixed endpoints $a, b : A$ and $n \in \mathbb{N}$, define

$$S_n(a, b) := \{p : \mathsf{Path}(a, b) \mid \mathsf{sc}(p) = n\}.$$

So $\mathsf{Path}(a, b)$ decomposes as a disjoint union of strata:

$$\mathsf{Path}(a, b) = \bigsqcup_{n \in \mathbb{N}} S_n(a, b).$$

Proposition 12.53 (Cross-stratum inequality). *If $p \in S_m(a, b)$ and $q \in S_n(a, b)$ with $m \neq n$, then $p \neq q$.*

Proof. Immediate from Corollary 12.28. □

Remark 12.54 (Practical payoff). Proposition 12.53 reduces many non-identity proofs to arithmetic in $\mathbb{N}$. Instead of manipulating full path terms, one compares integers. This explains why step count appears repeatedly in formal proofs of non-UIP and non-K.

Proposition 12.55 (Stratum shift under reflexive extension). *For each p : $\mathsf{Path}(a, a)$ in $S_n(a, a)$,*

$$\mathsf{trans}\big(p, \mathsf{stepLoop}(a)\big) \in S_{n+1}(a, a).$$

Proof. By additivity,

$$\mathsf{sc}\big(\mathsf{trans}(p, \mathsf{stepLoop}(a))\big) = \mathsf{sc}(p) + 1 = n + 1.$$

□

Corollary 12.56 (No fixed-point stratum for extension). *There is no n such that*

$$\mathsf{trans}(-, \mathsf{stepLoop}(a)) : S_n(a, a) \to S_n(a, a).$$

Proof. By Proposition 12.55, the codomain is always S_{n+1}. □

Remark 12.57 (Interpretation as graded action). Loop extension by $\mathsf{stepLoop}(a)$ acts like a degree-$+1$ operator on the graded family $(S_n(a, a))_{n \in \mathbb{N}}$. This "grading" viewpoint is often more informative than viewing loops as an undifferentiated set.

Example 12.58 (A graded non-cancellation pattern). Let p, q : $\mathsf{Path}(a, a)$ with $\mathsf{sc}(p) = \mathsf{sc}(q)$. Even if $p \neq q$, extension by one step sends both to the same higher stratum:

$$\mathsf{sc}\big(\mathsf{trans}(p, \mathsf{stepLoop}(a))\big) = \mathsf{sc}\big(\mathsf{trans}(q, \mathsf{stepLoop}(a))\big) = \mathsf{sc}(p) + 1.$$

Thus the grading alone does not detect all distinctions, but it still tracks coarse movement through complexity levels.

Remark 12.59 (Why keep coarse invariants?). Coarse invariants are valuable because they are robust and cheap. Fine invariants may distinguish more cases but usually require significantly more proof effort. This chapter deliberately uses the minimal invariant that already forces the key logical conclusions.

12.4 The Proof-Relevance / Proof-Irrelevance Dialectic

The preceding sections have established that computational paths are emphatically proof-relevant: UIP fails, K fails, loop spaces are infinite, and the step-count invariant makes these distinctions measurable. Yet the underlying propositional equality in the ambient type theory is proof-irrelevant. How can both be true?

The resolution is that these claims concern *different layers* of the framework. Computational paths form an intensional layer that tracks rewrite history; propositional equality forms an extensional layer that records only truth. The forgetful projection $\pi : \mathsf{Path}(a, b) \to \mathsf{Id}_A(a, b)$ connects the two, discarding step data and retaining only the equality witness. This section makes the interplay precise.

12.4.1 Projection-Level UIP

Definition 12.60 (Forgetful projection to propositional equality). For each $a, b : A$, define the *projection map*

$$\pi_{a,b} : \mathsf{Path}(a, b) \to \mathsf{Id}_A(a, b)$$

as the function that forgets step data and retains only the underlying equality witness. When endpoints are clear from context, we write simply $\pi(p)$.

Although paths themselves are proof-relevant, their projections to propositional equality satisfy UIP. This is the precise sense in which extensional reasoning remains available.

Theorem 12.61 (Projection-level UIP). *For any parallel paths $p, q : \mathsf{Path}(a, b)$,*

$$\pi(p) = \pi(q).$$

That is, all paths between the same endpoints project to the same propositional equality witness.

Proof. In the ambient equality layer, identity proofs between fixed endpoints are proof-irrelevant by assumption. Since $\pi(p)$ and $\pi(q)$ both inhabit $\mathsf{Id}_A(a, b)$, they must be equal. ☐

Projection-level UIP is strictly weaker than full $\mathsf{UIP}(A)$: the pair $(\mathsf{refl}_a, \mathsf{stepLoop}(a))$ demonstrates the gap, since their projections agree but the paths are distinct. The difference between p and q lies entirely in step structure, not in endpoint behaviour.

Example 12.62 (The fundamental pair). For any $a : A$:

$$\mathsf{refl}_a \neq \mathsf{stepLoop}(a), \qquad \text{yet} \qquad \pi(\mathsf{refl}_a) = \pi(\mathsf{stepLoop}(a)).$$

The two loops agree *extensionally* (as equality proofs) but differ *intensionally* (as computational paths with different step content).

12.4.2 Transport and Substitution

Theorem 12.63 (Transport is projection-dependent). *Let* $D : A \to \mathsf{Type}$*, let* $x : D(a)$*, and let* $p, q : \mathsf{Path}(a, b)$*. Then*

$$\mathsf{transport}^{D}_{\pi(p)}(x) = \mathsf{transport}^{D}_{\pi(q)}(x).$$

Proof. By Theorem 12.61, $\pi(p) = \pi(q)$ in $\mathsf{Id}_A(a, b)$. Transport along equal equality proofs yields equal transported terms. □

This does not imply $p = q$; it says only that dependent transport uses the propositional equality content, not the full path history. More generally, any operation defined solely through the projected equality $\pi(p)$ is insensitive to replacing p by a parallel path q.

12.4.3 The Two-Level Architecture

The dialectic can be summarised by separating two questions: the *extensional* question (are the endpoints equal?) and the *intensional* question (by which computational history are they equal?). The projection π answers the first and forgets the second. Computational paths are designed to retain both.

For quick reference, Table 12.1 contrasts the two levels.

Aspect	Path level (Path)	Projection level (Id)
Question answered	how equality is realised	whether endpoints are equal
Primary data	rewrite history	propositional equality witness
UIP status	fails	holds at fixed endpoints
K status	fails	reflexive collapse valid
Typical invariant	step count / step list	trivialised by irrelevance
Role in this book	coherence and higher cells	extensional transport

Table 12.1: Two-level equality structure in the computational-path framework.

The right column is not a rival theory but a quotient-like shadow of the left: it retains the truth of equality and forgets the history.

From a homotopical viewpoint, UIP amounts to requiring that every identity type be set-truncated. Its failure indicates the presence of genuinely higher-dimensional information—information that in our setting is concretely represented by step sequences and their compositions. This chapter supplies the negative half of a positive programme: negatively, UIP and K fail; positively, the failure is controlled by invariants and compatibility laws, in the form needed for the comparison chapters that follow.

12.4.4 Design Principles from the Dialectic

The chapter suggests a useful design doctrine for equality-based formalisms: distinguish a *validation layer* (extensional correctness) from a *trace layer* (intensional provenance). In computational paths, π extracts the first while retaining access to the second through the original path object.

Proposition 12.64 (Layered adequacy). *The computational-path framework is adequate in the following sense:*

1. *every path validates an extensional equality via* π*;*
2. *extensional constructions (transport/substitution) depend only on* π*;*
3. *intensional distinctions survive whenever path invariants differ.*

Proof. (1) is by definition of the projection map. (2) follows from Theorems 12.61 and 12.63. (3) follows from count-based separation (Corollary 12.28). □

For implementations, layered adequacy means one may optimise or reason extensionally without discarding proof traces. For mathematics, it explains why higher structure can coexist with classical extensional arguments: transport lemmas proceed proof-irrelevantly while nontrivial loop composition, 2-cells (Chapter 7), and coherence (Chapter 8) use the richer intensional layer.

12.4.5 A Final Structural Statement

We close with a theorem that consolidates the chapter's main results into a single structural characterisation of computational paths.

Theorem 12.65 (Proof-relevance principle for computational paths). *For every nonempty type* A*:*

1. $\neg\mathsf{UIP}(A)$ *and* $\neg\mathsf{PathK}(A)$ *(proof-relevance);*
2. *each loop space* $\mathsf{Path}(a, a)$ *is infinite (rich structure);*
3. *projection to* $\mathsf{Id}_A(a, a)$ *is constant on parallel paths (extensional compatibility).*

Proof. Item (1) follows from Theorems 12.10 and 12.18. Item (2) is Corollary 12.43. Item (3) is Theorem 12.61. □

The significance of this theorem is not merely that UIP fails—it is that the failure is *organised.* It comes equipped with invariants (step count), stability under operations (composition, symmetry, congruence), and a controlled interface to proof-irrelevant equality (the projection π). This organization makes proof-relevance mathematically useful rather than merely exotic.

Summary

Chapter 12 can be summarised in one sentence: *computational paths separate equality into truth and history, and only the first is proof-irrelevant.*

The non-UIP theorem and failure of K establish that rewrite histories cannot be collapsed without losing genuine mathematical structure. The step-count invariant shows that this non-collapse is measurable, pervasive, and computationally tractable. The projection results demonstrate that extensional reasoning remains fully available when history does not matter.

This two-level architecture prepares the ground for Part III's remaining chapters, which connect computational paths with external mathematical domains. When we compare with homotopical and categorical semantics (Chapters 14 and 15), we do so from a clear structural position: path spaces are intrinsically proof-relevant, yet they project naturally to the familiar proof-irrelevant world of propositional equality.

Historical Notes.

The question treated here belongs to the central modern debate over the meaning of identity in type theory. The intensional identity type originates in Martin-Löf's type theory [109, 111], where identity proofs are first-class objects and elimination is governed by path-induction principles. The groupoid interpretation of Hofmann and Streicher [86] established that identity witnesses naturally support higher-dimensional structure. The computational-path approach of de Queiroz and collaborators [53, 45, 30] gives a direct syntactic account of this structure via rewrite histories. From the perspective of modern HoTT [150], the failure of UIP is the expected behaviour of non-set-level types; here it is exhibited concretely by step-level invariants.

Formalization Notes. The results of this chapter are formalised in the following files of the accompanying Lean development (`https://github.com/Arthur742Ramos/ComputationalPathsLean`):

- `Path/Basic/UIP.lean` — UIP and K definitions, non-UIP/non-K theorems, step counting, and projection-level results
- `Comparison/UnivalenceAnalog.lean` — comparison material explaining why full univalence is incompatible with the computational-path framework
- `Comparison/HoTTComparison.lean` — the bridge to HoTT-style identity types used for the surrounding discussion

Key definitions. The relevant definitions and theorems appear in the following groups:

UIP and K failure (`Path/Basic/UIP.lean`):

- `UIP`, `PathK` — the principal definitions
- `not_uip_of_nonempty` — Theorem 12.10
- `not_pathK_of_nonempty` — Theorem 12.18

```
/-- UIP: any two paths with same endpoints are equal -/
noncomputable def UIP (A : Type u) : Prop :=
  forall {a b : A}, forall (p q : Path a b), p = q

/-- UIP fails for all nonempty types -/
theorem not_uip_of_nonempty {A : Type u} (hA : Nonempty A
   ) :
    not (UIP A) := by
  obtain (|a|) := hA
  intro h
  have := h (p := Path.refl a) (q := stepLoop a)
  exact refl_ne_stepLoop a this
```

Step counting (`Path/Basic/UIP.lean`):

- `stepCount`, `stepCount_trans` — the invariant and additivity
- `ne_of_stepCount_ne` — Corollary 12.28
- `stepCount_nfoldLoop` — Lemma 12.41
- `loop_space_infinite` — Corollary 12.43

```
/-- Step count is additive under composition -/
theorem stepCount_trans (p : Path a b) (q : Path b c) :
    stepCount (Path.trans p q) =
      stepCount p + stepCount q := by
  simp [stepCount, Path.trans]
```

Projection-level results (`Path/Basic/UIP.lean`):

- `toEq_unique`, `weak_uip` — Theorem 12.61
- `transport_path_irrelevant` — Theorem 12.63

Chapter 13

Fundamental Groupoids of Combinatorial Spaces

"The fundamental group is the first glimpse of the algebraic soul of a space. Compute it, and you begin to understand the space."

—An algebraic topologist's credo

This chapter brings the theory back to a familiar mathematical task: computing fundamental groups. After twelve chapters developing the theory of computational paths—from the basic definitions through rewriting systems, groupoid structure (Chapter 6), higher cells, coherence (Chapter 8), ω-groupoids (Chapter 9), and the Eckmann–Hilton argument (Chapter 10)—we now put that machinery to work on concrete spaces. The circle, torus, Klein bottle, projective plane, bouquets, and lens spaces all yield their classical fundamental groups when analysed through the computational-path perspective.

The approach is *combinatorial.* Rather than starting with point-set topology and continuous maps, we define spaces via generators and relations: a space is specified by a set of points, a set of path generators ("edges"), and a set of path relations ("faces"). This is the perspective of *CW-complexes* [79] or, in type-theoretic language, *higher inductive types* [150]. The fundamental groupoid emerges directly from the presentation, and the fundamental group at a basepoint is computed by reduction to normal forms. The use of groupoid methods for computing fundamental groups follows Brown [15, 18].

The chapter is organised as follows. Section 13.1 introduces the framework of combinatorial spaces via generators and relations. Section 13.2 computes $\pi_1(S^1) \cong \mathbb{Z}$, the paradigmatic example. Section 13.3 handles the torus T^2 and obtains $\pi_1(T^2) \cong \mathbb{Z}\times\mathbb{Z}$. Section 13.4 treats the Klein bottle with its non-abelian semidirect product $\pi_1(K) \cong \mathbb{Z}\rtimes\mathbb{Z}$. Section 13.5 computes the projective plane's fundamental group $\pi_1(\mathbb{R}P^2) \cong \mathbb{Z}/2\mathbb{Z}$. Section 13.6 introduces bouquets of circles and free groups. Section 13.7 applies the Seifert–van Kampen theorem to the figure-eight. Section 13.8 extends to lens spaces $L(p, q)$.

Throughout, we emphasize how the computational-path framework makes these classical computations explicit and mechanizable [55, 131, 58]. Each fundamental group emerges from a systematic analysis of how path expressions reduce under the rewrite rules of Chapter 4. The Seifert–van Kampen theorem is treated via computational paths in [135].

13.1 Combinatorial Spaces via Generators and Relations

Before computing fundamental groups, we need a precise language for defining spaces combinatorially. The key idea is to present a space by listing its cells: points (0-cells), paths between points (1-cells), and higher relations (2-cells and beyond). This is the combinatorial analogue of building a CW-complex [79] by attaching disks along their boundary.

13.1.1 The Idea of Cell Attachments

Consider how one might build the circle S^1 from scratch. The minimal data consists of:

- One point, which we call $\star$ (the basepoint).
- One path from $\star$ to itself, which we call ℓ (the loop).

There are no relations imposed on ℓ: any sequence of concatenations and inversions of ℓ is a valid loop, and two such sequences are equal only if they are algebraically equivalent (e.g., $\ell \cdot \ell^{-1} = \mathsf{refl}_\star$). The absence of additional relations makes ℓ freely generate the loop space.

By contrast, the projective plane $\mathbb{R}P^2$ has:

- One point $\star$.
- One loop ℓ at $\star$.
- One 2-cell witnessing that $\ell \cdot \ell = \mathsf{refl}_\star$.

The 2-cell "fills in" the doubled loop, forcing ℓ to have order 2.

This pattern generalises: a combinatorial space is specified by cell attachments at each dimension, with higher cells imposing relations on lower-dimensional structures.

13.1.2 Formal Definition

Definition 13.1 (Combinatorial space presentation). A *combinatorial space presentation* $\mathcal{P}$ consists of:

(i) A set P_0 of *points* (0-cells).

(ii) A set P_1 of *path generators* (1-cells), together with source and target maps $s, t : P_1 \to P_0$.

(iii) A set P_2 of *2-cell relations*, each specifying that two parallel paths (composed from generators using reflexivity, transitivity, and symmetry) are to be identified.

Higher-dimensional cells $P_3, P_4, \ldots$ may be added analogously, though for fundamental group computations we typically stop at dimension 2.

Remark 13.2 (Connection to HITs). In homotopy type theory, the data of Definition 13.1 corresponds exactly to a *higher inductive type* (HIT); see Chapter 14, Section 14.3, for a detailed comparison. The points are the point constructors, the path generators are path constructors, and the 2-cell relations are higher path constructors. The computational paths framework provides a concrete implementation of this idea: path expressions are syntactic objects that can be manipulated and normalised.

Definition 13.3 (Path expression over a presentation). Given a combinatorial space presentation $\mathcal{P}$, the *path expressions* over $\mathcal{P}$ are generated inductively:

(i) For each $g \in P_1$ with $s(g) = a$ and $t(g) = b$, there is a path expression $\widehat{g} : a \rightsquigarrow b$.

(ii) For each $a \in P_0$, there is a reflexivity path $\mathsf{refl}_a : a \rightsquigarrow a$.

(iii) If $p : a \rightsquigarrow b$ is a path expression, then $\mathsf{symm}(p) : b \rightsquigarrow a$ is a path expression.

(iv) If $p : a \rightsquigarrow b$ and $q : b \rightsquigarrow c$ are path expressions, then $\mathsf{trans}(p, q) : a \rightsquigarrow c$ is a path expression.

We write $\mathrm{Expr}(a, b)$ for the set of path expressions from a to b. When $a = b$, a path expression is called a *loop expression* at a; we write $\mathrm{Loop}(a) := \mathrm{Expr}(a, a)$. Throughout this chapter, "path expression" refers to an element of $\mathrm{Expr}(a, b)$ for possibly distinct a and b, while "loop expression" always means $a = b$.

The 2-cell relations specify when two path expressions should be considered equal. Beyond the intrinsic groupoid identities (associativity, unit laws, inverse laws), a presentation may impose additional equalities.

Definition 13.4 (Rewrite equivalence for presentations)**.** Given a combinatorial space presentation $\mathcal{P}$, the *rewrite equivalence* $\sim$ on path expressions is a presentation-specific instance of the general relation $\sim_{\mathsf{rw}}$ from earlier chapters. It is the smallest congruence containing:

(R1) The groupoid identities G1–G8 from Chapter 6.

(R2) For each 2-cell $r \in P_2$ relating p and q, the equation $p \sim q$.

Definition 13.5 (Fundamental groupoid of a presentation)**.** The *fundamental groupoid* $\Pi_1(\mathcal{P})$ of a combinatorial space presentation $\mathcal{P}$ has:

- Objects: the points P_0.
- Morphisms from a to b: equivalence classes $\mathrm{Expr}(a, b)/\sim$.
- Composition, identities, and inverses: induced from path operations.

For a chosen basepoint $\star \in P_0$, the *fundamental group* $\pi_1(\mathcal{P}, \star)$ is the automorphism group $\mathrm{Hom}_{\Pi_1(\mathcal{P})}(\star, \star)$.

Remark 13.6 (Normal forms and decidability)**.** For many presentations, the rewrite equivalence admits a confluent and terminating reduction strategy (cf. Chapter 5). In such cases, each equivalence class has a unique normal form, and the word problem for π_1 becomes decidable: two expressions are equivalent if and only if they share the same normal form. This computational character is one of the main advantages of the combinatorial approach.

13.1.3 The General Strategy

The remainder of this chapter applies the framework to specific spaces. The strategy is uniform:

1. **Define the presentation**: specify points, path generators, and 2-cell relations.
2. **Identify the loop space**: describe the path expressions from the basepoint to itself.
3. **Compute normal forms**: reduce loop expressions using the groupoid rules and presentation relations.
4. **Extract the group structure**: show that normal forms correspond bijectively to elements of a known group.

Circle S^1: one point, one free loop

$\mathbb{R}P^2$: one loop with $\ell^2 = \mathsf{refl}$

Figure 13.1: Combinatorial presentations of S^1 and $\mathbb{R}P^2$. The circle has a single free loop generator. The projective plane imposes the relation $\ell \cdot \ell = \mathsf{refl}$.

13.2 The Circle: $\pi_1(S^1) \cong \mathbb{Z}$

The circle is the simplest non-trivial example: a single point with a single loop generator and no 2-cell relations. The fundamental group is the integers $\mathbb{Z}$, with the loop generator corresponding to the integer 1. This classical result [79] emerges cleanly from the computational-path framework.

13.2.1 The Presentation

Definition 13.7 (Circle presentation). The *circle* S^1 is the combinatorial space with:

- One point: $P_0 = \{\star\}$.
- One path generator: $P_1 = \{\ell\}$, with $s(\ell) = t(\ell) = \star$.
- No 2-cell relations: $P_2 = \varnothing$.

The path expressions from $\star$ to $\star$ are built from ℓ, $\mathsf{refl}_\star$, symmetry, and transitivity. Since there is only one point, every path expression is automatically a loop.

13.2.2 Loop Expressions and Winding Numbers

A loop expression is a formal word in the generators ℓ and ℓ^{-1}. The groupoid identities allow us to:

- Cancel adjacent ℓ and ℓ^{-1} (inverse laws).
- Eliminate refl from compositions (unit laws).
- Reassociate parentheses freely (associativity).

Definition 13.8 (Winding number). The *winding number* of a loop expression p is defined inductively:

$$\begin{aligned} w(\ell) &= 1, \\ w(\mathsf{refl}_\star) &= 0, \\ w(\mathsf{symm}(p)) &= -w(p), \\ w(\mathsf{trans}(p, q)) &= w(p) + w(q). \end{aligned}$$

Lemma 13.9 (Winding number is invariant). *If $p \sim q$ under the groupoid rewrite equivalence, then $w(p) = w(q)$.*

Proof. We verify that each groupoid identity preserves the winding number.

- **Associativity**: $w(\mathsf{trans}(\mathsf{trans}(p, q), r)) = w(p) + w(q) + w(r) = w(\mathsf{trans}(p, \mathsf{trans}(q, r)))$.
- **Left unit**: $w(\mathsf{trans}(\mathsf{refl}, p)) = 0 + w(p) = w(p)$.
- **Right unit**: $w(\mathsf{trans}(p, \mathsf{refl})) = w(p) + 0 = w(p)$.
- **Left inverse**: $w(\mathsf{trans}(\mathsf{symm}(p), p)) = -w(p) + w(p) = 0 = w(\mathsf{refl})$.
- **Right inverse**: $w(\mathsf{trans}(p, \mathsf{symm}(p))) = w(p) - w(p) = 0 = w(\mathsf{refl})$.

Since $\sim$ is generated by these identities and their congruences, the winding number is constant on equivalence classes. □

Proposition 13.10 (Normal forms for circle loops). *Every loop expression at $\star$ is rewrite-equivalent to exactly one of:*

- $\ell^n := \underbrace{\ell \cdot \ell \cdot \cdots \cdot \ell}_{n}$ *for $n > 0$ (right-associated),*
- $\mathsf{refl}_\star$ *for $n = 0$,*
- $(\ell^{-1})^{|n|}$ *for $n < 0$.*

In particular, the normal form is uniquely determined by the winding number.

Proof. By confluence of the groupoid rewrite system (Chapter 5), every expression has a unique normal form. By Lemma 13.9, expressions with different winding numbers cannot be equivalent. A direct induction shows that every expression with winding number n reduces to the stated normal form. □

13.2.3 The Isomorphism

Theorem 13.11 ($\pi_1(S^1) \cong \mathbb{Z}$)**.** *The fundamental group of the circle is isomorphic to the integers:*

$$\pi_1(S^1, \star) \cong \mathbb{Z}.$$

The isomorphism sends the equivalence class $[\ell]$ *to* $1 \in \mathbb{Z}$*.*

Proof. Define maps:

$$\phi : \pi_1(S^1, \star) \to \mathbb{Z}, \quad [p] \mapsto w(p),$$
$$\psi : \mathbb{Z} \to \pi_1(S^1, \star), \quad n \mapsto [\ell^n],$$

where ℓ^n is the appropriate power (with $\ell^0 = \mathsf{refl}$ and $\ell^{-n} = (\ell^{-1})^n$).

By Lemma 13.9, ϕ is well-defined. It is a homomorphism since $w(\mathsf{trans}(p, q)) = w(p) + w(q)$.

The map ψ is a homomorphism since $\ell^{m+n} \sim \mathsf{trans}(\ell^m, \ell^n)$ by associativity.

We have $\phi \circ \psi(n) = w(\ell^n) = n$ and $\psi \circ \phi([p]) = [\ell^{w(p)}] = [p]$ by Proposition 13.10. □

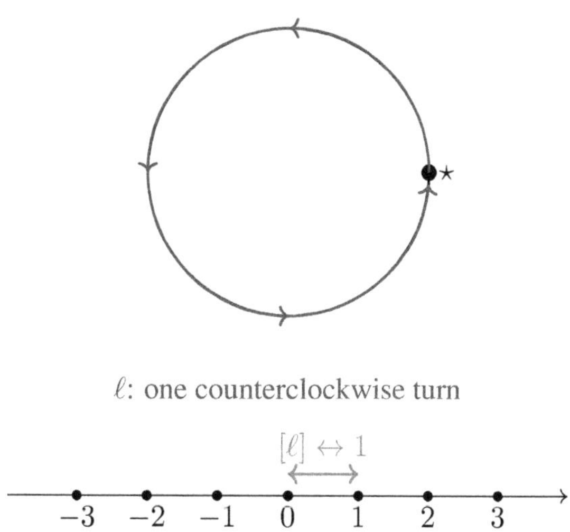

Figure 13.2: The correspondence $\pi_1(S^1) \cong \mathbb{Z}$. The loop generator ℓ maps to 1; its n-fold iteration ℓ^n maps to n.

Remark 13.12 (Computational decidability)**.** The winding number provides an effective decision procedure for path equivalence on the circle. Given two loop expressions p and q, compute their winding numbers: $p \sim q$ if and only if $w(p) = w(q)$. This is a simple linear-time algorithm in the size of the expressions.

Formalization Notes.

Formalization 13.13 (Circle fundamental group)**.** In the Lean formalisation, the circle is defined in `ComputationalPaths.Path.CompPath.CircleCompPath`, while the quotient-level interface for the fundamental group is packaged in `ComputationalPaths.Path.CompPath.CircleStep`. The distinguished base point is `circleBase`, the loop generator is `circleLoop`, the winding-number map is `circlePiOneEncode`, and the equivalence `circlePiOneEquivInt` establishes $\pi_1(S^1) \cong \mathbb{Z}$.

Key definitions:

- `circleLoop`: the loop generator ℓ.
- `circlePiOneEncode`: winding number on $\pi_1(S^1)$.
- `circleDecode`: reconstruction of loop classes from integers.
- `circlePiOneEquivInt`: the equivalence $\pi_1(S^1) \simeq \mathbb{Z}$.

```
@[simp] noncomputable def circlePiOneEncode : circlePiOne
    -> Int := ...
@[simp] noncomputable def circleDecode : Int ->
   circlePiOne := ...

noncomputable abbrev circlePiOneEquivInt :
    SimpleEquiv circlePiOne Int := ...
```

13.3 The Torus: $\pi_1(T^2) \cong \mathbb{Z} \times \mathbb{Z}$

The torus T^2 is the product of two circles: $T^2 = S^1 \times S^1$. Topologically, it is the familiar doughnut-shaped surface; combinatorially, it arises from a square with opposite edges identified. The fundamental group is $\mathbb{Z} \times \mathbb{Z}$, reflecting the two independent directions one can wind around the torus.

13.3.1 The Polygon Model

The standard construction of the torus identifies opposite edges of a square. If we label the horizontal edges a and the vertical edges b, and traverse the boundary starting from a corner, we obtain the word $a \cdot b \cdot a^{-1} \cdot b^{-1}$. This is the *commutator* $[a, b] = aba^{-1}b^{-1}$, and gluing the square into a torus imposes the relation $[a, b] = 1$, i.e., $ab = ba$.

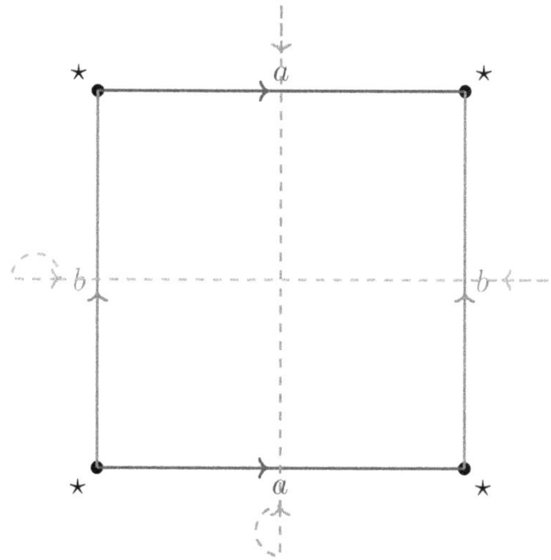

Identification: $aba^{-1}b^{-1} = 1$

Figure 13.3: The torus as a square with opposite edges identified. Both pairs of opposite edges are glued with matching orientations, yielding the identification word $aba^{-1}b^{-1}$.

13.3.2 The Presentation

Definition 13.14 (Torus presentation)**.** The *torus* T^2 is the combinatorial space with:

- One point: $P_0 = \{\star\}$.
- Two path generators: $P_1 = \{a, b\}$, both loops at $\star$.
- One 2-cell relation: $P_2 = \{r\}$, where r asserts $\mathsf{trans}(a, b) \sim \mathsf{trans}(b, a)$.

Equivalently, the 2-cell can be expressed as

$$\mathsf{trans}(\mathsf{trans}(\mathsf{trans}(a, b), \mathsf{symm}(a)), \mathsf{symm}(b)) \sim \mathsf{refl}_\star,$$

asserting that the commutator is trivial.

13.3.3 Normal Forms and the Group Structure

Loop expressions on the torus are words in a, a^{-1}, b, b^{-1}. The groupoid rules allow cancellation of adjacent inverses and reassociation, while the 2-cell relation permits commuting a past b.

Proposition 13.15 (Normal forms for torus loops). *Every loop expression on the torus is rewrite-equivalent to a unique expression of the form $a^m \cdot b^n$ for integers $m, n \in \mathbb{Z}$, where:*

- *a^m means a composed m times (or $(a^{-1})^{|m|}$ if $m < 0$, or* refl *if $m = 0$).*
- *Similarly for b^n.*

Proof. By the commutativity relation $ab \sim ba$, any word can be reordered to collect all as before all bs. The groupoid rules then reduce the a-block to a^m and the b-block to b^n for appropriate integers. Uniqueness follows from confluence: if $a^{m_1}b^{n_1} \sim a^{m_2}b^{n_2}$, then projecting to each factor (using the standard homomorphisms to $\mathbb{Z}$) gives $m_1 = m_2$ and $n_1 = n_2$. □

Theorem 13.16 ($\pi_1(T^2) \cong \mathbb{Z} \times \mathbb{Z}$). *The fundamental group of the torus is:*

$$\pi_1(T^2, \star) \cong \mathbb{Z} \times \mathbb{Z}.$$

The isomorphism sends $[a] \mapsto (1, 0)$ and $[b] \mapsto (0, 1)$.

Proof. Define:

$$\phi : \pi_1(T^2, \star) \to \mathbb{Z} \times \mathbb{Z}, \quad [a^m b^n] \mapsto (m, n),$$
$$\psi : \mathbb{Z} \times \mathbb{Z} \to \pi_1(T^2, \star), \quad (m, n) \mapsto [a^m b^n].$$

By Proposition 13.15, ϕ is well-defined. It is a homomorphism since $[a^{m_1}b^{n_1}] \cdot [a^{m_2}b^{n_2}] = [a^{m_1}b^{n_1}a^{m_2}b^{n_2}] = [a^{m_1+m_2}b^{n_1+n_2}]$ by commutativity. Similarly, ψ is a homomorphism, and ϕ, ψ are mutual inverses. □

Remark 13.17 (Abelianization). The fundamental group $\pi_1(T^2)$ is abelian precisely because the 2-cell relation enforces commutativity of the generators. In general, if a space is presented with commutator relations for all generator pairs, the fundamental group will be the free abelian group on those generators. The abelianness of $\pi_1(T^2)$ can also be seen as a shadow of the Eckmann–Hilton argument (Chapter 10), which shows that higher loop compositions are necessarily commutative.

Formalization Notes.

Formalization 13.18 (Torus fundamental group). The Lean formalisation defines the torus in `ComputationalPaths.Path.CompPath.Torus` and `ComputationalPaths.Path.CompPath.TorusStep`. Here the torus is represented concretely as `Torus := Circle × Circle`. The two generating loops are `torusLoop1` and `torusLoop2`, and the quotient-level encode/decode interface is packaged by `torusPiOneEncode`, `torusDecode`, and `torusPiOneEquivIntProdSimple`, establishing $\pi_1(T^2) \cong \mathbb{Z} \times \mathbb{Z}$.

Key definitions:

- `torusLoop1`, `torusLoop2`: the generators a and b.
- `torusDecode`: constructing loops from pairs (m, n).
- `torusPiOneEncode`: extracting pairs from loop classes.
- `torusPiOneEquivIntProdSimple`: the final quotient-level equivalence.

```
abbrev Torus : Type u := Prod Circle Circle

noncomputable def torusLoop1 : Path torusBase torusBase
    := ...
noncomputable def torusLoop2 : Path torusBase torusBase
    := ...

noncomputable def torusPiOneEquivIntProdSimple :
    SimpleEquiv torusPiOne (Prod Int Int) := ...
```

13.4 The Klein Bottle: $\pi_1(K) \cong \mathbb{Z} \rtimes \mathbb{Z}$

The Klein bottle K is a non-orientable surface that cannot be embedded in $\mathbb{R}^3$ without self-intersection. Like the torus, it is built from a square by edge identification, but with a twist: one pair of edges is glued with *reversed* orientation.

Why include the Klein bottle in this survey? The circle and torus yield abelian fundamental groups ($\mathbb{Z}$ and $\mathbb{Z} \times \mathbb{Z}$), and the projective plane yields a finite group ($\mathbb{Z}/2\mathbb{Z}$). All three are relatively tame from the group-theoretic standpoint. The Klein bottle is the simplest surface whose fundamental group is *non-abelian*: it forces the reader to confront the semidirect product structure that arises when edge identifications break orientability. It therefore serves as a test case for the computational-path framework's ability to handle non-commutative normal-form reductions, and it previews the kind of algebraic complexity that appears in more general surfaces.

13.4.1 The Polygon Model

Traversing the boundary of the Klein bottle's fundamental polygon yields $a \cdot b \cdot a \cdot b^{-1}$ (not $a \cdot b \cdot a^{-1} \cdot b^{-1}$ as for the torus). The relation is thus $aba = b$, or equivalently $ba = a^{-1}b$.

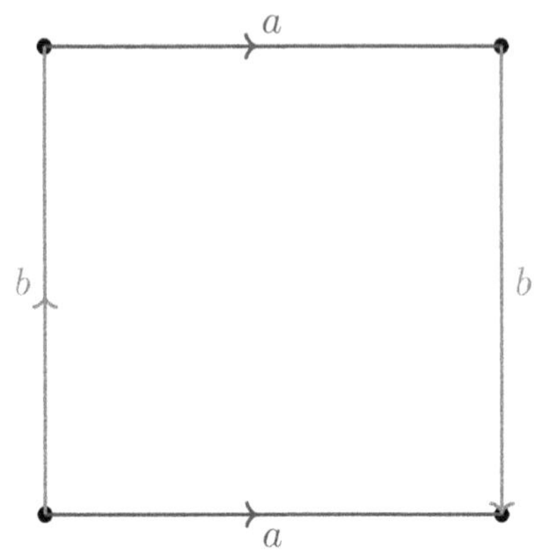

Identification: $aba = b$

Figure 13.4: The Klein bottle: opposite vertical edges are identified with *reversed* orientation (note the arrow directions). The relation becomes $aba = b$, or equivalently $ba = a^{-1}b$.

13.4.2 The Presentation and Group Structure

Definition 13.19 (Klein bottle presentation). The *Klein bottle* K is the combinatorial space with:

- One point: $P_0 = \{\star\}$.
- Two path generators: $P_1 = \{a, b\}$, both loops at $\star$.
- One 2-cell relation: $\mathsf{trans}(\mathsf{trans}(a, b), a) \sim b$, i.e., $aba = b$, equivalently $ba = a^{-1}b$.

The relation $ba = a^{-1}b$ implies $bab^{-1} = a^{-1}$: conjugating a by b yields its inverse. This is the characteristic feature of the *semidirect product* $\mathbb{Z} \rtimes \mathbb{Z}$.

Definition 13.20 (Semidirect product $\mathbb{Z} \rtimes \mathbb{Z}$). The semidirect product $\mathbb{Z} \rtimes \mathbb{Z}$ (with respect to the inversion action) is the group with underlying set $\mathbb{Z} \times \mathbb{Z}$ and multiplication:

$$(m_1, n_1) \cdot (m_2, n_2) = (m_1 + (-1)^{n_1} m_2, n_1 + n_2).$$

The identity is $(0, 0)$, and $(m, n)^{-1} = ((-1)^{n+1} m, -n)$.

Proposition 13.21 (Normal forms for Klein bottle loops). *Every loop expression on the Klein bottle is rewrite-equivalent to $a^m b^n$ for unique integers $m, n \in \mathbb{Z}$.*

Proof. Using the relation $ba = a^{-1}b$, any occurrence of b can be "pushed past" an a at the cost of inverting that a. More precisely:

$$\begin{aligned} ba &= a^{-1}b, \\ ba^{-1} &= ab, \\ b^{-1}a &= a^{-1}b^{-1}, \\ b^{-1}a^{-1} &= ab^{-1}. \end{aligned}$$

These relations allow moving all as to the left and all bs to the right, yielding a normal form $a^m b^n$.

For uniqueness, observe that projecting to the abelianization (obtained by additionally imposing $ab = ba$, which collapses to $\mathbb{Z}/2\mathbb{Z} \times \mathbb{Z}$) distinguishes different $(m \bmod 2, n)$ pairs. A finer analysis distinguishes all (m, n). □

Theorem 13.22 ($\pi_1(K) \cong \mathbb{Z} \rtimes \mathbb{Z}$)**.** *The fundamental group of the Klein bottle is:*

$$\pi_1(K, \star) \cong \mathbb{Z} \rtimes \mathbb{Z}.$$

The isomorphism sends $[a] \mapsto (1, 0)$ *and* $[b] \mapsto (0, 1)$.

Proof. Define $\phi([a^m b^n]) = (m, n)$. To verify this is a homomorphism, compute:

$$\begin{aligned} &[a^{m_1} b^{n_1}] \cdot [a^{m_2} b^{n_2}] \\ &= [a^{m_1} b^{n_1} a^{m_2} b^{n_2}] \\ &= [a^{m_1} a^{(-1)^{n_1} m_2} b^{n_1} b^{n_2}] \\ &\qquad \text{(commuting } b^{n_1} \text{ past } a^{m_2}\text{)} \\ &= [a^{m_1 + (-1)^{n_1} m_2} b^{n_1 + n_2}]. \end{aligned}$$

This matches the semidirect product multiplication. The inverse map $(m, n) \mapsto [a^m b^n]$ is clearly a homomorphism, and the two are mutual inverses by Proposition 13.21. □

Remark 13.23 (Non-abelian fundamental group)**.** The Klein bottle is the simplest surface with a non-abelian fundamental group. The non-commutativity arises from the twisted identification: traversing the "a-direction" and then the "b-direction" is not the same as traversing them in the opposite order.

13.5 Real Projective Plane: $\pi_1(\mathbb{R}P^2) \cong \mathbb{Z}/2\mathbb{Z}$

The real projective plane $\mathbb{R}P^2$ is obtained from a disk by identifying antipodal boundary points. Equivalently, it is a Möbius band with its boundary circle capped off by a disk. The fundamental group is $\mathbb{Z}/2\mathbb{Z}$: there is a single non-trivial loop that "wraps halfway around," and going around twice returns to the trivial loop.

13.5.1 The Presentation

Definition 13.24 ($\mathbb{R}P^2$ presentation)**.** The *real projective plane* $\mathbb{R}P^2$ is the combinatorial space with:

- One point: $P_0 = \{\star\}$.
- One path generator: $P_1 = \{\ell\}$, a loop at $\star$.
- One 2-cell relation: $P_2 = \{r\}$, asserting $\mathsf{trans}(\ell, \ell) \sim \mathsf{refl}_\star$, i.e., $\ell^2 = 1$.

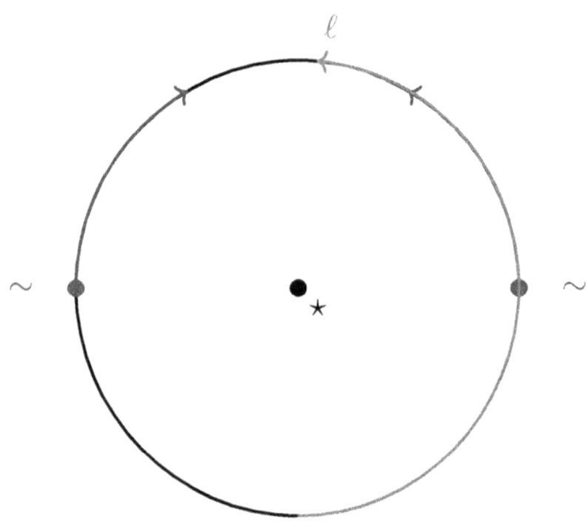

$\ell^2 = 1$: going around twice is contractible

Figure 13.5: The projective plane: a disk with antipodal boundary identification. The generator ℓ traverses half the boundary; ℓ^2 is the full boundary, which bounds the disk and hence is contractible.

13.5.2 Normal Forms and the Group

Proposition 13.25 (Normal forms for $\mathbb{R}P^2$ loops)**.** *Every loop expression on $\mathbb{R}P^2$ is rewrite-equivalent to either* $\mathsf{refl}_\star$ *or* ℓ.

Proof. Any loop expression is a word in ℓ and ℓ^{-1}. Since $\ell^2 = 1$, we have $\ell = \ell^{-1}$ (multiply both sides of $\ell^2 = 1$ by ℓ^{-1}). Thus every word reduces to ℓ^n

for some n, which further reduces to $\ell^{n \bmod 2}$: either refl (if n is even) or ℓ (if n is odd). □

Theorem 13.26 ($\pi_1(\mathbb{R}P^2) \cong \mathbb{Z}/2\mathbb{Z}$). *The fundamental group of the projective plane is:*

$$\pi_1(\mathbb{R}P^2, \star) \cong \mathbb{Z}/2\mathbb{Z}.$$

The isomorphism sends $[\ell] \mapsto 1 \in \mathbb{Z}/2\mathbb{Z}$.

Proof. By Proposition 13.25, there are exactly two equivalence classes of loops: $[\mathsf{refl}]$ and $[\ell]$. The group operation satisfies $[\ell] \cdot [\ell] = [\ell^2] = [\mathsf{refl}]$, so $\{[\mathsf{refl}], [\ell]\}$ with path composition is isomorphic to $\mathbb{Z}/2\mathbb{Z}$. □

Remark 13.27 (Geometric interpretation). Geometrically, ℓ represents the path that traverses half the boundary of the fundamental disk. Since antipodal points are identified, this is a closed loop. Traversing it twice covers the full boundary, which bounds the disk and is therefore null-homotopic.

13.6 Bouquets and Free Groups

A *bouquet* (or *wedge sum*) of circles is obtained by taking several circles and gluing them together at a single point. The fundamental group of a bouquet of n circles is the *free group* on n generators: the group with n generators and *no* relations.

13.6.1 Definition

Definition 13.28 (Bouquet of circles). The *bouquet of n circles*, denoted $\bigvee_{i=1}^{n} S^1$ or $S^1 \vee S^1 \vee \cdots \vee S^1$, is the combinatorial space with:

- One point: $P_0 = \{\star\}$.
- n path generators: $P_1 = \{\ell_1, \ell_2, \ldots, \ell_n\}$, all loops at $\star$.
- No 2-cell relations: $P_2 = \varnothing$.

13.6.2 The Free Group

Definition 13.29 (Free group). The *free group F_n* on n generators $g_1, \ldots, g_n$ is the group of all reduced words in the alphabet $\{g_1, g_1^{-1}, g_2, g_2^{-1}, \ldots, g_n, g_n^{-1}\}$, where a *reduced word* is one with no adjacent $g_i g_i^{-1}$ or $g_i^{-1} g_i$ pairs. The group operation is concatenation followed by reduction (canceling adjacent inverse pairs).

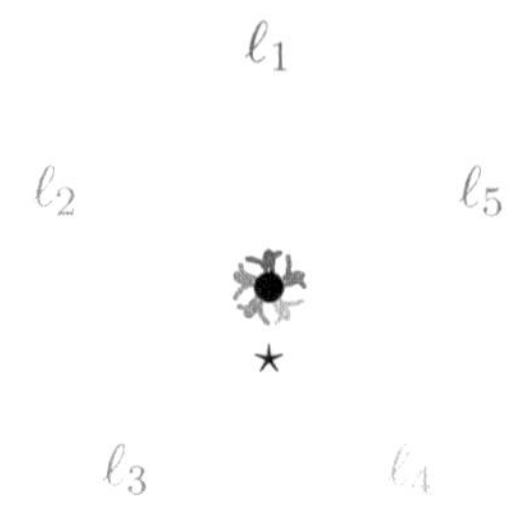

Bouquet of 5 circles: $\pi_1 \cong F_5$

Figure 13.6: A bouquet of five circles. Each loop generator contributes a free generator to the fundamental group.

Theorem 13.30 ($\pi_1(\bigvee_{i=1}^n S^1) \cong F_n$)**.** *The fundamental group of a bouquet of n circles is the free group on n generators:*

$$\pi_1\Big(\bigvee_{i=1}^{n} S^1, \star\Big) \cong F_n.$$

Proof. Path expressions on the bouquet are words in $\ell_1, \ell_1^{-1}, \ldots, \ell_n, \ell_n^{-1}$. The only rewrites available are the groupoid identities, which reduce words to reduced form (no adjacent inverse pairs) but impose no further relations. The set of reduced words, with concatenation-then-reduce as multiplication, is precisely F_n. □

Remark 13.31 (Rank and generators)**.** The number n is called the *rank* of the free group. Any two free groups of the same rank are isomorphic, and the rank is determined by the cardinality of any generating set. In particular, $F_1 \cong \mathbb{Z}$ (the integers under addition), and F_0 is the trivial group.

13.6.3 Normal Forms via Word Reduction

The computational-path framework provides an explicit algorithm for normal forms in free groups: the word reduction algorithm from the rewrite system of Chapter 4.

Proposition 13.32 (Word problem for free groups)**.** *Given two loop expressions p, q on a bouquet, the question of whether $[p] = [q]$ in π_1 is decidable: reduce both to normal form and compare.*

The formal development implements this by iteratively cancelling adjacent inverse pairs to produce a canonical representative for each equivalence class.

13.7 The Figure-Eight and Seifert–van Kampen

The *figure-eight space* $S^1 \vee S^1$ is the bouquet of two circles. Its fundamental group is the free group F_2 on two generators. This section revisits this computation using the *Seifert–van Kampen theorem*, a central tool for computing fundamental groups of spaces built from simpler pieces.

13.7.1 The Seifert–van Kampen Theorem

Theorem 13.33 (Seifert–van Kampen [138, 154, 15])**.** *Let $X = U \cup V$ be a space that is the union of two path-connected open sets U and V, with $U \cap V$ also path-connected and containing the basepoint $\star$. Then*

$$\pi_1(X, \star) \cong \pi_1(U, \star) *_{\pi_1(U \cap V, \star)} \pi_1(V, \star),$$

where $*_G$ denotes the amalgamated free product *over the common subgroup (the image of $\pi_1(U \cap V)$ in both $\pi_1(U)$ and $\pi_1(V)$).*

Remark 13.34 (Amalgamated free product)**.** The amalgamated free product $A *_C B$ is the group generated by A and B subject to the relations identifying elements of C in both copies. If C is trivial, this reduces to the ordinary free product $A * B$.

13.7.2 Application to the Figure-Eight

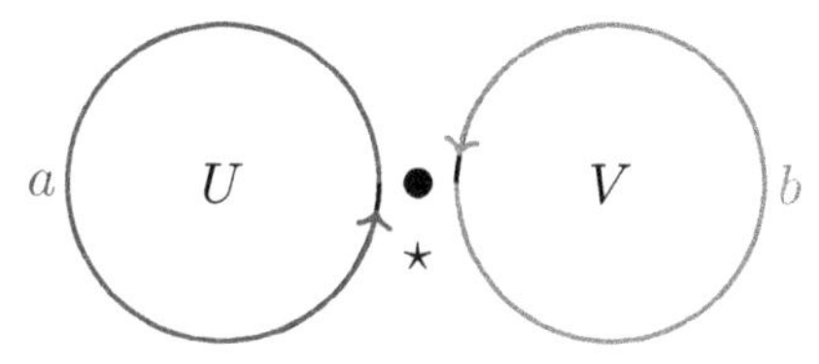

$$\pi_1(U) = \langle a \rangle \cong \mathbb{Z}, \quad \pi_1(V) = \langle b \rangle \cong \mathbb{Z}, \quad U \cap V = \{\star\}$$

Figure 13.7: The figure-eight decomposed for Seifert–van Kampen. Each circle contributes a copy of $\mathbb{Z}$; the intersection is a point with trivial fundamental group.

Consider the figure-eight $S^1 \vee S^1$. Decompose it as $U \cup V$ where:

- U is the left circle (with a small neighborhood of $\star$).
- V is the right circle (with a small neighborhood of $\star$).

- $U \cap V$ is a small contractible neighborhood of $\star$.

We have:

- $\pi_1(U) \cong \mathbb{Z}$, generated by a (the left loop).
- $\pi_1(V) \cong \mathbb{Z}$, generated by b (the right loop).
- $\pi_1(U \cap V) \cong 1$ (trivial, since $U \cap V$ is contractible).

By Seifert–van Kampen:

$$\pi_1(S^1 \vee S^1) \cong \mathbb{Z} *_1 \mathbb{Z} = \mathbb{Z} * \mathbb{Z} = F_2.$$

Corollary 13.35 ($\pi_1(S^1 \vee S^1) \cong F_2$)**.** *The fundamental group of the figure-eight is the free group on two generators:*

$$\pi_1(S^1 \vee S^1, \star) \cong F_2 = \langle a, b \mid \varnothing \rangle.$$

13.7.3 Computational Interpretation

In the computational-path framework, the Seifert–van Kampen (SVK) theorem is witnessed by the construction of path expressions. A loop in $X = U \cup V$ can be decomposed into segments alternating between U and V, with transitions at points of $U \cap V$. The amalgamation identifies loops in $U \cap V$ when viewed as elements of $\pi_1(U)$ or $\pi_1(V)$.

Formalization 13.36 (Seifert–van Kampen)**.** The Lean formalisation implements SVK via pushouts: `ComputationalPaths.Path.HIT.Pushout` defines the pushout construction, and the fundamental group of a pushout is computed via the amalgamated free product. The key theorem is `Pushout.pi1_equiv_amalg` establishing the SVK isomorphism.

For the figure-eight specifically, `ComputationalPaths.Path.CompPath.FigureEight` defines the space and `figureEightPiOneEquivFreeGroup` establishes $\pi_1 \cong F_2$.

```
/-- The 2-sphere as suspension of S^1 -/
abbrev Sphere2 : Type u := Suspension Circle

/-- SVK for suspension: pi_1(Sigma X) = 1 when both
    cone tips are contractible -/
theorem sphere2_isPathConnected :
    IsPathConnected Sphere2 :=
  Pushout.isPathConnected_of_components
    punit_isPathConnected
    punit_isPathConnected
    inferInstance
```

13.8 Lens Spaces

Lens spaces are a family of 3-dimensional manifolds with cyclic fundamental groups. They demonstrate a fundamental limitation of π_1: spaces with isomorphic fundamental groups need not be homeomorphic.

13.8.1 Definition

Definition 13.37 (Lens space). For integers $p > 0$ and $0 < q < p$ with $\gcd(p, q) = 1$, the *lens space* $L(p, q)$ is the quotient of the 3-sphere S^3 by the $\mathbb{Z}/p\mathbb{Z}$-action:

$$\zeta \cdot (z_1, z_2) = (\zeta z_1, \zeta^q z_2),$$

where $\zeta = e^{2\pi i/p}$ is a primitive p-th root of unity and $(z_1, z_2) \in S^3 \subset \mathbb{C}^2$.

Remark 13.38 (Special cases).

- $L(1, 1) \cong S^3$ (the 3-sphere).
- $L(2, 1) \cong \mathbb{R}P^3$ (real projective 3-space).

13.8.2 Combinatorial Presentation

A lens space can be built combinatorially by gluing two solid tori along their boundary. This gives a presentation:

Definition 13.39 (Lens space presentation). The lens space $L(p, q)$ has the combinatorial presentation:

- One point: $P_0 = \{\star\}$.
- One path generator: $P_1 = \{\ell\}$, a loop at $\star$.
- One 2-cell relation: $\ell^p \sim \mathsf{refl}_\star$.

The parameter q does not appear in the fundamental group but distinguishes the homeomorphism type of the space.

13.8.3 The Fundamental Group

Theorem 13.40 ($\pi_1(L(p, q)) \cong \mathbb{Z}/p\mathbb{Z}$). *For any lens space $L(p, q)$:*

$$\pi_1(L(p, q), \star) \cong \mathbb{Z}/p\mathbb{Z}.$$

Proof. The presentation has one generator ℓ with the single relation $\ell^p = 1$. By the same argument as for $\mathbb{R}P^2$ (Proposition 13.25), every loop reduces to ℓ^k for some $0 \leq k < p$. These p elements form a cyclic group of order p. □

Remark 13.41 (Lens spaces with same π_1)**.** The lens spaces $L(7,1)$ and $L(7,2)$ both have fundamental group $\mathbb{Z}/7\mathbb{Z}$. They are homotopy equivalent, but not homeomorphic, so π_1 does not determine a 3-manifold up to homeomorphism, even within the class of lens spaces. The complete classification of lens spaces uses *Reidemeister torsion* [79], a more refined invariant.

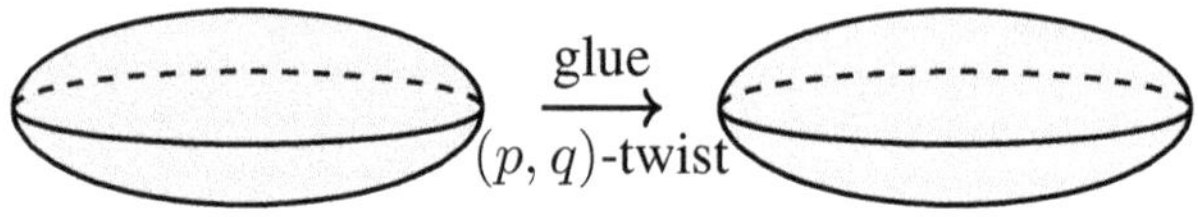

Figure 13.8: A lens space is built by gluing two solid tori with a (p,q)-twist. The fundamental group is $\mathbb{Z}/p\mathbb{Z}$, determined by p; the parameter q affects the topology but not π_1.

Summary

This chapter brought the computational-path framework to bear on a classical mathematical task: computing the fundamental groups of concrete spaces. We began with a combinatorial language for defining spaces via generators and relations, then applied it systematically to a sequence of examples of increasing algebraic complexity.

The circle, with its single free loop, produced the integers $\mathbb{Z}$—the paradigmatic computation showing that winding numbers classify loop equivalence classes. The torus, built from two commuting generators, yielded $\mathbb{Z} \times \mathbb{Z}$, illustrating how a commutativity relation turns a free group into an abelian one. The Klein bottle introduced genuine non-commutativity: its twisted identification forced the semidirect product $\mathbb{Z} \rtimes \mathbb{Z}$, requiring a more delicate normal-form argument that tracks how generators interact under conjugation. The projective plane, with its single relation $\ell^2 = 1$, gave the finite cyclic group $\mathbb{Z}/2\mathbb{Z}$, while bouquets of circles produced free groups of arbitrary rank. The Seifert–van Kampen theorem then showed how to decompose more complex spaces, as illustrated by the figure-eight. Finally, lens spaces demonstrated a fundamental limitation: spaces with the same π_1 need not be homeomorphic.

For reference, the results are collected in the following table:

Space	Presentation	π_1
Circle S^1	$\langle \ell \mid \varnothing \rangle$	$\mathbb{Z}$
Torus T^2	$\langle a, b \mid ab = ba \rangle$	$\mathbb{Z} \times \mathbb{Z}$
Klein bottle K	$\langle a, b \mid aba = b \rangle$	$\mathbb{Z} \rtimes \mathbb{Z}$
Projective plane $\mathbb{R}P^2$	$\langle \ell \mid \ell^2 = 1 \rangle$	$\mathbb{Z}/2\mathbb{Z}$
Bouquet $\bigvee_n S^1$	$\langle \ell_1, \ldots, \ell_n \mid \varnothing \rangle$	F_n
Figure-eight $S^1 \vee S^1$	$\langle a, b \mid \varnothing \rangle$	F_2
Lens space $L(p, q)$	$\langle \ell \mid \ell^p = 1 \rangle$	$\mathbb{Z}/p\mathbb{Z}$

Across all these examples, four themes recur. First, combinatorial presentations encode topology: a space is determined by its generators and relations. Second, normal forms compute π_1: the rewrite system reduces loop expressions to canonical representatives, and two loops are equivalent precisely when their normal forms coincide. Third, group structure emerges from rewrites: multiplication, inverses, and the identity arise directly from path operations rather than being imposed externally. Fourth, the Seifert–van Kampen theorem handles decompositions: complex spaces can be analysed by breaking them into simpler pieces whose fundamental groups are already known.

Remark 13.42 (Higher homotopy groups)**.** The framework extends to higher homotopy groups π_n via the ω-groupoid structure of Chapter 9. The computation of $\pi_2(S^2) \cong \mathbb{Z}$ requires techniques beyond the scope of this chapter but follows similar principles.

Remark 13.43 (Computational decidability)**.** For the spaces in this chapter, the word problem (deciding when two loops represent the same element of π_1) is decidable:

- Free groups: reduce to normal form and compare.
- Abelian groups: compute invariants (winding numbers, coordinates).
- Cyclic groups: reduce exponents modulo the order.

The semidirect product $\mathbb{Z} \rtimes \mathbb{Z}$ (Klein bottle) also has a decidable word problem via normal forms. In general, the word problem for finitely presented groups is undecidable, but the special structure of surface groups makes them tractable.

These calculations are not merely topological arguments rewritten in a different notation. The computational-path framework turns them into explicit algorithms. Rather than reasoning about continuous deformations, we manipulate symbolic expressions according to rewrite rules. The fundamental group emerges as the quotient of the expression space by rewrite equivalence—a structure that is at once mathematically natural and computationally effective. The categorical perspective on these constructions, including the connection to free groupoids and computads, is developed in Chapter 15.

Formalization Notes. The results of this chapter are formalised in the following files of the accompanying Lean development (`https://github.com/Arthur742Ramos/ComputationalPathsLean`):

- `ComputationalPaths/Path/CompPath/CircleCompPath.lean` and `ComputationalPaths/Path/CompPath/CircleStep.lean` — the computational circle, winding number, and the equivalence $\pi_1(S^1) \cong \mathbb{Z}$
- `ComputationalPaths/Path/CompPath/Torus.lean` and `ComputationalPaths/Path/CompPath/TorusStep.lean` — the torus as `Circle × Circle` and the equivalence $\pi_1(T^2) \cong \mathbb{Z} \times \mathbb{Z}$
- `ComputationalPaths/Path/CompPath/FigureEight.lean` — the figure-eight and $\pi_1 \cong F_2$
- `ComputationalPaths/Path/HIT/Pushout.lean` and `ComputationalPaths/Path/CompPath/PushoutPaths.lean`. These files provide the pushout construction and path decompositions.
- `ComputationalPaths/Path/Homotopy/SeifertVanKampen.lean` — the Seifert–van Kampen equivalence for pushouts

Chapter 14

Connections to Homotopy Type Theory

"The identity type is the heart of homotopy type theory. But that heart can beat in different ways: some systems make it pump computational content, while others let it carry geometric meaning."

—Inspired by Voevodsky's vision

Homotopy Type Theory (HoTT) is a research programme initiated by Vladimir Voevodsky and collaborators, establishing a profound correspondence between type theory and homotopy theory. At its centre lies the idea that the identity type of Martin-Löf type theory behaves like a space of paths, and that type-theoretic constructions naturally carry higher-dimensional structure.

The computational-path framework developed in this book moves in the same general region, but with different foundational commitments [129, 124]. Both frameworks organize identity proofs into higher structure; both exhibit groupoid-like behaviour. The purpose of this chapter, however, is comparison rather than assimilation. We do not claim that computational paths reproduce univalence, higher inductive types, or the full expressive power of HoTT. What we do claim is more precise: within the equality calculus developed earlier, computational paths provide *explicit rewrite witnesses* and *effective normalisation tests for rewrite equivalence*, whereas HoTT brings in genuinely stronger foundational principles.

This chapter provides a detailed comparison. Section 14.1 establishes a translation dictionary mapping concepts between the two frameworks. Section 14.2 explains what computational paths add within the present rewrite-theoretic setting: explicit step traces, decidable rewrite equivalence, and proof-relevant syntactic witnesses. Section 14.3 reverses the perspective, explaining what

HoTT adds: univalence and higher inductive types that are not available in our setting. Section 14.4 compares computational paths to cubical type theory, a computational refinement of HoTT. Finally, Section 14.5 discusses the open problem of whether computational paths can model univalence in some form.

The central distinction. Throughout this chapter, we maintain a crucial distinction between the two frameworks:

> **Computational paths work over an ambient equality satisfying UIP**, while **HoTT explicitly rejects UIP**.

In our framework, the ambient propositional equality $a = b$ satisfies the Uniqueness of Identity Proofs: any two equality proofs of $a = b$ are themselves equal (while computational path values can still differ—see Chapter 12). HoTT, by contrast, allows genuinely distinct identity proofs at every level, which is essential for univalence. This difference has profound consequences for what each system can express. It should be read as a difference in ambient equality, not as a claim that one framework subsumes the other.

Remark 14.1 (Two levels of identity)**.** The reader may wonder how UIP can hold here when Chapter 12 demonstrated its *failure*. The resolution is that two different equality notions are in play:

- The *ambient equality* $a = b$, used in this chapter for the semantic projection and for transport, is proof-irrelevant and therefore satisfies UIP.
- The *computational path type* $\mathsf{Path}(a, b)$, studied in Chapter 12, is proof-relevant: distinct rewrite traces can determine genuinely distinct path values.

These claims are compatible. Saying that UIP holds in the ambient theory does not collapse the computational path type; it says only that after one passes from a path to its underlying ambient equality witness, that witness is unique. The proof-relevant structure therefore lives in the path values themselves, not in the ambient equality between them.

14.1 Translation Dictionary: HoTT $\leftrightarrow$ Computational Paths

Before exploring the differences between the two frameworks, we establish their common vocabulary. Both systems arise from the same source—Martin-Löf's intensional type theory—and many concepts translate directly.

The reader already familiar with HoTT will find this section valuable for mapping known concepts to the computational paths terminology used throughout this book. Conversely, readers who have followed the development of computational paths can use this section as a guide when consulting the HoTT literature.

14.1.1 Basic Concepts

The two frameworks share the same fundamental intuition: identity proofs are *paths* connecting terms, and these paths carry algebraic structure. The following table summarises how the key concepts translate:

Concept	**HoTT**	**Computational Paths**
Identity type	$\mathsf{Id}_A(a,b)$ or $a =_A b$	$\mathsf{Path}(a,b)$
Reflexivity	$\mathsf{refl}_a : a = a$	refl_a
Symmetry	$p^{-1} : b = a$	$\mathsf{symm}(p)$
Transitivity	$p \cdot q : a = c$	$\mathsf{trans}(p,q)$
Congruence	$\mathsf{ap}_f(p) : f(a) = f(b)$	$\mathsf{congrArg}(f,p)$
Transport	$\mathsf{transport}^P(p,x)$	$\mathsf{transport}(P,p,x)$
J-eliminator	$\mathsf{J}(C, c_{\mathsf{refl}}, p)$	Derived, see Chapter 11
Path between paths	$p =_{a=b} q$	$\mathsf{Derivation}_2(p,q)$
Higher paths	iterated =	$\mathsf{Derivation}_n$
Proof irrelevance	Fails (in general)	Holds at the ambient equality layer
Rewrite eq. decidable	No (in general)	Yes via normalisation

Remark 14.2 (Naming conventions)**.** In the HoTT literature, following [150], the identity type is often written $a = b$ with the type A implicit. Transitivity is denoted $p \cdot q$ or $p \cdot q$, pronounced "p concatenated with q." The inverse is p^{-1} or $!p$. The application of f to a path p is $\mathsf{ap}_f(p)$, read as "ap of f along p."

In our notation, we use trans, symm, and congrArg to emphasize the rewriting viewpoint, but the operations are mathematically identical to their HoTT counterparts.

14.1.2 The J-Eliminator: A Point of Contact

The J-eliminator (path induction) is fundamental to both systems and provides the clearest point of contact between them. Intuitively, J says: "to prove something about all paths, it suffices to prove it for the reflexivity path."

In HoTT, J is typically taken as a primitive rule of the type theory:

Definition 14.3 (J-eliminator, HoTT style)**.** For a type A and type family $C : \prod_{a,b:A}(a = b) \to \mathcal{U}$, the J-eliminator has type:

$$\mathsf{J} : \Big(\prod_{a:A} C(a, a, \mathsf{refl}_a)\Big) \to \prod_{a,b:A} \prod_{p:a=b} C(a, b, p)$$

with the computation rule (sometimes called the "β-rule for J"):

$$\mathsf{J}(c, a, a, \mathsf{refl}_a) \equiv c(a).$$

In words: if we have a proof $c(a)$ for the reflexivity case at each point a, then J lifts this to a proof for any path $p : a = b$.

In Chapter 11, we *derived* J from the contractibility of the based path space. The derivation shows that J is a consequence of our groupoid operations and the rewrite rules, not an axiom we must accept on faith.

Proposition 14.4 (J from computational paths)**.** *Given the operations* refl, trans, symm, congrArg, *and* transport *of computational paths, together with the groupoid laws (proved via the 77 rewrite rules of Chapter 4), the J-eliminator is definable.*

Proof. See Theorem 11.25 in Chapter 11. The key ingredient is the contractibility of the based path space $\sum_{b:A} \mathsf{Path}(a, b)$. ☐

This derivation is significant: it means that in computational paths, J is not a black box but a constructed operation whose computational behaviour we can trace step by step.

14.1.3 Groupoid Laws: Same Theorems, Different Proofs

Both frameworks establish the standard groupoid laws G1–G8 already listed in Chapter 6. The difference lies not in the statements but in how they are obtained. In HoTT, these laws are proved by path induction and auxiliary lemmas. In computational paths, they appear as rewrite equivalences, each witnessed by a rule or short reduction sequence from Chapter 4. So the real contrast here is one of proof technology rather than algebraic content.

Example 14.5 (G3 in both systems)**.** The associativity law G3 illustrates the different proof styles.

In HoTT: A standard proof proceeds by path induction:

1. Induct on p; it suffices to prove when $p \equiv \mathsf{refl}_a$.

2. Then $(\mathsf{refl} \cdot q) \cdot r = q \cdot r$ by the computation rule for concatenation.

3. And $\mathsf{refl} \cdot (q \cdot r) = q \cdot r$ similarly.
4. Hence both sides equal $q \cdot r$, so they are equal.

In computational paths: G3 is a single rewrite rule:

$$\mathsf{trans}(\mathsf{trans}(p, q), r) \longrightarrow \mathsf{trans}(p, \mathsf{trans}(q, r)).$$

The proof *is* the rewrite step itself. There is no induction; the rule applies uniformly to all paths.

14.1.4 The Functor Laws for Congruence

Both systems prove that ap_f (HoTT notation) or $\mathsf{congrArg}(f, \cdot)$ (our notation) is functorial—it respects the groupoid operations:

$$\mathsf{ap}_f(\mathsf{refl}_a) = \mathsf{refl}_{f(a)} \tag{F1}$$

$$\mathsf{ap}_f(\mathsf{trans}(p, q)) = \mathsf{trans}(\mathsf{ap}_f(p), \mathsf{ap}_f(q)) \tag{F2}$$

$$\mathsf{ap}_f(\mathsf{symm}(p)) = \mathsf{symm}(\mathsf{ap}_f(p)) \tag{F3}$$

$$\mathsf{ap}_{g \circ f}(p) = \mathsf{ap}_g(\mathsf{ap}_f(p)) \tag{F4}$$

$$\mathsf{ap}_{\mathsf{id}}(p) = p \tag{F5}$$

In HoTT, F1 is definitional (it holds by the computation rule), while F2–F5 require proofs by path induction. In computational paths, all five are rewrite rules or immediate consequences thereof, making the functoriality computationally evident.

14.1.5 Higher Structure: 2-Cells and Beyond

Both systems organize paths into higher-dimensional structures. Understanding how each achieves this reveals a key architectural difference.

In HoTT. A path between paths $p = q$ is itself an element of an identity type. Higher paths arise by further iteration:

$$\mathsf{Id}_{a=b}(p, q), \qquad \mathsf{Id}_{p=q}(\alpha, \beta), \qquad \ldots$$

The result is a tower of identity types, each connected to the next by reflexivity, symmetry, transitivity, and congruence operations inherited from the general identity type. This uniformity is elegant: the same rules apply at every level.

In Computational Paths. We introduced $\mathsf{Derivation}_2$ for 2-cells (paths between paths) and $\mathsf{Derivation}_n$ for higher cells in Chapters 7 and 9. These are explicitly defined as rewrite sequences between path expressions, carrying step-by-step computational traces.

The key difference: in HoTT, the higher structure emerges *automatically* from the uniform treatment of identity types. In computational paths, each level is constructed *explicitly* with its own rewrite rules and coherence conditions.

Level	HoTT	Computational Paths
1-cells (paths)	$a = b$	$\mathsf{Path}(a, b)$
2-cells	$p =_{a=b} q$	$\mathsf{Derivation}_2(p, q)$
3-cells	$\alpha =_{p=q} \beta$	$\mathsf{Derivation}_3(\alpha, \beta)$
n-cells	iterated $=$	$\mathsf{Derivation}_n$
Contractibility	Fails in general	Holds at level ≥ 3

14.1.6 The ω-Groupoid Structure

The ω-groupoid construction itself is developed in detail in Chapter 9; Section 9.6 there gives a first comparison with HoTT identity types. Here we place that comparison in the broader context of the two frameworks.

Both frameworks yield weak ω-groupoid structures, a fundamental result connecting type theory to higher category theory.

In HoTT, this theorem is due to Lumsdaine [105] and van den Berg–Garner [153]: the iterated identity types of any type form a weak ω-groupoid in the sense of Batanin or Grothendieck.

In computational paths, we proved the analogous result in Chapter 9: the tower

$$\mathsf{Ob} \xrightarrow{\mathsf{Path}} \mathsf{Mor}_1 \xrightarrow{\mathsf{Derivation}_2} \mathsf{Mor}_2 \xrightarrow{\mathsf{Derivation}_3} \cdots$$

forms a weak ω-groupoid (Chapter 9) with explicit coherence witnesses through dimension 3 (Chapter 8) and stabilised higher fillers thereafter, all constructed from the rewrite rules.

The structures are abstractly isomorphic, but the presentations differ:

- HoTT's ω-groupoid is *implicit*—it emerges from the uniform rules for identity types.
- Computational paths' ω-groupoid is *explicit*—each level is constructed with specific rewrite rules and coherence cells.

14.2 What Computational Paths Add

Having established the common ground, we now examine what computational paths provide that HoTT's standard presentation does not. The key additions are *explicit rewrite witnesses*, *decidable path equality*, and *concrete computational content*.

These features are not merely technical conveniences; they enable entirely new applications in proof automation and program verification.

14.2.1 Explicit Rewrite Witnesses

In HoTT, a path $p : a = b$ is an abstract proof term. We know it exists and satisfies certain rules, but the internal structure—*how* the path was constructed—is not directly accessible. The path is a certificate that a and b are equal, but it does not tell us the sequence of steps that establishes this equality.

In computational paths, every path carries an explicit rewrite trace together with an ambient equality witness:

$$\mathsf{Path}(a, b) = \{\, (s, h) \mid s \text{ is a rewrite trace from } a \text{ to } b,\ h : a = b\,\}.$$

The trace component records the sequence of atomic rewrite operations that justify the path. This is analogous to recording the sequence of moves in a chess game, rather than just noting the final position.

Example 14.6 (Two paths, same endpoints, different traces)**.** Consider the function $g(x) = (\lambda y.\, y)\, x$ on natural numbers and the path $p : \mathsf{Path}(g(n), n)$ given by the β-reduction collapsing $g(n)$ to n. We can construct two paths from $g(g(n))$ to n:

$$\begin{aligned} p_1 &= \mathsf{trans}(\mathsf{congrArg}(g, p), p) \\ p_2 &= \mathsf{trans}(q, p) \end{aligned}$$

where $q : \mathsf{Path}(g(g(n)), g(n))$ is the path obtained by reducing the outer redex first. Both have the same endpoints and the same underlying ambient equality witness (by proof irrelevance at that level). But they have *different step traces* recording different sequences of rewrites: p_1 first reduces the inner occurrence of $g(n)$, while p_2 reduces the outer redex and only then the remaining one.

This explicit structure enables several capabilities not available in standard HoTT:

Step counting. We can count how many atomic steps a path contains. This provides a concrete measure of "proof complexity" that has no direct analogue in HoTT, where paths are opaque.

Trace inspection. We can examine the specific sequence of rules applied. This enables debugging ("why did the proof fail here?"), optimisation ("is there a shorter path?"), and analysis of proof structure.

Computational behaviour. The rewrite rules have explicit computational meaning. We can *run* the rewrite system to simplify paths, obtaining normal forms. In HoTT, such computation is generally not available.

14.2.2 Decidable Rewrite Equivalence

Perhaps the most significant technical advantage of computational paths is a **decidable normalisation test for rewrite equivalence** of paths.

Theorem 14.7 (Decidability of rewrite equivalence)**.** *For any paths* $p, q : \mathsf{Path}(a, b)$*, it is decidable whether* $p \sim_{\mathsf{rw}} q$ *(where* $\sim_{\mathsf{rw}}$ *is rewrite equivalence* RwEq*).*

Proof. By the confluence and termination results of Chapter 5:

1. Compute $\mathsf{normalize}(p)$ and $\mathsf{normalize}(q)$.
2. Compare the normal forms for syntactic equality.
3. $p \sim_{\mathsf{rw}} q$ if and only if $\mathsf{normalize}(p) = \mathsf{normalize}(q)$.

Each step is computable, so the overall procedure is decidable. □

In HoTT, path equality is *not* generally decidable. Given $p, q : a = b$, deciding whether $p = q$ (in the identity type $p =_{a=b} q$) would require solving the word problem for the ω-groupoid structure, which is undecidable in general.

Remark 14.8 (The price of decidability)**.** This decidability comes with a trade-off: computational paths use UIP at the meta-level. The ambient equality witness carried by a path is proof-irrelevant in the ambient extensional foundation, meaning all *meta-level equality proofs* of $a = b$ are considered equal—while the path values themselves remain distinct whenever their step traces differ (see Remark 14.1). This is why we can quotient by RwEq and obtain a decidable quotient equality.

In HoTT, different proofs of $a = b$ can be genuinely distinct—this is essential for capturing homotopy-theoretic phenomena but precludes this kind of normalisation-based comparison. The trade-off is fundamental: one cannot have both univalence and the present normalisation-based decision procedure for path comparison.

14.2.3 The Two-Level Structure

Computational paths employ a *two-level structure* that is key to achieving both proof-relevance (for step traces) and decidability (via UIP at the base):

1. **Proof-relevant level:** The path expressions $\mathsf{Path}(a, b)$ with their step traces, organised into $\mathsf{Derivation}_2$, $\mathsf{Derivation}_3$, etc. Different traces are distinguished; this level carries computational information.

2. **Proof-irrelevant level:** The underlying ambient equality $a = b$, whose witness is unique by UIP in the ambient foundation. This level grounds the path in actual equality.

This two-level design allows us to have computational content (the step traces) while maintaining decidability (via the proof-irrelevant base).

$$\begin{array}{ccc} \mathsf{Path}(a,b) & \xrightarrow{\text{ambient equality}} & a = b \\ {\scriptstyle\text{rewrite trace}}\Big\downarrow & & \Big\downarrow{\scriptstyle\text{UIP}} \\ \text{rewrite traces} & \xrightarrow[\text{quotient by } \sim_{\mathsf{rw}}]{} & \mathsf{Path}(a,b)/{\sim_{\mathsf{rw}}} \end{array}$$

The horizontal arrow forgets the trace and remembers only the ambient equality; the vertical arrow records the computational content and then passes to rewrite-equivalence classes. The quotient identifies paths that differ only by rewrite equivalence, yielding decidable equality.

14.2.4 Concrete Computation Examples

To make these ideas concrete, let us trace through some computations.

Example 14.9 (Simplifying a complex path)**.** Consider the path expression:

$$p = \mathsf{trans}(\mathsf{trans}(\mathsf{refl}_a, \mathsf{refl}_a), \mathsf{symm}(\mathsf{refl}_a))$$

Applying the rewrite rules step by step:

$$\begin{aligned} p &\longrightarrow \mathsf{trans}(\mathsf{refl}_a, \mathsf{symm}(\mathsf{refl}_a)) && \text{(G1: left unit)} \\ &\longrightarrow \mathsf{trans}(\mathsf{refl}_a, \mathsf{refl}_a) && \text{(G7: symm of refl)} \\ &\longrightarrow \mathsf{refl}_a && \text{(G1: left unit)} \end{aligned}$$

The normal form is refl_a. The step trace records exactly which rules were applied and in what order.

Example 14.10 (Distinguishing paths by trace). Let $p = \mathsf{refl}_a$ and $q = \mathsf{trans}(\mathsf{refl}_a, \mathsf{refl}_a)$. Both determine the same underlying ambient equality witness $\mathsf{rfl} : a = a$. But:

$$\mathsf{normalize}(q) = \mathsf{refl}_a = \mathsf{normalize}(p)$$

So $p \sim_{\mathsf{rw}} q$ under rewrite equivalence. However, *before normalisation*, p and q have different step traces. The rewrite system records that q required one application of the left-unit rule to reach p.

14.2.5 Computational Meaning of Groupoid Laws

The 77 rewrite rules of Chapter 4 give computational meaning to the groupoid laws. Each rule is an *oriented* rewrite:

$$\mathsf{trans}(\mathsf{refl}_a, p) \longrightarrow p$$

This directionality is significant. It says: "when you encounter a path that starts with refl composed with something, simplify by removing the refl."

In HoTT, the analogous statement $\mathsf{refl} \cdot p = p$ is an equality without preferred direction. The computational paths version chooses a canonical simplification direction, enabling the normalisation algorithm that yields decidable path equality.

14.2.6 Summary: Computational Paths Advantages

To summarise, computational paths add:

(i) **Explicit step traces:** Every path records its construction history.

(ii) **Decidable equality:** $p \sim_{\mathsf{rw}} q$ is decidable via normalisation.

(iii) **Oriented rewriting:** The 77 rules give computational meaning to path operations.

(iv) **Normal forms:** Every path has a canonical normal form.

(v) **Complexity measures:** Step counts and trace inspection are available.

These features make computational paths suitable for applications requiring concrete computation with paths, such as:

- Proof assistants where path equality must be checked automatically.
- Program verification where rewrite traces justify transformations.
- Term rewriting systems where confluence must be established.

14.3 What HoTT Adds: Univalence and Higher Inductive Types

Having seen what computational paths provide, we now turn to what HoTT offers that our framework does not directly support. The two key features are the *univalence axiom* and *higher inductive types*.

14.3.1 The Univalence Axiom

Univalence is the central innovation of homotopy type theory, proposed by Voevodsky around 2006. The idea is deceptively simple: *equivalent types should be considered equal.*

To make this precise, we first define what it means for two types to be equivalent.

Definition 14.11 (Equivalence of types). An *equivalence* between types A and B consists of:

- A function $f : A \to B$ (the forward map);
- A function $g : B \to A$ (the inverse);
- A homotopy $\eta : \prod_{x:A} g(f(x)) = x$ (left inverse);
- A homotopy $\epsilon : \prod_{y:B} f(g(y)) = y$ (right inverse);
- A coherence datum $\tau : \prod_{x:A} \mathsf{ap}_f(\eta_x) = \epsilon_{f(x)}$, witnessing that the two ways of identifying $f(g(f(x)))$ with $f(x)$ agree.

The datum τ is the *half-adjoint coherence condition.* Without it, the type of bi-invertible maps (f, g, η, ϵ) is not a mere proposition: there can be distinct choices of η and ϵ that are not related. Adding τ cuts down the data to a contractible fibre, ensuring that $A \simeq B$ is well-behaved as a type.

We write $A \simeq B$ for the type of equivalences from A to B. Intuitively, $A \simeq B$ means that A and B "have the same structure" in a very strong sense: they can be put in perfect bijection, and this bijection is compatible with all type-theoretic operations.

There is a canonical map that turns an identity into an equivalence:

Definition 14.12 (idtoequiv). For any types $A, B : \mathcal{U}$, there is a canonical function

$$\mathsf{idtoequiv} : (A = B) \to (A \simeq B)$$

defined by path induction: $\mathsf{idtoequiv}(\mathsf{refl}_A)$ is the identity equivalence on A (the equivalence whose forward and backward maps are both the identity function).

Here the domain $A = B$ is the ambient equality of types; below, this distinction from the computational path type $\mathsf{Path}(A, B)$ will be crucial.

The key observation is that idtoequiv is not obviously surjective. Just because two types are equivalent, must they be equal? Classical mathematics would say no—$\mathbb{N}$ and $\mathbb{Z}_{>0}$ are equivalent but not literally identical. Univalence says yes:

Axiom 14.13 (Univalence)**.** For all types $A, B : \mathcal{U}$, the function idtoequiv : $(A = B) \to (A \simeq B)$ is itself an equivalence.

In other words, identities between types correspond *exactly* to equivalences between types. This has profound consequences:

- Equivalent types are interchangeable in *all* contexts.
- The universe $\mathcal{U}$ is not a set; it has non-trivial higher structure (automorphisms of types give loops in $\mathcal{U}$).
- Function extensionality follows from univalence.

14.3.2 Why Univalence Lies Outside the Present Framework

Univalence is **incompatible** with the computational paths framework as presented in this book. The reason is our use of UIP at the meta-level.

The intuition is straightforward: univalence demands that genuinely different equivalences between types (such as the identity on Bool and boolean negation) give rise to genuinely different paths at the type level. But our framework enforces UIP for the ambient propositional equality Eq, which means that any two inhabitants of Bool = Bool are identified. There is simply no room in the ambient equality for two distinct witnesses, and therefore no way to reflect the non-trivial equivalence not as a distinct path.

Theorem 14.14 (Full univalence fails)**.** *In the computational paths framework, there is no function* $\mathsf{ua} : (A \simeq B) \to \mathsf{Path}(A, B)$ *such that, for every equivalence* e*, applying* idtoequiv *to the ambient equality carried by* $\mathsf{ua}(e)$ *recovers* e*.*

Proof. We exhibit a specific equivalence that cannot come from any path. Consider the equivalence not : Bool $\simeq$ Bool given by boolean negation (swapping true and false).

For any path $p : \mathsf{Path}(\mathsf{Bool}, \mathsf{Bool})$, we have $\mathsf{toEq}(p) : \mathsf{Bool} = \mathsf{Bool}$. By UIP for the ambient equality, $\mathsf{toEq}(p) = \mathsf{refl}_{\mathsf{Bool}}$. Therefore transport along the ambient equality extracted from any such path acts trivially on Bool.

But if such a function ua existed, then idtoequiv applied to the ambient equality underlying $\mathsf{ua}(\mathsf{not})$ would have to recover the equivalence not, whose forward map is boolean negation. We would therefore need transport along that ambient equality to send true to false. Yet we have just shown that transport along any ambient equality extracted from a computational path is the identity. This yields $\mathsf{true} = \mathsf{false}$, which is absurd. □

Remark 14.15 (A boundary of the present design)**.** This theorem identifies a structural boundary of the framework developed in this book.

Univalence requires *non-trivial paths at the type level*—paths between types that are not reflexivity. In HoTT, $\mathsf{Bool} = \mathsf{Bool}$ has at least two distinct elements: the identity equivalence and the one induced by negation.

In our setting, UIP *forces all ambient type equalities to be trivial*: $\mathsf{Bool} = \mathsf{Bool}$ has only one inhabitant (reflexivity) at the ambient equality level.

One therefore cannot simultaneously keep this ambient UIP-based equality and also recover full univalence. The present design chooses explicit rewrite witnesses and decidable normalisation over non-trivial type-level paths.

Remark 14.16 (Practical consequence)**.** The failure of univalence means that in computational paths:

- equivalences between types are handled externally, not encoded as non-trivial ambient equalities;
- transporting along an ambient type equality is always trivial;
- the ambient universe $\mathcal{U}$ behaves like a set (all paths are reflexivity).

This should be read as a boundary of the current design rather than as a defect. The same simplifying choice that blocks univalence is what keeps the ambient equality compatible with the normalisation and decidability results of Chapters 4 and 5.

14.3.3 Higher Inductive Types

Higher inductive types (HITs) are another key feature of HoTT that computational paths do not directly support. While ordinary inductive types are specified by point constructors (elements of the type), HITs also include path constructors (equalities between elements) and potentially higher path constructors.

Definition 14.17 (Higher inductive type, informally)**.** A *higher inductive type* is an inductive type specified by:

- **Point constructors:** elements of the type;
- **Path constructors:** paths between points;

- **Higher path constructors:** paths between paths, etc.

The elimination principle for a HIT requires specifying where each constructor lands, including coherence conditions for the paths.

Example 14.18 (The circle S^1)**.** The circle is the paradigmatic higher inductive type:

- One point constructor: $\mathsf{base} : S^1$
- One path constructor: $\mathsf{loop} : \mathsf{base} = \mathsf{base}$

The elimination principle says: to define $f : S^1 \to A$, provide $b : A$ and $\ell : b = b$; then $f(\mathsf{base}) = b$ and $\mathsf{ap}_f(\mathsf{loop}) = \ell$. This captures the homotopy-theoretic idea that the circle has one point and one non-trivial loop.

Example 14.19 (The interval I)**.** The interval is the simplest non-trivial HIT:

- Two point constructors: $0, 1 : I$
- One path constructor: $\mathsf{seg} : 0 = 1$

Despite its simplicity, the interval is contractible and can be used to construct function extensionality.

Example 14.20 (Quotients as HITs)**.** Quotient types arise naturally as HITs. Given a type A and an equivalence relation $R : A \to A \to \mathsf{Prop}$, the quotient A/R is a HIT with:

- Point constructor: $[a] : A/R$ for each $a : A$
- Path constructor: for each $a, b : A$ and $r : R(a, b)$, a path $[a] = [b]$
- Truncation: A/R is a set (all paths between paths are equal)

This provides a clean type-theoretic treatment of quotients that is difficult to achieve without HITs.

14.3.4 HITs and Computational Paths: A Comparison

In Chapter 13, we computed fundamental groups of combinatorial spaces using *generators and relations*. Our approach was:

1. Postulate atomic steps corresponding to path constructors.
2. Define relations as paths between paths (2-cells).
3. Compute the groupoid presented by these generators and relations.

This is reminiscent of HITs, but there are crucial differences:

HITs are built into the type theory. In HoTT, the circle S^1 is a genuine type with a genuine eliminator. Functions out of S^1 are defined by the elimination principle, and the computation rules are part of the type theory.

Computational paths use external postulates. We postulate atomic steps and relations but do not have a type-theoretic elimination principle. The computation is performed *within* the path algebra, not by type-theoretic recursion.

Example: $\pi_1(S^1) \cong \mathbb{Z}$ in both systems. Both HoTT and computational paths can prove that the fundamental group of the circle is the integers, but the proofs have different character:

- In HoTT, it follows from the elimination principle for S^1 and univalence [101]. The proof uses the universal property of S^1.
- In computational paths, it follows from analysing the structure of the path groupoid on a space with one point and one non-trivial loop. The proof is combinatorial.

14.3.5 Summary: HoTT Advantages

To summarise, HoTT provides:

(i) **Univalence:** Equivalent types are equal, enabling transport of structure.

(ii) **Higher inductive types:** Built-in support for spaces with non-trivial homotopy.

(iii) **Function extensionality:** Provable from univalence (or as a consequence of HITs).

(iv) **Quotient types:** Clean construction via HITs.

(v) **Synthetic homotopy theory:** Homotopy groups, fibrations, etc., can be defined and studied type-theoretically.

These features make HoTT suitable for:

- Foundations of mathematics where structure should transfer across equivalences.
- Synthetic homotopy theory and the study of homotopy types.
- Formalization of category theory with natural identification of isomorphic objects.

14.4 Cubical Type Theory and Computational Paths

Cubical type theory is a refinement of HoTT that gives computational content to univalence. It provides an illuminating comparison point for computational paths, as both frameworks aim to make equality "computational" but in different ways.

14.4.1 What Is Cubical Type Theory?

Cubical type theory, developed by Cohen–Coquand–Huber–Mörtberg (CCHM) [23] and in variant forms [4], is a type theory where paths are functions out of an "interval" type. The key ideas are:

- Paths are literally functions $\mathbb{I} \to A$.
- The interval $\mathbb{I}$ has two endpoints $0, 1 : \mathbb{I}$.
- Univalence has computational content via "Glue types."

Definition 14.21 (Path type in cubical TT). In cubical type theory, a path from a to b in type A is a function from the interval that agrees with a at 0 and b at 1:

$$\mathsf{Path}_A(a, b) := (i : \mathbb{I}) \to A\, [\, i = 0 \mapsto a,\ i = 1 \mapsto b\,]$$

The notation $[\, i = 0 \mapsto a,\ i = 1 \mapsto b\,]$ indicates boundary constraints. The path p is a continuous family of elements $p(i) : A$ parametrised by the interval, with $p(0) = a$ and $p(1) = b$.

14.4.2 Comparison: Cubical vs. Computational Paths

The following table highlights the key differences between the two approaches to computational paths:

Feature	**Cubical TT**	**Computational Paths**
Path representation	Function $\mathbb{I} \to A$	Step traces + proof
Interval	Primitive type $\mathbb{I}$	Not needed
Reflexivity	$\lambda i.\, a$	refl_a (atomic)
Symmetry	$\lambda i.\, p(1 - i)$	$\mathsf{symm}(p)$ (rewrite rule)
Transitivity	Connection squares	$\mathsf{trans}(p, q)$ (rewrite rules)
Univalence	Computes via Glue types	Fails (meta-level UIP)
HITs	Supported	External postulates
Decidable rewrite eq.	No (in general)	Yes (via normalisation)

14.4.3 The Interval: Same Name, Different Objects

Both cubical type theory and computational paths use the word "path," but the underlying objects are fundamentally different:

Cubical paths are functions. A cubical path $p : \mathsf{Path}_A(a, b)$ is a function $p : \mathbb{I} \to A$. The path *is* the function; there is no separate "step trace." Evaluating $p(i)$ gives a point of A for any $i : \mathbb{I}$. The path literally describes a continuous family of points interpolating between a and b.

Computational paths are structured records. A computational path $p : \mathsf{Path}(a, b)$ has two components: a list of atomic steps (the trace) and a proof that the endpoints are equal. There is no "intermediate point"—the path records *how* a and b were shown equal, not a continuous interpolation between them.

Remark 14.22 (Geometric vs. algebraic perspectives)**.** One might characterise the difference as follows:

- Cubical paths are *geometric*: they parametrise a continuous family of points, like a path in topology.
- Computational paths are *algebraic*: they record a sequence of rewrite operations, like a word in a group presentation.

In topology, these correspond to two classical views of paths:

- *Geometric:* A path is a continuous map $[0, 1] \to X$.
- *Algebraic:* A path is a word in the fundamental groupoid.

Both perspectives are valuable, and each captures different aspects of the notion of "connection" between points.

14.4.4 How Cubical TT Achieves Computational Univalence

The key innovation of cubical type theory is *Glue types*: type formers that allow "gluing" a type to an equivalent type along boundaries. This is what makes univalence compute.

Definition 14.23 (Glue types, informally)**.** If ϕ is a face formula (a condition on interval variables) and T is a type with $e : T \simeq A$ when ϕ holds, then:

$$\mathsf{Glue}\ [\phi \mapsto (T, e)]\ A$$

is a type that equals T when ϕ holds and A otherwise, with the equivalence e providing the "gluing map" along the face.

Using Glue types, one can define a path witnessing any equivalence:

$$\mathsf{ua}(e) := \lambda i.\, \mathsf{Glue}\ [(i = 0) \mapsto (A, e)]\ B$$

This is a path from A to B: at $i = 0$ we get A, at $i = 1$ we get B.

Theorem 14.24 (Univalence computes in cubical TT). *In cubical type theory, if $e : A \simeq B$ and $x : A$, then*

$$\mathsf{transport}(\mathsf{ua}(e), x) = e.\mathsf{toFun}(x)$$

definitionally (or up to a simple path, depending on the variant).

This is the sense in which cubical type theory "computes univalence": transporting along the path induced by an equivalence actually applies the equivalence function. The transport is not a black box; it has explicit computational behaviour.

14.4.5 What Computational Paths Cannot Achieve (and Why)

Computational paths cannot achieve computational univalence. The obstacle is not a limitation of our rewrite rules but a fundamental incompatibility with our design choice to use an extensional equality supporting UIP at the ambient level.

The chain of reasoning is (cf. Remark 14.1 and Chapter 12):

1. We use an ambient equality that satisfies UIP.
2. All *meta-level* equality proofs of $A = B$ are equal; consequently every path in $\mathsf{Path}(A, B)$ carries the same ambient equality witness, namely reflexivity. Note: this does *not* mean that distinct path values are identified—they can still differ in their step traces.
3. Transport along such a path is the identity map (since $\mathsf{transport}(\mathsf{rfl}, x) = x$).
4. Therefore transport cannot compute a non-trivial equivalence.

To achieve computational univalence, we would need:

- A custom equality type that does *not* satisfy UIP.
- Type-level paths that genuinely differ from reflexivity.
- A transport operation that consults the path structure to compute non-trivially.

This would require rebuilding the framework in a different type theory (e.g., implementing cubical TT in an appropriate proof assistant or using a cubical proof assistant like Cubical Agda or Arend).

14.4.6 Cubical Ideas That Could Transfer

Despite the fundamental incompatibility regarding univalence, several cubical ideas could inspire extensions of computational paths:

Composition structures. Cubical TT has explicit "composition" operations (the comp and hcomp primitives) that fill higher-dimensional cubes. Our rewrite rules play a similar role but are less systematic. A more systematic treatment of higher compositions, perhaps inspired by the cubical composition laws, might simplify our coherence proofs.

Face lattice structure. The interval $\mathbb{I}$ in cubical TT comes with a rich structure of faces and boundaries, organised into a De Morgan algebra. Some of this structure could be mirrored in how we organize $\mathsf{Derivation}_n$ cells.

Kan operations. Cubical TT requires all types to have "Kan operations" (fillings for open boxes). Our higher-filler results (Chapters 8 and 9) are related—they guarantee the existence of fillers for certain configurations—but the connection is only partial. In the current formal development, the main level-3 route proceeds by isolating the inverse loop determined by two parallel 2-cells, normalising that loop, and comparing it structurally inside the derivation tower. However, the exported core connector still retains a residual fallback through propositional equality of the underlying rewrite equivalences; this is not a cubical Kan operation, and levels 4 and above are instead filled by generic higher-dimensional fillers.

14.5 Open Problem: Can Computational Paths Model Univalence?

We conclude this chapter by discussing an open question: is there a variant of computational paths that can model univalence while retaining the advantages of explicit rewrite witnesses and decidable rewrite equivalence?

14.5.1 The Challenge

The challenge is to find a design that achieves both:

- **Decidability:** Rewrite equivalence (or an equally usable quotient-level path comparison) should be decidable, enabling automatic simplification and proof checking.

- **Univalence:** Equivalent types should have non-trivial paths between them, enabling transport of structure.

These desiderata appear to be in tension. Decidability typically requires normalisation, which requires unique normal forms, which requires collapsing enough paths. But univalence requires paths to remain distinct when they correspond to distinct equivalences.

14.5.2 Partial Univalence Outside the Present Framework

Within the framework developed in this book, Chapter 12 has already shown that every nonempty type carries genuinely distinct computational paths. So there is no substantial internal class of nonempty "set-truncated" types on which one could hope to impose a restricted univalence principle. Any partial-univalence proposal must therefore be read as a target for a *modified* framework rather than as a live option inside the present one.

Definition 14.25 (Partial univalence for sets)**.** We say *partial univalence for sets* holds if, for any set-truncated types A and B (types where all paths between elements are equal):

- Every equivalence $e : A \simeq B$ induces a path $\mathsf{ua}(e) : A = B$.
- $\mathsf{idtoequiv}(\mathsf{ua}(e))$ gives back e up to RwEq-equivalence on the component paths.

Here $A = B$ denotes the *ambient (strict) equality* of types, not the computational path type $\mathsf{Path}(A, B)$. The distinction is crucial: Theorem 14.14 shows that full univalence fails precisely because the ambient equality, satisfying UIP, is too coarse to carry the information of a non-trivial equivalence.

This definition should therefore be read counterfactually: it describes what one might ask for in a variant of the theory with a genuine 0-truncated fragment. In such a setting the obstruction from Theorem 14.14 would be less severe. In the current framework, however, even this weaker principle fails because of UIP at the type level (the ambient equality between types admits only reflexivity; see Remark 14.1).

14.5.3 Alternative Approaches

Several alternative approaches might be worth exploring in future work:

Stratified universes. Use different universes for "computational" types (with decidable equality and UIP) and "homotopical" types (with univalence but without decidability). Paths within computational types have traces; paths between homotopical types behave abstractly. This mirrors how mathematicians sometimes work with both "concrete" and "abstract" equality.

Quotiented paths. Define a quotient of the path type that identifies paths up to some coarser equivalence than syntactic equality. The quotient might have decidable equality while the full path type retains homotopical information.

Directed type theory. Replace symmetric paths with directed paths (morphisms that may not be invertible). This is the setting of directed type theory [100], where univalence takes a different form (the "transport" of structure becomes functorial action).

Two-level type theory (2LTT). Voevodsky's two-level type theory [157] distinguishes "strict" and "non-strict" types. Computational paths could live at the strict level (with UIP); homotopical phenomena at the non-strict level (with univalence).

14.5.4 What Would Be Needed

To properly model univalence while retaining some form of decidability, we would likely need:

1. **A custom identity type:** Not the ambient built-in equality (which satisfies UIP), but a custom type that can have non-trivial inhabitants at the type level.
2. **Type-level step traces:** Just as paths between terms have step traces recording rewrites, paths between types would need traces recording the equivalence that justifies the identification.
3. **Normalization for type-level paths:** A rewrite system for type-level paths that is confluent and terminating, at least for some fragment.
4. **Transport that uses traces:** Transport along a type-level path should consult the trace to determine the action on elements, not simply be the identity.

This is a substantial undertaking. It would essentially require implementing a cubical-style type theory with explicit rewrite rules, combining the computational content of both approaches.

14.5.5 Current Status

As of this writing, the question remains open. The current formal development includes:

- A proof that full univalence fails for the present framework.
- Partial results on restricted univalence principles.
- A path-equivalence structure that mimics HoTT equivalences.

Whether a satisfactory synthesis of decidability and univalence is possible remains an interesting direction for future research.

14.5.6 A Philosophical Remark

The tension between computational paths and univalence reflects a genuine dichotomy in the foundations of mathematics:

- **The constructive perspective:** Proofs carry computational content; equality is decidable where possible; different constructions are distinguishable.
- **The extensional perspective:** Equivalent things are equal; structure transfers across equivalences; only observable differences matter.

Classical set-theoretic mathematics dissolves the tension by treating equality as a two-valued relation without computational content. HoTT commits to the extensional view, collapsing equivalence into equality. Computational paths commit to the constructive view, expanding equality into explicit computation.

The most promising future direction is a framework that accommodates both views, letting the user choose the appropriate level of abstraction for each application. Such a framework would distinguish "observable equality" (extensional, univalent) from "computational equality" (intensional, decidable), with controlled passage between them.

Summary

This chapter compared computational paths to homotopy type theory, making clear both the common ground and the decisive differences. The comparison matters precisely because the two frameworks are close enough to invite confusion and far enough apart to force real choices.

Common Ground. The overlap is substantial. Both frameworks:

- Organize identity proofs into higher structure (groupoids, ω-groupoids).
- Support the J-eliminator and path induction.
- Prove the groupoid laws and functoriality of congruence.
- Connect type theory to homotopy theory.

What Computational Paths Add. Computational paths make proof history explicit:

- Explicit rewrite witnesses (step traces).
- Decidable rewrite equivalence via normalisation.
- Concrete computational content for path operations.
- A two-level structure separating computational traces from proof-irrelevant meta-level equality (Remark 14.1).

What HoTT Adds. Homotopy type theory, by contrast, expands the expressive horizon:

- The univalence axiom (equivalent types are equal).
- Higher inductive types (types with path constructors).
- A uniform treatment of all levels of identity types.
- Synthetic homotopy theory.

The Fundamental Incompatibility. Computational paths accept UIP at the meta-level—meaning the underlying ambient equality witness is unique—while HoTT rejects UIP, allowing genuinely distinct identity proofs (see Chapter 12 for how path *values* nonetheless remain distinct). This is not a bug but a design choice: meta-level UIP enables decidable rewrite equivalence and quotient-level equality, but precludes non-trivial type-level paths.

Cubical Type Theory. Cubical TT is a computational refinement of HoTT that achieves computational univalence via interval types and Glue types. It represents a different point in the design space: geometric (paths as functions) rather than algebraic (paths as rewrite traces).

Open Problem. Whether there exists a variant of computational paths that supports some form of univalence while retaining decidability remains an open question, and one of the most natural directions for future research.

Historical Notes.

Homotopy type theory emerged from Voevodsky's programme on univalent foundations, beginning around 2006. The first systematic presentation was the HoTT Book [150], a collaborative effort by dozens of researchers published in 2013. Key early papers include Voevodsky's notes on univalent foundations [156] and Awodey–Warren's interpretation of identity types in homotopy theory [7].

The ω-groupoid interpretation of type theory was developed independently by Lumsdaine [105] and van den Berg–Garner [153]. This work showed that the identity types of any type form a weak ω-groupoid, a result we paralleled in Chapter 9 for computational paths.

Cubical type theory originated with Bezem–Coquand–Huber [13] and was refined in the CCHM paper [23] and cartesian variants [4]. It is implemented in Cubical Agda [155] and the Arend proof assistant.

The relationship between HoTT and computational approaches to equality has been explored from several directions. Altenkirch–Kaposi [1] developed quotient inductive types that combine ideas from both traditions. Angiuli et al. [3] showed that cubical type theory provides a computational interpretation of homotopy type theory.

Two-level type theory was proposed by Voevodsky [157] as a way to combine strict and homotopical reasoning, and has since been formalised and studied by various researchers.

Formalization Notes. Several central comparison constructions from this chapter are formalised in the following files of the accompanying Lean development (`https://github.com/Arthur742Ramos/ComputationalPathsLean`):

- `Comparison/HoTTComparison.lean` — The translation dictionary between HoTT-style identity types and computational paths. Includes `HoTTId`, `hottJ`, `CompPathId`, and comparison functions. Also defines `CompPathsStrengths`, recording what computational paths add (explicit traces, decidability, rewriting theory).
- `Comparison/UnivalenceAnalog.lean` — The `PathEquiv` structure (computational-paths analogue of equivalence), the `idToEquiv` function, and `full_univalence_fails`, the proof that full univalence is incompatible with our framework. Contains partial univalence results for sets.

- `Path/TypeTheory/UnivalencePaths.lean` — transport- and equivalence-based infrastructure for the univalence comparison
- `Path/Homotopy/HoTT.lean` — HoTT-style transport, homotopy, equivalence, and weak function-extensionality infrastructure

Key definitions. The HoTT J-eliminator is derivable from pattern matching:

```
/-- HoTT-style J eliminator -/
noncomputable def hottJ {A : Type u} {a : A}
    (C : (b : A) -> HoTTId a b -> Sort v)
    (crefl : C a rfl) {b : A} (p : HoTTId a b) : C b p :=
        by
  cases p
  exact crefl
```

The path-equivalence structure for computational paths:

```
/-- Path-equivalence: maps with comp-path round-trips -/
structure PathEquiv (alpha beta : Type u) : Type (u + 1)
   where
  toFun   : alpha -> beta
  invFun : beta -> alpha
  left_inv   : forall x : alpha, Path (invFun (toFun x)) x
  right_inv : forall y : beta, Path (toFun (invFun y)) y
```

The formalisation demonstrates both the connections (J-elimination, groupoid laws transfer) and the divergence (univalence failure, decidability vs. homotopical content) between the two approaches.

Chapter 15

Connections to Category Theory and Higher Algebra

"Category theory is the mathematics of mathematics."

—Eugenia Cheng

Throughout this book, we have developed the theory of computational paths from the inside: paths arise from rewrite steps, organise themselves into groupoids (Chapter 6), extend to 2-cells and 3-cells (Chapters 7–8), and ultimately form a weak ω-groupoid (Chapter 9). That development has been largely self-contained, requiring only the language of term rewriting and some basic categorical notions.

The present chapter steps outside and looks back. The constructions developed earlier are not ad hoc. They participate in well-established categorical frameworks, and recognising this participation yields concrete benefits—it clarifies the abstract status of what the core chapters have proved, it connects the rewrite theory to a large body of existing tools and coherence results, and it identifies precisely where further categorical development remains to be done.

Each identification below anchors a different layer of the path structure—generators, hom-objects, composition directions, finiteness conditions, and higher-dimensional completion—in a categorical concept that has its own body of theorems. The payoff is that results proved once in the abstract setting transfer immediately to computational paths:

(i) The path groupoid $\Pi_1(A)$ is the *free groupoid on a computad* (Section 15.1)—a presentation by generators and relations that category theorists use to build groupoids from combinatorial data.

(ii) The hom-objects $\mathrm{Hom}(a, b)$ carry groupoid structure themselves, making the path structure a *2-groupoid* (Section 15.2)—a category whose hom-objects are themselves groupoids.

(iii) The rewrite theory also admits a formal finite package inspired by *Squier's theorem* on finite derivation types (Section 15.3), packaging finite generating data for steps, critical pairs, and homotopy-basis bounds.

(iv) The double-source structure of 2-cells admits a *double category* and *double groupoid* interpretation (Section 15.4)—capturing the two natural directions in which 2-cells can be composed.

(v) The weak ω-groupoid and its stabilised higher structure connect to the homotopy hypothesis and ∞-categorical foundations (Section 15.5)—the modern framework for higher category theory.

Why does any of this matter? Because category theory is, at bottom, the study of how structured things relate to other structured things—and computational paths are structured things par excellence. At a practical level, the categorical identifications provide *independent verification*: our constructions instantiate well-studied abstract patterns, and the fact that they do so lends confidence in their correctness. More substantially, they *unlock existing tools*. Coherence theorems, classification results, and computational techniques from higher category theory become directly available for reasoning about computational paths—one need not re-derive them from scratch. Finally, the categorical viewpoint *suggests generalisations* that would be difficult to see from the rewrite-theoretic side alone, such as directed paths (where inverses are not required) and higher-dimensional analogues beyond the type-theoretic setting.

15.1 The Path Groupoid as a Free Groupoid on a Computad

The groupoid $\Pi_1(A)$ of Chapter 6 arose from quotienting raw computational paths by rewrite equivalence. That construction was entirely internal to type theory: we defined objects, morphisms, composition, and inverses, and verified the groupoid laws one by one. But there is a question the internal construction leaves unanswered: *what kind* of groupoid is $\Pi_1(A)$? Is it an arbitrary groupoid that happens to satisfy the axioms, or does it occupy a special position in the landscape of all groupoids?

The answer is the latter. The groupoid $\Pi_1(A)$ is *free*: it is generated by a collection of atomic edges (the elementary rewrite steps) and relations (the 77 rewrite rules), with no additional identifications imposed. We find this characterisation particularly illuminating because it explains, at a single stroke, why the word problem for paths is decidable, why induction on generators is a valid proof strategy, and why the combinatorial data of the rewrite system suffice to pin down the groupoid completely. Freeness is not a structural curiosity—it is the reason the entire machinery works.

The explanation of how this construction instantiates a general categorical pattern—*free groupoids on computads* (also called 1-polygraphs or 1-computads)—is the business of the present section.

15.1.1 Graphs, Categories, and Groupoids

We begin by recalling the "algebraicist's" hierarchy of structures built from directed graphs. This progression—from graphs to categories to groupoids—is fundamental to understanding how computational paths fit into the broader categorical landscape. Each step in the hierarchy adds more algebraic structure, and each step corresponds to a familiar operation on computational paths.

A *graph* is the simplest possible structure: just objects (vertices) and arrows (edges) between them. There is no composition, no identities, no algebraic structure whatsoever.

Definition 15.1 (Graph). A *graph* consists of a set V of vertices and a set E of edges, together with source and target functions $s, t : E \to V$ assigning to each edge its starting and ending vertices.

Given a graph, we can freely generate a category by allowing edges to be composed. The morphisms of this free category are *paths in the graph*: finite sequences of edges where each edge starts where the previous one ended.

Definition 15.2 (Free category on a graph). Given a graph (V, E, s, t), the *free category* $F(V, E)$ has:

(i) Objects: the vertices V.

(ii) Morphisms from a to b: finite sequences of composable edges

$$a = v_0 \xrightarrow{e_1} v_1 \xrightarrow{e_2} \cdots \xrightarrow{e_n} v_n = b.$$

(iii) Composition: concatenation of sequences.

(iv) Identities: empty sequences.

The category axioms (associativity, unitality) hold *strictly* because concatenation is associative and the empty sequence is a two-sided unit for concatenation.

A groupoid goes one step further: it allows edges to be traversed in *either* direction. Formally, we adjoin an inverse for every edge and impose the groupoid equations.

Definition 15.3 (Free groupoid on a graph). The *free groupoid* $G(V, E)$ on a graph (V, E, s, t) is constructed similarly, but edges may be traversed in either direction. Formally, we adjoin a formal inverse $e^{-1} : t(e) \to s(e)$ for each edge e, and morphisms are equivalence classes of sequences under the relations

$$e \cdot e^{-1} \sim \mathrm{id}_{s(e)},$$
$$e^{-1} \cdot e \sim \mathrm{id}_{t(e)}.$$

This construction is universal: any graph homomorphism $\phi : (V, E) \to \mathcal{G}$ into a groupoid $\mathcal{G}$ extends uniquely to a functor $\tilde{\phi} : G(V, E) \to \mathcal{G}$. This universal property is the hallmark of "free" constructions in category theory.

15.1.2 Computads (1-polygraphs)

The free groupoid on a graph, however, is not the end of the story. A free groupoid on a graph treats *every* pair of distinct paths as genuinely different. But in our setting, the 77 rewrite rules declare certain pairs of paths to be equivalent. To capture this, we need a structure that packages not only the generators (edges) but also the relations (equivalences between paths). This is precisely what a computad provides.

A *computad* (or *1-polygraph*) is a graph equipped with additional 2-dimensional data: relations on paths. Just as a group presentation consists of generators and relations, a computad presents a groupoid by specifying its edges and the identifications between paths.

Definition 15.4 (Computad). A *computad* $\mathcal{C} = (V, E, R, \sigma, \tau)$ consists of:

(i) A graph (V, E, s, t) (the 0-cells and 1-cells).

(ii) A set R of *relations* (the 2-cells).

(iii) Functions $\sigma, \tau : R \to \mathrm{Paths}(V, E)$ assigning to each relation its *source* and *target* paths, where both paths have the same endpoints.

The relations in a computad say: "these two paths, though perhaps syntactically different, should be treated as equivalent." The free groupoid on the computad then identifies all such pairs.

Example 15.5 (The circle as a computad). Consider the computad with one vertex $*$, one edge $\ell : * \to *$ (the loop), and no relations. The free groupoid on this computad is $\mathbf{B}\mathbb{Z}$: the groupoid with one object whose automorphism group is $\mathbb{Z}$. Each integer n corresponds to going around the loop $|n|$ times, forward if $n > 0$, backward if $n < 0$.

Example 15.6 (The torus as a computad). The torus can be presented by the computad with one vertex $*$, two edges $a, b : * \to *$, and one relation $r : a \cdot b \sim b \cdot a$. This relation says that going around the a-loop then the b-loop is equivalent to going around the b-loop then the a-loop. The resulting groupoid has $\pi_1 \cong \mathbb{Z} \times \mathbb{Z}$—the fundamental group of the torus.

Definition 15.7 (Free groupoid on a computad). Given a computad $\mathcal{C} = (V, E, R, \sigma, \tau)$, the *free groupoid* $G(\mathcal{C})$ is the quotient of the free groupoid $G(V, E)$ by the relations

$$\sigma(r) \sim \tau(r) \qquad \text{for all } r \in R.$$

This is the "most general" groupoid built from the given generators (edges) and relations.

15.1.3 Computational Paths as a Computad

We now arrive at the key observation of this section: the computational path structure on a type A is precisely a computad, and the fundamental groupoid $\Pi_1(A)$ is the free groupoid on this computad.

Fix a type A with its term-rewriting theory.

Construction 15.8 (The computational path computad). The *computational path computad* $\mathcal{C}_A$ has:

(i) **0-cells:** Terms of type A.

(ii) **1-cells:** Atomic rewrite steps $a \to b$ arising from the base reductions (β, η, etc.) and their contexts.

(iii) **2-cells:** The 77 rewrite rules of Chapter 4, each identifying a pair of parallel paths.

The generators (1-cells) are the basic computational moves, and the relations (2-cells) are the rewrite rules that tell us when two paths should be considered equivalent.

Proposition 15.9 (Path groupoid as free groupoid on computad). *The fundamental groupoid $\Pi_1(A)$ from Chapter 6 is isomorphic to the free groupoid on the computad $\mathcal{C}_A$:*

$$\Pi_1(A) \cong G(\mathcal{C}_A).$$

Proof. Start with the underlying directed graph of $\mathcal{C}_A$: its vertices are the terms of type A, and its edges are the elementary rewrite steps. By Theorem 6.48, the groupoid generated from these edges using identities, inverses, and composition is the free groupoid on that graph.

Passing from the free groupoid to $G(\mathcal{C}_A)$ imposes exactly the 2-cells of the computad, namely the 77 rewrite rules. But this is also how the fundamental groupoid $\Pi_1(A)$ is defined: its morphisms are computational paths modulo the congruence generated by those same rewrite rules. Therefore both constructions have the same objects and the same hom-sets, namely paths built from elementary steps modulo the same relations.

The identity-on-objects map sending a computational path to its class in $G(\mathcal{C}_A)$ is therefore well-defined and bijective on every hom-set. It respects composition because both sides compose by concatenating path expressions and then passing to the quotient by the same congruence. Hence it is an isomorphism of groupoids. □

15.1.4 Consequences of the Free Groupoid Perspective

The identification of $\Pi_1(A)$ with a free groupoid on a computad has several important consequences that connect our type-theoretic development to mainstream categorical methods.

Example 15.10 (The circle's path groupoid as free groupoid on a computad)**.** To see Proposition 15.9 in action, consider the circle S^1 as presented in Chapter 13, Section 13.2: one point $\star$ and one loop generator $\ell : \star \to \star$, with no 2-cell relations.

The computational path computad $\mathcal{C}_{S^1}$ has:

(i) one 0-cell: $\star$;

(ii) one 1-cell: ℓ;

(iii) no 2-cells (beyond the groupoid identities).

This is exactly the computad of Example 15.5 above. The free groupoid $G(\mathcal{C}_{S^1})$ has one object and its automorphism group is the free group on one generator—that is, $\mathbb{Z}$. On the other hand, the fundamental group computation of Chapter 13 yields $\pi_1(S^1, \star) \cong \mathbb{Z}$ by explicit normal-form analysis: every loop expression reduces to ℓ^n for a unique $n \in \mathbb{Z}$.

Proposition 15.9 explains *why* the normal-form argument works: the computational path groupoid is *free* on the computad, so its automorphism group at $\star$ is the free group on the generators, which is $\mathbb{Z}$. The confluence and termination machinery of Chapter 5 then makes this freeness computationally effective by providing a decision procedure for the word problem.

The same pattern governs every space in Chapter 13: the torus computad (Example 15.6) has two loops and one commutativity relation, yielding $\pi_1 \cong \mathbb{Z} \times \mathbb{Z}$; the projective-plane computad has one loop and a torsion relation $\ell \cdot \ell \sim \mathrm{id}_\star$, yielding $\pi_1 \cong \mathbb{Z}/2\mathbb{Z}$. In each case, the combinatorial presentation of Chapter 13 is precisely the computad data, and the fundamental group is the automorphism group of the free groupoid on that computad.

Corollary 15.11 (Universal property). *For any groupoid $\mathcal{G}$ and any assignment of morphisms in $\mathcal{G}$ to the generators (rewrite steps) of $\mathcal{C}_A$ that respects the 77 relations, there exists a unique functor $\Pi_1(A) \to \mathcal{G}$ extending the assignment.*

This universal property is the abstract encapsulation of "proofs by induction on path structure." When we prove a theorem about paths by case analysis on how the path was constructed, we are implicitly using this universal property.

Corollary 15.12 (Decidability of the word problem). *Since the rewrite system on paths is confluent and terminating (Chapter 5), the word problem for $\Pi_1(A)$ is decidable: given two paths $p, q : a \rightsquigarrow b$, we can effectively determine whether $[p] = [q]$ in $\Pi_1(A)$ by reducing both to normal form and comparing.*

This is a strong property not shared by all finitely presented groupoids. The word problem for groups is famously undecidable in general; our computational paths escape this trap because of the confluence and termination properties of the rewrite system.

Corollary 15.13 (Finite presentation). *If the base type theory has finitely many reduction rules and the type A is finitely presented (finitely many constructors and definitional equalities), then $\Pi_1(A)$ is finitely presented as a groupoid.*

15.2 Enrichment and 2-Groupoid Structure

The previous section showed that $\Pi_1(A)$ is a free groupoid on a computad—but in doing so it deliberately threw information away. It treated the hom-sets $\mathrm{Hom}(a, b)$ as bare sets, discarding the 2-cell structure that Chapters 7 and 8 worked so hard to develop. This was a useful simplification for the computad story, which captures only the 1-dimensional skeleton. But any account that stops here misrepresents the richness of the actual path algebra.

To bring the higher cells back into view, we need a framework in which the hom-*sets* are replaced by hom-*objects* carrying their own algebraic structure. This is precisely what enriched category theory provides. When the hom-objects are themselves groupoids—as they are for computational paths, where the 2-cells between any two parallel paths form a groupoid under vertical composition—the resulting structure is called a *2-groupoid*. The payoff for the reader is a single coherent framework that organises all three levels (terms, paths, 2-cells) into one algebraic package, rather than treating them as three loosely related layers.

Chapter 6 showed that computational paths form a groupoid. Chapter 7 then revealed that paths between paths (2-cells) also carry groupoid structure. The natural categorical framework for capturing both levels simultaneously is that of *enriched categories* and *2-groupoids*.

15.2.1 Enriched Categories

Ordinary categories have hom-*sets*: for each pair of objects a, b, the collection $\mathrm{Hom}(a, b)$ is a set. An *enriched category* replaces these sets by objects of some ambient monoidal category $\mathcal{V}$, allowing the hom-data to carry structure of its own. The richer the ambient category, the richer the questions one can ask.

For instance, enriching in metric spaces lets one measure how "close" two morphisms are; enriching in groupoids—the case relevant to us—lets one compare morphisms to each other via invertible 2-cells. The general definition makes this precise.

Definition 15.14 ($\mathcal{V}$-enriched category)**.** Let $(\mathcal{V}, \otimes, I)$ be a monoidal category. A *$\mathcal{V}$-enriched category $\mathcal{C}$* consists of:

(i) A collection $\mathrm{Ob}(\mathcal{C})$ of objects.

(ii) For each pair a, b, a *hom-object* $\mathcal{C}(a, b) \in \mathrm{Ob}(\mathcal{V})$.

(iii) For each triple a, b, c, a *composition morphism*

$$\circ_{a,b,c} : \mathcal{C}(b, c) \otimes \mathcal{C}(a, b) \to \mathcal{C}(a, c)$$

in $\mathcal{V}$.

(iv) For each object a, an *identity morphism* $\mathrm{id}_a : I \to \mathcal{C}(a, a)$ in $\mathcal{V}$.

These data satisfy associativity and unitality axioms expressed as commutative diagrams in $\mathcal{V}$.

Example 15.15 (Categories enriched in sets)**.** When $\mathcal{V} = \mathbf{Set}$ with Cartesian product, a $\mathbf{Set}$-enriched category is an ordinary (locally small) category. This is the "baseline" case—ordinary category theory is enriched category theory over sets.

The case that matters for computational paths is enrichment over groupoids. Recall that the paths from a to b can be compared by 2-cells, and that these 2-cells compose and invert (Chapter 7). This is exactly the data of a hom-groupoid, so the path structure is naturally a category enriched in $\mathbf{Gpd}$.

Example 15.16 (Categories enriched in groupoids). When $\mathcal{V} = \mathbf{Gpd}$ (the category of groupoids) with Cartesian product, a $\mathbf{Gpd}$-enriched category is called a *2-groupoid*: each hom-object $\mathcal{C}(a, b)$ is itself a groupoid, and composition is functorial.

15.2.2 The 2-Groupoid of Computational Paths

We now make precise the claim that computational paths on a type A form a 2-groupoid. The construction packages the data from Chapters 6 and 7 into the enriched category framework.

Definition 15.17 (The weak 2-groupoid $\widetilde{\Pi}_2(A)$). For a type A, we define the *weak 2-groupoid* $\widetilde{\Pi}_2(A)$ by assembling the data from Chapters 6 and 7 into the enriched-category framework.

(i) **Objects:** Terms $a : A$.

(ii) **Hom-groupoid $\widetilde{\Pi}_2(A)(a, b)$:** For each pair of terms, the hom-groupoid has paths $p : a \rightsquigarrow b$ as its objects and 2-cells $\alpha : p \Rrightarrow q$ as its morphisms. Vertical composition provides the groupoid multiplication; reflexivity 2-cells serve as identities; and every 2-cell is invertible by Proposition 7.8.

(iii) **Composition functor:**

$$\circ : \widetilde{\Pi}_2(A)(b, c) \times \widetilde{\Pi}_2(A)(a, b) \to \widetilde{\Pi}_2(A)(a, c)$$

On objects (1-cells), this is path concatenation $(q, p) \mapsto q \cdot p$. On morphisms (2-cells), it is horizontal composition $(\beta, \alpha) \mapsto \beta * \alpha$.

(iv) **Identity functor:** $\mathrm{id}_a = \mathrm{refl}_a$ with identity 2-cell.

Theorem 15.18 ($\widetilde{\Pi}_2(A)$ is a 2-groupoid). *The structure $\widetilde{\Pi}_2(A)$ defined above is a 2-groupoid: the composition functor is associative up to natural isomorphism (the associator 2-cells), and identities are units up to natural isomorphism (the unitor 2-cells).*

Proof. This follows from the results of Chapters 7 and 8. Specifically:

(i) Each hom-groupoid is indeed a groupoid by Proposition 7.8.

(ii) Horizontal composition is functorial by Proposition 7.18.

(iii) Associativity and unitality hold up to 2-cells, which are the coherence data of Chapter 8.

□

15.2.3 Strict vs. Weak 2-Groupoids

An important distinction arises between *strict* and *weak* 2-groupoids. Understanding this distinction is crucial for appreciating the relationship between syntactic and semantic structures.

In a *weak* 2-groupoid, the associativity law $(p \cdot q) \cdot r = p \cdot (q \cdot r)$ holds only "up to isomorphism"—there is a specified 2-cell (the *associator*) witnessing the equivalence, but the two sides are not literally equal. This is the natural state of affairs when working with concrete syntax: the expressions $(p \cdot q) \cdot r$ and $p \cdot (q \cdot r)$ are syntactically distinct, even if they represent the same abstract path.

Definition 15.19 (Strict 2-groupoid). A *strict 2-groupoid* is a 2-groupoid in which associativity and unitality hold as *equalities*, not merely up to isomorphism.

The 2-groupoid $\widetilde{\Pi}_2(A)$ is *not* strict: the associativity $(p \cdot q) \cdot r = p \cdot (q \cdot r)$ holds only up to a rewrite-equivalence 2-cell, not as a literal equality of paths. However, after quotienting by rewrite equivalence, we obtain a strict 2-groupoid.

Proposition 15.20 (Strictification). *Let $\Pi_2(A)$ denote the quotient of $\widetilde{\Pi}_2(A)$ by rewrite equivalence at both the 1-cell and 2-cell levels. Then $\Pi_2(A)$ is a strict 2-groupoid.*

The passage from weak to strict is an instance of a general coherence theorem (Mac Lane–Paré coherence for bicategories). The weak structure $\widetilde{\Pi}_2(A)$ is the "natural" one arising from the syntax; the strict quotient $\Pi_2(A)$ is computationally simpler but loses some proof-relevant information.

15.2.4 The Gray Tensor Product

In a strict 2-category, the *interchange law* holds exactly:

$$(\alpha * \beta) \cdot (\gamma * \delta) = (\alpha \cdot \gamma) * (\beta \cdot \delta).$$

This says that we can "slide" horizontal and vertical compositions past each other. But in many natural examples—including computational paths—this law holds only up to a higher cell.

The *Gray tensor product* provides the right framework for this situation. It is a way of "tensoring" 2-categories that weakens the interchange law.

Definition 15.21 (Gray-category). A *Gray-category* is a category enriched in the monoidal category $(\mathbf{2\text{-}Cat}, \otimes_{\mathrm{Gray}}, \mathbf{1})$ where $\otimes_{\mathrm{Gray}}$ is the Gray tensor product. Concretely, a Gray-category has:

(i) Objects, 1-morphisms, and 2-morphisms as in a 2-category.

(ii) The interchange law

$$(\alpha * \beta) \cdot (\gamma * \delta) = (\alpha \cdot \gamma) * (\beta \cdot \delta)$$

holds up to a specified invertible 3-cell (the *interchanger*), rather than strictly.

Theorem 15.22 (Path structure as Gray-category). *Computational paths on a type A carry the structure of a Gray-category: the interchange law holds up to the explicit 3-cells constructed in Chapter 8.*

Proof. This follows from Theorem 7.32 and the construction of interchange witnesses as 3-cells. The interchanger 3-cells satisfy the required coherence conditions by the contractibility results of Chapter 8. □

The formalisation includes an explicit `GrayCategory` structure built from computational paths, with pentagon and triangle coherence 3-cells witnessing the required axioms.

The enriched perspective just developed tells us *what* algebraic structure the path groupoid carries: it is a 2-groupoid whose hom-objects are themselves groupoids, with interchange governed by the Gray tensor product. But it does not address a different question: *how much data* is needed to specify this structure? Is the path 2-groupoid, with its infinitely many paths and 2-cells, determined by some finite combinatorial core?

This is precisely the question that Squier's theorem answers. Where the enriched viewpoint organises the path structure *algebraically*, Squier's theorem analyses it *combinatorially*, showing that the confluence and termination properties of the rewrite system guarantee that a finite package of steps and critical-pair fillers suffices to generate all of the higher coherence data. The two perspectives are complementary: enrichment tells us the structure is a 2-groupoid, while Squier tells us it is a *finitely presented* one.

15.3 Relation to Squier's Theorem

One of the deepest connections between term rewriting and algebraic topology is Squier's theorem. It addresses a question that may seem surprising at first: what does the *shape* of proof space have to do with whether two expressions can be shown equivalent? The answer—and we find it a beautiful one—is that confluence and termination, properties we established for our system by hard combinatorial work in Chapter 5, are not merely computational conveniences. They have topological consequences: they guarantee that the space of all possible proofs of equivalence is "tame" in a precise homological sense.

For computational paths, this means that the entire infinite tower of 2-cells, 3-cells, and higher coherence data is controlled by a finite combinatorial skeleton. The reader gains from this perspective a concrete bound on the complexity of the path theory: although the rewrite system generates infinitely many paths and derivations, all of them are assembled from finitely many building blocks.

What follows is a self-contained treatment of the theorem and its relevance to computational paths.

15.3.1 Finite Derivation Type

Consider a monoid M presented by generators and relations. When we ask whether two words represent the same element, we are asking whether there is a *derivation*—a chain of rewrite steps—connecting them. But derivations themselves have structure: two derivations connecting the same pair of words might be "homotopic" (deformable into each other) or not.

The *derivation type* of M captures the "shape" of all possible proofs that two words represent the same element. Think of it as the topological complexity of proof space.

Definition 15.23 (Finite derivation type (FDT)). A monoid presentation (X, R) has *finite derivation type* (FDT) if there exists:

(i) A finite set of generating 1-cells: the relations R.

(ii) A finite homotopy basis of generating 2-cells: a finite family of critical-pair resolutions and coherence fillers from which every parallel derivation is generated up to composition.

Formally, FDT means that the associated Squier complex admits a finite homotopy basis, not merely that the presentation complex has a finite 2-skeleton.

The intuition is this: a presentation has FDT if the "proof space" can be triangulated with finitely many 2-cells. Any proof that two words are equal can be decomposed into finitely many basic pieces.

Theorem 15.24 (Squier, 1987). *If a monoid M admits a finite complete (= confluent and terminating) rewrite system, then M has finite derivation type.*

This is a remarkable result: it says that good computational properties (confluence and termination) imply good topological properties (finite derivation type). The converse is false: there exist finitely presented monoids with FDT that admit no finite complete rewrite system.

15.3.2 Homological Consequences

The significance of FDT lies in its homological implications. Homological finiteness conditions measure how "tame" an algebraic structure is from the perspective of homological algebra.

Definition 15.25 (Type FP_n)**.** A group (or monoid) G is of *type* FP_n if there exists a projective resolution

$$\cdots \to P_n \to P_{n-1} \to \cdots \to P_1 \to P_0 \to \mathbb{Z} \to 0$$

of the trivial $\mathbb{Z}[G]$-module $\mathbb{Z}$ with P_i finitely generated for $i \leq n$.

Intuitively, type FP_n means that the first n levels of algebraic complexity are "finite." Type FP_1 says the monoid is finitely generated; type FP_2 adds that the relations are finitely generated; type FP_3 adds that the "syzygies" (relations between relations) are finitely generated.

Theorem 15.26 (Squier–Groves)**.** *If a monoid presentation has finite derivation type, then the monoid is of type* FP_3.

15.3.3 Computational Paths and Squier's Theorem

The confluence and termination results of Chapter 5 make a Squier-style analysis natural for computational paths. This section repackages, rather than replaces, the explicit 77-rule presentation: the earlier chapters provide the concrete normalisation engine, while the present argument extracts from it the finite step data and critical-pair fillers relevant to Squier's framework.

Theorem 15.27 (Finite Squier-style package for computational paths)**.** *The computational-path rewrite system admits a finite Squier-style generation package. Specifically:*

(i) the one-step reduction relation is generated by a finite catalogue of primitive rewrite steps refining the rule presentation of Chapters 4 and 5;

(ii) the critical-pair analysis yields an explicit finite list of 33 *non-trivial overlap cases, each equipped with a chosen join;*

(iii) every rewrite-equivalence witness is obtained by composing primitive steps and their formal inverses, while the chosen critical-pair joins induce finitely many coherence fillers between the resulting parallel derivations.

Consequently the computational-path system carries finite proof-relevant coherence data in the Squier sense: a finite step catalogue together with a finite family of critical-pair fillers controlling higher rewriting.

Proof. For (i), the rewrite system of Chapters 4 and 5 is presented by finitely many primitive rule schemas, so its one-step reductions are generated by a finite catalogue of basic steps.

Clause (ii) is exactly the outcome of the local-confluence analysis in Chapter 5: among all possible rule overlaps, only 33 non-trivial schemas remain, and each is resolved there by an explicit common reduct.

For the first assertion in (iii), rewrite-equivalence is by definition the closure of one-step reduction under identity, inverse, and composition. Therefore any rewrite-equivalence witness decomposes into a finite composite of primitive steps and their formal inverses.

For the second assertion in (iii), each chosen join from (ii) determines two parallel derivations with a common target, and Theorem 8.37 turns that join into a proof-relevant coherence filler between those derivations. Since there are only finitely many critical-pair schemas, these fillers also form a finite family.

Taken together, the finite step catalogue and the finite critical-pair fillers constitute the claimed Squier-style generation package. □

Remark 15.28 (Relation to the 77-rule presentation)**.** The finite package above is an extracted summary of the concrete rewrite theory, not a second presentation replacing it. Chapters 4 and 5 remain responsible for the explicit reductions, normal forms, and joins; the present section records the finite generative data that those arguments expose.

15.3.4 The Homotopy Basis

Following Squier's framework, we can extract a *homotopy basis*: a finite set of 2-cells that generate all others. This is analogous to how a vector space basis generates all vectors, but in the higher-dimensional setting of 2-cells.

Definition 15.29 (Homotopy basis)**.** A *homotopy basis* for a rewrite system is a finite set $\mathcal{H}$ of 2-cells such that every 2-cell (derivation between paths with the same endpoints) can be expressed as a composition of cells in $\mathcal{H}$ and their inverses.

Proposition 15.30 (Homotopy basis for computational paths)**.** *The current formalisation packages the homotopy basis as the explicit list of* 33 *critical-pair cases. A uniform two-position overlap estimate then yields the numerical upper bound* 66.

The formalized analysis records this basis as an explicit list of the critical-pair cases together with a verified bound on its size.

15.3.5 Homological Shadow of FP_3

The current formalisation also packages an abstract shadow of the classical Squier–Groves implication:

Corollary 15.31 (Finite-data FP_3 shadow). *The formal development proves an abstract finiteness predicate expressing that the chosen homotopy basis admits a global numerical bound.*

This should be read as a finite-data shadow of the classical homological statement: it records bounded 2-dimensional coherence data, but it should not yet be read as a formal proof that the actual path monoid has a projective resolution with finitely generated terms through dimension 3.

15.4 Double Categories and Double Groupoids

The categorical frameworks developed so far—free groupoids on computads, enriched 2-groupoids, and Squier's finite generation packages—all organise the path structure along a single "vertical" axis: objects, then morphisms between objects, then 2-cells between morphisms. But the 2-cells of Chapter 7 have a feature that this one-dimensional hierarchy does not fully capture: they admit *two* independent directions of composition, vertical and horizontal, and these two directions interact through the interchange law.

Double categories are the categorical structure designed precisely for this situation. Where a 2-category stacks its cells in a single tower, a double category lays them out on a grid, with horizontal edges, vertical edges, and squares that mediate between the two. The resulting picture is richer and, for computational paths, more faithful to the actual algebra of 2-cells.

To visualise the idea, think of a chessboard where the horizontal and vertical edges represent different kinds of morphisms, and each square represents a 2-cell relating the two ways of getting from one corner to the opposite corner.

15.4.1 Double Categories

Definition 15.32 (Double category). A *double category* $\mathbb{D}$ is specified by the following data:

(i) A set of *objects* (0-cells).

(ii) Two kinds of *morphisms*:

- *Horizontal morphisms* $f : a \to b$, forming a category $\mathbb{D}_H$.
- *Vertical morphisms* $u : a \to c$, forming a category $\mathbb{D}_V$.

(iii) *Squares* (2-cells) α of the form

$$\begin{array}{ccc} a & \xrightarrow{f} & b \\ {\scriptstyle u}\big\downarrow & & \big\downarrow{\scriptstyle v} \\ c & \xrightarrow[g]{} & d \end{array}$$

with four boundary edges.

(iv) *Horizontal composition* of squares (stacking left-right).

(v) *Vertical composition* of squares (stacking top-bottom).

These compositions satisfy interchange and unitality axioms.

The key feature is that there are *two* independent directions of composition for the 2-cells, corresponding to the two directions on the chessboard.

Definition 15.33 (Double groupoid)**.** A *double groupoid* is a double category in which:

(i) Every horizontal morphism is invertible.

(ii) Every vertical morphism is invertible.

(iii) Every square is invertible (in both directions).

15.4.2 The Double Groupoid of Computational Paths

Computational paths naturally organize into a double groupoid structure. The two directions of composition correspond to the vertical and horizontal composition of 2-cells from Chapter 7.

Construction 15.34 (The path double groupoid $\mathbb{P}(A)$)**.** For a type A, define:

(i) **Objects:** Terms $a : A$.

(ii) **Horizontal morphisms:** Paths $p : a \rightsquigarrow b$.

(iii) **Vertical morphisms:** Also paths $u : a \rightsquigarrow c$, since the type A furnishes both directions.

(iv) **Squares:**

$$\begin{array}{ccc} a & \xrightarrow{p} & b \\ {\scriptstyle u}\big\downarrow & & \big\downarrow{\scriptstyle v} \\ c & \xrightarrow[q]{} & d \end{array}$$

where the square is filled by a 2-cell witnessing $p \cdot v \sim u \cdot q$ (both paths go from a to d).

Remark 15.35 (Thin structure)**.** In the path double groupoid, the horizontal and vertical categories are the *same* (both are $\Pi_1(A)$). This makes $\mathbb{P}(A)$ a *symmetric* or *edge-symmetric* double groupoid.

Theorem 15.36 (Double groupoid structure)**.** *$\mathbb{P}(A)$ is a double groupoid:*

(i) Horizontal and vertical composition of squares correspond to the two whiskering operations from Chapter 7.

(ii) Interchange follows from the interchange theorem (Theorem 7.32).

(iii) All morphisms and squares are invertible by the groupoid and 2-groupoid structures.

15.4.3 Connection to Crossed Modules

Double groupoids are closely related to *crossed modules*, an algebraic structure that captures the first two levels of a homotopy type. This connection is important because crossed modules are computationally tractable algebraic objects that encode significant topological information.

Intuitively, a crossed module packages a group G (think: fundamental group), a group H (think: second homotopy group), and the data describing how they interact.

Definition 15.37 (Crossed module)**.** A *crossed module* is given by groups G and H, a group homomorphism $\partial : H \to G$ (the "boundary map"), and an action of G on H by automorphisms, subject to two conditions:

(i) The *Peiffer identity*: $\partial(g \cdot h) = g\,\partial(h)\,g^{-1}$.

(ii) The *crossed module identity*: $\partial(h_1) \cdot h_2 = h_1\,h_2\,h_1^{-1}$.

Theorem 15.38 (Brown–Spencer)**.** *There is an equivalence of categories between:*

(i) Connected double groupoids with one object.

(ii) Crossed modules.

This theorem says that double groupoids and crossed modules are two different ways of packaging the same algebraic data. For computational paths based at a point a, we obtain:

Corollary 15.39 (Crossed module from paths)**.** *The loop space $\Omega(A, a)$ and second loop space $\Omega^2(A, a)$ form a crossed module:*

- *$G = \pi_1(A, a)$ (the fundamental group).*

- $H = \pi_2(A, a)$ *(the second homotopy group, which is abelian by Eckmann–Hilton).*
- $\partial : \pi_2 \to \pi_1$ *is the boundary map (trivial for the path groupoid).*
- *The action is the standard* π_1*-action on* π_2.

15.4.4 Higher Double Structures

The double category perspective extends to higher dimensions. A *triple category* would capture the three-fold composition structure of 3-cells; a *quadruple category* would handle 4-cells; and so on. The contractibility results of Chapter 8 imply that these higher structures become progressively simpler (degenerate) as dimension increases—a reflection of the fact that higher coherence data becomes increasingly constrained.

15.5 Sketch: Path ∞-Groupoids and ∞-Cosmoi

The preceding sections have situated computational paths within 1-dimensional (computads), 2-dimensional (enriched 2-groupoids, double categories), and finitary (Squier) categorical frameworks. Each of these captures part of the picture, but none of them accounts for the *full* tower of higher cells—the 3-cells, 4-cells, and beyond—that the ω-groupoid of Chapter 9 provides.

This is where honest accounting is called for. The ω-groupoid of Chapter 9 was constructed by explicit combinatorial means: we built cells level by level, verified coherence through dimension 3, and proved stabilisation from dimension 4 onward. Modern higher category theory has developed several frameworks that handle cells at all levels in a single sweep—quasicategories, Kan complexes, complete Segal spaces, and the model-independent language of ∞-cosmoi—but the connection between our globular, rewrite-based construction and these simplicial or model-categorical frameworks has not yet been made fully precise. We sketch the landscape here, noting both the points of contact that we can identify with confidence and the gaps that remain open.

15.5.1 The Homotopy Hypothesis

The *homotopy hypothesis*, proposed by Grothendieck in *Pursuing Stacks* (1983), is one of the foundational conjectures of higher category theory. It asserts that ∞-groupoids are equivalent to spaces (in the sense of homotopy theory).

The intuition is compelling, and it runs parallel to the story of this book: a space X has points (0-cells), paths between points (1-cells), homotopies between paths (2-cells), and so on, all the way up. All of these are "reversible" in an appropriate sense, so we get a groupoid structure at every level. The homotopy hypothesis says that *every* ∞-groupoid arises this way, and conversely, every space can be recovered from its ∞-groupoid of points, paths, homotopies, etc. If the hypothesis is correct, then the ω-groupoid of Chapter 9 is not merely an algebraic artefact of our rewrite system—it *is* a space, in the sense that matters to homotopy theory.

Conjecture 15.40 (Homotopy hypothesis)**.** There is an equivalence of $(\infty, 1)$-categories:

$$\infty\text{-}\mathbf{Gpd} \simeq \mathbf{Spaces}.$$

Under this equivalence, the fundamental ∞-groupoid $\Pi_\infty(X)$ of a space X corresponds to X itself, and the geometric realisation $|G|$ of an ∞-groupoid G is a space with $\Pi_\infty(|G|) \simeq G$.

In this broad model-independent form, the homotopy hypothesis is best read as a guiding slogan rather than as a single theorem with one canonical formulation. Different models of ∞-groupoids lead to different precise statements. For globular ω-groupoids (as in Chapter 9), precise versions of the equivalence were established by several authors including Ara–Maltsiniotis and Bourke–Gurski.

15.5.2 (∞, n)-Categories

The notion of ∞-groupoid is part of a more general family. An (∞, n)-*category* is a higher category where:

- k-cells for $k > n$ are all invertible.
- k-cells for $k \leq n$ may be non-invertible.

The number n controls "how much non-invertibility" is allowed.

Example 15.41 (Low-dimensional cases)**.**

- $(\infty, 0)$-categories = ∞-groupoids. All cells at all levels are invertible. This is the structure of computational paths.
- $(\infty, 1)$-categories = ∞-categories in the sense of Joyal–Lurie. 1-cells (morphisms) may be non-invertible, but all higher cells are invertible.
- $(\infty, 2)$-categories = 1-cells and 2-cells may be non-invertible, but 3-cells and higher are invertible.

Computational paths form an $(\infty, 0)$-category (i.e., an ∞-groupoid) because all paths and higher cells are invertible. This reflects the symmetric nature of definitional equality: if $a = b$, then $b = a$.

15.5.3 Quasicategories and the Joyal Model

The most widely used model for $(\infty, 1)$-categories is Joyal's *quasicategories*: simplicial sets satisfying a weak Kan condition. This model has become the foundation for much of modern higher category theory, particularly through Lurie's work.

Definition 15.42 (Quasicategory). A *quasicategory* is a simplicial set X such that every inner horn $\Lambda^n_k \to X$ (for $0 < k < n$) admits a filler $\Delta^n \to X$.

The horn-filling condition is a weak form of composition: it says that whenever we have a "partially specified" composition, we can complete it. The fundamental ∞-groupoid of computational paths *ought* to be representable as a Kan complex (a quasicategory where *all* horns, not just inner ones, have fillers), but the translation between the globular cells of our construction and the simplicial cells of a Kan complex has not been carried out in detail for the computational-path setting. The general globular-to-simplicial comparison is well-understood in the abstract [6], and we see no obstruction to instantiating it here, but the verification remains a concrete piece of future work.

15.5.4 ∞-cosmoi

Riehl and Verity's ∞-*cosmoi* provide a model-independent framework for ∞-category theory. Rather than committing to a specific model (quasicategories, complete Segal spaces, etc.), they axiomatise the properties that any "category of ∞-categories" should satisfy. For our purposes the appeal is clear: if computational path ω-groupoids can be shown to live inside an ∞-cosmos, the entire Riehl–Verity toolkit becomes available without choosing a simplicial model first.

Definition 15.43 (∞-Cosmos (informal)). An ∞-*cosmos* $\mathcal{K}$ is a category enriched in quasicategories, equipped with:

- A class of *isofibrations* (functors with good lifting properties).
- A class of *trivial fibrations*.
- Sufficient limits (products, pullbacks, cotensors).

The category of quasicategories, Segal categories, complete Segal spaces, and other models each form an ∞-cosmos. Whether the category of computational-path ω-groupoids itself forms an ∞-cosmos—or embeds into one of the standard ones in a well-behaved way—is an open question that we find genuinely interesting. The stabilisation beyond dimension 3 (Chapter 9) suggests that the answer should be positive, since the contractibility of higher coherence data mimics the horn-filler conditions, but the details have not yet been worked out.

15.5.5 Directed Computational Paths

Perhaps the most tantalising generalisation suggested by the categorical viewpoint is to drop the invertibility requirement on paths. Throughout this book, every path has an inverse—reflecting the fact that definitional equality is symmetric. But rewriting is not inherently symmetric: a β-reduction $(\lambda x. M)\, N \to M[N/x]$ has a preferred direction. Taking that directionality seriously would yield *directed computational paths* forming an $(\infty, 1)$-category rather than an ∞-groupoid.

Open Problem 15.44 (Directed paths)**.** Develop a theory of *directed computational paths* where:

(i) 1-cells are oriented rewrite sequences $a \Rightarrow b$ without formal inverses.

(ii) The groupoid laws G1–G8 are replaced by category laws.

(iii) Higher cells remain invertible, yielding $(\infty, 1)$-structure.

This would connect computational paths to the rich theory of ∞-categories with non-invertible morphisms, including the applications to stable homotopy theory and derived algebraic geometry.

15.5.6 Comparison with HoTT

The contrast with Homotopy Type Theory (HoTT) is sharpened by the categorical perspective. HoTT takes the homotopy hypothesis as a *design principle*: identity types are stipulated to behave like path spaces, and univalence is postulated to make this behaviour productive. Computational paths take the opposite approach: the ω-groupoid structure is *derived* from the term rewriting system, not postulated. Neither approach is superior in the abstract; they answer different questions. HoTT asks "what can we do if we assume homotopical structure?" while computational paths ask "where does homotopical structure come from?"

The following table summarises the key differences:

	HoTT	**Comp. Paths**
Axiom	Univalence	None beyond ordinary type theory
Identity	Iterated identity types	Explicit rewrite-carrying paths
Higher str.	By iterated identity-type structure	By rewriting plus higher derivations
Decidability	Undecidable in general	Decidable (confluence)

The computational paths approach demonstrates that ω-groupoid structure can arise purely from syntactic considerations—from rewrite rules on proof terms—without appeal to topological intuition or univalence axioms. This is, we believe, the deepest lesson of the categorical perspective: the algebra is already there in the syntax, waiting to be recognised.

Formalization Notes. The results of this chapter are formalised in the following files of the accompanying Lean development (`https://github.com/Arthur742Ramos/ComputationalPathsLean`):

- `ComputationalPaths/Path/Rewrite/SquierDeep.lean` — Squier-style finite package for steps, critical pairs, and abstract `TypeFP3`
- `ComputationalPaths/HigherCategory/TwoCategory.lean` — path-based 2-category interface
- `ComputationalPaths/HigherCategory/GrayCategory.lean` — Gray-style 3- and 4-cell packaging
- `ComputationalPaths/Path/DoubleCategoryPaths.lean` — double-category interface on path squares
- `ComputationalPaths/Path/OmegaGroupoid.lean` — the underlying derivation tower and contractibility maps

Key definitions.

Squier-style finite package.
The file `Path/Rewrite/SquierDeep.lean` packages the Squier-style analysis:

- `SquierFiniteDerivationType`: structure recording the rewrite generators together with critical pair resolvers.
- `squierFDT`: canonical finite package built from the finite list of step generators and the listed critical-pair resolvers.
- `squier_generates_all_rweq`: theorem that every `RwEq` derivation decomposes into basic generators.
- `fdt_implies_typeFP3`: theorem bridging the file's abstract `HasFDT` predicate to its abstract `TypeFP3` predicate.

Gray-category structure.
The file `HigherCategory/GrayCategory.lean` defines:

- `Hom`, `TwoCell`, `ThreeCell`, and `FourCell`: the cell interfaces used by the Gray scaffold.
- `pentagon3` and `triangle3`: canonical coherence 3-cells.
- `pentagonContraction` and `triangleContraction`: 4-cells contracting arbitrary witnesses to the canonical ones.

Double-category packaging.
The file `Path/DoubleCategoryPaths.lean` makes the double-category structure explicit:

- `PathSquare`: squares represented by rewrite-equality 2-cells.
- `DoubleCategory`: the path-based double-category interface.
- `pathDoubleCategory`: the canonical double category of computational paths.
- `pathDoubleCategory_to_twoCategory`: compatibility with the 2-category interchange law.

Key definitions. The free groupoid structure on computational paths:

```
/-- Minimal groupoid interface -/
structure CPGroupoid where
  Obj : Type u
  Hom : Obj -> Obj -> Type v
  id : (a : Obj) -> Hom a a
  comp : {a b c : Obj} -> Hom a b -> Hom b c -> Hom a c
  inv : {a b : Obj} -> Hom a b -> Hom b a
  comp_assoc : forall (f : Hom a b) (g : Hom b c) (h :
     Hom c d),
    comp (comp f g) h = comp f (comp g h)
  -- ... groupoid laws

/-- Free groupoid: morphisms are paths mod RwEq -/
noncomputable def FreeGroupoid (A : Type u) : CPGroupoid
   where
  Obj := A
  Hom := fun a b => PathRwQuot A a b
  id := fun a => PathRwQuot.refl a
  comp := fun x y => PathRwQuot.trans x y
  inv := fun x => PathRwQuot.symm x
  comp_assoc := fun x y z => free_comp_assoc x y z
```

2-groupoid enrichment. The hom-groupoids and their composition are developed in `HigherCategory/TwoCategory.lean` and `Path/OmegaGroupoid.lean`. The main exported notions here are `Hom`, `TwoCell`, `hcomp`, `interchangeIdentity`, `pentagonIdentity`, and `triangleIdentity`.

Formalisation status. The following aspects remain informal:

- The identification of $\Pi_1(A)$ with the free groupoid on a computad (statement made precise in this chapter).
- The equivalence with crossed modules and related algebraic comparison theorems are discussed in the text, but are not yet packaged as dedicated Lean interfaces.
- Connections to quasicategories and ∞-cosmoi (beyond the scope of the current formalisation).

Summary

This chapter has one principal aim: to show that the theory of computational paths is not an isolated development but a natural inhabitant of several well-established categorical frameworks. These identifications are not merely decorative; they clarify the abstract character of what has been built. In particular, the path groupoid $\Pi_1(A)$ appears as the free groupoid on a computad (Section 15.1), tying the generators-and-relations presentation of computational paths to the combinatorial foundations of higher category theory. The hom-objects carry their own groupoid structure, so the path theory is naturally 2-groupoidal (Section 15.2), and the non-strict interchange law is captured precisely by the Gray tensor product, yielding a Gray-category. Likewise, the confluence and termination results of the rewrite system connect to a Squier-style finite generating package (Section 15.3), providing an independent combinatorial check on the finiteness of the step and critical-pair data.

The double-source structure of 2-cells was then organised as a double category interface (Section 15.4), while the related double-groupoid and crossed-module interpretations supply algebraic guidance for the surrounding discussion. Finally, the sketch of connections to ∞-groupoid models (Section 15.5)—quasicategories, Kan complexes, and ∞-cosmoi—shows that the weak ω-groupoid of Chapter 9 belongs naturally within the modern landscape of ∞-category theory. Taken together, these identifications do more than corroborate the earlier constructions: they make available a wider stock of coherence theorems and classification results, and they point toward generalisations—such as directed computational paths—that reach beyond the present type-theoretic setting.

15.6 Historical Notes and Further Reading

The references collected here mark the principal categorical lineages on which this chapter draws.

Computads and polygraphs. The notion of computad was introduced by Street [145] in his work on categorical coherence. The equivalent notion of n-polygraph was developed by Burroni [19] from the perspective of higher-dimensional rewriting. The modern treatment by Métayer [115] connects polygraphs to algebraic topology.

Enriched categories. The theory of enriched categories was systematized by Kelly [92], building on earlier work by Eilenberg, Kelly, and others. The specific case of 2-categories and 2-groupoids is treated in Lack's survey [95].

Squier's theorem. Squier's original paper [141] established the connection between finite complete rewrite systems and homological finiteness. The theory was extended by Pride [123] and Kobayashi [94]. A modern treatment appears in the book by Otto [14].

Double categories. Double categories were introduced by Ehresmann [65]. The connection to crossed modules and 2-groupoids was established by Brown and Spencer [17]. Grandis and Paré [76] developed the theory systematically.

∞-categories. The quasicategory model is due to Joyal [89], building on Boardman and Vogt's work on homotopy coherent diagrams. Lurie's *Higher Topos Theory* [106] and *Higher Algebra* [107] provide comprehensive treatments. The ∞-cosmos approach is developed in Riehl–Verity [137].

Homotopy hypothesis. Grothendieck's *Pursuing Stacks* [77] introduced the homotopy hypothesis. Various proofs for specific models appear in Ara [6] (for Grothendieck ω-groupoids) and Joyal [90] (for quasicategories).

In the next chapter, the categorical picture developed here gives way to open problems about extension, univalence, and applications to program verification.

Chapter 16

Perspectives and Open Problems

> *"The value of a theory is measured not by the problems it solves but by the problems it reveals."*
>
> —David Hilbert (paraphrased)

Throughout this book we have developed the theory of computational paths as an explicit rewrite-theoretic account of equality in typed systems. Parts I and II built the mathematical core: the labelled-deduction background, the path calculus, the 77-rule rewrite theory, its normalisation and confluence results, and the higher structure culminating in the weak ω-groupoid theorem. Part III then examined some of the principal interfaces of that core with path induction, UIP, fundamental groupoids, HoTT, and category theory.

This final chapter adds no new structural theorem to that core. Its task is different: to separate the main next questions into a few concrete directions. These include extension to dependent types and universes, the status of univalence-like principles, directed variants of rewriting, applications to program verification, and broader homotopical interpretations of formal systems. Some of these problems remain close to the existing rewrite engine; others are deliberately speculative. It is part of the purpose of a concluding chapter to keep that distinction explicit.

16.1 Extension to Dependent Types and Universes

The computational paths framework developed in this book operates on *simple types*: the simply-typed λ-calculus extended with products, coproducts, and a base of ground types. The central question is whether this foundation can be extended to full dependent type theory with universes. This section explains why such an extension matters, where the obstacles lie, and what future work must resolve.

16.1.1 The Challenge of Dependency

In dependent type theory, types themselves may depend on terms. A dependent function type $\Pi(x : A).B(x)$ binds x in the type B; the type of the output depends on the value of the input. Similarly, dependent pair types $\Sigma(x : A).B(x)$ bundle a term with a type that depends on it.

For computational paths, the core challenge is immediate: if $p : \mathsf{Path}(a, a')$ in type A, and $B : A \to \mathsf{Type}$, then we need a notion of *path transport*

$$\mathsf{transport}_B(p) : B(a) \to B(a')$$

that respects the path structure. We introduced transport in Chapter 11 as a consequence of the J-eliminator. But in our framework, paths also carry explicit step information, so the question becomes whether that information survives and remains meaningful in dependent types.

Intuitively, we want a "path over a path": given a base path p witnessing $a = a'$ in A, a dependent path should witness $b = b'$ "along" p, recording how the rewrite steps in p lift to the fiber B.

Open Problem 16.1 (Dependent paths with step preservation)**.** The computational paths framework (Chapters 3–9) equips each path $p : \mathsf{Path}(a, b)$ with a list of rewrite steps recording *how* the equality $a = b$ is derived. In the simply-typed setting, these steps live in a single fiber. The problem is to extend this to dependent types, where $B : A \to \mathsf{Type}$ and a base path p in A should lift to paths in the varying fibers $B(a)$ and $B(a')$.

Define a notion of *dependent computational path* $\mathsf{DPath}_B(p)(b, b')$ for $p : \mathsf{Path}(a, a')$, $b : B(a)$, $b' : B(a')$ such that:

(i) Step information is preserved: $\mathsf{DPath}_B(p)(b, b')$ carries a step list derived from p.

(ii) The groupoid laws extend: dependent paths compose, invert, and satisfy associativity and unit laws.

(iii) J-elimination follows: the standard path induction principle is derivable from the structure of dependent paths.

16.1.2 Universes and Type-Level Paths

The challenge of dependency becomes even more acute when we move to *universes*. In a system with a universe $\mathsf{Type}_0 : \mathsf{Type}_1$, we can form paths between types themselves: $\mathsf{Path}(A, B)$ where $A, B : \mathsf{Type}_0$. Preliminary formal work exists on this, but the full picture remains unclear.

The key observation is that computational paths between types behave differently from computational paths between terms. If $p : \mathsf{Path}(A, B)$ at the type level, then p carries step information, but the ambient equality witness it determines is a term of $A = B$ in the ambient extensional logic, which is proof-irrelevant. This creates a mismatch: the step information in p is "invisible" to the type system's transport mechanisms.

Theorem 16.2 (Type-level transport is step-blind). *For any type-level path $p : \mathsf{Path}(\mathsf{Type}, A, B)$ and any $x : A$, the transport $\mathsf{transport}(p)(x) : B$ depends only on the ambient equality carried by p and not on its rewrite trace. In particular, if p and q determine the same underlying equality but have different step lists, then $\mathsf{transport}(p)(x) = \mathsf{transport}(q)(x)$.*

Proof. This follows from the fact that transport is defined solely from the underlying ambient equality carried by p, so it ignores the step list. □

This theorem reveals a fundamental tension: at the term level, step information matters (paths with different steps are distinguished), but at the type level, step information is erased by transport. Resolving this tension is essential for a satisfactory extension to universes.

Open Problem 16.3 (Step-sensitive type transport). Is there a variant of type transport that *does* depend on step information? Such a mechanism would need to:

(i) Track which rewrite rules were used to transform A into B.

(ii) Produce different transports for paths with different step lists.

(iii) Remain coherent: transports along $\sim_{\mathsf{rw}}$-equivalent paths should agree.

A positive answer to Problem 16.3 would require moving beyond the ambient proof-irrelevant equality to a custom equality type. This is technically feasible (one would define an inductive type `Path'` that carries computational content) but would sacrifice integration with existing rewriting infrastructure.

16.1.3 Stratified Universes and Consistency

A more modest extension would stratify the treatment: use computational paths with full step information at lower levels, but allow step erasure at the universe level. This mirrors the two-level type theory approach of Voevodsky [157], where an "outer" layer of strict equality coexists with an "inner" layer of homotopy-theoretic identity.

Conjecture 16.4 (Stratified computational paths)**.** There exists a consistent extension of the computational paths framework to dependent types with a stratified universe hierarchy $\mathsf{Type}_0 : \mathsf{Type}_1 : \mathsf{Type}_2 : \cdots$ such that:

(i) Paths at level n are computational paths with step lists.

(ii) Transport from level n to level n preserves step information.

(iii) Transport crossing universe levels, e.g. via $\mathsf{Path}(\mathsf{Type}_n, A, B)$, may erase step information.

(iv) The groupoid laws hold at each level.

(v) The system is conservative over the simply-typed fragment.

Establishing this conjecture would require a careful analysis of universe polymorphism and its interaction with the 77 rewrite rules. The main danger is introducing inconsistency through Girard-style paradoxes if the interaction between paths and universes is not properly controlled.

This tension between step information and proof irrelevance at the type level leads directly to the most consequential question for computational paths: their relationship to the univalence axiom.

16.2 The Univalence Question

The univalence axiom, introduced by Voevodsky [156, 150], asserts that equivalence of types is equivalent to identity of types:

$$(A \simeq B) \simeq (A = B).$$

This axiom has profound consequences: it implies function extensionality, justifies the practice of treating equivalent types as interchangeable, and provides a computational interpretation of the informal mathematical practice of "identifying isomorphic structures."

The central question for computational paths is: *can our framework model univalence, and if so, what would it take?*

16.2.1 Why Full Univalence Fails

The formal development includes a proof that *full univalence is inconsistent* with the computational paths framework.

To state the result precisely, we need three definitions. The first adapts the HoTT notion of equivalence to the computational paths setting.

Definition 16.5 (Path-equivalence)**.** Two types A and B are *path-equivalent* if there exist functions $f : A \to B$ and $g : B \to A$ with computational path witnesses:

$$\forall x : A.\, \mathsf{Path}(g(f(x)), x) \qquad \text{and} \qquad \forall y : B.\, \mathsf{Path}(f(g(y)), y).$$

We write $\mathsf{PathEquiv}(A, B)$ for the type of such equivalences.

The second definition converts a type-level path into a path-equivalence.

Definition 16.6 (idToEquiv)**.** The function $\mathsf{idToEquiv} : \mathsf{Path}(A, B) \to \mathsf{PathEquiv}(A, B)$ converts a type-level path into a path-equivalence via transport:

$$\mathsf{idToEquiv}(p) = \big(\mathsf{transport}(p), \mathsf{transport}(\mathsf{symm}\, p), \text{witnesses}\big).$$

The third definition states what a full univalence principle would look like in our setting.

Definition 16.7 (Full univalence)**.** The *full univalence principle* for computational paths would assert:

(i) There exists $\mathsf{ua} : \mathsf{PathEquiv}(A, B) \to \mathsf{Path}(A, B)$.

(ii) $\mathsf{idToEquiv}(\mathsf{ua}(e)) = e$ for all $e : \mathsf{PathEquiv}(A, B)$.

Theorem 16.8 (Full univalence fails)**.** *The full univalence principle is inconsistent with computational paths.*

Proof. Consider the type Bool and the path-equivalence $\mathsf{boolNegEquiv} : \mathsf{PathEquiv}(\mathsf{Bool}, \mathsf{Bool})$ induced by negation:

$$\mathsf{boolNegEquiv} = (\neg, \neg, \text{involutivity witnesses}).$$

If full univalence held, there would exist $p : \mathsf{Path}(\mathsf{Bool}, \mathsf{Bool})$ such that $\mathsf{idToEquiv}(p) = \mathsf{boolNegEquiv}$. In particular, $(\mathsf{idToEquiv}(p)).\mathsf{toFun} = \neg$.

But for any $p : \mathsf{Path}(\mathsf{Bool}, \mathsf{Bool})$, Theorem 16.2 implies that $\mathsf{transport}(p)$ is the identity function (since the underlying ambient equality is refl by proof irrelevance for $\mathsf{Bool} = \mathsf{Bool}$). Thus $(\mathsf{idToEquiv}(p)).\mathsf{toFun} = \mathsf{id} \neq \neg$. Contradiction. □

Remark 16.9 (The obstruction is proof irrelevance)**.** The failure of full univalence is not a defect of computational paths *per se*, but rather reflects the underlying proof irrelevance of the ambient equality type. In a system where type equality itself carried computational content, the situation might differ.

This obstruction runs deep. Computational paths owe their decidability to operating *above* an underlying equality that is proof-irrelevant, yet that same feature prevents them from modelling univalence. Univalence demands that distinct equivalences between types give rise to distinct paths, but proof irrelevance collapses all type-level equalities to a single witness, leaving transport blind to the step structure. The tension between decidability and higher-dimensional richness is therefore fundamental, not an incidental limitation that a cleverer encoding might circumvent.

The situation evokes other "no-go" results in logic and type theory. Just as Gödel's incompleteness theorems reveal that certain desirable properties of formal systems cannot coexist, Theorem 16.8 reveals a choice point in the computational paths framework: between the decidability of path equivalence and the full strength of the univalence principle. Any future extension aspiring to reconcile these goals will need to introduce genuinely new structure—whether cubical, higher-inductive, or of some other kind—rather than merely refining the existing rewrite rules.

16.2.2 Partial Univalence for 0-Types

Although full univalence fails, a *partial* form succeeds. For 0-truncated types—set-like types whose parallel paths collapse to a single rewrite-equivalence class—a restricted univalence principle holds.

Definition 16.10 (0-truncation)**.** A type A is *0-truncated* (a *set-like type*) if any two parallel paths are $\sim_{\mathsf{rw}}$-equivalent:

$$\forall a, b : A.\, \forall p, q : \mathsf{Path}(a, b).\, p \sim_{\mathsf{rw}} q.$$

Intuitively, a 0-type is one where all computational witnesses between fixed endpoints become indistinguishable after quotienting by rewrite equivalence.

Theorem 16.11 (Partial univalence for 0-types)**.** *If* Type_0 *is 0-truncated (i.e., any two parallel type-level paths are* $\sim_{\mathsf{rw}}$*-equivalent), then* idToEquiv *is well-defined on* $\sim_{\mathsf{rw}}$*-classes. Specifically, for paths* $p, q : \mathsf{Path}(A, B)$ *with* $p \sim_{\mathsf{rw}} q$*:*

$$(\mathsf{idToEquiv}(p)).\mathsf{toFun} = (\mathsf{idToEquiv}(q)).\mathsf{toFun}$$
$$(\mathsf{idToEquiv}(p)).\mathsf{invFun} = (\mathsf{idToEquiv}(q)).\mathsf{invFun}.$$

This theorem shows that the rewrite quotient of paths respects the equivalence structure: the $\sim_{\mathsf{rw}}$-class of a type-level path uniquely determines the pair of functions (up to functional extensionality).

16.2.3 Towards a Univalent Extension

The partial result above suggests a modular approach. To achieve full univalence, we identify three distinct strategies, each balancing theoretical depth against practical feasibility.

Approach 16.12 (Higher inductive types)**.** In HoTT, the circle S^1 is defined as a higher inductive type with a point base and a path $\mathsf{loop} : \mathsf{Path}(\mathsf{base}, \mathsf{base})$. One could attempt to define a univalent universe as a higher inductive type with:

- Points for each type A
- Paths $\mathsf{ua}(e) : \mathsf{Path}(A, B)$ for each equivalence $e : A \simeq B$
- Higher paths ensuring coherence

The challenge is ensuring that the step structure of computational paths is preserved through the higher inductive construction.

Approach 16.13 (Cubical structure)**.** Cubical type theory [23, 4] achieves computational univalence by adding an interval type $\mathbb{I}$ and Kan operations. One could investigate whether computational paths can be given a cubical interpretation:

- Paths as functions $\mathbb{I} \to A$ (or more precisely, as step-decorated such functions)
- Composition via Kan filling
- Univalence as a derived operation on the cubical structure

This approach would require significant extensions to the rewrite system.

Approach 16.14 (Step-carrying equality)**.** The most direct approach is to replace the ambient built-in equality with a custom inductive type that carries step information at all levels:

$$\begin{aligned}
&\textbf{inductive}\ \mathsf{Path}' : A \to A \to \mathsf{Type}\ \textbf{where} \\
&\quad \mid \mathsf{refl} : \mathsf{Path}'(a, a) \\
&\quad \mid \mathsf{step} : (s : \mathsf{Step}) \to \mathsf{Path}'(a, b) \to \mathsf{Path}'(a, b)
\end{aligned}$$

Transport would then be defined by recursion on the step list, potentially producing different results for different step sequences. The challenge is ensuring confluence and coherence in this richer setting.

Open Problem 16.15 (Univalent computational paths)**.** The univalence axiom (Section 16.2) asserts that equivalent types are identical; it fails in the current framework because transport is step-blind (Theorem 16.8). The problem is to find a modification that reconciles the rewrite-theoretic core with univalence.

Concretely, is there a modification of the computational paths framework that:

(i) Retains the 77 rewrite rules and confluence/termination results

(ii) Supports full dependent types with universes

(iii) Validates the univalence axiom

(iv) Provides a computational interpretation of ua

A solution to Problem 16.15 would represent a significant advance, providing a rewriting-based approach to univalent foundations.

To appreciate the difficulty, consider what a concrete reconciliation might require. In cubical type theory [23], univalence is witnessed by a Kan composition that fills a cube whose faces encode the equivalence and its inverse. Transposing this into the computational paths setting would mean: given an equivalence $e : A \simeq B$, one must construct a path $\mathsf{ua}(e) : \mathsf{Path}(A, B)$ together with a step list recording *how* A is transformed into B—not at the level of elements, but at the level of types. The 77 rewrite rules would then need to be augmented with rules governing how these type-level steps compose, invert, and interact with term-level paths. The resulting system would be substantially more complex: preliminary estimates suggest an additional 20–30 rules for the universe level alone, and the confluence proof would need to account for critical pairs between term-level and type-level rewrites.

A further obstacle is that decidability, which is one of the framework's most distinctive features, may not survive the extension. The decidability of path equivalence (Theorem 4.50 of Chapter 4) relies on the fact that every path reduces to the single-step path determined by its ambient equality witness, and that this underlying equality is proof-irrelevant. A step-sensitive transport mechanism (Problem 16.3) would introduce type-level paths whose normal forms depend on the step list, potentially making the word problem for path equivalence undecidable at universe level. Whether a restricted form of decidability can be preserved—perhaps by stratifying the system so that decidability holds within each universe level while cross-level reasoning is semi-decidable—is itself an open question.

16.2.4 The Decidability–Univalence Spectrum

Before discussing why the univalence question matters, it is useful to map out the logical space of possible answers. The tension between decidability and higher-dimensional richness suggests a spectrum of positions, each with distinct consequences.

Position 1. *Full univalence is achievable with modified paths.* The obstruction identified in Theorem 16.8 stems from the fact that `transport` sees only the ambient equality carried by a path and not the step list. If one could design a variant of `Path` where transport is step-sensitive (Problem 16.3), the obstruction would dissolve. The cost would be abandoning the ambient built-in equality and its attendant automation.

Position 2. *Partial univalence is the correct level.* Perhaps the right analogy is not with full HoTT univalence but with the observational type theory of Altenkirch, McBride, and Swierstra [2], where function extensionality holds but full univalence need not. On this view, the partial univalence result (Theorem 16.11) is already the strongest statement one should expect, and the framework is complete at the 0-truncated level.

Position 3. *A cubical hybrid is needed.* The framework might be extended with an interval type $\mathbb{I}$ and Kan composition operations, yielding a system where paths carry both step information (from the rewrite rules) and cubical structure (from the interval). The interaction between these two sources of path structure is unexplored and potentially delicate: the 77 rewrite rules would need to be shown compatible with Kan filling.

Each position leads to a different research programme. Position 1 requires a radical redesign of the path type. Position 2 calls for a careful study of which mathematical structures can be captured within the 0-truncated framework. Position 3 demands a synthesis of two hitherto separate traditions—term rewriting and cubical type theory—that has not been attempted. We regard Position 3 as the most promising, though also the most technically demanding.

A key test case for any proposed resolution is the *structure identity principle* [150]: given two groups $(G, \cdot)$ and $(H, *)$ that are isomorphic as groups, a univalent framework should provide a path $\mathsf{Path}(G, H)$ whose transport respects the group structure. In our current framework, such a path would exist (since both groups inhabit the same type) but its transport would be step-blind. Making transport respect the group isomorphism would require precisely the step-sensitive mechanism of Position 1 or the cubical structure of Position 3.

16.2.5 Why the Univalence Question Matters

The question of univalence is not merely a technical puzzle; it lies at the heart of the relationship between computational paths and modern foundations of mathematics. We highlight three reasons.

Equivalence invariance. In mathematical practice, mathematicians routinely treat equivalent structures as interchangeable: they speak of "the" real numbers without specifying a construction (Dedekind cuts vs. Cauchy sequences). Univalence justifies this practice type-theoretically: if $A \simeq B$, then A and B are interchangeable in all contexts. For computational paths, achieving this property would mean that path-equivalent types are substitutable—a powerful reasoning principle for verified software.

Decidability trade-offs. A striking feature of the computational paths framework is the *decidability* of path equivalence (Theorem 4.50 in Chapter 4), which relies on confluence and termination. Full univalence, as currently understood in HoTT, does *not* enjoy this property: checking whether two terms are path-equal in a univalent setting is in general undecidable. Any successful univalent extension of computational paths would therefore need to navigate this tension carefully—perhaps by restricting univalence to decidable fragments or by accepting a weaker form of decidability.

Connection to cubical type theories. The most successful computational interpretations of univalence [23, 4] use cubical structure. Approach 16.13 above suggests investigating a cubical interpretation of computational paths. A concrete sub-problem is whether the step-decorated paths of our framework can be recast as cubical functions $\mathbb{I} \to A$ in a way that preserves the 77-rule rewrite system. If so, the Kan operations of cubical type theory would provide a natural mechanism for defining ua [23].

Taken together, these considerations reveal the univalence question as a nexus where several of the deepest themes in the foundations of mathematics converge: the nature of equality, the role of computation in proofs, and the boundary between decidable and undecidable reasoning. For computational paths, resolving this question is not simply a matter of adding a new axiom; it requires rethinking how step-level information propagates through type-level constructions. The partial univalence result (Theorem 16.11) shows that the framework already captures the right intuition for 0-truncated types, where parallel paths collapse to a single rewrite-equivalence class. The challenge is to extend this intuition to higher levels without sacrificing the concrete, computable character of the rewrite system that gives the theory its distinctive power.

The preceding sections probed the *type-theoretic depth* of computational paths—their foundational constraints and possibilities. We now turn to their *structural breadth*: extensions that enlarge the classes of systems the framework can model.

16.3 Directed Computational Paths

Throughout this book, we have treated computational paths as inherently *reversible*. Every path $p : \mathsf{Path}(a, b)$ has an inverse $\mathsf{symm}\, p : \mathsf{Path}(b, a)$, and this symmetry is fundamental to the groupoid structure. But in many computational settings, transitions are *not* reversible: a program executes forward, a reduction sequence proceeds from redex to reduct, and time flows in one direction.

This observation motivates the study of *directed computational paths*, where the symmetry operation is absent or restricted. The resulting structures resemble categories rather than groupoids, and open the door to connections with directed type theory and $(\infty, 1)$-categories.

16.3.1 From Groupoids to Categories

Recall that our computational paths carry an inverse (symm), which is what makes them form a *groupoid* rather than a mere category. Dropping the inverse operation from a groupoid leaves a category. Dropping it from an ω-groupoid leaves an ω-category [106, 99]—or more precisely, an $(\infty, 1)$-category if we retain invertibility at dimension ≥ 1. The following definition captures this idea.

Definition 16.16 (Directed path structure). A *directed path structure* on a type A consists of:

(i) A type $\mathsf{DirPath}(a, b)$ of directed paths for each $a, b : A$

(ii) Identity: $\mathsf{refl}_a : \mathsf{DirPath}(a, a)$

(iii) Composition: $\mathsf{trans} : \mathsf{DirPath}(a, b) \to \mathsf{DirPath}(b, c) \to \mathsf{DirPath}(a, c)$

(iv) Associativity and unit laws (up to higher paths)

Note the absence of a symmetry/inverse operation.

Example 16.17 (Reduction paths). In a term rewriting system, define $\mathsf{DirPath}(t, u)$ to be the type of reduction sequences $t \to t_1 \to \cdots \to t_n = u$. This is naturally directed: we can compose reduction sequences, but we cannot (in general) reverse a reduction. The reflexive path is the empty sequence.

Example 16.18 (Computation traces). For a programming language with operational semantics, define $\mathsf{DirPath}(\sigma, \sigma')$ to be the type of execution traces from state σ to state σ'. These are inherently directed by the flow of time.

16.3.2 Directed Type Theory

The examples above show that directed paths arise naturally in practice. Recent work on *directed type theory* (DTT) [136, 137, 100, 159] provides a type-theoretic foundation for $(\infty, 1)$-categories. The key idea is to distinguish:

- **Covariant** types, where transport is forward-only
- **Contravariant** types, where transport goes backward
- **Invariant** types, where transport is unavailable

Open Problem 16.19 (Directed computational paths)**.** The standard computational paths of this book are *invertible*: every path has a symmetry (Chapter 6), making them groupoidal rather than merely categorical. Many computational settings—term reduction, program execution, irreversible state transitions—lack this invertibility. The problem is to develop a directed counterpart of the theory.

Specifically:

(i) Define $\mathsf{DirPath}(a, b)$ as a rewrite sequence without symmetry.

(ii) Establishes confluence for directed paths (a directed path has a unique normal form)

(iii) Characterizes the categorical structure: when do directed paths form an $(\infty, 1)$-category?

(iv) Connects to directed type theory: is there a translation between DTT and directed computational paths?

16.3.3 Mixed Structures

In practice, the most interesting setting is neither purely invertible nor purely directed, but a mixture of both. In such a *mixed* structure:

- Some paths are invertible (e.g., renamings, commutative rewrites)
- Others are one-way (e.g., β-reduction, genuine computation)
- The interaction between the two must be carefully controlled

Definition 16.20 (Mixed path structure)**.** A *mixed path structure* consists of:

(i) A type $\mathsf{MixPath}(a, b)$ of mixed paths

(ii) A predicate $\mathsf{isInvertible} : \mathsf{MixPath}(a, b) \to \mathsf{Prop}$

(iii) Identity and composition (always available)

(iv) Symmetry: $\mathsf{symm} : (p : \mathsf{MixPath}(a,b)) \to \mathsf{isInvertible}(p) \to \mathsf{MixPath}(b,a)$

(v) Coherence conditions relating directed and invertible parts

In the computational paths framework, a natural mixed structure arises from distinguishing:

- **Invertible steps**: Those corresponding to rules that are symmetric (e.g., $\mathsf{trans}(\mathsf{refl}, p) \leftrightarrow p$)
- **Directed steps**: Those corresponding to oriented reductions (e.g., β-reduction $(\lambda x.t)\,u \to t[u/x]$)

Conjecture 16.21 (Directed substructure)**.** The 77 rewrite rules of Chapter 4 can be partitioned into:

(i) A set R_{sym} of "symmetric" rules that admit inverse rules in the system

(ii) A set R_{dir} of "directed" rules that do not

The directed subset R_{dir} forms a confluent, terminating rewrite system in its own right, giving rise to a directed path structure.

A concrete test case for Conjecture 16.21 is the groupoid core (Category A). Rules R1–R6 are genuinely directed: they simplify expressions by eliminating redundant structure. Rules R7 (contravariance) and R8 (associativity) are re-association rules whose direction is chosen by convention. A directed variant of the groupoid core would retain R1–R6 but might leave R7 and R8 as unoriented equations, yielding a *directed groupoid* in which forward paths compose and simplify but cannot be reversed.

The preceding sections have focused on extending the *mathematical* theory of computational paths. We now turn to a complementary direction: applying the existing framework to problems in software engineering.

16.4 Applications to Program Verification

One of the original motivations for computational paths was to provide a foundation for reasoning about program equivalence. This section turns from foundations to applications: if two programs are connected by an explicit path of rewrite steps, they should behave identically in all observable contexts.

16.4.1 Refactoring as Path Construction

Refactoring is ubiquitous in software engineering: code is restructured while its behaviour is preserved. From the computational-paths perspective, a refactoring is nothing other than the construction of a path.

Example 16.22 (Inlining as a path). Consider the refactoring that inlines a function call:

```
let f = λx.t in ... f(u) ...
    ⇝ let f = λx.t in ... t[u/x] ...
```

This is justified by β-reduction. The computational path witnessing this refactoring contains the step $\mathsf{betaStep}((\lambda x.t), u)$.

Example 16.23 (Loop fusion). The optimisation that fuses two loops:

```
map g (map f xs)  ⇝  map (g ∘ f) xs
```

corresponds to a path built from the functor law for `map`. This path witnesses that the optimisation is semantics-preserving.

Open Problem 16.24 (Refactoring certificates). Develop a system where:

(i) Refactoring tools output computational paths as certificates

(ii) The paths can be mechanically verified against the rewrite rules

(iii) Invalid refactorings are rejected (no valid path exists)

(iv) Path composition tracks chains of refactorings

16.4.2 Program Equivalence Proofs

Beyond refactoring, computational paths can serve as proofs of program equivalence more generally. The idea is simple: if two programs are connected by a path, they must behave identically in all contexts.

Definition 16.25 (Observational equivalence). Two programs P and Q of the same type are *observationally equivalent* if they cannot be distinguished by any context:

$$P \approx Q \iff \forall C.\, C[P] \Downarrow v \iff C[Q] \Downarrow v$$

where $C[-]$ ranges over contexts and $\Downarrow$ denotes termination with value v.

Conjecture 16.26 (Paths as equivalence proofs). For a suitably rich rewrite system, the existence of a computational path $p : \mathsf{Path}(P, Q)$ implies observational equivalence $P \approx Q$.

The conjecture requires care: the rewrite rules must be *sound*—they must not identify observationally distinct programs. The 77 rules of Chapter 4 are designed for the path algebra and may not directly apply to general program transformations. Extending them to richer programming languages is an important direction.

Open Problem 16.27 (Sound rewrite rules for realistic languages). For a realistic programming language (e.g., a subset of ML or Haskell):

(i) Identify a set of sound rewrite rules (ones that preserve observational equivalence)

(ii) Prove confluence and termination for the rule set

(iii) Characterize which observational equivalences are captured by paths

16.4.3 Differential Reasoning

Software systems are built from components, and reasoning about the whole often reduces to reasoning about the parts. An intriguing application of computational paths is *differential* reasoning: given paths $p : \mathsf{Path}(P, P')$ and $q : \mathsf{Path}(Q, Q')$ for individual components, construct a path witnessing the equivalence of the combined systems.

Example 16.28 (Compositional paths). If $p : \mathsf{Path}(f, f')$ and $q : \mathsf{Path}(g, g')$, then congrArg yields:

$$\mathsf{congrArg}(\circ, p, q) : \mathsf{Path}(f \circ g, f' \circ g')$$

This allows reasoning about composed systems from reasoning about their parts.

The functorial properties of congrArg (Chapter 3) provide the infrastructure for such compositional reasoning. A full-fledged system for differential program verification would build on this foundation.

16.4.4 Security Applications

Program equivalence has direct applications to security: many security properties can be phrased as the requirement that two programs—one with secrets, one without—be indistinguishable. If a program transformation is *secure* (does not introduce information leaks), this can be witnessed by a path satisfying additional constraints.

Example 16.29 (Constant-time equivalence). Two programs are *constant-time equivalent* if their execution time depends only on public inputs. A path $p : \mathsf{Path}(P, Q)$ witnesses constant-time equivalence if every step in p preserves timing behaviour.

Open Problem 16.30 (Security-typed paths). Develop a variant of computational paths where:

(i) Steps are labeled with security levels

(ii) Paths respect a noninterference property

(iii) The existence of a well-typed path implies secure equivalence

The applications above—refactoring certificates, equivalence proofs, compositional reasoning, and security—only hint at the broader landscape. The core insight is that the explicit, structured nature of computational-path witnesses makes them a natural substrate for verified program transformation tools.

A particularly promising near-term application is *proof repair*: when a library evolves (types change, functions are refactored), the computational path between old and new versions enables automatic transformation of downstream proofs. If the refactoring is witnessed by a path p, then transport along p carries proofs from the old version to the new one. The step list in p records exactly which changes were made, enabling targeted repair rather than wholesale re-verification.

16.5 The Homotopy Theory of Formal Systems

We conclude with a broader view. The computational paths framework, and more generally the connection between type theory and homotopy theory revealed by HoTT, suggest a unifying paradigm: the *homotopy theory of formal systems*.

16.5.1 Formal Systems as Spaces

The central insight of this section is simple but far-reaching: a formal system—whether a logic, a type theory, or a programming language—can be viewed as a mathematical space. The points of this space are terms, proofs, or programs; the paths are derivations, rewrites, or reductions. From this perspective, classical tools from algebraic topology become immediately relevant.

Definition 16.31 (The classifying space of a formal system). Given a formal system $\mathcal{F}$, its *classifying space* $B\mathcal{F}$ is the ω-groupoid (or $(\infty, 1)$-category in the directed case) whose:

- 0-cells are well-typed terms of $\mathcal{F}$
- 1-cells are single-step rewrites
- 2-cells are rewrites between rewrite sequences

- n-cells are $(n-1)$-cell equivalences

The computational paths of this book provide a concrete construction of $B\mathcal{F}$ for the simply-typed λ-calculus with the 77 rewrite rules.

Theorem 16.32 (Higher structural stabilisation)**.** *The results proved in Chapters 8 and 9 show that the* structural *higher-cell comparison problems in $B\mathcal{F}$ stabilise after the genuine 3-dimensional coherence layer, so from dimension 4 onward. In particular:*

- *structural 3-cell diagrams commute up to 4-cells (Theorem 8.25);*
- *every structural boundary problem from dimension 4 onward has at least one higher filler;*
- *any two such structural fillers are connected by one more structural cell (Theorem 9.61).*

This is the part of the truncation picture currently justified by the book and by the current mechanised development. It shows that $B\mathcal{F}$ already behaves much like an object close to a 3-type on its designated coherence data. A full theorem that $B\mathcal{F}$ is literally 3-coskeletal, or that $\pi_n(B\mathcal{F}, a) = 0$ for all $n \geq 4$, is not established here and remains an open problem.

Remark 16.33 (About the current mechanisation of truncation)**.** The theorem is correct in the extensional setting of this book, but the current mechanised proof of the level-3 step now splits into two layers. On the raw path tower, the main comparison route factors through inverse-loop normalisation and structural 3-cell contraction, yet the exported connector still retains one localized transport along propositional equality. On the explicit-syntax/polygraphic side, the same critical-pair analysis already yields a fully constructive proof-relevant coherent presentation through dimension 3 and acyclicity above dimension 3. What remains open is a direct transport-free comparison between that explicit-syntax theorem and the raw path-level hierarchy.

16.5.2 Homotopy Invariants of Formal Systems

If formal systems are spaces, then classical invariants from algebraic topology—homotopy groups, homology, cohomology—become invariants of formal systems. The homotopy groups $\pi_n(B\mathcal{F})$ are such invariants. Different formal systems may have different homotopy types, and the invariants may encode deep properties of the logic.

Question 16.34 (Homotopy classification)**.** Can formal systems be classified by their homotopy type? More precisely:

(i) Which homotopy types arise as classifying spaces of formal systems?

(ii) Do "equivalent" formal systems (in some suitable sense) have equivalent classifying spaces?

(iii) What do the homotopy groups tell us about the expressive power of a formal system?

Example 16.35 (Simply-typed vs. polymorphic)**.** The simply-typed λ-calculus and System F (polymorphic λ-calculus) have different classifying spaces. System F has richer structure at the type level, which should manifest as different homotopy groups. Making this precise is an open problem.

16.5.3 Morphisms Between Formal Systems

Just as continuous maps between topological spaces induce homomorphisms on homotopy groups, translations between formal systems should induce maps between classifying spaces.

Definition 16.36 (Homotopy-theoretic comparison data)**.** A *translation* $\Phi : \mathcal{F} \to \mathcal{G}$ between formal systems may be compared homotopy-theoretically at several different strengths via the induced map $B\Phi : B\mathcal{F} \to B\mathcal{G}$. Depending on the question, one may ask that $B\Phi$ be:

(i) A weak homotopy equivalence (isomorphism on all π_n), or

(ii) A fibration-like map preserving designated lifting data, or

(iii) At minimum, a continuous map that preserves the chosen path data.

Example 16.37 (Curry–Howard as a target comparison theorem)**.** A long-term aspiration would be to formulate the Curry–Howard correspondence between intuitionistic propositional logic and the simply-typed λ-calculus as a weak equivalence between suitable classifying spaces. Propositions would correspond to types, proofs to terms, and proof transformations to rewrite sequences. Making this precise, however, requires a careful choice of the spaces and comparison maps, and is not proved in the present book.

16.5.4 A Research Program

The ideas above—classifying spaces, homotopy invariants, and maps between formal systems—fit together into a coherent vision. We therefore propose a long-term research programme: *develop the homotopy theory of formal systems as a unifying framework for understanding the structure and relationships of logics, type theories, and programming languages.*

1. **Foundations**: Establish precise definitions of classifying spaces for broad classes of formal systems (logics, type theories, rewrite systems, automata).

2. **Invariants**: Compute homotopy invariants (π_n, homology, cohomology) for specific formal systems. Connect these to syntactic/semantic properties.

3. **Classification**: Develop classification theorems: which homotopy types arise? When are formal systems homotopy-equivalent?

4. **Constructions**: Study how formal system constructions (products, coproducts, exponentials, quotients) affect homotopy type.

5. **Applications**: Apply homotopy-theoretic methods to problems in logic, type theory, and programming languages: decidability, expressiveness, compilation, optimisation.

Open Problem 16.38 (Homotopy theory of rewrite systems)**.** A term rewriting system $\mathcal{R}$ (Appendix A) determines a space whose points are terms, whose 1-cells are rewrite steps, and whose higher cells record relationships between rewrite sequences. The problem is to connect the homotopy type of this space to established invariants from rewriting theory.

Concretely, given a term rewriting system $\mathcal{R}$:

(i) Define the classifying ω-groupoid of $\mathcal{R}$

(ii) Compute its homotopy groups in terms of critical pairs and confluence properties

(iii) Relate the homotopy type to Squier's theorem on finite derivation type [141, 78]

Problem 16.38 connects to Chapter 15, where we noted Squier's result that a finitely presented monoid has finite derivation type (a finiteness condition on 2-cells) if and only if it has finite homological type. This is precisely a homotopy-theoretic statement about rewrite systems.

16.5.5 The Synthetic Perspective

Finally, we note that the homotopy theory of formal systems can itself be formalised *within* a sufficiently powerful formal system—this is the "synthetic" approach championed in HoTT [150].

In this view, formal systems are objects of a universe, and their classifying spaces are constructed type-theoretically. The homotopy theory of formal systems becomes part of the internal mathematics of the metatheory.

Conjecture 16.39 (Synthetic formalisation)**.** In a univalent type theory with higher inductive types, the classifying space of a formal system $\mathcal{F}$ can be constructed as a higher inductive type. The homotopy groups $\pi_n(B\mathcal{F})$ are then definable internally, and homotopy-theoretic reasoning about $\mathcal{F}$ becomes ordinary type-theoretic reasoning.

This closes a circle: we began by using type theory to study computational paths, and we end by imagining computational paths as part of the homotopy theory of type theories themselves.

Formalization Notes. The results of this chapter are formalised in the following files of the accompanying Lean development (`https://github.com/Arthur742Ramos/ComputationalPathsLean`):

- `ComputationalPaths/Comparison/UnivalenceAnalog.lean` — The PathEquiv structure, idToEquiv, Theorem 16.2 (step-blind transport), Theorem 16.8 (failure of full univalence), and partial univalence for 0-types (Theorem 16.11).
- `ComputationalPaths/Comparison/HoTTComparison.lean` — The translation dictionary referenced in Section 16.2 and used in the proof of the univalence obstruction.

Key definitions. The full univalence axiom is incompatible with computational paths:

```
/-- Univalence axiom: every path-equivalence arises from
    a type-level path -/
noncomputable def FullUnivalence : Prop :=
  forall (alpha beta : Type u), IsEquiv (@idToEquiv alpha
      beta)

/-- Counterexample: two distinct paths between Bool and
    Bool -/
theorem full_univalence_fails : not FullUnivalence := by
  intro univ
  -- The identity and swap equivalences give different
     paths
  -- but idToEquiv cannot distinguish them at the step
     level
  have h := step_blind_transport (D := fun _ => Bool)
  exact absurd h path_steps_matter
```

The open problems and conjectures stated in this chapter (Problems 16.1–16.38 and Conjectures 16.4–16.39) are not formalised and represent directions for future work.

16.6 Conclusion

Let us step back. The book has established a proof-relevant calculus of equality whose mathematical core can be stated in three layers.

- **Explicit rewrite theory.** Computational paths record equality proofs as stepwise derivations, and the 77-rule system gives a normalising and confluent calculus for comparing them.
- **Higher structure.** Paths, paths between paths, and higher witnesses assemble into groupoid, 2-cell, and 3-cell structures, culminating in a weak ω-groupoid with explicit low-dimensional coherence and stabilisation from dimension 4 onward.
- **Interfaces.** Path induction, the analysis of UIP, fundamental groupoids, HoTT comparisons, and categorical reformulations show how this core interacts with neighbouring frameworks without collapsing into any one of them.

The open problems collected in this chapter fall into two broad classes. Some are internal extensions of the existing theory: dependent types, universes, directed rewrites, and verification-oriented semantics. Others are external comparison questions: univalence, geometric models, and a possible homotopy theory of formal systems. The distinction matters, because the former ask for new structure on the proved core, whereas the latter ask how far that core can be taken.

The right conclusion is therefore this. Computational paths do not settle every question about equality, identity types, or higher structure. What they provide is a concrete rewrite-theoretic and higher-algebraic object against which those questions can now be posed with precision. What remains open is substantial; it is no longer vague.

Appendix A

Term Rewriting Systems

> *"Rewriting is a very powerful method for dealing with equations. Essentially, equations are used only in one direction: they are turned into rewrite rules. By restricting the way equations are used, one obtains a more goal-oriented proof method."*
>
> —Franz Baader & Tobias Nipkow [8]

This appendix provides a self-contained introduction to the theory of term rewriting systems (TRS) for readers who may not be familiar with this area. The reader who already knows what confluence and termination mean, and who is comfortable with critical pairs and recursive path orderings, may skip directly to Section A.5, which explains how these concepts specialise to the path calculus. For everyone else, the appendix offers the background needed to follow the metatheory of Chapters 4 and 5 without consulting external references.

Term rewriting is one of the foundational tools of theoretical computer science and mathematical logic, providing both a computational paradigm for working with equational theories and a powerful framework for studying the decidability of word problems.

The computational path calculus presented in this book—particularly the 77-rule rewrite system of Chapter 4 and its metatheory in Chapter 5—is an instance of a term rewriting system. Understanding the general theory illuminates why certain properties (confluence, termination) are crucial for our purposes and provides context for the techniques used in proving these properties.

We begin with the basic definitions of terms, signatures, and rewrite rules in Section A.1. Section A.2 discusses termination and the well-founded orderings used to prove it. Section A.3 covers confluence and the critical pair method. Section A.4 treats normal forms and the decidability of word problems. Finally, Section A.5 connects these concepts to our specific application: the rewrite system for computational paths.

A.1 Basic Definitions

We begin with the fundamental concepts: the syntactic objects that rewriting operates on, and the rules that govern their transformation.

A.1.1 Signatures and Terms

Definition A.1 (Signature). A *signature* Σ is a set of *function symbols*, each equipped with a fixed *arity* $n \geq 0$. We write $f : n$ or f/n to indicate that f has arity n. Function symbols of arity 0 are called *constants*.

Example A.2. The signature for groups consists of:

- A constant $e : 0$ (the identity element)
- A unary symbol $\mathsf{inv} : 1$ (inversion)
- A binary symbol $\cdot : 2$ (multiplication)

We write this as $\Sigma_{\mathsf{Group}} = \{e/0, \mathsf{inv}/1, \cdot/2\}$.

Example A.3. The signature for computational paths, used throughout this book, includes:

- Constants $\mathsf{refl}_a/0$ for reflexivity paths (one for each term a)
- A unary symbol $\mathsf{symm}/1$ for symmetry
- A binary symbol $\mathsf{trans}/2$ for transitivity
- A binary symbol $\mathsf{congrArg}/2$ for congruence
- Additional symbols for transport, products, sigma types, etc.

Definition A.4 (Terms). Let Σ be a signature and $\mathcal{V}$ be a countably infinite set of *variables*. The set $\mathcal{T}(\Sigma, \mathcal{V})$ of *terms* over Σ and $\mathcal{V}$ is defined inductively:

1. Every variable $x \in \mathcal{V}$ is a term.
2. If $f \in \Sigma$ has arity n and $t_1, \ldots, t_n$ are terms, then $f(t_1, \ldots, t_n)$ is a term.

A term is *ground* if it contains no variables. We write $\mathcal{T}(\Sigma)$ for the set of ground terms.

The structure of terms can be visualised as labeled trees, where each node is labeled by a function symbol or variable, and the children of a node labeled f/n are the n subterms.

Definition A.5 (Subterms and Positions)**.** A *position* in a term is a sequence of positive integers specifying a path from the root to a node. We write ϵ for the empty sequence (the root position). For a term t:

- $\mathsf{Pos}(t)$ denotes the set of all positions in t.
- $t|_p$ denotes the subterm of t at position p.
- $t[s]_p$ denotes the term obtained by replacing the subterm at position p with s.

Example A.6. Consider the term $t = f(g(a, b), c)$, where $f/2$, $g/2$, $a/0$, $b/0$, and $c/0$. The positions and corresponding subterms are:

- ϵ: the whole term $f(g(a, b), c)$
- 1: the first argument $g(a, b)$
- 2: the second argument c
- 1.1: the first argument of the first argument, namely a
- 1.2: the second argument of the first argument, namely b

A.1.2 Substitution and Matching

Definition A.7 (Substitution)**.** A *substitution* is a mapping $\sigma : \mathcal{V} \to \mathcal{T}(\Sigma, \mathcal{V})$ that is the identity on all but finitely many variables. We write substitutions as $\sigma = \{x_1 \mapsto t_1, \ldots, x_n \mapsto t_n\}$, listing only the non-identity mappings.

The application of a substitution σ to a term t, written $t\sigma$ or $\sigma(t)$, is defined inductively:

$$\begin{aligned} x\sigma &= \sigma(x) \quad \text{for } x \in \mathcal{V} \\ f(t_1, \ldots, t_n)\sigma &= f(t_1\sigma, \ldots, t_n\sigma) \end{aligned}$$

Example A.8. Let $\sigma = \{x \mapsto g(a), y \mapsto b\}$ and $t = f(x, h(y, x))$. Then:

$$t\sigma = f(g(a), h(b, g(a)))$$

Definition A.9 (Composition of Substitutions)**.** The *composition* of substitutions σ and τ, written $\sigma\tau$ or $\tau \circ \sigma$, is defined by:

$$x(\sigma\tau) = (x\sigma)\tau$$

Composition is associative: $(\sigma\tau)\rho = \sigma(\tau\rho)$.

Definition A.10 (Matching). A term ℓ *matches* a term t if there exists a substitution σ such that $\ell\sigma = t$. In this case, σ is called a *matcher* (or *match substitution*) of ℓ to t.

When it exists, the matcher is unique up to variables not occurring in ℓ.

Example A.11. Consider the term pattern $\ell = f(x, g(x))$ and the term $t = f(a, g(a))$. The substitution $\sigma = \{x \mapsto a\}$ is a matcher: $\ell\sigma = t$.

However, ℓ does not match $t' = f(a, g(b))$, since this would require both $x \mapsto a$ and $x \mapsto b$, which is inconsistent.

Matching is a restricted form of equation solving where only one side contains variables. The more general problem of *unification* allows variables on both sides.

Definition A.12 (Unification). Two terms s and t are *unifiable* if there exists a substitution σ (called a *unifier*) such that $s\sigma = t\sigma$. A unifier σ is a *most general unifier* (mgu) if every other unifier τ factors through σ: there exists ρ such that $\tau = \sigma\rho$.

Theorem A.13 (Robinson, 1965). *If two terms are unifiable, then they have a most general unifier, which is unique up to variable renaming. There is an algorithm to compute the mgu or determine that no unifier exists.*

A.1.3 Rewrite Rules and Reduction

Definition A.14 (Rewrite Rule). A *rewrite rule* is a pair of terms (ℓ, r), written $\ell \to r$, subject to the following conditions:

1. ℓ is not a variable.

2. Every variable occurring in r also occurs in ℓ.

The term ℓ is called the *left-hand side* (lhs) and r the *right-hand side* (rhs). The first condition prevents rules like $x \to a$ that would rewrite anything; the second prevents rules from introducing "free" variables.

Definition A.15 (Term Rewriting System). A *term rewriting system* (TRS) is a pair $\mathcal{R} = (\Sigma, R)$ where Σ is a signature and R is a set of rewrite rules over Σ.

Informally, a TRS is a machine for simplifying symbolic expressions. Given a term, we scan it for subterms that match the left-hand side of some rule; when we find one, we replace it by the corresponding right-hand side. The fundamental questions are: does this process always terminate? And if there are multiple choices of which subterm to rewrite first, do we always end up at the same result? These are the questions of *termination* and *confluence*, developed in the next two sections.

Definition A.16 (Reduction)**.** Let $\mathcal{R}$ be a TRS and $\ell \to r$ a rule in $\mathcal{R}$. A term s *reduces* to t by the rule $\ell \to r$ at position p, written $s \to_{\mathcal{R}} t$ or simply $s \to t$, if:

1. The subterm $s|_p$ matches the left-hand side: $s|_p = \ell\sigma$ for some substitution σ.

2. The term t is obtained by replacing this subterm with the instantiated right-hand side: $t = s[r\sigma]_p$.

The subterm $s|_p = \ell\sigma$ is called a *redex* (reducible expression).

Example A.17. Consider the group theory TRS with rules:

$$\begin{aligned} e \cdot x &\to x \\ \mathsf{inv}(x) \cdot x &\to e \\ (x \cdot y) \cdot z &\to x \cdot (y \cdot z) \end{aligned}$$

Starting from $(\mathsf{inv}(a) \cdot a) \cdot b$, we can reduce:

$$\begin{aligned} (\mathsf{inv}(a) \cdot a) \cdot b &\to e \cdot b \quad \text{(by rule 2 at position 1)} \\ &\to b \quad \text{(by rule 1)} \end{aligned}$$

Definition A.18 (Reduction Relations)**.** Let $\to$ be the one-step reduction relation. We define:

- $\to^*$: the reflexive-transitive closure of $\to$ (zero or more steps)
- $\to^+$: the transitive closure of $\to$ (one or more steps)
- $\leftrightarrow$: the symmetric closure of $\to$ (one step in either direction)
- $\leftrightarrow^*$: the reflexive-transitive-symmetric closure, called *conversion* or *equational equivalence*

The relation $\leftrightarrow^*$ is exactly the equational theory generated by treating the rewrite rules as equations. Two terms s and t satisfy $s \leftrightarrow^* t$ if and only if they are provably equal using the rules as bidirectional equations.

A.2 Termination

A term rewriting system *terminates* (or is *strongly normalising*) if there is no infinite reduction sequence $t_0 \to t_1 \to t_2 \to \cdots$. Termination is a crucial property: it guarantees that the rewriting process always halts, which is necessary for using rewriting as a decision procedure.

A.2.1 Well-Founded Orderings

The key tool for proving termination is the concept of a well-founded ordering.

Definition A.19 (Well-Founded Ordering)**.** A strict partial ordering $\succ$ on a set S is *well-founded* if there is no infinite descending chain $s_0 \succ s_1 \succ s_2 \succ \cdots$.

Equivalently, $\succ$ is well-founded if and only if every non-empty subset of S has a minimal element (an element with no smaller elements in the subset).

Theorem A.20 (Termination via Well-Founded Orderings)**.** *A TRS $\mathcal{R}$ terminates if and only if there exists a well-founded ordering $\succ$ on terms such that $s \to_{\mathcal{R}} t$ implies $s \succ t$ for all terms s and t.*

Proof. If such an ordering exists, any reduction sequence $t_0 \to t_1 \to \cdots$ would give a descending chain $t_0 \succ t_1 \succ \cdots$, contradicting well-foundedness. Conversely, if the TRS terminates, we can take $\succ$ to be the transitive closure of $\to$. □

Example A.21. The standard ordering $>$ on $\mathbb{N}$ is well-founded. A TRS terminates if we can assign each term a natural number "weight" that decreases with each reduction.

For more complex TRSs, simple weight functions are insufficient. We need more sophisticated orderings that can handle rules where the right-hand side is not obviously "simpler" than the left-hand side.

A.2.2 Recursive Path Orderings (RPO)

The *recursive path ordering* (RPO), introduced by Dershowitz [61], is one of the most widely used methods for proving termination. It is based on a *precedence*—a well-founded ordering on function symbols—and extends this to an ordering on terms.

Definition A.22 (Precedence)**.** A *precedence* is a strict partial ordering $>_F$ on the function symbols in a signature Σ.

Definition A.23 (Recursive Path Ordering)**.** Given a precedence $>_F$, the *recursive path ordering* $>_{\mathsf{rpo}}$ is defined inductively: $s = f(s_1, \ldots, s_m) >_{\mathsf{rpo}} t$ if one of the following holds:

(RPO1) $s_i \geq_{\mathsf{rpo}} t$ for some $i \in \{1, \ldots, m\}$. (A subterm of s dominates t.)

(RPO2) $t = g(t_1, \ldots, t_n)$ with $f >_F g$, and $s >_{\mathsf{rpo}} t_j$ for all $j \in \{1, \ldots, n\}$. (The root symbol of s dominates that of t, and s recursively dominates all subterms of t.)

(RPO3) $t = f(t_1, \ldots, t_n)$ (same root symbol), and $(s_1, \ldots, s_m) >^{\mathsf{lex}}_{\mathsf{rpo}} (t_1, \ldots, t_n)$ or $(s_1, \ldots, s_m) >^{\mathsf{mul}}_{\mathsf{rpo}} (t_1, \ldots, t_n)$, and $s >_{\mathsf{rpo}} t_j$ for all j. (Same root, but arguments decrease lexicographically or multiset-wise.)

Here $\geq_{\mathsf{rpo}}$ means $>_{\mathsf{rpo}}$ or syntactic equality.

The choice between lexicographic (lex) and multiset (mul) comparison in (RPO3) can be made independently for each function symbol, resulting in variants called LPO (lexicographic path ordering) and MPO (multiset path ordering).

Theorem A.24 (Dershowitz, 1982). *For any well-founded precedence $>_F$, the recursive path ordering $>_{\mathsf{rpo}}$ is well-founded.*

Example A.25. Consider the group theory rules with precedence $\cdot >_F \mathsf{inv} >_F e$. The rule $e \cdot x \to x$ is oriented by RPO: $e \cdot x >_{\mathsf{rpo}} x$ because x is a subterm of $e \cdot x$ (case RPO1).

The rule $\mathsf{inv}(\mathsf{inv}(x)) \to x$ is oriented because $\mathsf{inv}(x)$ is a subterm of $\mathsf{inv}(\mathsf{inv}(x))$ and x is a subterm of $\mathsf{inv}(x)$, so $\mathsf{inv}(\mathsf{inv}(x)) >_{\mathsf{rpo}} \mathsf{inv}(x) >_{\mathsf{rpo}} x$ by repeated application of RPO1.

A.2.3 Knuth-Bendix Ordering (KBO)

The *Knuth-Bendix ordering* (KBO), introduced by Knuth and Bendix [93], combines a weight-based approach with a precedence-based tiebreaker.

Definition A.26 (Weight Function). A *weight function* for a signature Σ consists of:

- A weight $w_0 > 0$ for variables.
- A weight $w(f) \geq 0$ for each function symbol $f \in \Sigma$.
- The constraint that if f is unary with $w(f) = 0$, then f must be the greatest element in the precedence.

The weight of a term is defined by:

$$|x| = w_0 \quad \text{for variables } x$$
$$|f(t_1, \ldots, t_n)| = w(f) + |t_1| + \cdots + |t_n|$$

Definition A.27 (Knuth-Bendix Ordering). Given a weight function and a precedence $>_F$, the *Knuth-Bendix ordering* $>_{\mathsf{kbo}}$ is defined by: $s >_{\mathsf{kbo}} t$ if the following conditions hold:

1. $|s| > |t|$, or

2. $|s| = |t|$ and one of the following:
   (a) $s = f^n(x)$ and $t = x$ for some variable x and $n \geq 1$, where f is the unique unary symbol with $w(f) = 0$.
   (b) $s = f(s_1, \ldots, s_m)$, $t = g(t_1, \ldots, t_n)$, and either:
      - $f >_F g$, or
      - $f = g$ and $(s_1, \ldots, s_m) >_{\mathsf{kbo}}^{\mathsf{lex}} (t_1, \ldots, t_n)$.

Additionally, for all variables x, the number of occurrences of x in s must be at least the number in t.

Theorem A.28 (Knuth-Bendix, 1970). *For any valid weight function and well-founded precedence, the Knuth-Bendix ordering $>_{\mathsf{kbo}}$ is well-founded.*

KBO has the advantage of being simple to compute: first compare weights, then use the precedence as a tiebreaker. However, it cannot orient certain rules that RPO can handle (and vice versa).

A.2.4 Polynomial Interpretations

Polynomial interpretations provide another powerful method for proving termination. The idea is to interpret terms as polynomials over $\mathbb{N}$ and show that reduction decreases the polynomial value.

Definition A.29 (Polynomial Interpretation). A *polynomial interpretation* assigns to each n-ary function symbol f a polynomial $[f] \in \mathbb{N}[x_1, \ldots, x_n]$ such that:

1. Each $[f]$ is *weakly monotone*: for each argument position, $[f]$ is weakly increasing in that argument.

2. The interpretation of any term t under any assignment of natural numbers to variables is a natural number.

The interpretation extends to terms by: $[x] = x$ for variables, and $[f(t_1, \ldots, t_n)] = [f]([t_1], \ldots, [t_n])$.

Theorem A.30. *A TRS terminates if there exists a polynomial interpretation such that for every rule $\ell \to r$: $[\ell] > [r]$ for all assignments of natural numbers ≥ 1 to variables.*

Example A.31. Consider the TRS with rules:

$$A(0, y) \to S(y)$$
$$A(S(x), 0) \to A(x, S(0))$$
$$A(S(x), S(y)) \to A(x, A(S(x), y))$$

This computes the Ackermann function. Termination can be proved using the interpretation $[0] = 1$, $[S](x) = x + 1$, and $[A](x, y) = 2^x(y + 1)$ (note that this goes beyond polynomial interpretations, illustrating that richer methods are sometimes needed).

Polynomial interpretations are particularly useful for TRSs where reduction increases term size or rearranges subterms, as can occur with associativity rules.

A.2.5 Lexicographic Combinations

When no single termination method suffices, *lexicographic combinations* can be employed.

Theorem A.32 (Lexicographic Termination). *Let $\succ_1$ and $\succ_2$ be well-founded orderings on terms. The lexicographic combination $\succ_{\mathsf{lex}}$, defined by*

$$s \succ_{\mathsf{lex}} t \iff s \succ_1 t \text{ or } (s \not\succ_1 t \text{ and } t \not\succ_1 s \text{ and } s \succ_2 t)$$

is well-founded.

More generally, one can partition the rules into layers and use different orderings for different layers, provided the interaction between layers is well-controlled. This technique is used in Chapter 5 to prove termination of the 77-rule system.

A.3 Confluence

Confluence is the property that ensures rewriting is deterministic in a strong sense: although multiple reduction sequences may be possible from a given term, they all lead to the same result. This property is essential for using rewriting as a computational mechanism.

A.3.1 The Church-Rosser Property

Definition A.33 (Confluence). A reduction relation $\to$ is *confluent* (or has the *Church-Rosser property*) if for all terms t, s_1, and s_2:

$$s_1 \;{}^*\!\!\leftarrow t \to^* s_2 \quad \Longrightarrow \quad \exists u.\ s_1 \to^* u \;{}^*\!\!\leftarrow s_2$$

In words: if a term t can reduce to both s_1 and s_2 (in any number of steps), then s_1 and s_2 can both reduce to some common term u.

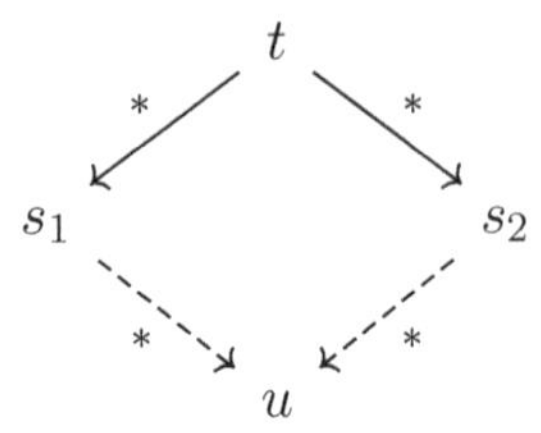

Figure A.1: The confluence (Church-Rosser) property.

Theorem A.34 (Church-Rosser Equivalence). *A relation $\rightarrow$ is confluent if and only if:*

$$s \leftrightarrow^* t \quad \Longrightarrow \quad \exists u.\ s \rightarrow^* u \ {}^*\!\leftarrow t$$

That is, convertible terms have a common reduct.

Corollary A.35. *If $\rightarrow$ is confluent and $s \leftrightarrow^* t$, and both s and t are in normal form (no further reductions possible), then $s = t$.*

This corollary is the key to using rewriting for decision procedures: if we can reduce any term to a normal form and the system is confluent, then two terms are equivalent if and only if they have the same normal form.

A.3.2 Local Confluence

Definition A.36 (Local Confluence). A relation $\rightarrow$ is *locally confluent* (or has the *weak Church-Rosser property*) if for all terms t, s_1, and s_2:

$$s_1 \leftarrow t \rightarrow s_2 \quad \Longrightarrow \quad \exists u.\ s_1 \rightarrow^* u \ {}^*\!\leftarrow s_2$$

Local confluence is a weaker property than confluence: it only requires that *single-step* divergences can be rejoined. However, a remarkable theorem due to Newman shows that for terminating systems, local confluence implies full confluence.

Theorem A.37 (Newman's Lemma, 1942). *If a relation $\rightarrow$ is terminating (strongly normalising) and locally confluent, then it is confluent.*

Proof. Let $<$ be the well-founded ordering supplied by termination. We prove confluence by well-founded induction on the source term t. Thus fix t and assume as induction hypothesis that every term strictly smaller than t is confluent. We must show that whenever

$$s_1 \ {}^*\!\leftarrow t \rightarrow^* s_2,$$

the terms s_1 and s_2 are joinable.

If either reduction sequence is empty, there is nothing to prove. So we may write

$$t \to t_1 \to^* s_1, \qquad t \to t_2 \to^* s_2.$$

By local confluence, the one-step peak $t_1 \leftarrow t \to t_2$ is joinable, so there exists u with

$$t_1 \to^* u \qquad \text{and} \qquad t_2 \to^* u.$$

Since $t_1 < t$ and $t_2 < t$, the induction hypothesis applies to both t_1 and t_2.

Apply it first to the pair of reductions $t_1 \to^* s_1$ and $t_1 \to^* u$. We obtain v such that

$$s_1 \to^* v \qquad \text{and} \qquad u \to^* v.$$

Apply it next to the pair of reductions $t_2 \to^* s_2$ and $t_2 \to^* u$. We obtain w such that

$$s_2 \to^* w \qquad \text{and} \qquad u \to^* w.$$

Now $u < t$, so the induction hypothesis applied to $u \to^* v$ and $u \to^* w$ yields some z with

$$v \to^* z \qquad \text{and} \qquad w \to^* z.$$

Composing the reduction sequences gives $s_1 \to^* v \to^* z$ and $s_2 \to^* w \to^* z$, so s_1 and s_2 are joinable. Hence the relation is confluent. □

Newman's Lemma is enormously practical: it reduces the problem of proving confluence (which involves arbitrary reduction sequences) to the problem of proving local confluence (which involves only single steps) plus termination.

A.3.3 Critical Pairs

The *critical pair* method, due to Knuth and Bendix [93], provides an algorithmic approach to proving local confluence.

Definition A.38 (Overlap). An *overlap* between rules $\ell_1 \to r_1$ and $\ell_2 \to r_2$ (with variables renamed to be disjoint) occurs when:

- A non-variable subterm of ℓ_1 at position p unifies with ℓ_2.

Let σ be the most general unifier of $\ell_1|_p$ and ℓ_2. The overlap produces the term $\ell_1\sigma$, which can be reduced in two ways:

1. By the first rule at the root: $\ell_1\sigma \to r_1\sigma$.

2. By the second rule at position p: $\ell_1\sigma \to \ell_1[r_2]_p\sigma$.

Definition A.39 (Critical Pair). A *critical pair* is the pair of terms $(r_1\sigma, \ell_1[r_2]_p\sigma)$ arising from an overlap. A critical pair is *trivial* if its two components are identical.

Example A.40. Consider the group theory rules:

$$\ell_1 \to r_1 : \mathsf{inv}(x) \cdot x \to e$$
$$\ell_2 \to r_2 : (x \cdot y) \cdot z \to x \cdot (y \cdot z)$$

The term $(\mathsf{inv}(x) \cdot x) \cdot z$ instantiates ℓ_2 with the subterm $\mathsf{inv}(x) \cdot x$ at position 1 matching ℓ_1. This gives the critical pair:

$$(\mathsf{inv}(x) \cdot (x \cdot z), \quad e \cdot z)$$

Definition A.41 (Joinable)**.** Two terms s and t are *joinable*, written $s \downarrow t$, if they have a common reduct: $\exists u.\ s \to^* u \;{}^*\!\!\leftarrow t$.

A critical pair (s, t) is *joinable* if $s \downarrow t$.

A rewrite rule is *left-linear* if no variable occurs more than once in its left-hand side. A TRS is left-linear if every rule is left-linear.

Theorem A.42 (Critical Pair Lemma)**.** *A left-linear TRS is locally confluent if and only if all its critical pairs are joinable.*

Proof. ($\Rightarrow$) Every critical pair is, by construction, a particular one-step peak. If the system is locally confluent, every such peak is joinable, hence every critical pair is joinable.

($\Leftarrow$) Assume the TRS is left-linear and that all critical pairs are joinable. Consider an arbitrary local peak

$$s_1 \leftarrow t \to s_2.$$

Let the two steps contract redexes at positions p and q of t. There are three cases.

If p and q are disjoint, then the two replacements occur in different subterms, so they commute. Replacing the subterm at p and then the one at q yields the same term as performing the two replacements in the opposite order. Hence the peak is joinable.

If one position is contained in the other, say $q = pr$, then either the inner redex lies in a non-variable subterm of the outer left-hand side or it lies inside a variable occurrence. In the first situation we have a genuine overlap, so after renaming variables apart the peak is an instance of a critical pair and is therefore joinable by assumption. In the second situation, left-linearity is essential: the variable in question occurs only once in the outer left-hand side, so reducing inside its substituted instance does not create interference elsewhere in the rule. One may therefore first perform the inner reduction inside the substitution and then apply the outer rule, or apply the outer rule first and then propagate the same substitutional reduction to every corresponding occurrence in the right-hand side. The two branches again have a common reduct.

Finally, if $p = q$, the two rules overlap at the same occurrence, so the peak is precisely an instance of a critical pair and is joinable by assumption.

Thus every one-step peak is joinable, which is local confluence. □

Combining Newman's Lemma with the Critical Pair Lemma gives:

Corollary A.43. *A terminating left-linear TRS is confluent if and only if all its critical pairs are joinable.*

This provides an effective procedure for checking confluence of terminating systems: enumerate all critical pairs (a finite set for finite TRSs) and check that each pair is joinable.

A.3.4 The Knuth-Bendix Completion Procedure

The Knuth-Bendix completion procedure [93] is an algorithm that attempts to transform a set of equations into an equivalent confluent and terminating TRS.

Algorithm: Knuth-Bendix Completion.

Input: A set E of equations and a reduction ordering $>$.
Output: A confluent TRS R equivalent to E, or failure.

1. Initialize $R := \emptyset$.
2. For each equation $s = t$ in E:
   - If $s > t$, add the rule $s \to t$ to R.
   - If $t > s$, add the rule $t \to s$ to R.
   - If neither, report failure.
3. Compute all critical pairs of R.
4. For each non-trivial critical pair (s, t):
   - Reduce s and t to normal forms s' and t'.
   - If $s' \neq t'$, add the equation $s' = t'$ to E and go to step 2.
5. If no new equations are added, return R.

The procedure may fail (if an equation cannot be oriented) or may not terminate (if critical pairs generate infinitely many new rules). However, when it succeeds, it produces a canonical TRS for the equational theory.

Example A.44. Applying Knuth-Bendix completion to the group axioms:

$$e \cdot x = x \qquad x \cdot e = x$$
$$\mathsf{inv}(x) \cdot x = e \qquad x \cdot \mathsf{inv}(x) = e$$
$$(x \cdot y) \cdot z = x \cdot (y \cdot z)$$

yields a confluent TRS after computing and orienting several critical pairs. The completed system includes rules such as $\mathsf{inv}(\mathsf{inv}(x)) \to x$ and $\mathsf{inv}(x \cdot y) \to \mathsf{inv}(y) \cdot \mathsf{inv}(x)$.

A.4 Normalization

When a TRS is both confluent and terminating, every term has a unique normal form. This section explores the implications for decidability and computational practice.

A.4.1 Normal Forms

Definition A.45 (Normal Form). A term t is in *normal form* (with respect to a TRS $\mathcal{R}$) if no rewrite rule applies to it—that is, if there is no term s such that $t \to_{\mathcal{R}} s$.

For any term t, a *normal form of* t is a term t' in normal form such that $t \to^* t'$.

Theorem A.46 (Existence and Uniqueness of Normal Forms). *Let $\mathcal{R}$ be a TRS.*

1. *If $\mathcal{R}$ is terminating, then every term has at least one normal form.*
2. *If $\mathcal{R}$ is confluent, then every term has at most one normal form.*
3. *If $\mathcal{R}$ is both terminating and confluent, then every term has a unique normal form.*

Proof.

1. Termination implies that every reduction sequence eventually halts. The final term of any maximal reduction sequence is a normal form.
2. Suppose $t \to^* s_1$ and $t \to^* s_2$ with s_1 and s_2 both in normal form. By confluence, there exists u with $s_1 \to^* u$ and $s_2 \to^* u$. Since s_1 and s_2 are in normal form, we must have $s_1 = u = s_2$.
3. Immediate from (1) and (2). □

We write $t{\downarrow}$ or $\mathsf{nf}(t)$ for the unique normal form of t (when it exists).

A.4.2 Unique Normal Forms

The property of having unique normal forms (UN) is fundamental to using TRSs as decision procedures.

Definition A.47 (UN Property). A TRS has the *UN property* (unique normal forms) if whenever s and t are normal forms with $s \leftrightarrow^* t$, then $s = t$.

Theorem A.48. *If a TRS $\mathcal{R}$ is confluent, then it has the UN property. If $\mathcal{R}$ is both confluent and terminating, then every term reduces to a unique normal form. Equivalently, whenever $s \leftrightarrow^* t$ and $s \rightarrow^* s'$, $t \rightarrow^* t'$ with s' and t' normal forms, one has $s' = t'$.*

When a TRS is confluent and terminating, it provides a *canonical* representation for each equivalence class of terms: the unique normal form. Two terms are equivalent if and only if they reduce to the same normal form.

A.4.3 Decidability of Word Problems

Definition A.49 (Word Problem). The *word problem* for an equational theory E is the problem of deciding, given two terms s and t, whether $s =_E t$ (whether s and t are provably equal using the equations in E).

The word problem is undecidable in general—even for finitely presented semigroups. However, term rewriting provides a powerful technique for those cases where decidability can be achieved.

Theorem A.50 (Decidability via Rewriting). *If an equational theory E has an equivalent TRS $\mathcal{R}$ that is both terminating and confluent, then the word problem for E is decidable.*

Proof. Given terms s and t:

1. Compute $\mathsf{nf}(s)$ and $\mathsf{nf}(t)$ (possible by termination).
2. Check whether $\mathsf{nf}(s) = \mathsf{nf}(t)$ (syntactic equality).
3. Return "yes" if they are equal, "no" otherwise.

By confluence, $s \leftrightarrow^* t$ if and only if $\mathsf{nf}(s) = \mathsf{nf}(t)$, so this procedure is correct. □

The complexity of the decision procedure depends on the complexity of normalisation. For many practical TRSs, normal forms can be computed in polynomial time.

Example A.51. The Knuth-Bendix completion of group theory produces a confluent, terminating TRS. Therefore, the word problem for group theory is decidable: given two group-theoretic expressions, we can determine whether they are equal by reducing both to normal form.

A.4.4 Rewriting Strategies

When multiple redexes are present, different *reduction strategies* can be employed.

Definition A.52 (Reduction Strategy). A *reduction strategy* specifies which redex to reduce at each step. Common strategies include:

- *Leftmost-innermost*: reduce the leftmost redex that contains no other redexes.
- *Leftmost-outermost*: reduce the leftmost redex that is not contained in another redex.
- *Parallel-outermost*: reduce all outermost redexes simultaneously.

For confluent TRSs, all strategies lead to the same normal form (when they terminate). For terminating TRSs, all strategies terminate. However, for TRSs that are only confluent or only terminating, the choice of strategy can affect whether normalisation succeeds.

Theorem A.53 (Strategy Independence). *If a TRS is confluent and terminating, then every reduction strategy reaches the (unique) normal form, and the choice of strategy affects only the length of the reduction sequence, not the result.*

A.5 Connection to Computational Paths

We now connect the general theory of term rewriting systems to the specific application developed in this book: the rewrite system for computational paths.

A.5.1 The Path Rewrite System as a TRS

The 77-rule system presented in Chapter 4 is a term rewriting system in the technical sense defined in Section A.1. The signature consists of the path constructors:

Definition A.54 (Path Signature). The signature Σ_{Path} for computational paths includes:

- Constants refl_a for each term a (reflexivity paths)
- Unary symbols: symm (symmetry)
- Binary symbols: trans (transitivity), $\mathsf{congrArg}$ (congruence), $\mathsf{congrFun}$ (function congruence)

- Additional constructors for products, sigma types, transport, etc.

The terms of this TRS are *path expressions*—syntactic representations of computational paths built from the primitive constructors. The 77 rewrite rules specify how these expressions can be simplified.

Example A.55. Some rules from Chapter 4, written in TRS notation:

$$\begin{aligned}
\mathsf{trans}(\mathsf{refl}, p) &\to p && \text{(Left identity)}\\
\mathsf{trans}(p, \mathsf{refl}) &\to p && \text{(Right identity)}\\
\mathsf{trans}(\mathsf{symm}(p), p) &\to \mathsf{refl} && \text{(Left inverse)}\\
\mathsf{trans}(p, \mathsf{symm}(p)) &\to \mathsf{refl} && \text{(Right inverse)}\\
\mathsf{symm}(\mathsf{symm}(p)) &\to p && \text{(Involution)}\\
\mathsf{trans}(\mathsf{trans}(p, q), r) &\to \mathsf{trans}(p, \mathsf{trans}(q, r)) && \text{(Associativity)}
\end{aligned}$$

These correspond to Rules 1–6 and 8 from the 77-rule system.

The system is *many-sorted* in practice, since paths have source and target types that constrain composition. However, the TRS framework handles this through implicit typing constraints.

A.5.2 The 77 Rules as a TRS

The 77 rules of Chapter 4 are organised into eight categories, corresponding to different aspects of path algebra:

1. **Groupoid core (Rules 1–8)**: These establish the basic algebraic structure—identity, inverse, and associativity. They form a TRS for the free groupoid on path generators.

2. **Type constructor rules (Rules 9–25)**: These govern how paths interact with product types, sigma types, sum types, and function types. They extend the groupoid structure to account for type-theoretic constructions.

3. **Transport rules (Rules 26–32)**: These handle the interaction of paths with transport in dependent type families.

4. **Context rules (Rules 33–48)**: These specify how paths propagate through one-hole contexts, implementing the functoriality of path application.

5. **Dependent context rules (Rules 49–60)**: The dependent counterpart, handling contexts where the codomain varies with the input.

6. **Binary context rules (Rules 61–68)**: Congruence for two-argument functions, both in dependent and non-dependent settings.

7. **Partial map rules (Rules 69–72)**: Interactions between left-partial and right-partial congruences in binary contexts.
8. **Structural completion (Rules 73–77)**: Closure rules that ensure the system is closed under path constructors.

From the TRS perspective, each category adds rewrite rules for additional function symbols while maintaining compatibility with the existing rules.

A.5.3 Why Confluence and Termination Matter

The central results of Chapter 5—that the 77-rule system is both confluent and terminating—have profound implications for the theory of computational paths:

Decidability of Path Equality. By Theorem A.50, confluence and termination imply that path equivalence is decidable. Given two path expressions p and q, we can determine whether they represent the same computational path by reducing both to normal form and checking syntactic equality.

This is not merely a theoretical nicety. In proof assistants and type checkers, equality of proofs often needs to be determined algorithmically. The decidability of path equality means this can be done without ad-hoc heuristics or incomplete approximations.

Canonical Representatives. Every equivalence class of paths has a unique normal form representative. This provides a canonical "simplest" expression for any path, eliminating redundancy in representations.

Consistency. Confluence ensures that the equational theory of paths is consistent: different reduction strategies cannot lead to contradictory results. This is essential for the coherence of the path algebra.

Computability of Path Simplification. Termination guarantees that path simplification always succeeds—it cannot loop indefinitely. Combined with confluence, this means that the normal form can be computed by any reduction strategy.

A.5.4 Techniques Used in the Confluence Proof

The confluence proof in Chapter 5 employs the techniques described in this appendix:

Termination Proof. The termination of the 77-rule system is established using a *lexicographic combination* of measures (Section A.2.5). The primary measure is a weight function (similar to those used in KBO; see Section A.2.3) that assigns values to path constructors. A secondary measure based on term structure handles rules that preserve weight.

Critical Pair Analysis. Local confluence in Chapter 5 is established by a refined critical-pair analysis. The genuine overlap families are enumerated and shown joinable, while disjoint and nested contextual peaks are handled by separate commutation and lifting lemmas. Because some computational-path rules repeat variables on the left-hand side, the argument is not a bare application of Theorem A.42; rather, it follows the same strategy with the additional peak cases made explicit.

Newman's Lemma. Once termination and local confluence are established, global confluence follows immediately from Newman's Lemma (Theorem A.37).

A.5.5 The Path Algebra as a Completed System

From the perspective of Knuth-Bendix completion (Section A.3.4), the 77-rule system can be viewed as a *completed* rewrite system for the equational theory of computational paths.

The starting point would be the fundamental laws that computational paths are expected to satisfy:

- Groupoid laws: identity, inverse, associativity
- Functoriality of congruence
- Coherence conditions for type constructors
- Transport laws

The 77 rules represent the result of "completing" these equations into a confluent, terminating TRS. Some rules (like associativity) come directly from the axioms; others (like the structural completion rules 73–77) arise from resolving critical pairs during completion.

This perspective explains the structure of the 77-rule system: it is not an arbitrary collection of identities, but a principled completion of the fundamental path laws into a decision procedure.

A.5.6 Comparison with Other Equational Theories

The computational path system shares features with several classical equational theories:

Group Theory. Like group theory, the path system has identity, inverse, and associativity. The groupoid core rules (1–8) are analogous to the completed TRS for groups. However, paths form a *groupoid* (category where all morphisms are invertible) rather than a group: composition is only defined when endpoints match.

Category Theory. Paths can be viewed as morphisms in a category, with composition given by transitivity. The functoriality rules capture how paths interact with functors (type constructors). The coherence conditions relate to coherence theorems in category theory.

Lambda Calculus. The interaction of paths with function types has parallels with the lambda calculus. Congruence in function arguments corresponds to beta-reduction; function extensionality corresponds to eta-expansion.

Type Theory. The path system is ultimately a fragment of intensional type theory, capturing the computational content of equality proofs. The termination and confluence results establish that this fragment has good computational properties.

A.6 Further Reading

For readers wishing to explore term rewriting systems in greater depth, we recommend the following resources:

Comprehensive Textbooks.

- Baader and Nipkow's *Term Rewriting and All That* [8] is the standard modern textbook, covering the theory thoroughly with careful proofs.
- Terese's *Term Rewriting Systems* [148] is an encyclopedic reference compiled by a consortium of researchers.

Foundational Papers.

- Knuth and Bendix's original paper [93] introduced the completion procedure and the KBO ordering.
- Newman's 1942 paper [116] established the fundamental lemma connecting local and global confluence.
- Dershowitz's papers [61, 62] developed the recursive path ordering and termination theory.

Connections to Type Theory.

- The relationship between rewriting and type theory is explored in the context of proof normalisation for dependent types [24].
- The connection to homotopy type theory is developed in the HoTT Book [150].

Summary

This appendix has provided the necessary background on term rewriting systems for understanding the technical machinery of Chapters 4 and 5. The key points are:

1. **Basic concepts**: A TRS consists of a signature (function symbols) and rewrite rules (directed equations). Reduction replaces instances of left-hand sides with corresponding right-hand sides.
2. **Termination**: A TRS terminates if all reduction sequences are finite. This is proved by well-founded orderings such as RPO and KBO, or by polynomial interpretations.
3. **Confluence**: A TRS is confluent if divergent reductions can always rejoin. Newman's Lemma reduces this to local confluence for terminating systems. For left-linear systems, the Critical Pair Lemma provides an algorithmic check for local confluence; more generally, critical-pair analysis is often supplemented by separate commutation arguments for nested or contextual peaks.
4. **Normalization**: Termination and confluence together imply unique normal forms, enabling decidability of the word problem.

5. **Computational paths**: The 77-rule system of Chapter 4 is a TRS for path expressions. Its confluence and termination (Chapter 5) establish decidability of path equality and provide canonical representatives for path equivalence classes.

The theory of term rewriting thus provides not only a framework for understanding the path calculus, but also the tools for proving its fundamental metatheoretic properties.

Appendix B

Groupoids and Higher Groupoids

"The notion of groupoid is one of the most transparent and flexible of category theory's contributions to mathematics."

—Ronald Brown [15]

This appendix provides the category-theoretic background necessary for understanding the higher algebraic structure of computational paths. The reader gains two things from this material. First, a precise vocabulary—categories, functors, natural transformations, coherence conditions—for describing the structures that arise in Part II without ad-hoc circumlocution. Second, and more importantly, an understanding of *why* the weak ω-groupoid theorem (Chapter 9) is the natural destination of the theory: computational paths do not merely satisfy groupoid laws by coincidence, but instantiate a well-studied categorical pattern that extends to all dimensions.

While Chapter 6 introduced groupoids in the context of paths and Chapter 9 developed the weak ω-groupoid structure, we did not pause to give the full categorical context. Here we remedy that omission, presenting the theory of groupoids and their higher-dimensional analogues in a self-contained manner.

The material progresses through increasing levels of abstraction. Section B.1 covers the basic theory of categories and groupoids, including the fundamental examples that motivate the definitions. Section B.2 introduces 2-categories and 2-groupoids, developing the notion of morphisms between morphisms. Section B.3 extends the pattern to n-groupoids and ω-groupoids, discussing both strict and weak variants. Section B.4 addresses coherence theorems, which explain when diagrams of higher cells necessarily commute. Finally, Section B.5

connects this categorical machinery to computational paths, explaining how the structures developed in Part II fit into the broader landscape.

B.1 Categories and Groupoids

We begin with the foundational notion of category, then specialise to groupoids—categories in which every morphism is invertible. The reader who has worked through Chapter 6 will find some repetition here, but the emphasis differs: there we focused on computational paths, while here we develop the abstract theory.

B.1.1 Basic Definitions

Definition B.1 (Category)**.** A *category* $\mathcal{C}$ consists of:

(i) A class $\mathrm{Ob}(\mathcal{C})$ of *objects*.

(ii) For each pair of objects $A, B \in \mathrm{Ob}(\mathcal{C})$, a class $\mathrm{Hom}_{\mathcal{C}}(A, B)$ of *morphisms* (or *arrows*) from A to B. We write $f : A \to B$ to indicate $f \in \mathrm{Hom}_{\mathcal{C}}(A, B)$.

(iii) For each object A, an *identity morphism* $\mathrm{id}_A : A \to A$.

(iv) For each triple of objects A, B, C, a *composition* operation

$$\circ : \mathrm{Hom}_{\mathcal{C}}(B, C) \times \mathrm{Hom}_{\mathcal{C}}(A, B) \to \mathrm{Hom}_{\mathcal{C}}(A, C),$$

writing $g \circ f$ for the composite of $f : A \to B$ followed by $g : B \to C$.

These data must satisfy three axioms:

Associativity. For all $f : A \to B$, $g : B \to C$, and $h : C \to D$,

$$h \circ (g \circ f) = (h \circ g) \circ f.$$

Left unit. For all $f : A \to B$,

$$\mathrm{id}_B \circ f = f.$$

Right unit. For all $f : A \to B$,

$$f \circ \mathrm{id}_A = f.$$

A *small category* is one where both $\mathrm{Ob}(\mathcal{C})$ and all hom-classes are sets. A *locally small category* is one where each $\mathrm{Hom}_{\mathcal{C}}(A, B)$ is a set, though the collection of objects may be a proper class.

Definition B.2 (Isomorphism). A morphism $f : A \to B$ in a category $\mathcal{C}$ is an *isomorphism* if there exists a morphism $g : B \to A$ such that

$$g \circ f = \mathrm{id}_A \quad \text{and} \quad f \circ g = \mathrm{id}_B.$$

In this case, g is called an *inverse* of f and is denoted f^{-1}. The inverse, when it exists, is unique.

Definition B.3 (Groupoid). A *groupoid* is a category in which every morphism is an isomorphism. Equivalently, a groupoid is a category $\mathcal{G}$ equipped with an *inverse operation* $(-)^{-1} : \mathrm{Hom}(A, B) \to \mathrm{Hom}(B, A)$ satisfying

$$f^{-1} \circ f = \mathrm{id}_A, \tag{B.1}$$

$$f \circ f^{-1} = \mathrm{id}_B \tag{B.2}$$

for all $f : A \to B$.

Intuitively, a groupoid is a "multi-object group": the algebraic structure of a group—composition, identity, inverses—is present, but there are potentially many objects rather than just one, and two morphisms can only be composed when the target of the first matches the source of the second. This is precisely the situation for computational paths, where a path from a to b can be composed with a path from b to c but not with one from d to e.

Proposition B.4 (Groupoid inverse laws). *In any groupoid:*

(i) $(\mathrm{id}_A)^{-1} = \mathrm{id}_A$.

(ii) $(f^{-1})^{-1} = f$.

(iii) $(g \circ f)^{-1} = f^{-1} \circ g^{-1}$.

Proof. (i) By uniqueness of inverses: $\mathrm{id}_A \circ \mathrm{id}_A = \mathrm{id}_A$, so id_A is its own inverse.
(ii) We verify that f is an inverse of f^{-1}: $f \circ f^{-1} = \mathrm{id}_B$ and $f^{-1} \circ f = \mathrm{id}_A$.
(iii) We compute:

$$\begin{aligned}(f^{-1} \circ g^{-1}) \circ (g \circ f) &= f^{-1} \circ (g^{-1} \circ g) \circ f \\ &= f^{-1} \circ \mathrm{id}_B \circ f \\ &= f^{-1} \circ f = \mathrm{id}_A.\end{aligned}$$

Similarly for the other composite. □

B.1.2 Examples of Groupoids

We now present the principal examples that motivate the abstract definition.

Example B.5 (Groups as one-object groupoids). Every group $(G, \cdot, e, (-)^{-1})$ determines a groupoid $\mathbf{B}G$ with a single object $*$, where $\mathrm{Hom}_{\mathbf{B}G}(*, *) = G$. Composition is the group operation, the identity morphism is e, and inverses are group inverses.

Conversely, every groupoid with exactly one object is essentially a group: the set of morphisms forms a group under composition. Thus groupoids generalise groups from one object to many—or equivalently, groups are "groupoids with a single object."

This perspective illuminates the role of groupoids in symmetry: a group captures the symmetries of a single object (the automorphisms), while a groupoid captures symmetries and identifications among many objects.

Example B.6 (Fundamental groupoid of a topological space). Let X be a topological space. The *fundamental groupoid* $\Pi_1(X)$ has:

- **Objects:** points of X.
- **Morphisms:** homotopy classes of continuous paths. Specifically, $\mathrm{Hom}_{\Pi_1(X)}(x, y)$ consists of continuous maps $\gamma : [0,1] \to X$ with $\gamma(0) = x$ and $\gamma(1) = y$, modulo homotopy relative to endpoints.
- **Identity:** the constant path at x.
- **Composition:** concatenation of paths. Given $\gamma : x \to y$ and $\delta : y \to z$, define $\delta \circ \gamma$ by

$$(\delta \circ \gamma)(t) = \begin{cases} \gamma(2t) & \text{if } 0 \leq t \leq 1/2, \\ \delta(2t - 1) & \text{if } 1/2 \leq t \leq 1. \end{cases}$$

- **Inverse:** path reversal. Given $\gamma : x \to y$, define $\gamma^{-1}(t) = \gamma(1 - t)$.

The groupoid axioms hold only up to homotopy—for instance, associativity of path concatenation requires a reparametrization—but since we quotient by homotopy, we obtain a strict groupoid.

The *fundamental group* $\pi_1(X, x_0)$ is recovered as the automorphism group $\mathrm{Hom}_{\Pi_1(X)}(x_0, x_0)$. Working with the full groupoid rather than the group at a basepoint has technical advantages: it avoids the arbitrary choice of basepoint, and it captures all path-connectedness information in a single structure.

Example B.7 (Action groupoid). Let G be a group acting on a set X from the left, with the action written $g \cdot x$ for $g \in G$ and $x \in X$. The *action groupoid* (or *transformation groupoid*) $X /\!\!/ G$ has:

- **Objects:** elements of X.
- **Morphisms:** pairs (g, x) with $g \in G$ and $x \in X$, representing a morphism from x to $g \cdot x$.
- **Identity:** $(e, x) : x \to x$.
- **Composition:** $(h, g \cdot x) \circ (g, x) = (hg, x)$.
- **Inverse:** $(g, x)^{-1} = (g^{-1}, g \cdot x)$.

The morphisms from x to y are precisely the group elements g such that $g \cdot x = y$. Two objects are in the same connected component of the groupoid if and only if they lie in the same orbit of the group action. The automorphism group at x is the *stabilizer* (or *isotropy group*) of x:

$$\mathrm{Aut}(x) = \mathrm{Stab}_G(x) = \{g \in G : g \cdot x = x\}.$$

Action groupoids play a central role in equivariant topology and the theory of orbifolds.

Example B.8 (Equivalence relations as groupoids)**.** Let $\sim$ be an equivalence relation on a set X. Define a groupoid with objects X and

$$\mathrm{Hom}(x, y) = \begin{cases} \{(x, y)\} & \text{if } x \sim y, \\ \varnothing & \text{otherwise.} \end{cases}$$

Reflexivity gives identities, symmetry gives inverses, and transitivity gives composition.

This groupoid is *thin*: each hom-set has at most one element. Conversely, every thin groupoid determines an equivalence relation. Thus equivalence relations are precisely "groupoids with at most one morphism between any two objects."

In type theory, such structures are called *setoids*—sets equipped with an equivalence relation playing the role of equality. The setoid model of type theory interprets types as groupoids of this form.

Example B.9 (Pair groupoid)**.** The *pair groupoid* (or *codiscrete groupoid*) on a set X has X as objects and exactly one morphism between any two objects: $\mathrm{Hom}(x, y) = \{(x, y)\}$. Composition is $(y, z) \circ (x, y) = (x, z)$, and the inverse of (x, y) is (y, x).

This is the "maximally connected" groupoid on X: all objects are isomorphic. It corresponds to the total equivalence relation where $x \sim y$ for all x, y.

B.1.3 Functors and Natural Transformations

Groupoids form a category (indeed, a 2-category), and understanding maps between groupoids is essential.

Definition B.10 (Functor). A *functor* $F : \mathcal{C} \to \mathcal{D}$ between categories consists of:

(i) A function $F : \mathrm{Ob}(\mathcal{C}) \to \mathrm{Ob}(\mathcal{D})$.

(ii) For each pair A, B, a function $F : \mathrm{Hom}_{\mathcal{C}}(A, B) \to \mathrm{Hom}_{\mathcal{D}}(F(A), F(B))$.

These must satisfy:

$$F(\mathrm{id}_A) = \mathrm{id}_{F(A)}, \tag{B.3}$$

$$F(g \circ f) = F(g) \circ F(f). \tag{B.4}$$

Proposition B.11. *A functor between groupoids automatically preserves inverses:* $F(f^{-1}) = F(f)^{-1}$.

Proof. $F(f) \circ F(f^{-1}) = F(f \circ f^{-1}) = F(\mathrm{id}) = \mathrm{id}$, and similarly for the other composite. By uniqueness of inverses, $F(f^{-1}) = F(f)^{-1}$. □

Definition B.12 (Natural transformation). Let $F, G : \mathcal{C} \to \mathcal{D}$ be functors. A *natural transformation* $\alpha : F \Rightarrow G$ consists of a family of morphisms $\alpha_A : F(A) \to G(A)$ in $\mathcal{D}$, one for each object $A \in \mathcal{C}$, such that for every morphism $f : A \to B$ in $\mathcal{C}$, the following square commutes:

$$\begin{array}{ccc} F(A) & \xrightarrow{\alpha_A} & G(A) \\ {\scriptstyle F(f)}\downarrow & & \downarrow{\scriptstyle G(f)} \\ F(B) & \xrightarrow[\alpha_B]{} & G(B) \end{array} \tag{B.5}$$

That is, $G(f) \circ \alpha_A = \alpha_B \circ F(f)$.

Definition B.13 (Natural isomorphism). A natural transformation $\alpha : F \Rightarrow G$ is a *natural isomorphism* if each component α_A is an isomorphism. In this case, we write $F \cong G$ and say that F and G are *naturally isomorphic*.

Between groupoids, a natural transformation is automatically a natural isomorphism: since every morphism in a groupoid is invertible, each component α_A has an inverse.

B.2 2-Categories and 2-Groupoids

The notion of natural transformation suggests that categories themselves have "morphisms between morphisms." This leads to the concept of a 2-category, where we have objects, 1-morphisms (morphisms between objects), and 2-morphisms (morphisms between 1-morphisms).

B.2.1 2-Categories

Definition B.14 (Strict 2-category)**.** A *strict 2-category* $\mathcal{C}$ consists of:

(i) A class of *objects* (or *0-cells*).

(ii) For each pair of objects A, B, a category $\mathcal{C}(A, B)$ whose objects are called *1-morphisms* (or *1-cells*) and whose morphisms are called *2-morphisms* (or *2-cells*).

(iii) For each object A, an identity 1-morphism $\mathrm{id}_A \in \mathcal{C}(A, A)$.

(iv) For each triple A, B, C, a composition functor

$$\circ : \mathcal{C}(B, C) \times \mathcal{C}(A, B) \to \mathcal{C}(A, C).$$

These must satisfy the usual category axioms strictly: associativity and unit laws hold as equalities.

The composition functor $\circ$ acts on both 1-morphisms and 2-morphisms. On 1-morphisms, it gives ordinary composition. On 2-morphisms, it gives *horizontal composition*: if $\alpha : f \Rightarrow f'$ and $\beta : g \Rightarrow g'$ are 2-morphisms with $f, f' : A \to B$ and $g, g' : B \to C$, then $\beta \circ \alpha : g \circ f \Rightarrow g' \circ f'$.

Within each hom-category $\mathcal{C}(A, B)$, we have *vertical composition* of 2-morphisms: if $\alpha : f \Rightarrow g$ and $\beta : g \Rightarrow h$ are 2-cells with the same source and target objects, their vertical composite is $\beta \cdot \alpha : f \Rightarrow h$.

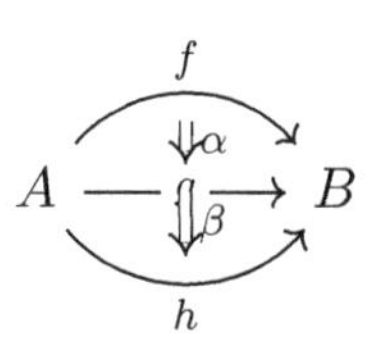

Proposition B.15 (Interchange law)**.** *In any 2-category, horizontal and vertical composition satisfy the* interchange law*: given 2-cells*

$$\alpha : f \Rightarrow f', \quad \alpha' : f' \Rightarrow f'', \quad \beta : g \Rightarrow g', \quad \beta' : g' \Rightarrow g''$$

with appropriate types, we have

$$(\beta' \cdot \beta) \circ (\alpha' \cdot \alpha) = (\beta' \circ \alpha') \cdot (\beta \circ \alpha). \tag{B.6}$$

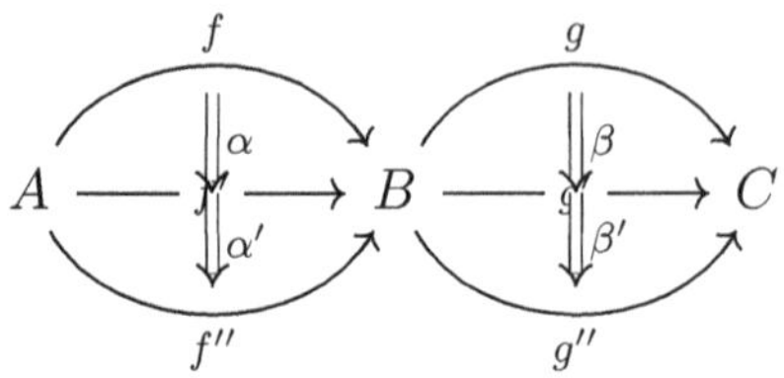

The interchange law says we can compute the composite 2-cell from $g \circ f$ to $g'' \circ f''$ in two ways: first horizontally then vertically, or first vertically then horizontally.

Example B.16 ($\mathbf{Cat}$: the 2-category of categories)**.** The paradigmatic example of a 2-category is $\mathbf{Cat}$:

- **Objects:** small categories.
- **1-morphisms:** functors.
- **2-morphisms:** natural transformations.

Vertical composition is composition of natural transformations; horizontal composition is whiskering.

Example B.17 ($\mathbf{Gpd}$: the 2-category of groupoids)**.** The 2-category $\mathbf{Gpd}$ is the full sub-2-category of $\mathbf{Cat}$ consisting of groupoids. Since natural transformations between functors of groupoids are automatically natural isomorphisms, every 2-morphism in $\mathbf{Gpd}$ is invertible.

B.2.2 Strict 2-Groupoids

Definition B.18 (Strict 2-groupoid)**.** A *strict 2-groupoid* is a strict 2-category in which:

(i) Every 1-morphism $f : A \to B$ has a strict inverse $f^{-1} : B \to A$ with $f^{-1} \circ f = \mathrm{id}_A$ and $f \circ f^{-1} = \mathrm{id}_B$.

(ii) Every 2-morphism is invertible.

Equivalently, each hom-category $\mathcal{C}(A, B)$ is a groupoid, and every 1-morphism is strictly invertible.

B.2.3 Bicategories: Weak 2-Categories

In many situations, the axioms of a 2-category hold only up to coherent isomorphism rather than strictly. This leads to the notion of a bicategory.

Definition B.19 (Bicategory)**.** A *bicategory* $\mathcal{B}$ consists of:

(i) A class of objects.

(ii) For each pair A, B, a category $\mathcal{B}(A, B)$.

(iii) For each object A, a 1-morphism $I_A : A \to A$.

(iv) For each triple A, B, C, a composition functor $\circ : \mathcal{B}(B, C) \times \mathcal{B}(A, B) \to \mathcal{B}(A, C)$.

(v) For each quadruple A, B, C, D and 1-morphisms $f : A \to B, g : B \to C, h : C \to D$, an *associator* 2-isomorphism

$$\alpha_{h,g,f} : (h \circ g) \circ f \overset{\cong}{\Rightarrow} h \circ (g \circ f).$$

(vi) For each $f : A \to B$, *unitor* 2-isomorphisms

$$\lambda_f : I_B \circ f \overset{\cong}{\Rightarrow} f, \qquad \rho_f : f \circ I_A \overset{\cong}{\Rightarrow} f.$$

The associators and unitors must be natural in all arguments and satisfy the *coherence conditions* described below.

The coherence conditions for a bicategory are:

1. **Pentagon identity.** For composable 1-morphisms f, g, h, k, the following diagram of 2-isomorphisms commutes:

$$\begin{array}{ccccc} & & ((k \circ h) \circ g) \circ f & & \\ & \swarrow^{\alpha_{k,h,g} \circ \mathrm{id}_f} & & \searrow^{\alpha_{k \circ h,g,f}} & \\ (k \circ (h \circ g)) \circ f & & & & (k \circ h) \circ (g \circ f) \\ \downarrow{\scriptstyle \alpha_{k,h \circ g,f}} & & & & \downarrow{\scriptstyle \alpha_{k,h,g \circ f}} \\ k \circ ((h \circ g) \circ f) & & \xrightarrow[\mathrm{id}_k \circ \alpha_{h,g,f}]{} & & k \circ (h \circ (g \circ f)) \end{array} \tag{B.7}$$

2. **Triangle identity.** For composable 1-morphisms $f : A \to B$ and $g : B \to C$:

$$\begin{array}{ccc} (g \circ I_B) \circ f & \xrightarrow{\alpha_{g,I_B,f}} & g \circ (I_B \circ f) \\ & {\scriptstyle \rho_g \circ \mathrm{id}_f} \searrow \quad \swarrow {\scriptstyle \mathrm{id}_g \circ \lambda_f} & \\ & g \circ f & \end{array} \tag{B.8}$$

Remark B.20. A strict 2-category is a bicategory where all associators and unitors are identity 2-morphisms. The generalisation from strict to weak is crucial: many naturally occurring 2-categorical structures are bicategories but not strict 2-categories.

Example B.21 (Bimodules)**.** Let $\mathbf{Bimod}$ have rings as objects. A 1-morphism from R to S is an (R, S)-bimodule. A 2-morphism is a bimodule homomorphism. Composition is tensor product over the intermediate ring. This forms a bicategory but not a strict 2-category: the tensor product is associative only up to isomorphism.

Example B.22 (Spans)**.** The bicategory $\mathbf{Span}(\mathcal{C})$ has objects of $\mathcal{C}$ as objects. A 1-morphism from A to B is a span $A \leftarrow M \rightarrow B$. Composition is by pullback, which is associative only up to isomorphism.

B.3 n-Groupoids and ω-Groupoids

The pattern continues: just as 2-categories have 2-morphisms between 1-morphisms, 3-categories have 3-morphisms between 2-morphisms, and so on. The infinite-dimensional limit gives ω-categories. When all cells are invertible, we obtain n-groupoids and ω-groupoids.

B.3.1 The Pattern

Definition B.23 (Informal: strict n-category)**.** A *strict n-category* has:

- 0-cells (objects)
- 1-cells (morphisms between 0-cells)
- 2-cells (morphisms between parallel 1-cells)

$\vdots$

-
- n-cells (morphisms between parallel $(n-1)$-cells)

with composition operations in each dimension satisfying associativity and unit laws strictly.

Definition B.24 (Informal: strict n-groupoid)**.** A *strict n-groupoid* is a strict n-category in which every k-cell (for $k \geq 1$) is invertible.

Making these definitions precise requires care, which leads to the notion of globular sets.

B.3.2 Globular Sets

Definition B.25 (Globular set)**.** A *globular set* G consists of:

(i) Sets G_n for each $n \in \mathbb{N}$ (the set of n-cells).

(ii) Source and target maps $s_n, t_n : G_n \to G_{n-1}$ for $n \geq 1$.

These must satisfy the *globular identities*: for $n \geq 2$,

$$s_{n-1} \circ s_n = s_{n-1} \circ t_n, \tag{B.9}$$
$$t_{n-1} \circ s_n = t_{n-1} \circ t_n. \tag{B.10}$$

Remark B.26. The globular identities express that an n-cell goes between *parallel* $(n-1)$-cells—cells with the same source and target. Pictorially:

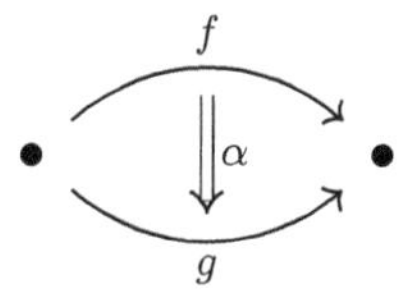

A 2-cell α has $s_2(\alpha) = f$ and $t_2(\alpha) = g$, with $s_1(f) = s_1(g)$ and $t_1(f) = t_1(g)$.

Definition B.27 (Reflexive globular set)**.** A *reflexive globular set* is a globular set equipped with identity maps $i_n : G_n \to G_{n+1}$ satisfying $s_{n+1} \circ i_n = t_{n+1} \circ i_n = \mathrm{id}_{G_n}$.

B.3.3 Strict ω-Categories

Definition B.28 (Strict ω-category)**.** A *strict ω-category* is a reflexive globular set G equipped with composition operations

$$\circ_k : G_n \times_{G_k} G_n \to G_n$$

for all $n > k \geq 0$, where $G_n \times_{G_k} G_n$ is the set of k-composable pairs:

$$G_n \times_{G_k} G_n = \{(x, y) \in G_n \times G_n : t_n^k(x) = s_n^k(y)\}.$$

Here $s_n^k = s_{k+1} \circ \cdots \circ s_n$ and $t_n^k = t_{k+1} \circ \cdots \circ t_n$.

These operations must satisfy:

(S1) **Source/target.** $s_n(y \circ_k x) = s_n(x)$ if $k = n-1$, otherwise $s_n(y \circ_k x) = s_n(y) \circ_k s_n(x)$. Similarly for t_n.

(S2) **Associativity.** $(z \circ_k y) \circ_k x = z \circ_k (y \circ_k x)$.

(S3) **Units.** $i_{n-1}(t_{n-1}(x)) \circ_{n-1} x = x = x \circ_{n-1} i_{n-1}(s_{n-1}(x))$.

(S4) **Interchange.** $(y' \circ_k y) \circ_j (x' \circ_k x) = (y' \circ_j x') \circ_k (y \circ_j x)$ when $j < k$ and compositions are defined.

Definition B.29 (Strict ω-groupoid). A *strict ω-groupoid* is a strict ω-category equipped with inverse operations $(-)^{-1}_k : G_n \to G_n$ for all $n > k \geq 0$, satisfying

$$x_k^{-1} \circ_k x = i_k(s_n^k(x)), \qquad x \circ_k x_k^{-1} = i_k(t_n^k(x)).$$

B.3.4 Weak ω-Categories and Weak ω-Groupoids

Strict ω-categories are algebraically tractable but too rigid for many purposes. In topology, for instance, the composition of paths is associative only up to homotopy, not on the nose. This motivates weak ω-categories, where the laws hold up to coherent higher cells.

Definition B.30 (Informal: weak ω-category). A *weak ω-category* is an ω-categorical structure where:

- Associativity, unit, and interchange laws hold up to invertible higher cells.
- These coherence cells satisfy their own coherence conditions, again up to invertible cells one dimension higher.
- This continues at all levels.

Making this precise is notoriously difficult. We mention two approaches.

Batanin's Definition

Batanin [11] defined weak ω-categories using the machinery of *globular operads* (see also Penon [120] for an alternative approach and Brown–Higgins [16] for the algebra of cubes and crossed complexes). The idea is:

1. Define a *globular operad* $\mathcal{O}$ that encodes all possible composition operations and their coherence data.
2. A weak ω-category is an algebra over a suitable contractible globular operad.
3. The contractibility condition ensures that any two ways of composing cells that "should" be equal are connected by a unique coherence cell.

This approach is technically sophisticated but provides a clean universal property.

Grothendieck's Homotopy Hypothesis

In his manuscript *Pursuing Stacks* [77], Grothendieck proposed that weak ω-groupoids should model homotopy types. This is the *homotopy hypothesis*:

Conjecture B.31 (Grothendieck's homotopy hypothesis)**.** There is an equivalence of homotopy theories between:

(i) The homotopy theory of topological spaces (or CW-complexes).

(ii) The homotopy theory of weak ω-groupoids.

More precisely, the fundamental ω-groupoid construction should give an equivalence from spaces to ω-groupoids, with a geometric realisation functor going the other way.

The homotopy hypothesis has been verified for several definitions of weak ω-groupoid, including Kan complexes (which are one model for weak ω-groupoids in the simplicial setting) [106, 90].

Remark B.32. The homotopy hypothesis connects algebra to topology in a profound way: the algebraic structure of ω-groupoids exactly captures the homotopy-theoretic structure of spaces. This connection underlies homotopy type theory, where types are interpreted as spaces and identity types as path spaces.

B.3.5 Truncation

Definition B.33 (n-truncation)**.** A weak ω-groupoid G is *n-truncated* (or is a *weak n-groupoid*) if no new homotopical information appears above dimension n: for every $k > n$ and every pair of parallel $(k-1)$-cells, the comparison problem is weakly contractible one dimension higher. Equivalently, all cells above dimension n serve only as coherent fillers rather than as new independent data.

Examples:

- A 0-groupoid is a set.
- A 1-groupoid is an ordinary groupoid.
- A 2-groupoid has 0-, 1-, and 2-cells with non-trivial structure, but all higher cells are determined.

The fundamental groupoid $\Pi_1(X)$ is 1-truncated: it captures only the 1-dimensional homotopy information (paths modulo homotopy). The fundamental 2-groupoid $\Pi_2(X)$ captures homotopies between paths as well.

B.4 Coherence

Coherence theorems address the question: when do all diagrams of a certain type commute? In categorical structures with non-strict composition, multiple ways of associating and applying units can lead to different formal expressions. Coherence theorems assert that, despite this apparent ambiguity, all such expressions are canonically equivalent.

B.4.1 Mac Lane's Coherence Theorem

The prototypical coherence theorem is Mac Lane's result for monoidal categories [98] (see also Selinger [139] for a survey of graphical languages for monoidal categories).

Definition B.34 (Monoidal category). A *monoidal category* $(\mathcal{C}, \otimes, I, \alpha, \lambda, \rho)$ consists of:

- A category $\mathcal{C}$.
- A bifunctor $\otimes : \mathcal{C} \times \mathcal{C} \to \mathcal{C}$.
- A unit object I.
- Natural isomorphisms:

$$\begin{aligned} \alpha_{A,B,C} &: (A \otimes B) \otimes C \xrightarrow{\cong} A \otimes (B \otimes C), \\ \lambda_A &: I \otimes A \xrightarrow{\cong} A, \\ \rho_A &: A \otimes I \xrightarrow{\cong} A. \end{aligned}$$

These must satisfy the pentagon and triangle identities (analogous to (B.7) and (B.8)).

Theorem B.35 (Mac Lane's coherence theorem). *In a monoidal category, any two morphisms built from α, λ, ρ, and their inverses, with the same source and target, are equal.*

Corollary B.36. *Every monoidal category is monoidally equivalent to a* strict *monoidal category, where α, λ, and ρ are identity morphisms.*

This means we can "pretend" that associativity and unit laws hold strictly when working with monoidal categories, as long as we only care about results up to equivalence.

B.4.2 Stasheff Associahedra

Coherence for associativity can be visualised geometrically using *Stasheff associahedra* (or *associahedra*) [142].

The n-th associahedron K_n is a convex polytope whose vertices correspond to the different ways of fully parenthesizing a product of n terms. Edges connect parenthesizations that differ by a single application of the associativity law.

- K_2 is a point (only one way to parenthesize ab).
- K_3 is a line segment, with vertices $(ab)c$ and $a(bc)$.
- K_4 is a pentagon, with vertices $((ab)c)d$, $(a(bc))d$, $a((bc)d)$, $a(b(cd))$, $(ab)(cd)$.
- K_5 is a 3-dimensional polytope with 14 vertices.

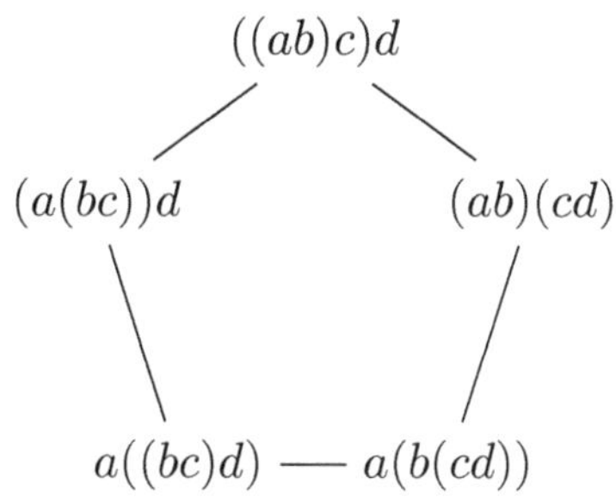

The pentagon identity (equation (B.7)) is precisely the assertion that the boundary of K_4 commutes. Higher associahedra encode higher coherence conditions.

Theorem B.37 (Stasheff)**.** *The coherence conditions for an A_∞-space (a space with multiplication associative up to all higher homotopies) are encoded by the associahedra: the n-th condition involves the faces of K_n.*

This connects coherence to topology: the associahedra are the "shapes" of the coherence conditions.

B.4.3 Coherence for Bicategories

The coherence theorem for bicategories is more subtle than for monoidal categories.

Theorem B.38 (Coherence for bicategories)**.** *Every bicategory is biequivalent to a strict 2-category.*

However, unlike the monoidal case, we cannot always strictify "in place": the strict 2-category may have different objects. What we can say is:

Theorem B.39. *In a bicategory, any diagram of 2-cells built from associators, unitors, and their inverses, with the same source and target, commutes.*

This is the bicategorical analogue of Mac Lane's theorem: the coherence isomorphisms, though non-identity, are uniquely determined.

B.4.4 Coherence for ω-Groupoids

For weak ω-groupoids, coherence takes a different form. Instead of asserting that diagrams commute, we assert that any diagram is filled by a higher cell, uniquely up to further coherence.

Theorem B.40 (Coherence for weak ω-groupoids)**.** *In a weak ω-groupoid, any two parallel n-cells that are "formally equivalent" (i.e., can be connected by a sequence of coherence isomorphisms) are connected by an $(n+1)$-cell, and any two such connectors agree up to higher coherence.*

This is the infinite-dimensional analogue of coherence: rather than asserting equality, we assert canonical connection at the next level, controlled by the higher cells above it.

B.5 Connection to Computational Paths

We now connect the abstract theory to the computational paths developed in Part II of this book.

B.5.1 Computational Paths as a Weak ω-Groupoid

The main theorem of Chapter 9 asserts that computational paths form a weak ω-groupoid. Let us revisit this in light of the categorical background.

Definition B.41 (Path globular set)**.** For a type A, define a globular set $\mathcal{P}(A)$ by:

- $\mathcal{P}(A)_0 = A$ (terms of type A).
- $\mathcal{P}(A)_1 = \{p : \mathsf{Path}(a, b) \mid a, b \in A\}$ (paths).
- $\mathcal{P}(A)_2 = \{D : p \Rrightarrow q\}$ (derivations/2-cells).
- $\mathcal{P}(A)_n$: the n-cells (derivations between $(n-1)$-cells).

Source and target maps are the evident projections.

Theorem B.42 (Part II, main result). *$\mathcal{P}(A)$ carries the structure of a weak ω-groupoid:*

(i) Composition at level 1 is trans.

(ii) Inverses at level 1 come from symm.

(iii) The groupoid laws (G1–G8) hold up to 2-cells.

(iv) Composition at level 2 is vertical composition of derivations.

(v) The 2-categorical laws hold up to 3-cells.

(vi) This pattern continues at all levels.

The key insight is that the rewrite rules of Chapter 4 provide exactly the coherence cells needed for a weak ω-groupoid structure. Each rewrite rule is a 2-cell witnessing a coherence equation.

B.5.2 Stabilisation Above Dimension 3

A striking feature of computational paths is that the ω-groupoid structure becomes "trivial" above a certain dimension.

Theorem B.43 (Contractibility, Chapter 9). *For computational paths, the ω-groupoid $\mathcal{P}(A)$ is 3-truncated in the weak sense: 3-cells carry genuine coherence data, but from dimension 4 onward the higher comparison problems are canonically contractible. In particular, any two parallel 2-cells $D, D' : p \Rrightarrow q$ are connected by a 3-cell, any two parallel 3-cells are connected by a 4-cell, and the same pattern continues in all higher dimensions.*

This means computational paths form a genuinely higher weak ω-groupoid whose non-trivial homotopical content is concentrated in dimensions 0, 1, 2, and 3 rather than a fully non-degenerate tower in every dimension. The contrast with homotopy theory is instructive:

Structure	**Comp. Paths**	**Homotopy Theory**
0-cells (points/terms)	Terms of A	Points of X
1-cells (paths)	$\mathsf{Path}(a, b)$	Paths $x \rightsquigarrow y$
2-cells (homotopies)	Derivations	Homotopies
3-cells	**Genuine coherence data**	Generally non-trivial
n-cells ($n \geq 4$)	**Contractible fillers**	May be non-trivial

B.5.3 Comparison with Topological ω-Groupoids

The fundamental ω-groupoid of a topological space X has non-trivial structure in all dimensions (unless X is homotopically trivial). For instance:

- $\pi_2(S^2) \cong \mathbb{Z}$ (2-cells are non-trivial).
- $\pi_3(S^2) \cong \mathbb{Z}$ (3-cells are non-trivial).
- The homotopy groups $\pi_n(S^2)$ are non-trivial for infinitely many n.

In contrast, computational paths (as developed in this book) are 3-truncated in the weak sense: all the genuinely independent structure is concentrated in dimensions 0, 1, 2, and 3. This corresponds to the fact that:

- Paths between terms are non-trivial.
- Derivations (rewrites between paths) are non-trivial.
- The basic 3-dimensional coherence witnesses are also non-trivial.
- From dimension 4 onward, higher comparison problems are canonically contractible.

This 3-truncation has important consequences:

1. It means computational paths cannot directly model all homotopy types—their stabilised tower behaves like a 3-type rather than an arbitrary space with non-trivial homotopy in every dimension.
2. It reflects the "syntactic" nature of paths: path identity is determined by a finite rewrite system, unlike the infinite complexity of topological homotopy.
3. It gives decidability properties: determining whether two paths are equal (up to the relevant rewrite equivalence) is algorithmically tractable, while still leaving room for genuine low-dimensional coherence.

B.5.4 The Fundamental Groupoid

Just as topological spaces have fundamental groupoids, types have fundamental groupoids via computational paths.

Definition B.44 (Fundamental groupoid of a type)**.** For a type A, the *fundamental groupoid* $\Pi_1(A)$ has:

- **Objects:** terms $a : A$.

- **Morphisms:** equivalence classes $[p]$ of paths $p : \mathsf{Path}(a, b)$, where $p \sim q$ if there exists a 2-cell $D : p \Rrightarrow q$.

By the coherence results of Part II, $\Pi_1(A)$ is a strict groupoid: the equivalence relation $\sim$ identifies paths that are "provably equal" via the rewrite system, and the groupoid laws hold strictly on equivalence classes.

Remark B.45 (Type-theoretic/topological comparison)**.** When a geometric realisation $|A|$ is available and is compatible with the path structure under discussion, one expects $\Pi_1(A)$ to agree with the topological fundamental groupoid $\Pi_1(|A|)$. Making this precise requires a specific realisation functor together with a theorem identifying computational paths with continuous paths up to homotopy, so we record the comparison only as guiding intuition here.

This comparison connects the type-theoretic and topological notions of fundamental groupoid. The computational paths of type theory correspond to the continuous paths of topology, and the rewrite equivalence corresponds to homotopy equivalence.

B.5.5 Summary

The categorical framework of ω-groupoids provides the right language for understanding computational paths:

1. **Paths are 1-cells** in a globular structure.
2. **Derivations are 2-cells**, witnessing the groupoid laws.
3. **Coherence** ensures that all reasonable diagrams commute.
4. **Weak 3-truncation** distinguishes computational paths from fully homotopical structures: genuine coherence persists through dimension 3, while higher comparison problems stabilise above that.
5. **The fundamental groupoid** connects type-theoretic and topological perspectives.

This completes our survey of groupoids and higher groupoids. The reader interested in pursuing these topics further may consult Leinster [99] for an accessible introduction to higher categories, or the original sources [11, 77] for the technical foundations of weak ω-categories.

Summary

This appendix provided the category-theoretic foundations underlying the higher algebraic structure of computational paths. We covered standard definitions of categories and groupoids, then ascended to 2-categories, bicategories, and the coherence theorems that govern them. The progression continued through n-groupoids and weak ω-groupoids in the sense of Batanin and Leinster, culminating in the connection between these abstract structures and the computational paths developed in the main text.

Appendix C

The Lean 4 Formalization

The core constructions discussed in this appendix have corresponding Lean 4 developments. This appendix documents the structure of the `ComputationalPathsLean` *library, explains its key design decisions, and records how the current formalisation realises the higher coherence tower explicitly as Type-valued derivations, together with the normalisation and diamond-filler arguments that prove stabilisation above dimension 3.*

The Lean 4 formalisation accompanying this book serves multiple purposes [132, 133]. First, it provides machine-checked verification of a substantial core of the theory, including the main constructions and many of the major theorems used throughout the manuscript, ensuring that the intricate coherence arguments and higher-dimensional constructions are free from subtle errors. The Lean 4 proof assistant [28] was chosen for its efficient kernel, support for dependent types, and a metaprogramming framework that we use extensively for automation. Second, it demonstrates that the computational paths framework is practical—the path-level operations and rewrite procedures are computable, while the higher coherence layer is represented by explicit Type-valued derivations `Derivation2`, `Derivation3`, `Derivation4`, and `DerivationHigh`. The development proves the low-dimensional coherences directly and establishes stabilisation through normalisation and diamond fillers, without adding axioms, while isolating the remaining `Prop`-level transport to a single legacy fallback in the raw-`Path` level-3 connector. It integrates smoothly with Mathlib, the community-maintained library of formalised mathematics. Third, it serves as a reference implementation that readers can use to experiment with their own extensions and applications.

This appendix is organised into five sections. Section C.1 records the companion repository and the reproducibility conventions for the formalisation.

Section C.2 describes the core architecture and module organization. Section C.3 discusses the key design decisions that shape the formalisation. Section C.4 provides guidance for navigating the proofs and finding specific theorems. Section C.5 explains how to extend the formalisation with new rewrite rules or coherence results.

C.1 Getting Started

This section records how the companion formalisation is packaged, cited, and reproduced, without attempting to duplicate the repository's own toolchain instructions.

C.1.1 Companion repository and reproducibility

The Lean formalisation lives in the external companion repository `ComputationalPathsLean`. Because toolchain instructions, caches, and editor integrations change more quickly than the mathematics, this appendix does not attempt to serve as a step-by-step installation manual. The authoritative build instructions are maintained with the repository itself. For the purposes of the book, the important reproducibility fact is that the development is version-pinned through `lean-toolchain` and `lakefile.lean`, so a checked-out release reproduces the cited theorems and definitions.

C.1.2 Accessing the companion repository

The repository is hosted on GitHub:

`https://github.com/Arthur742Ramos/ComputationalPathsLean`

Readers who wish to inspect or rebuild the development should consult the README of that companion repository, which records the current workflow, editor support, and release procedure. In the present appendix we focus on the mathematical architecture of the formalisation rather than on environment-specific setup steps.

C.1.3 Project Structure Overview

The Lean formalisation lives in the external companion repository `ComputationalPathsLean`; it is not vendored into the book repository itself. Its top-level layout is:

```
ComputationalPathsLean/
  ComputationalPaths.lean
  Main.lean
  lakefile.lean
  lean-toolchain
  ComputationalPaths/
    Basic.lean
    Path.lean
    Path/
      Basic/
      Rewrite/
      Homotopy/
      CompPath/
      OmegaGroupoid/
      Polygraph/
      Algebra/
      Category/
      Logic/
      Topology/
    HigherCategory/
    Comparison/
    Examples/
    [many domain trees]
  docs/
  docbuild/
  paper/
  Scripts/
```

The `lean-toolchain` and `lakefile.lean` files pin the development to Lean 4.24.0 and Mathlib v4.24.0. The core library is organised around `ComputationalPaths.lean` and `ComputationalPaths/Path/`, while many additional top-level directories collect broader application-oriented developments. In practice, readers working directly with the material in this book will spend most of their time inside the `Path/`, `HigherCategory/`, `Comparison/`, and `Examples/` subtrees.

C.2 Core Architecture

The formalisation implements the theoretical framework developed in this book through four main layers: the `Path` type and basic operations, the rewrite system, the derivation types (2-cells), and the higher coherence structures.

C.2.1 Path.lean: The Path Type and Basic Operations

The fundamental definition is the `Path` structure, which packages a propositional equality proof with a trace of rewrite steps:

```
/-- A computational path from a to b. The equality
    semantics
     come solely from `proof : a = b`; the `steps` list is
```

```
    metadata recording a rewrite trace. -/
structure Path {A : Type u} (a b : A) where
  steps : List (Step A)
  proof : a = b
```

Here `Step A` records an atomic rewrite:

```
/-- An elementary rewrite step between two elements of A.
    -/
structure Step (A : Type u) where
  src : A
  tgt : A
  proof : src = tgt
```

The basic path operations are defined straightforwardly. These correspond directly to the path constructors introduced in Chapter 3: `refl` is Definition 3.19, `trans` is Definition 3.20, and `symm` is Definition 3.22.

```
/-- Reflexive path (empty sequence of rewrites). -/
@[simp] noncomputable def refl (a : A) : Path a a :=
  Path.mk [] rfl

/-- Compose two paths, concatenating their step lists. -/
@[simp] noncomputable def trans (p : Path a b) (q : Path
    b c) :
    Path a c :=
  Path.mk (p.steps ++ q.steps) (p.proof.trans q.proof)

/-- Reverse a path by reversing and inverting each step.
    -/
@[simp] noncomputable def symm (p : Path a b) : Path b a
    :=
  Path.mk (p.steps.reverse.map Step.symm) p.proof.symm
```

The crucial design point is that the `proof` field carries the semantic content (propositional equality), while the `steps` field preserves the computational trace. This separation allows us to work within Lean's standard type theory while tracking rewrite provenance.

In the concrete Lean representation, the basic groupoid laws hold definitionally for paths:

```
@[simp] theorem trans_refl_left (p : Path a b) :
    trans (refl a) p = p := by
  cases p; simp [trans, refl]

@[simp] theorem trans_refl_right (p : Path a b) :
    trans p (refl b) = p := by
  cases p; simp [trans, refl]
```

```
@[simp] theorem trans_assoc (p : Path a b) (q : Path b c)
    (r : Path c d) :
    trans (trans p q) r = trans p (trans q r) := by
  cases p; cases q; cases r
  simp [trans, List.append_assoc]
```

These definitional equalities correspond to Rules R3, R4, and R8 from the 77-rule rewrite system (Chapter 4, Appendix D). They simplify many proofs, as the type checker automatically recognises that `trans (trans p q) r` and `trans p (trans q r)` are the same path. This is a feature of the chosen Lean representation; the raw path calculus in the main text acquires the corresponding laws only after quotienting by the rewrite/coherence relation.

C.2.2 Step.lean: The Rewrite Relation

The rewrite layer introduces a second object, also named `Step` in its own namespace, which defines one-step rewrites between computational paths. To avoid confusion, one should distinguish this rewrite-level `Step` from the earlier trace-step structure `Step A`. Each constructor of the rewrite-level type corresponds to one of the computational path axioms:

```
/-- A single rewrite step between computational paths. -/
inductive Step :
  {A : Type u} -> {a b : A} -> Path a b -> Path a b ->
      Prop

  -- Basic Path Algebra
  | symm_refl (a : A) :
      Step (symm (refl a)) (refl a)
  | symm_symm (p : Path a b) :
      Step (symm (symm p)) p
  | trans_refl_left (p : Path a b) :
      Step (trans (refl a) p) p
  | trans_refl_right (p : Path a b) :
      Step (trans p (refl b)) p
  | trans_symm (p : Path a b) :
      Step (trans p (symm p)) (refl a)
  | symm_trans (p : Path a b) :
      Step (trans (symm p) p) (refl b)
  | symm_trans_congr (p : Path a b) (q : Path b c) :
      Step (symm (trans p q)) (trans (symm q) (symm p))
  | trans_assoc (p : Path a b) (q : Path b c) (r : Path c
      d) :
      Step (trans (trans p q) r) (trans p (trans q r))
```

```
  -- Congruence closure
  | symm_congr {p q : Path a b} :
      Step p q -> Step (symm p) (symm q)
  | trans_congr_left (q : Path b c) {p p' : Path a b} :
      Step p p' -> Step (trans p q) (trans p' q)
  | trans_congr_right (p : Path a b) {q q' : Path b c} :
      Step q q' -> Step (trans p q) (trans p q')

  -- Product types, sigma types, functions, transport...
  -- (approximately 60 constructors total)
```

The constructors are organised into categories matching the eight rule categories of Chapter 4 and Appendix D: basic path algebra (Rules 1–8, Category A), product types (Rules 9–16), sigma types (Rules 17–22), sum types (Rules 23–24), function types (Rules 25–28), dependent application (Rule 29), transport (Rules 30–36, Category C), context rules (Rules 37–52, Categories D–E), and structural congruence rules (Rules 53–56).

Every step preserves the semantic equality:

```
theorem step_toEq {p q : Path a b} (h : Step p q) :
    p.toEq = q.toEq := by
  induction h <;> simp [toEq]
```

C.2.3 RwEq.lean: Rewrite Equivalence

The reflexive-symmetric-transitive closure of `Step` defines when two paths are *rewrite equivalent*:

```
/-- Symmetric reflexive-transitive closure of Step. -/
inductive RwEq {A : Type u} {a b : A} :
    Path a b -> Path a b -> Type (u + 1)
  | refl (p : Path a b) : RwEq p p
  | step {p q : Path a b} : Step p q -> RwEq p q
  | symm {p q : Path a b} : RwEq p q -> RwEq q p
  | trans {p q r : Path a b} :
      RwEq p q -> RwEq q r -> RwEq p r
```

Note that `RwEq` is a *Type*, not a `Prop`. This is crucial: we need to track *how* two paths are related, not just *whether* they are. A `Prop`-valued version is available when only existence matters:

```
/-- Prop-valued wrapper for quotients. -/
abbrev RwEqProp {A : Type u} {a b : A} (p q : Path a b) :
    Prop :=
  Nonempty (RwEq p q)
```

The RwEq type supports the expected operations: congruence under path operations, normalisation, and conversion to propositional equality.

C.2.4 OmegaGroupoid.lean: Higher Cells and Coherence

The OmegaGroupoid module implements the weak ω-groupoid structure [11, 99, 105]. The key definitions are the Derivation types at each level:

```
/-- 2-cells: Rewrite derivations between paths -/
inductive Derivation2 {a b : A} :
    Path a b -> Path a b -> Type (u + 1) where
  | refl (p : Path a b) : Derivation2 p p
  | step {p q : Path a b} : Step p q -> Derivation2 p q
  | inv {p q : Path a b} :
      Derivation2 p q -> Derivation2 q p
  | vcomp {p q r : Path a b} :
      Derivation2 p q -> Derivation2 q r -> Derivation2 p
          r
```

The Derivation₂ type tracks explicit derivation trees, recording which rewrite steps were applied in which order. This corresponds to the 2-cells (derivations between paths) of Chapter 7; in the terminology of that chapter, a value of type Derivation2 p q is a witness $D : p \Rrightarrow q$. This is essential for the higher coherence structure: two derivations of the same fact may be genuinely different 2-cells.

At level 3, we have meta-derivations between 2-cells:

```
/-- Meta-steps at level 3: primitive 3-cells -/
inductive MetaStep3 : {a b : A} -> {p q : Path a b} ->
    Derivation2 p q -> Derivation2 p q -> Type (u + 2)
        where
  -- Groupoid laws
  | vcomp_refl_left (d : Derivation2 p q) :
      MetaStep3 (.vcomp (.refl p) d) d
  | vcomp_refl_right (d : Derivation2 p q) :
      MetaStep3 (.vcomp d (.refl q)) d
  | vcomp_assoc (d1 : Derivation2 p q)
      (d2 : Derivation2 q r) (d3 : Derivation2 r s) :
      MetaStep3 (.vcomp (.vcomp d1 d2) d3)
                (.vcomp d1 (.vcomp d2 d3))
  | inv_inv (d : Derivation2 p q) :
      MetaStep3 (.inv (.inv d)) d
  -- ...

  -- Pentagon coherence
  | pentagon (f : Path a b) (g : Path b c)
```

```
      (h : Path c d) (k : Path d e) :
      MetaStep3 <pentagon-left-side> <pentagon-right-side
         >

  -- Triangle coherence
  | triangle (f : Path a b) (g : Path b c) :
      MetaStep3 <triangle-left-side> <triangle-right-side
         >

  -- Interchange law
  | interchange (alpha : Derivation2 f f')
      (beta : Derivation2 g g') :
      MetaStep3 <interchange-left> <interchange-right>
```

The `pentagon`, `triangle`, and `interchange` constructors encode the key coherences for bicategories [12, 98]. These arise naturally from critical pairs in the rewrite system [141] (see Chapter 8).

The `Derivation`$_3$ type closes under reflexivity, symmetry, and transitivity:

```
/-- 3-cells: Meta-derivations between 2-cells -/
inductive Derivation3 {a b : A} {p q : Path a b} :
    Derivation2 p q -> Derivation2 p q -> Type (u + 2)
        where
  | refl (d : Derivation2 p q) : Derivation3 d d
  | step {d1 d2 : Derivation2 p q} :
      MetaStep3 d1 d2 -> Derivation3 d1 d2
  | inv {d1 d2 : Derivation2 p q} :
      Derivation3 d1 d2 -> Derivation3 d2 d1
  | vcomp {d1 d2 d3 : Derivation2 p q} :
      Derivation3 d1 d2 -> Derivation3 d2 d3 ->
         Derivation3 d1 d3
```

The next stages are defined in the same style: `Derivation4` compares parallel 3-cells, and `DerivationHigh` packages the stable higher-dimensional case for all levels above 4. None of these levels is represented by `Unit` or `PUnit`; each is a genuine inductive family carrying identities, inverses, and composition.

C.3 Key Design Decisions

The formalisation makes several deliberate design choices that distinguish it from other approaches to identity types and higher category theory. Understanding these decisions is essential for working with the code.

C.3.1 Why Path Uses UIP (Intentionally Not HoTT)

The computational paths framework *intentionally* works within a type theory where the uniqueness of identity proofs (UIP) holds [86]. In Lean 4, the `Eq` type lives in `Prop`, and proofs are definitionally irrelevant. This is reflected in the `Path` structure:

```
structure Path {A : Type u} (a b : A) where
  steps : List (Step A)
  proof : a = b   -- Lives in Prop, hence proof-
     irrelevant
```

The `proof` field provides the semantic content, while the `steps` field (which lives in `Type`) carries the computational trace. Two paths with different step sequences but the same endpoints are *not* definitionally equal, even though their underlying equality proofs are.

This design has several advantages:

1. **Compatibility with Mathlib.** The formalisation integrates seamlessly with Mathlib's extensive library of formalised mathematics, which assumes UIP throughout.

2. **Decidable equality.** With UIP, propositional equality between paths can often be decided by normalising the step sequences. In HoTT [150], such decidability is generally not available.

3. **Coherence proofs are simpler.** Many higher coherences in the semantic equality field are simplified by UIP, but the higher coherences between derivation traces remain genuine Type-valued data. The interesting structure lives in the step sequences and their higher derivations, not in duplicate equality proofs.

4. **Focus on computation.** The computational paths framework is fundamentally about *computational* aspects of equality, not about homotopy-theoretic identity. UIP reflects this focus.

This does *not* mean that computational paths are incompatible with HoTT [150]. Rather, the framework provides a *different* perspective: it extracts higher-categorical structure from the computational content of proof normalisation, rather than from the homotopy-theoretic structure of identity types.

C.3.2 Type-Valued 2-Cells

A crucial design decision is that `Derivation₂` (and every higher derivation type) are *Type*-valued, not `Prop`-valued:

```
inductive Derivation2 {a b : A} :
    Path a b -> Path a b -> Type (u + 1) where
  -- ...
```

This means derivations carry computational content: they record *which* sequence of rewrite steps connects two paths, not just *whether* such a connection exists. This is essential for the ω-groupoid structure:

- Two different derivations of the same rewrite equivalence are genuinely different 2-cells.
- The higher layers `Derivation3`, `Derivation4`, and `DerivationHigh` connect these different 2-, 3-, and higher cells.
- Without this distinction, the entire higher structure would collapse.

When only existence matters, a `Prop`-valued wrapper is available:

```
abbrev RwEqProp (p q : Path a b) : Prop := Nonempty (RwEq
    p q)
```

This is used, for example, when defining quotients by rewrite equivalence.

C.3.3 Stabilisation and Contractibility above Level 3

The weak ω-groupoid structure requires *contractibility* at sufficiently high dimensions: any two parallel cells should be connected by a cell one dimension higher. In the current formalisation, the resulting tower is genuinely non-trivial in dimensions 1–3 and stabilises from dimension 4 onward:

- **Level 2 (not contractible):** `Derivation`$_2$ `p q` is only inhabited when there is an actual sequence of rewrite steps from `p` to `q`. Parallel paths without such a connection have no derivation between them.
- **Level 3 (genuinely Type-valued):** `contractibility`$_3$ connects any two parallel `Derivation`$_2$ terms by isolating the inverse loop they determine, normalising that loop, and contracting it through structural `MetaStep`$_3$ fillers. In the raw-`Path` tower, a localized fallback to `strict_transport`$_3$ remains for the last unmatched strict-shape cases. Alongside this, the explicit-syntax polygraphic package now isolates a fully constructive proof-relevant 3-dimensional coherent presentation with ten 2-cell families and nine generating 3-cell families.
- **Level 4:** `contractibility`$_4$ compares parallel `Derivation`$_3$ terms by the analogous normalisation argument, now closed by `MetaStep`$_4$`.diamond_filler`.

- **Levels 5 and above:** `contractibilityHigh` packages the stable case for all higher dimensions using `normalize4` and `MetaStepHigh.diamond_filler`.

This asymmetry is *crucial* for preserving non-trivial fundamental groups. Consider the circle S^1:

```
-- In the circle, 'loop' and 'loop . loop' are different
   paths
-- with no derivation between them.
-- This is why pi_1(S^1) = Z, not trivial!
```

If contractibility held at level 2, all loops would be rewrite-equivalent to the identity, collapsing $\pi_1(S^1)$ to the trivial group.

The kernel-level interfaces for the stabilisation theorem are:

```
noncomputable def contractibility3 {p q : Path a b}
    (d1 d2 : Derivation2 p q) : Derivation3 d1 d2 :=
  by
    -- isolate the inverse loop d1 . d2^{-1},
    -- normalize it, and connect the endpoints
       structurally
    exact ...

noncomputable def contractibility4 {d1 d2 : Derivation2 p
    q}
    (Gamma1 Gamma2 : Derivation3 d1 d2) : Derivation4
       Gamma1 Gamma2 :=
  by
    -- normalize 3-cells and use MetaStep4.diamond_filler
    exact ...

noncomputable def contractibilityHigh {n : Nat}
    (Xi0 Xi1 : DerivationHigh n x y) :
    DerivationHigh (n + 1) Xi0 Xi1 :=
  by
    -- stable case: normalize4 + MetaStepHigh.
       diamond_filler
    exact ...
```

The decisive point is that the current formal picture now has two fronts. On the raw-`Path` side, the comparison maps are organised mainly inside the Type-valued derivation tower: the level-3 route goes through inverse-loop normalisation and structural comparison, and the final legacy fallback is still `strict_transport3 := MetaStep3.rweq_transport(...)`. On the explicit-syntax/polygraph side, the same critical-pair analysis now packages a fully constructive proof-relevant coherent presentation through dimension 3, with ten generating 2-cell families, nine generating 3-cell families, and an acyclicity theorem above dimension 3, so no generating 4-cells are needed there. By contrast, level 4 on the raw-`Path` side is handled by `MetaStep₄.diamond_filler`, and the higher levels by the uniform stable `MetaStepHigh.diamond_filler` construction. This is the precise formal sense in which the stabilised computational ω-groupoid is 3-truncated: dimensions 1–3 retain genuine proof-relevant content, while dimensions 4 and above are contractible.

C.3.4 MetaStep$_3$ Constructors for Coherences

The `MetaStep₃` inductive type includes explicit constructors for the pentagon, triangle, and interchange coherences:

```
inductive MetaStep3 : ... where
  -- Pentagon: critical pair when two trans_assoc overlap
  | pentagon (f : Path a b) (g : Path b c)
      (h : Path c d) (k : Path d e) :
      MetaStep3
        (.vcomp (.step (Step.trans_assoc (trans f g) h k)
            )
                (.step (Step.trans_assoc f g (trans h k))
                    ))
        (.vcomp (.vcomp (...) (...)) (...))

  -- Triangle: critical pair of trans_assoc and
      trans_refl_right
  | triangle (f : Path a b) (g : Path b c) :
      MetaStep3
        (.vcomp (.step (Step.trans_assoc f (refl b) g))
                (.step (Step.trans_congr_right f
                  (Step.trans_refl_left g))))
        (.step (Step.trans_congr_left g
          (Step.trans_refl_right f)))

  -- Interchange: (d1 . d2) * (d3 . d4) = (d1 * d3) . (d2
      * d4)
  | interchange {f f' : Path a b} {g g' : Path b c}
      (alpha : Derivation2 f f') (beta : Derivation2 g g
        ') :
      MetaStep3
        (.vcomp (whiskerRight alpha g) (whiskerLeft f'
          beta))
        (.vcomp (whiskerLeft f beta) (whiskerRight alpha
          g'))
```

Although `contractibility₃`, `contractibility₄`, and the stable higher fillers can compare any parallel cells in their respective ranges, keeping these designated coherence constructors explicit has several advantages:

1. **Categorical clarity:** The pentagon and triangle are the classical axioms of bicategories [98, 142], and having them as explicit generators makes the connection clear.
2. **Computational efficiency:** Explicit constructors are faster to work with than derived contractibility proofs.
3. **Genuine coherence content:** The fourteen named 3-cell coherence proofs were independently tested by replacing their witnesses with `Derivation3.refl`; Lean rejects every such replacement.
4. **Rewriting-theoretic motivation:** The constructors mirror important low-dimensional critical-pair patterns in the rewrite system [93, 88], making the Squier-style motivation [141] visible while remaining compatible with the normalisation-and-diamond-filler proof of higher contractibility.

C.4 Navigating the Proofs

The formalisation spans many modules. This section provides guidance for finding specific theorems and understanding the module organization.

C.4.1 Module Hierarchy

The module hierarchy mirrors the conceptual structure of the theory:

Core Definitions (`Path/Basic/`)

- `Core.lean`: `Path`, `Step` structures
- `Context.lean`: Contexts and substitution
- `Congruence.lean`: Congruence rules
- `UIP.lean`: Uniqueness of identity proofs
- `PathAlgebra.lean`: Path algebra operations

Rewrite System (`Path/Rewrite/`)

- `Step.lean`: Primitive rewrite-step relation

- `Rw.lean`: Reflexive-transitive closure
- `RwEq.lean`: Symmetric closure
- `Confluence.lean`: Confluence theorem
- `Normalization.lean`: Normal forms
- `PathTactic.lean`: Automation tactics

Homotopy Theory (`Path/Homotopy/`)

- `Loops.lean`: Loop spaces
- `FundamentalGroup.lean`: π_1 definition
- `EckmannHilton.lean`: Eckmann–Hilton argument
- `SeifertVanKampen.lean`: pushout-level π_1 computation
- `FundamentalGroupoid.lean`: groupoid packaging of path classes
- `HigherHomotopyGroups.lean`: π_n interfaces

Computational HIT-Style Constructions (`Path/CompPath/`)

- `CircleCompPath.lean`, `CircleStep.lean`: circle and $\pi_1(S^1)$
- `Torus.lean`, `TorusStep.lean`: torus and $\pi_1(T^2)$
- `FigureEight.lean`: wedge-of-circles example
- `PushoutCompPath.lean`, `PushoutPaths.lean`: pushouts and path codes
- `SeifertVanKampenDerived.lean`: standard corollaries of SVK

Higher Coherence (`Path/OmegaGroupoid/`)

- `OmegaGroupoid.lean`: main derivation tower and contractibility maps
- `GroupoidProofs.lean`: explicit low-dimensional coherences
- `PentagonProof.lean`: pentagon filler package
- `Stabilization.lean`: stabilized contractibility interfaces

- `TruncationProof.lean`: path-level compatibility theorem and constructive explicit-syntax wrappers

Higher-Categorical Packaging

- `HigherCategory/TwoCategory.lean`: path-based 2-category interface
- `HigherCategory/GrayCategory.lean`: Gray-style 3- and 4-cell packaging
- `Path/DoubleCategoryPaths.lean`: double-category interface on path squares
- `Path/Category/`: broader categorical developments

Comparison and Navigation

- `Comparison/UnivalenceAnalog.lean`: path-equivalence and univalence obstruction
- `Comparison/HoTTComparison.lean`: HoTT comparison layer
- `Path/Polygraph/ThreeCells.lean`: proof-relevant 3-cell structures
- `Examples/`: standalone worked examples such as circle π_1

C.4.2 Finding Specific Theorems

The Lean 4 language server provides powerful search capabilities. In VS Code or Emacs, you can:

1. **Search by name:** Use the command palette to search for definitions by name fragment (e.g., "confluence", "pentagon").
2. **Search by type:** Use `#check` or goal-directed search to find lemmas with a given type signature.
3. **Navigate definitions:** Ctrl-click (or M-.) on any identifier to jump to its definition.

Key theorems and their locations:

Theorem	Location
Confluence	`Path/Rewrite/Confluence.lean`
Normalization	`Path/Rewrite/Normalization.lean`
$\pi_1(S^1) \cong \mathbb{Z}$	`Path/CompPath/CircleStep.lean`
$\pi_1(T^2) \cong \mathbb{Z}^2$	`Path/CompPath/TorusStep.lean`
Eckmann–Hilton	`Path/Homotopy/EckmannHilton.lean`
Pentagon coherence	`Path/OmegaGroupoid/PentagonProof.lean`
Triangle coherence	`Path/OmegaGroupoid/GroupoidProofs.lean`
Contractibility$_3$	`Path/OmegaGroupoid.lean`
Constructive polygraph theorem	`Path/OmegaGroupoid/TruncationProof.lean`
Interchange law	`Path/DoubleCategoryPaths.lean`
3-cell structure	`Path/Polygraph/ThreeCells.lean`

C.4.3 Key Lemmas and Where They Live

The formalisation establishes several key normalisation interfaces. For the stabilisation construction, the most important ones are:

```
-- Normalize the comparison witness to strict normal form
noncomputable def normalize :
    Derivation2 p q -> StrictNormalForm p q := ...

-- Structural connector between strict normal forms
noncomputable def connect_core_strict_structural ... :
    Derivation3 d1 d2 := ...

-- 4-dimensional diamond filler
noncomputable def MetaStep4.diamond_filler ... :
    Derivation4 Gamma1 Gamma2 := ...

-- Stable higher-dimensional filler
noncomputable def MetaStepHigh.diamond_filler ... :
    DerivationHigh (n + 1) Xi0 Xi1 := ...
```

These definitions isolate the constructive heart of the current stabilisation proof: normalisation plus structural connection inside the derivation tower itself. The remaining caveat is that the public core connector still falls back, in unmatched strict-shape cases, to `strict_transport`$_3$ and hence to

MetaStep3.rweq_transport. Local confluence enters through the diamond fillers, which turn joinability data for critical pairs into ordinary higher-cell composition, while the harder level-3 comparison has now been reduced to this isolated legacy boundary.

For the fundamental group computations:

```
-- Circle encode/decode (Path/CompPath/CircleCompPath.
   lean)
noncomputable def circlePiOneEncode : circlePiOne -> Int
   := ...
noncomputable def circleDecode : Int -> circlePiOne :=
   ...

-- Round-trip equations
theorem circlePiOneEncode_circleDecode (z : Int) :
    circlePiOneEncode (circleDecode z) = z := ...
theorem circleDecode_circlePiOneEncode (x : circlePiOne)
   :
    circleDecode (circlePiOneEncode x) = x := ...
```

C.5 Extending the Formalization

The formalisation is designed to be extensible. This section explains how to add new rewrite rules, prove new coherence results, and contribute to the project.

C.5.1 Adding New Rewrite Rules

To add a new rewrite rule to the system:

1. **Define the rule.** Add a new constructor to the Step inductive type in Path/Rewrite/Step.lean:

```
inductive Step : ... where
  -- existing constructors...

  | my_new_rule {A : Type u} {a b : A} (p : Path a b
    ) :
      Step <left-hand-side> <right-hand-side>
```

2. **Prove preservation of equality.** Add a case to the step_toEq proof showing that your rule preserves the underlying propositional equality:

```
theorem step_toEq {p q : Path a b} (h : Step p q) :
    p.toEq = q.toEq := by
```

```
induction h with
-- existing cases...
| my_new_rule p => simp [toEq]
```

3. **Check confluence.** New rules may create new critical pairs. Analyse these pairs and add corresponding diamond fillers or 3-cell constructors as needed.

4. **Update normalisation.** If the rule affects normal forms, update the normalisation procedures in `Path/Rewrite/Normalization.lean`.

C.5.2 Proving New Coherence Results

To prove a new coherence result (e.g., a new commuting diagram in the bicategorical structure):

1. **Identify the cells involved.** Determine what 2-cells (derivations) are being compared.

2. **Use existing 3-cell constructors.** Check whether the coherence can be derived from the existing `MetaStep`$_3$ constructors using vertical composition and whiskering.

3. **Apply the appropriate contractibility theorem.** If the cells are parallel (same source and target), use `contractibility`$_3$, `contractibility`$_4$, or `contractibilityHigh` according to dimension; each works by normalisation and structural diamond fillers from Section C.3.3.

4. **Make explicit if needed.** For frequently-used coherences, consider adding an explicit constructor to `MetaStep`$_3$ for efficiency.

Example: proving a modified pentagon:

```
theorem modified_pentagon (f : Path a b) (g : Path b c)
    (h : Path c d) (k : Path d e) :
    Derivation3 <left-diagram> <right-diagram> := by
  -- Use existing pentagon and groupoid laws
  apply Derivation3.vcomp
  . exact Derivation3.step (MetaStep3.pentagon f g h k)
  . exact Derivation3.step (MetaStep3.vcomp_assoc ...)
```

C.5.3 Contributing Guidelines

Contributions to the formalisation are welcome. The project follows standard open-source practices:

1. **Fork and branch.** Fork the repository on GitHub and create a feature branch for your changes.
2. **Follow style conventions.** Match the existing code style: use `noncomputable` where needed, add documentation comments, and organize lemmas logically.
3. **Ensure CI passes.** The repository has continuous integration that checks all files compile. Current build commands and contribution workflow are maintained in the companion repository.
4. **Write documentation.** Add docstrings to new definitions and update module headers to describe new functionality.
5. **Submit a pull request.** Describe what your changes accomplish and reference any relevant issues or discussions.

Priority areas for contributions include:

- Additional higher inductive types (e.g., real projective spaces, Eilenberg–Mac Lane spaces)
- More derived coherences and their proofs
- Improved automation and tactics
- Documentation and examples
- Performance optimizations

Summary

The Lean 4 formalisation provides a machine-checked foundation for computational path theory. The key architectural decisions—UIP compatibility, the full Type-valued derivation tower, explicit coherence constructors, and the stabilisation theorem proving 3-truncation above dimension 3—combine to create a practical and mathematically rigorous implementation, while keeping the remaining raw-`Path` transport fallback explicit and local.

The module hierarchy follows the theoretical development: `Path/Basic/` for core definitions, `Path/Rewrite/` for the rewrite system, `Path/Homotopy/` for fundamental groups and higher homotopy, `Path/CompPath/` for higher inductive types, and `Path/OmegaGroupoid/` for the weak ω-groupoid structure.

For readers approaching the formalisation for the first time, we recommend:

1. Start with `Path/Basic/Core.lean` to understand the fundamental definitions.
2. Move to `Path/Rewrite/Step.lean` to see the rewrite rules.
3. Study `Path/CompPath/CircleStep.lean` for a complete worked example ($\pi_1(S^1) \cong \mathbb{Z}$).
4. Explore `Path/OmegaGroupoid.lean` for the higher coherence structure.

The formalisation demonstrates that computational paths provide a practical framework for higher-dimensional type theory, with the central algebraic constructions and current higher-cell comparison results checked by Lean's kernel.

Repository: `https://github.com/Arthur742Ramos/ComputationalPathsLean`

Appendix D

Complete List of Named Rewrite Clauses

This appendix serves as a reference card for the 77 named clauses that define the computational path rewrite relation. It is designed for two uses. First, as a **lookup table** when reading Chapters 4 and 5: if a proof invokes "Rule R37" or "L-WHISKER-INTRO," the reader can turn here to see the exact left- and right-hand sides. Second, as a **checklist** for anyone extending the system: the eight categories (A–H) and the summary at the end make it easy to verify that a proposed new rule fits into the existing structure without duplicating or conflicting with existing clauses.

For the motivation and detailed discussion behind each rule, see Chapter 4. For the confluence and termination proofs, see Chapter 5. The rules are also formalised in the Lean development (Appendix C, `Path/Rewrite/Step.lean`).

D.1 Rule Format and Notation

Each primitive rewrite clause has the form:

$$\textbf{LHS} \longrightarrow \textbf{RHS}$$

where the left-hand side (LHS) may be replaced by the right-hand side (RHS) in any context. The replacement is *directed*: only LHS to RHS reductions are permitted. When a clause is schematic (for example a context-compatibility clause), the displayed line should be read as generating the corresponding family of admissible instances.

D.1.1 Notation Summary

Symbol	Meaning
refl_a	Reflexivity path at a
$\mathsf{symm}(p)$	Symmetry (inverse) of path p
$\mathsf{trans}(p, q)$	Transitivity (composition) of paths p and q
$\mathsf{congrArg}(f, p)$	Congruence: apply function f to path p
$\mathsf{congrFun}(p, a)$	Function congruence: evaluate path p at a
$\mathsf{transport}(p, x)$	Transport x along path p
$C[p]$	Context C applied to path p (abbrev. for $\mathsf{congrArg}(C, p)$)
$L_C(r, p)$	Left whisker: r followed by $C[p]$
$R_C(p, t)$	Right whisker: $C[p]$ followed by t
$\mathsf{DC}[p]$	Dependent context applied to p
$DL_C(r, p)$	Dependent left whisker
$DR_C(p, t)$	Dependent right whisker
$\mathsf{sc}(h)$	Step constructor: lift propositional equality h to path
$\mathsf{toEq}(p)$	Semantic projection: extract equality from path p
$\mathsf{prodMk}(p, q)$	Product path from component paths
$\mathsf{sigmaMk}(p, q)$	Sigma path from component paths
$\mathsf{lamCongr}(p)$	Function extensionality path
$\mathsf{apd}_f(p)$	Dependent application of f along p
$\iota(h)$	Canonical coherence path from witness h

D.1.2 Type Conventions

Throughout the rules:

- p, q, r, s, t, u, v denote paths of appropriate types
- a, b, c, x, y denote terms
- f, g, h denote functions
- C denotes a (non-dependent) context
- DC denotes a dependent context
- K denotes a binary context
- A, B denote types; $B : A \to \mathsf{Type}$ denotes a type family
- T_p denotes the transport map along path p

D.2 Category A: Groupoid Core (Rules 1–8)

The groupoid core rules express the fundamental algebraic laws of path composition and inversion, ensuring that paths form a groupoid structure.

Rule	Name	LHS $\longrightarrow$ RHS	Description
R1	Symm-Refl	$\mathsf{symm}(\mathsf{refl}_a) \to \mathsf{refl}_a$	Inverse of identity is identity
R2	Symm-Invol	$\mathsf{symm}(\mathsf{symm}(p)) \to p$	Double inverse cancels
R3	Trans-Left-Id	$\mathsf{trans}(\mathsf{refl}_a, p) \to p$	Left identity law
R4	Trans-Right-Id	$\mathsf{trans}(p, \mathsf{refl}_b) \to p$	Right identity law
R5	Trans-Right-Inv	$\mathsf{trans}(p, \mathsf{symm}(p)) \to \mathsf{refl}_a$	Right inverse law
R6	Trans-Left-Inv	$\mathsf{trans}(\mathsf{symm}(p), p) \to \mathsf{refl}_b$	Left inverse law
R7	Symm-Trans	$\mathsf{symm}(\mathsf{trans}(p, q)) \to \mathsf{trans}(\mathsf{symm}(q), \mathsf{symm}(p))$	Contravariance of inverse
R8	Trans-Assoc	$\mathsf{trans}(\mathsf{trans}(p, q), r) \to \mathsf{trans}(p, \mathsf{trans}(q, r))$	Right-associativity

Type signatures:

$$\begin{aligned}
\text{R1:}&\quad a : A \\
\text{R2:}&\quad p : \mathsf{Path}(a, b) \\
\text{R3, R4:}&\quad p : \mathsf{Path}(a, b) \\
\text{R5:}&\quad p : \mathsf{Path}(a, b),\ \text{result} : \mathsf{Path}(a, a) \\
\text{R6:}&\quad p : \mathsf{Path}(a, b),\ \text{result} : \mathsf{Path}(b, b) \\
\text{R7:}&\quad p : \mathsf{Path}(a, b),\ q : \mathsf{Path}(b, c) \\
\text{R8:}&\quad p : \mathsf{Path}(a, b),\ q : \mathsf{Path}(b, c),\ r : \mathsf{Path}(c, d)
\end{aligned}$$

D.3 Category B: Type Constructor Rules (Rules 9–25)

These rules govern how paths interact with the type-forming operations: products, sigma types, coproducts, and function types.

D.3.1 Binary Congruence and Products (Rules 9–15)

R9 Map$_2$-Decomp: Binary congruence decomposes

$$\begin{aligned}
\mathsf{map}_2(f, p, q) \ \to\ &\mathsf{trans}(\mathsf{mapRight}_f(a_1, q), \\
&\quad \mathsf{mapLeft}_f(p, b_2))
\end{aligned}$$

R10 Prod-β_1: First projection of product path

$$\mathsf{congrArg}(\pi_1, \mathsf{map}_2(\langle\cdot,\cdot\rangle, p, q)) \ \to\ p$$

R11 Prod-β_2: Second projection of product path

$$\mathsf{congrArg}(\pi_2, \mathsf{map}_2(\langle\cdot,\cdot\rangle, p, q)) \;\to\; q$$

R12 Prod-Rec: Product recursor on product path

$$\mathsf{congrArg}(\mathsf{Prod.rec}(f), \mathsf{map}_2(\langle\cdot,\cdot\rangle, p, q)) \;\to\; \mathsf{map}_2(f, p, q)$$

R13 Prod-η: Product uniqueness

$$\mathsf{prodMk}(\mathsf{congrArg}(\pi_1, p),\ \mathsf{congrArg}(\pi_2, p)) \;\to\; p$$

R14 Prod-Symm: Inverse distributes to components

$$\mathsf{symm}(\mathsf{prodMk}(p, q)) \;\to\; \mathsf{prodMk}(\mathsf{symm}(p), \mathsf{symm}(q))$$

R15 Prod-Func: Componentwise function

$$\mathsf{congrArg}(\lambda x.\,(g(x.1), h(x.2)),\ \mathsf{prodMk}(p, q)) \;\to\; \mathsf{prodMk}(\mathsf{congrArg}(g, p), \mathsf{congrArg}(h, q))$$

D.3.2 Sigma Types (Rules 16–19)

R16 Sigma-β_1: First projection of sigma path

$$\mathsf{congrArg}(\Sigma.\mathsf{fst},\ \mathsf{sigmaMk}(p, q)) \;\to\; \mathsf{sc}(\mathsf{toEq}(p))$$

R17 Sigma-β_2: Second projection of sigma path

$$\mathsf{sigmaSnd}(\mathsf{sigmaMk}(p, q)) \;\to\; \mathsf{sc}(\mathsf{toEq}(q))$$

R18 Sigma-η: Sigma uniqueness

$$\mathsf{sigmaMk}(\mathsf{sigmaFst}(p),\ \mathsf{sigmaSnd}(p)) \;\to\; p$$

R19 Sigma-Symm: Inverse of sigma path

$$\mathsf{symm}(\mathsf{sigmaMk}(p, q)) \;\to\; \mathsf{sigmaMk}(\mathsf{symm}(p),\ \mathsf{symmSnd}(p, q))$$

D.3.3 Sum Types (Rules 20–21)

R20 Sum-Rec-Inl: Sum elimination on left injection

$$\mathsf{congrArg}(\mathsf{Sum.rec}(f, g),\ \mathsf{congrArg}(\mathsf{inl}, p)) \;\to\; \mathsf{congrArg}(f, p)$$

R21 Sum-Rec-Inr: Sum elimination on right injection

$$\mathsf{congrArg}(\mathsf{Sum.rec}(f, g),\ \mathsf{congrArg}(\mathsf{inr}, q)) \;\to\; \mathsf{congrArg}(g, q)$$

D.3.4 Function Types (Rules 22–25)

R22 FUN-β: Function extensionality at a point

$$\mathsf{congrFun}(\mathsf{lamCongr}(p), a) \rightarrow p(a)$$

R23 FUN-η: Function path uniqueness

$$\mathsf{lamCongr}(\lambda x.\, \mathsf{congrFun}(p, x)) \rightarrow p$$

R24 FUN-SYMM: Inverse of function path

$$\mathsf{symm}(\mathsf{lamCongr}(p)) \rightarrow \mathsf{lamCongr}(\lambda x.\, \mathsf{symm}(p(x)))$$

R25 APD-REFL: Dependent application on identity

$$\mathsf{apd}_f(\mathsf{refl}_a) \rightarrow \mathsf{refl}_{f(a)}$$

D.4 Category C: Transport Rules (Rules 26–32)

Transport rules govern the interaction of paths with the transport operation in dependent type families. Given a type family $B : A \rightarrow \mathsf{Type}$ and a path $p : \mathsf{Path}_A(a_1, a_2)$, transport yields a map $\mathsf{transport}(p, -) : B(a_1) \rightarrow B(a_2)$.

Rule	Name	LHS $\longrightarrow$ RHS
R26	TRANSPORT-REFL	$\mathsf{sc}(\mathsf{transport\_refl}(a, x)) \rightarrow \mathsf{refl}_x$
R27	TRANSPORT-TRANS	$\mathsf{sc}(\mathsf{transport\_trans}(p, q, x)) \rightarrow \iota(h_{27})$
R28	TRANSPORT-SYMM-L	$\mathsf{sc}(\mathsf{transport\_symm\_left}(p, x)) \rightarrow \iota(h_{28})$
R29	TRANSPORT-SYMM-R	$\mathsf{sc}(\mathsf{transport\_symm\_right}(p, y)) \rightarrow \iota(h_{29})$
R30	TRANSPORT-Σ-FST	$\mathsf{sc}(\mathsf{transport\_sigmaMk\_fst}(p, q, x)) \rightarrow \iota(h_{30})$
R31	TRANSPORT-Σ-DEP	$\mathsf{sc}(\mathsf{transport\_sigmaMk\_dep}(p, q, x)) \rightarrow \iota(h_{31})$
R32	SUBST-Σ-DEP	$\mathsf{sc}(\mathsf{subst\_sigmaMk\_dep}(p, q, x)) \rightarrow \iota(h_{32})$

Note: The terms $\iota(h_{27})$ through $\iota(h_{32})$ denote canonical coherence paths constructed from the underlying propositional equalities witnessing the relevant transport laws.

D.5 Category D: Context Rules (Rules 33–48)

Context rules govern how paths interact with contexts, that is, functions $C : A \to B$ through which paths can be propagated. The whisker operations L_C (left whisker) and R_C (right whisker) manage compositions involving context-lifted paths.

D.5.1 Basic Context Schemata (Rules 33–36)

Rule	Name	LHS $\longrightarrow$ RHS
R33	CTX-CONGR	$p \to q \Rightarrow C[p] \to C[q]$
R34	CTX-SYMM	$\mathsf{symm}(C[p]) \to C[\mathsf{symm}(p)]$
R35	CTX-ABSORB-L	$\mathsf{trans}(C[p], \mathsf{trans}(C[\mathsf{symm}(p)], v)) \to \mathsf{trans}(C[\mathsf{trans}(p, \mathsf{symm}(p))], v)$
R36	CTX-ABSORB-R	$\mathsf{trans}(\mathsf{trans}(v, C[p]), C[\mathsf{symm}(p)]) \to \mathsf{trans}(v, C[\mathsf{trans}(p, \mathsf{symm}(p))])$

Note. Rules R33–R36 are schemata rather than single first-order contractions: each line abbreviates the corresponding family of context-closure instances.

D.5.2 Whisker Introduction and Conversion (Rules 37–41)

Rule	Name	LHS $\longrightarrow$ RHS		Description
R37	L-WHISKER-INTRO	$\mathsf{trans}(r, C[p]) \to L_C(r, p)$		Left whisker introduction
R38	L-FROM-R	$\mathsf{trans}(r, R_C(p, \mathsf{refl}_{C(a_2)}))$ $L_C(r, p)$	$\to$	Left from right with trivial tail
R39	L-TO-R	$\mathsf{trans}(L_C(r, p), t)$ $\mathsf{trans}(r, R_C(p, t))$	$\to$	Left-to-right conversion
R40	R-WHISKER-INTRO	$\mathsf{trans}(C[p], t) \to R_C(p, t)$		Right whisker introduction
R41	R-WHISKER-ASSOC	$\mathsf{trans}(R_C(p, t), u)$ $R_C(p, \mathsf{trans}(t, u))$	$\to$	Right whisker associativity

D.5.3 Whisker Simplification (Rules 42–48)

Rule	Name	LHS $\longrightarrow$ RHS	Description
R42	L-WHISKER-REFL	$L_C(r, \mathsf{refl}_a) \to r$	Left whisker by identity
R43	L-WHISKER-REFL-PRE	$L_C(\mathsf{refl}_{C(a_1)}, p) \to C[p]$	Left whisker with identity prefix
R44	R-WHISKER-REFL	$R_C(\mathsf{refl}_a, r) \to r$	Right whisker by identity
R45	R-WHISKER-REFL-SUF	$R_C(p, \mathsf{refl}_{C(a_2)}) \to C[p]$	Right whisker with identity suffix
R46	L-WHISKER-IDEM	$L_C(L_C(r, \mathsf{refl}_{a_1}), p) \to L_C(r, p)$	Left whisker idempotence
R47	R-WHISKER-IDEM-1	$R_C(p, R_C(\mathsf{refl}_{a_2}, t)) \to R_C(p, t)$	Right whisker idempotence (prefix)
R48	R-WHISKER-IDEM-2	$R_C(\mathsf{refl}_{a_1}, R_C(p, t)) \to R_C(p, t)$	Right whisker idempotence (id prefix)

Type signatures for whisker operations:

$$L_C(r, p) : \mathsf{Path}(x, C(a_2)) \quad \text{where } r : \mathsf{Path}(x, C(a_1)),\ p : \mathsf{Path}(a_1, a_2)$$
$$R_C(p, t) : \mathsf{Path}(C(a_1), y) \quad \text{where } p : \mathsf{Path}(a_1, a_2),\ t : \mathsf{Path}(C(a_2), y)$$

D.6 Category E: Dependent Context Rules (Rules 49–60)

Dependent context rules are the dependent counterparts of Category D, for dependent functions $C : (a : A) \to B(a)$. Transport is required to align types in the dependent setting.

Rule	Name	LHS $\longrightarrow$ RHS	Description
R49	DC-CONGR	$p \to q \Rightarrow \mathsf{DC}[p] \to \mathsf{DC}[q]$	Dependent context congruence
R50	DC-SYMM	$\mathsf{symm}(\mathsf{DC}[p]) \to \mathsf{DC.symmMap}(p)$	Dependent symmetry
R51	DL-WHISKER-INTRO	$\mathsf{trans}(T_p(r), \mathsf{DC}[p]) \to DL_C(r, p)$	Dependent left whisker introduction
R52	DL-TO-DR	$\mathsf{trans}(DL_C(r, p), t) \to \mathsf{trans}(T_p(r), DR_C(p, t))$	Dependent left-to-right conversion
R53	DR-WHISKER-INTRO	$\mathsf{trans}(\mathsf{DC}[p], t) \to DR_C(p, t)$	Dependent right whisker introduction
R54	DR-WHISKER-ASSOC	$\mathsf{trans}(DR_C(p, t), u) \to DR_C(p, \mathsf{trans}(t, u))$	Dependent right whisker associativity
R55	DL-WHISKER-REFL	$DL_C(r, \mathsf{refl}_a) \to r$	Dependent left whisker by identity
R56	DL-WHISKER-REFL-PRE	$DL_C(\mathsf{refl}_{C(a_1)}, p) \to \mathsf{DC}[p]$	Dependent left whisker with id prefix
R57	DR-WHISKER-REFL	$DR_C(\mathsf{refl}_a, r) \to r$	Dependent right whisker by identity
R58	DR-WHISKER-REFL-SUF	$DR_C(p, \mathsf{refl}_{C(a_2)}) \to \mathsf{DC}[p]$	Dependent right whisker with id suffix
R59	DL-WHISKER-IDEM	$DL_C(DL_C(r, \mathsf{refl}_{a_1}), p) \to DL_C(r, p)$	Dependent left whisker idempotence
R60	DR-WHISKER-IDEM	$DR_C(p, DR_C(\mathsf{refl}_{a_2}, t)) \to DR_C(p, t)$	Dependent right whisker idempotence

Correspondence with non-dependent rules:

Operation	Non-dependent	Dependent
Congruence	R33	R49
Symmetry	R34	R50
Left whisker introduction	R37	R51
Left-to-right conversion	R39	R52
Right whisker introduction	R40	R53
Right whisker associativity	R41	R54
Left whisker refl	R42	R55
Left whisker refl prefix	R43	R56
Right whisker refl	R44	R57
Right whisker refl suffix	R45	R58
Left whisker idempotence	R46	R59
Right whisker idempotence	R47	R60

D.7 Category F: Binary Context Rules (Rules 61–68)

Binary context rules handle congruence for two-argument functions $K : A \to B \to C$. Both dependent (DBC) and non-dependent (BC) variants are provided.

D.7.1 Dependent Binary Context (Rules 61–64)

Rule	Name	LHS ⟶ RHS	Description
R61	DBC-MAPLEFT	$p \to q \Rightarrow \mathsf{DBC.mapLeft}_K(p, b) \to \mathsf{DBC.mapLeft}_K(q, b)$	Dep. binary left congruence
R62	DBC-MAPRIGHT	$p \to q \Rightarrow \mathsf{DBC.mapRight}_K(a, p) \to \mathsf{DBC.mapRight}_K(a, q)$	Dep. binary right congruence
R63	DBC-MAP₂-L	$p \to q \Rightarrow \mathsf{DBC.map}_2(K, p, r) \to \mathsf{DBC.map}_2(K, q, r)$	Dep. binary cong. (left pos.)
R64	DBC-MAP₂-R	$q \to r \Rightarrow \mathsf{DBC.map}_2(K, p, q) \to \mathsf{DBC.map}_2(K, p, r)$	Dep. binary cong. (right pos.)

D.7.2 Non-Dependent Binary Context (Rules 65–68)

Rule	Name	LHS ⟶ RHS	Description
R65	BC-MAPLEFT	$p \to q \Rightarrow \mathsf{BC.mapLeft}_K(p, b) \to \mathsf{BC.mapLeft}_K(q, b)$	Binary left congruence
R66	BC-MAPRIGHT	$p \to q \Rightarrow \mathsf{BC.mapRight}_K(a, p) \to \mathsf{BC.mapRight}_K(a, q)$	Binary right congruence
R67	BC-MAP₂-L	$p \to q \Rightarrow \mathsf{BC.map}_2(K, p, r) \to \mathsf{BC.map}_2(K, q, r)$	Binary cong. (left position)
R68	BC-MAP₂-R	$q \to r \Rightarrow \mathsf{BC.map}_2(K, p, q) \to \mathsf{BC.map}_2(K, p, r)$	Binary cong. (right position)

Note: Rules R61–R68 are all *compatibility rules* (congruence rules). They ensure that the rewrite relation is preserved by binary context operations but do not simplify expressions themselves.

D.8 Category G: Partial Map Rules (Rules 69–72)

Partial map rules govern left-partial and right-partial congruence operations that arise from binary functions.

R69 MAPLEFT-CONGR: Left map congruence

$$p \to q \;\Rightarrow\; \mathsf{mapLeft}_f(p, b) \to \mathsf{mapLeft}_f(q, b)$$

R70 MAPRIGHT-CONGR: Right map congruence

$$p \to q \;\Rightarrow\; \mathsf{mapRight}_f(a, p) \to \mathsf{mapRight}_f(a, q)$$

R71 MAPLEFT-COMP: Left map on step constructor

$$\mathsf{mapLeft}_f(\mathsf{sc}(h), b) \;\to\; \mathsf{sc}(\mathsf{congrArg}(\lambda x.\, f(x, b),\; h))$$

R72 MAPRIGHT-COMP: Right map on step constructor

$$\mathsf{mapRight}_f(a, \mathsf{sc}(h)) \;\to\; \mathsf{sc}(\mathsf{congrArg}(f(a, \cdot),\; h))$$

Explanation: For a binary function $f : A \to B \to C$:

- $\mathsf{mapLeft}_f(p, b)$ applies $f(\cdot, b)$ to path p, holding second argument fixed
- $\mathsf{mapRight}_f(a, p)$ applies $f(a, \cdot)$ to path p, holding first argument fixed

Rules R71–R72 reduce these operations when applied to step constructors, pushing the partial application through to the underlying equality.

D.9 Category H: Structural Completion (Rules 73–77)

The structural completion rules close the system under path constructors and provide cancellation laws.

D.9.1 Congruence Rules (Rules 73–75)

Rule	Name	LHS $\longrightarrow$ RHS	Description
R73	SYMM-CONGR	$p \to q \Rightarrow \mathsf{symm}(p) \to \mathsf{symm}(q)$	Congruence under symmetry
R74	TRANS-CONGR-L	$p \to q \Rightarrow \mathsf{trans}(p, r) \to \mathsf{trans}(q, r)$	Congruence under trans (left)
R75	TRANS-CONGR-R	$q \to r \Rightarrow \mathsf{trans}(p, q) \to \mathsf{trans}(p, r)$	Congruence under trans (right)

D.9.2 Cancellation Rules (Rules 76–77)

Rule	Name	LHS $\longrightarrow$ RHS	Description
R76	LEFT-CANCEL	$\mathsf{trans}(p, \mathsf{trans}(\mathsf{symm}(p), q)) \to q$	Left cancellation
R77	RIGHT-CANCEL	$\mathsf{trans}(\mathsf{symm}(p), \mathsf{trans}(p, q)) \to q$	Right cancellation

Note: Rules R76–R77 extend the basic inverse laws R5–R6. They handle cancellation in the *middle* of a longer composition, enabling more direct reduction than first applying associativity.

Summary: Rules by Category

Cat.	Name	Rules	Purpose
A	Groupoid Core	R1–R8	Fundamental algebraic laws
B	Type Constructors	R9–R25	Products, sums, sigma, functions
C	Transport	R26–R32	Dependent type families
D	Context	R33–R48	Substitution compatibility
E	Dependent Context	R49–R60	Dependent substitution
F	Binary Context	R61–R68	Two-argument congruence
G	Partial Maps	R69–R72	Partial application
H	Structural Completion	R73–R77	Closure and cancellation
	Total	**77**	

Rule types:

- **Simplification rules** (41 rules): R1–R8, R10–R24, R26, R42–R48, R55–R60, R71–R72, R76–R77

  These rules reduce the complexity of path expressions, replacing patterns with simpler forms.

- **Coherence rules** (6 rules): R27–R32

  These rules reduce transport expressions to canonical coherence paths.

- **Structural rules** (11 rules): R9, R25, R37–R41, R51–R54

  These rules introduce or convert between whisker forms, preparing expressions for further simplification.

- **Compatibility rules** (19 rules): R33–R36, R49–R50, R61–R70, R73–R75

  These rules (mostly meta-rules) ensure the rewrite relation is a congruence, propagating rewrites through all term constructors.

D.10 Quick Reference: Groupoid Rules

For convenient reference, here are the eight groupoid core rules in equation form:

$$\mathsf{symm}(\mathsf{refl}_a) \to \mathsf{refl}_a \tag{R1}$$

$$\mathsf{symm}(\mathsf{symm}(p)) \to p \tag{R2}$$

$$\mathsf{trans}(\mathsf{refl}_a, p) \to p \tag{R3}$$

$$\mathsf{trans}(p, \mathsf{refl}_b) \to p \tag{R4}$$

$$\mathsf{trans}(p, \mathsf{symm}(p)) \to \mathsf{refl}_a \tag{R5}$$

$$\mathsf{trans}(\mathsf{symm}(p), p) \to \mathsf{refl}_b \tag{R6}$$

$$\mathsf{symm}(\mathsf{trans}(p, q)) \to \mathsf{trans}(\mathsf{symm}(q), \mathsf{symm}(p)) \tag{R7}$$

$$\mathsf{trans}(\mathsf{trans}(p, q), r) \to \mathsf{trans}(p, \mathsf{trans}(q, r)) \tag{R8}$$

D.11 Quick Reference: Type Constructor Rules

D.11.1 Products

$$\mathsf{map}_2(f, p, q) \to \mathsf{trans}(\mathsf{mapRight}_f(a_1, q), \mathsf{mapLeft}_f(p, b_2)) \tag{R9}$$

$$\mathsf{congrArg}(\pi_1, \mathsf{map}_2(\langle\cdot,\cdot\rangle, p, q)) \to p \tag{R10}$$

$$\mathsf{congrArg}(\pi_2, \mathsf{map}_2(\langle\cdot,\cdot\rangle, p, q)) \to q \tag{R11}$$

$$\mathsf{prodMk}(\mathsf{congrArg}(\pi_1, p), \mathsf{congrArg}(\pi_2, p)) \to p \tag{R13}$$

$$\mathsf{symm}(\mathsf{prodMk}(p, q)) \to \mathsf{prodMk}(\mathsf{symm}(p), \mathsf{symm}(q)) \tag{R14}$$

D.11.2 Sigma Types

$$\mathsf{congrArg}(\Sigma.\mathsf{fst}, \mathsf{sigmaMk}(p, q)) \to \mathsf{sc}(\mathsf{toEq}(p)) \tag{R16}$$

$$\mathsf{sigmaSnd}(\mathsf{sigmaMk}(p, q)) \to \mathsf{sc}(\mathsf{toEq}(q)) \tag{R17}$$

$$\mathsf{sigmaMk}(\mathsf{sigmaFst}(p), \mathsf{sigmaSnd}(p)) \to p \tag{R18}$$

$$\mathsf{symm}(\mathsf{sigmaMk}(p, q)) \to \mathsf{sigmaMk}(\mathsf{symm}(p),\ \mathsf{symmSnd}(p, q)) \tag{R19}$$

D.11.3 Function Types

$$\mathsf{congrFun}(\mathsf{lamCongr}(p), a) \to p(a) \tag{R22}$$

$$\mathsf{lamCongr}(\lambda x.\, \mathsf{congrFun}(p, x)) \to p \tag{R23}$$

$$\mathsf{symm}(\mathsf{lamCongr}(p)) \to \mathsf{lamCongr}(\lambda x.\, \mathsf{symm}(p(x))) \tag{R24}$$

$$\mathsf{apd}_f(\mathsf{refl}_a) \to \mathsf{refl}_{f(a)} \tag{R25}$$

D.12 Quick Reference: Context and Whisker Rules

D.12.1 Non-Dependent Context

$$\mathsf{symm}(C[p]) \to C[\mathsf{symm}(p)] \tag{R34}$$

$$\mathsf{trans}(r, C[p]) \to L_C(r, p) \tag{R37}$$

$$\mathsf{trans}(C[p], t) \to R_C(p, t) \tag{R40}$$

$$\mathsf{trans}(L_C(r, p), t) \to \mathsf{trans}(r, R_C(p, t)) \tag{R39}$$

$$L_C(r, \mathsf{refl}_a) \to r \tag{R42}$$

$$L_C(\mathsf{refl}_{C(a_1)}, p) \to C[p] \tag{R43}$$

$$R_C(\mathsf{refl}_a, r) \to r \tag{R44}$$

$$R_C(p, \mathsf{refl}_{C(a_2)}) \to C[p] \tag{R45}$$

D.12.2 Dependent Context

$$\mathsf{symm}(\mathsf{DC}[p]) \to \mathsf{DC.symmMap}(p) \tag{R50}$$

$$\mathsf{trans}(T_p(r), \mathsf{DC}[p]) \to DL_C(r, p) \tag{R51}$$

$$\mathsf{trans}(\mathsf{DC}[p], t) \to DR_C(p, t) \tag{R53}$$

$$DL_C(r, \mathsf{refl}_a) \to r \tag{R55}$$

$$DL_C(\mathsf{refl}_{C(a_1)}, p) \to \mathsf{DC}[p] \tag{R56}$$

$$DR_C(\mathsf{refl}_a, r) \to r \tag{R57}$$

$$DR_C(p, \mathsf{refl}_{C(a_2)}) \to \mathsf{DC}[p] \tag{R58}$$

D.13 Quick Reference: Cancellation Rules

The following rules handle inverse cancellation:

Pattern	Result	Rule
$\mathsf{trans}(p, \mathsf{symm}(p))$	refl_a	R5
$\mathsf{trans}(\mathsf{symm}(p), p)$	refl_b	R6
$\mathsf{trans}(p, \mathsf{trans}(\mathsf{symm}(p), q))$	q	R76
$\mathsf{trans}(\mathsf{symm}(p), \mathsf{trans}(p, q))$	q	R77

D.14 Index of Rules by Name

APD-REFL R25
BC-MAP$_2$-L R67
BC-MAP$_2$-R R68
BC-MAPLEFT R65
BC-MAPRIGHT R66
CTX-ABSORB-L R35
CTX-ABSORB-R R36
CTX-CONGR R33
CTX-SYMM R34
DBC-MAP$_2$-L R63
DBC-MAP$_2$-R R64
DBC-MAPLEFT R61
DBC-MAPRIGHT R62
DC-CONGR R49
DC-SYMM R50
DL-TO-DR R52
DL-WHISKER-IDEM R59
DL-WHISKER-INTRO R51
DL-WHISKER-REFL R55
DL-WHISKER-REFL-PRE R56
DR-WHISKER-ASSOC R54
DR-WHISKER-IDEM R60
DR-WHISKER-INTRO R53
DR-WHISKER-REFL R57
DR-WHISKER-REFL-SUF R58
FUN-β R22
FUN-η R23
FUN-SYMM R24
L-FROM-R R38
L-TO-R R39
L-WHISKER-IDEM R46
L-WHISKER-INTRO R37
L-WHISKER-REFL R42
L-WHISKER-REFL-PRE R43
LEFT-CANCEL R76
MAP$_2$-DECOMP R9
MAPLEFT-COMP R71
MAPLEFT-CONGR R69
MAPRIGHT-COMP R72
MAPRIGHT-CONGR R70
PROD-β_1 .... R10
PROD-β_2 .... R11
PROD-η R13
PROD-FUNC R15
PROD-REC R12
PROD-SYMM R14
R-WHISKER-ASSOC R41
R-WHISKER-IDEM-1 .... R47
R-WHISKER-IDEM-2 .... R48
R-WHISKER-INTRO R40
R-WHISKER-REFL R44
R-WHISKER-REFL-SUF R45
RIGHT-CANCEL R77
SIGMA-β_1 .... R16
SIGMA-β_2 .... R17
SIGMA-η R18
SIGMA-SYMM R19
SUBST-Σ-DEP R32
SUM-REC-INL R20
SUM-REC-INR R21
SYMM-CONGR R73
SYMM-INVOL R2
SYMM-REFL R1
SYMM-TRANS R7
TRANS-ASSOC R8
TRANS-CONGR-L R74
TRANS-CONGR-R R75
TRANS-LEFT-ID R3

TRANS-LEFT-INV R6
TRANS-RIGHT-ID R4
TRANS-RIGHT-INV R5
TRANSPORT-REFL R26
TRANSPORT-Σ-DEP R31
TRANSPORT-Σ-FST R30
TRANSPORT-SYMM-L R28
TRANSPORT-SYMM-R R29
TRANSPORT-TRANS R27

Bibliography

[1] Thorsten Altenkirch and Ambrus Kaposi. Type theory in type theory using quotient inductive types. In *Proceedings of the 43rd Annual ACM SIGPLAN-SIGACT Symposium on Principles of Programming Languages (POPL)*, pages 18–29, 2016. doi: 10.1145/2837614.2837638.

[2] Thorsten Altenkirch, Conor McBride, and Wouter Swierstra. Observational equality, now! In *Proceedings of the 2007 Workshop on Programming Languages Meets Program Verification (PLPV)*, pages 57–68. ACM, 2007. doi: 10.1145/1292597.1292608.

[3] Carlo Angiuli, Kuen-Bang Hou (Favonia), and Robert Harper. Cartesian cubical computational type theory: Constructive reasoning with paths and equalities. In *Computer Science Logic (CSL 2018)*, volume 119 of *LIPIcs*, pages 6:1–6:17, 2018. doi: 10.4230/LIPIcs.CSL.2018.6.

[4] Carlo Angiuli, Guillaume Brunerie, Thierry Coquand, Robert Harper, Kuen-Bang Hou (Favonia), and Daniel R. Licata. Syntax and models of cartesian cubical type theory. *Mathematical Structures in Computer Science*, 31(4):424–468, 2021. doi: 10.1017/S0960129521000347.

[5] Danil Annenkov, Paolo Capriotti, Nicolai Kraus, and Christian Sattler. Two-level type theory and applications. *Mathematical Structures in Computer Science*, 33(8):688–743, 2023. doi: 10.1017/S0960129523000130. arXiv:1705.03307v3.

[6] Dimitri Ara. Higher quasi-categories vs higher Rezk spaces. *Journal of K-Theory*, 14(3):701–749, 2014. doi: 10.1017/S1865243315000021.

[7] Steve Awodey and Michael A. Warren. Homotopy theoretic models of identity types. *Mathematical Proceedings of the Cambridge Philosophical Society*, 146(1):45–55, 2009. doi: 10.1017/S0305004108001783.

[8] Franz Baader and Tobias Nipkow. *Term Rewriting and All That*. Cambridge University Press, Cambridge, 1998.

[9] Henk P. Barendregt. *The Lambda Calculus: Its Syntax and Semantics*, volume 103 of *Studies in Logic and the Foundations of Mathematics*. North-Holland, 1981.

[10] Henk P. Barendregt. *The Lambda Calculus: Its Syntax and Semantics*, volume 103 of *Studies in Logic and the Foundations of Mathematics*. North-Holland, Amsterdam, revised edition, 1984.

[11] Michael A. Batanin. Monoidal globular categories as a natural environment for the theory of weak n-categories. *Advances in Mathematics*, 136: 39–103, 1998. doi: 10.1006/aima.1998.1724.

[12] Jean Bénabou. Introduction to bicategories. In Jean Bénabou et al., editors, *Reports of the Midwest Category Seminar*, volume 47 of *Lecture Notes in Mathematics*, pages 1–77, Berlin, 1967. Springer-Verlag.

[13] Marc Bezem, Thierry Coquand, and Simon Huber. A model of type theory in cubical sets. In *19th International Conference on Types for Proofs and Programs (TYPES 2013)*, volume 26 of *LIPIcs*, pages 107–128, 2014. doi: 10.4230/LIPIcs.TYPES.2013.107.

[14] Ronald V. Book and Friedrich Otto. *String-Rewriting Systems*. Texts and Monographs in Computer Science. Springer, 1993. doi: 10.1007/978-1-4 613-9771-7.

[15] Ronald Brown. *Topology and Groupoids*. BookSurge, 2006. Third edition of Elements of Modern Topology (1968).

[16] Ronald Brown and Philip J. Higgins. On the algebra of cubes. *Journal of Pure and Applied Algebra*, 21:233–260, 1981.

[17] Ronald Brown and Christopher B. Spencer. Double groupoids and crossed modules. *Cahiers de Topologie et Géométrie Différentielle Catégoriques*, 17(4):343–362, 1976.

[18] Ronald Brown, Philip J. Higgins, and Rafael Sivera. *Nonabelian Algebraic Topology: Filtered Spaces, Crossed Complexes, Cubical Homotopy Groupoids*, volume 15 of *EMS Tracts in Mathematics*. European Mathematical Society, 2011.

[19] Albert Burroni. Higher-dimensional word problems with applications to equational logic. *Theoretical Computer Science*, 115(1):43–62, 1993. doi: 10.1016/0304-3975(93)90054-W.

[20] Philippe Le Chenadec. *Canonical Forms in Finitely Presented Algebras*. Research Notes in Theoretical Computer Science. Pitman, 1989.

[21] Alonzo Church. An unsolvable problem of elementary number theory. *American Journal of Mathematics*, 58(2):345–363, 1936. doi: 10.2307/2371045.

[22] Alonzo Church and J. Barkley Rosser. Some properties of conversion. *Transactions of the American Mathematical Society*, 39(3):472–482, 1936. doi: 10.2307/1989762.

[23] Cyril Cohen, Thierry Coquand, Simon Huber, and Anders Mörtberg. Cubical type theory: A constructive interpretation of the univalence axiom. In Tarmo Uustalu, editor, *21st International Conference on Types for Proofs and Programs (TYPES 2015)*, volume 69 of *LIPIcs*, pages 5:1–5:34. Schloss Dagstuhl – Leibniz-Zentrum für Informatik, 2018. doi: 10.4230/LIPIcs.TYPES.2015.5.

[24] Thierry Coquand. An algorithm for testing conversion in type theory. In Gérard Huet and Gordon Plotkin, editors, *Logical Frameworks*, pages 255–279. Cambridge University Press, 1991.

[25] Haskell B. Curry. Functionality in combinatory logic. *Proceedings of the National Academy of Sciences*, 20(11):584–590, 1934. doi: 10.1073/pnas.20.11.584.

[26] Haskell B. Curry and Robert Feys. *Combinatory Logic*, volume 1. North-Holland, 1958.

[27] Nicolaas Govert de Bruijn. AUTOMATH, a language for mathematics. Technical Report 68-WSK-05, Technological University Eindhoven, 1968.

[28] Leonardo de Moura and Sebastian Ullrich. The Lean 4 theorem prover and programming language. In André Platzer and Geoff Sutcliffe, editors, *Automated Deduction – CADE 28*, volume 12699 of *Lecture Notes in Computer Science*, pages 625–635. Springer, 2021.

[29] Anjolina G. de Oliveira and Ruy J.G.B. de Queiroz. A normalization procedure for the equational fragment of labelled natural deduction. *Logic Journal of the IGPL*, 7(2):173–215, 1999. doi: 10.1093/jigpal/7.2.173.

[30] Anjolina G. de Oliveira and Ruy J.G.B. de Queiroz. A new basic set of proof transformations. In Sergei Artemov et al., editors, *We Will Show Them: Essays in Honour of Dov Gabbay*, volume 2, pages 499–528. College Publications, London, 2005.

[31] Ruy J.G.B. de Queiroz. A proof-theoretic account of programming and the role of reduction rules. *Dialectica*, 42(4):265–282, 1988. doi: 10.1111/j.1746-8361.1988.tb00919.x.

[32] Ruy J.G.B. de Queiroz. The mathematical language and its semantics: To show the consequences of a proposition is to give its meaning. In Paul Weingartner and Gerhard Schurz, editors, *Reports of the Thirteenth International Wittgenstein Symposium 1988*, volume 18 of *Schriftenreihe der Wittgenstein-Gesellschaft*, pages 259–266. Hölder-Pichler-Tempsky, Vienna, 1989.

[33] Ruy J.G.B. de Queiroz. Meaning, function, purpose, usefulness, consequences – interconnected concepts (abstract). In *Abstracts of the Fourteenth International Wittgenstein Symposium (Centenary Celebration)*, page 20, 1989.

[34] Ruy J.G.B. de Queiroz. *Proof Theory and Computer Programming*. PhD thesis, Imperial College, University of London, 1990.

[35] Ruy J.G.B. de Queiroz. Meaning as grammar plus consequences. *Dialectica*, 45(1):83–86, 1991. doi: 10.1111/j.1746-8361.1991.tb00979.x.

[36] Ruy J.G.B. de Queiroz. Grundgesetze alongside Begriffsschrift (abstract). In *Abstracts of the Fifteenth International Wittgenstein Symposium*, pages 15–16, 1992.

[37] Ruy J.G.B. de Queiroz. Normalisation and language-games. *Dialectica*, 48(2):83–123, 1994. doi: 10.1111/j.1746-8361.1994.tb00107.x.

[38] Ruy J.G.B. de Queiroz. Meaning, function, purpose, usefulness, consequences – interconnected concepts. *Logic Journal of the IGPL*, 9(5): 693–734, 2001. doi: 10.1093/jigpal/9.5.693.

[39] Ruy J.G.B. de Queiroz. On reduction rules, meaning-as-use, and proof-theoretic semantics. *Studia Logica*, 90(2):211–247, 2008. doi: 10.1007/s11225-008-9150-5.

[40] Ruy J.G.B. de Queiroz. From Tractatus to later writings and back – new implications from Wittgenstein's Nachlass. *SATS - Northern European Journal of Philosophy*, 24(2):167–203, 2023. doi: 10.1515/sats-2022-0016.

[41] Ruy J.G.B. de Queiroz. From the Notebooks to the Investigations and beyond. *SATS - Northern European Journal of Philosophy*, 26(1):61–113, 2025. doi: 10.1515/sats-2024-0015.

[42] Ruy J.G.B. de Queiroz. Meaning as use, application, employment, purpose, usefulness. To appear in SATS - Northern European Journal of Philosophy. Preprint: arXiv:2506.07131, 2025.

[43] Ruy J.G.B. de Queiroz. Our view of language is a one-sided one. Draft dated 2025/26. Submitted for publication, 2026.

[44] Ruy J.G.B. de Queiroz and Anjolina G. de Oliveira. Term rewriting systems with labelled deductive systems. In *Proceedings of the Brazilian Symposium on Artificial Intelligence (SBIA'94)*, pages 59–72, 1994.

[45] Ruy J.G.B. de Queiroz and Anjolina G. de Oliveira. The functional interpretation of direct computations. *Electronic Notes in Theoretical Computer Science*, 269:19–40, 2011. doi: 10.1016/j.entcs.2011.03.003.

[46] Ruy J.G.B. de Queiroz and Anjolina G. de Oliveira. Natural deduction for equality: The missing entity. In Edward Hermann Haeusler, Luiz Carlos Pereira, and Valeria de Paiva, editors, *Advances in Natural Deduction*, Trends in Logic, pages 63–91. Springer, Dordrecht, 2014. doi: 10.1007/978-94-007-7548-0_4. Earlier pre-publication citations list this chapter as forthcoming (2008).

[47] Ruy J.G.B. de Queiroz and Dov M. Gabbay. Equality in labelled deductive systems and the functional interpretation of propositional equality. In *Proceedings of the 9th Amsterdam Colloquium*, pages 547–565, 1994.

[48] Ruy J.G.B. de Queiroz and Dov M. Gabbay. The functional interpretation of the existential quantifier. *Bulletin of the IGPL*, 3(2–3):243–290, 1995. doi: 10.1093/jigpal/3.2-3.243.

[49] Ruy J.G.B. de Queiroz and Dov M. Gabbay. The functional interpretation of modal necessity. In Maarten de Rijke, editor, *Advances in Intensional Logic*, pages 61–91. Kluwer Academic Publishers, 1997. doi: 10.1007/978-94-015-8879-9_3.

[50] Ruy J.G.B. de Queiroz and Dov M. Gabbay. Labelled natural deduction. In Hans Jürgen Ohlbach and Uwe Reyle, editors, *Logic, Language and Reasoning: Essays in Honour of Dov Gabbay*, pages 173–250. Kluwer Academic Publishers, 1999.

[51] Ruy J.G.B. de Queiroz and Thomas S.E. Maibaum. Proof theory and computer programming. *Zeitschrift für mathematische Logik und Grundlagen der Mathematik*, 36(5):389–414, 1990. doi: 10.1002/malq.19900360505.

[52] Ruy J.G.B. de Queiroz and Thomas S.E. Maibaum. Abstract data types and type theory: Theories as types. *Zeitschrift für mathematische Logik und Grundlagen der Mathematik*, 37(9–12):149–166, 1991. doi: 10.1002/malq.19910370904.

[53] Ruy J.G.B. de Queiroz, Anjolina G. de Oliveira, and Dov M. Gabbay. *The Functional Interpretation of Logical Deduction*, volume 5 of *Advances in Logic*. World Scientific, Singapore, 2011.

[54] Ruy J.G.B. de Queiroz, Anjolina G. de Oliveira, and Arthur F. Ramos. Propositional equality, identity types, and computational paths. *South American Journal of Logic*, 2(2):245–296, 2016. URL `https://www.sa-logic.org/sajl-v2-i2/05-De%20Queiroz-De%20Oliveira-Ramos-SAJL.pdf`.

[55] Tiago M.L. de Veras, Arthur F. Ramos, Ruy J.G.B. de Queiroz, and Anjolina G. de Oliveira. On the calculation of fundamental groups in homotopy type theory by means of computational paths. arXiv:1804.01413, 2018.

[56] Tiago M.L. de Veras, Arthur F. Ramos, Ruy J.G.B. de Queiroz, and Anjolina G. de Oliveira. A topological application of labelled natural deduction. arXiv:1906.09105, 2019.

[57] Tiago M.L. de Veras, Arthur F. Ramos, Ruy J.G.B. de Queiroz, and Anjolina G. de Oliveira. Computational paths – an approach in the LND_{EQ}-TRS_2 system. arXiv:2007.07769, 2020.

[58] Tiago M.L. de Veras, Arthur F. Ramos, Ruy J.G.B. de Queiroz, and Anjolina G. de Oliveira. Calculation of fundamental groups via computational paths. In *Encontro de Teoria da Computação (ETC 2021)*, 2021. doi: 10.5753/etc.2021.16370.

[59] Tiago M.L. de Veras, Arthur F. Ramos, Ruy J.G.B. de Queiroz, and Anjolina G. de Oliveira. Computational paths – a weak groupoid. *Journal of Logic and Computation*, 35(5), 2025. doi: 10.1093/logcom/exad071.

[60] Max Dehn. Über unendliche diskontinuierliche gruppen. *Mathematische Annalen*, 71:116–144, 1911. doi: 10.1007/BF01456932.

[61] Nachum Dershowitz. Orderings for term-rewriting systems. *Theoretical Computer Science*, 17(3):279–301, 1982.

[62] Nachum Dershowitz. Termination of rewriting. *Journal of Symbolic Computation*, 3(1–2):69–115, 1987.

[63] Michael Dummett. The philosophical basis of intuitionistic logic. In H.E. Rose and J.C. Shepherdson, editors, *Logic Colloquium '73*, pages 5–40. North-Holland, Amsterdam, 1975.

[64] Beno Eckmann and Peter J. Hilton. Group-like structures in general categories. I. Multiplications and comultiplications. *Mathematische Annalen*, 145:227–255, 1962. doi: 10.1007/BF01451367.

[65] Charles Ehresmann. Catégories structurées. *Annales scientifiques de l'École Normale Supérieure*, 80(4):349–426, 1963.

[66] Gottlob Frege. *Begriffsschrift, eine der arithmetischen nachgebildete Formelsprache des reinen Denkens*. Louis Nebert, Halle, 1879. English translation in van Heijenoort (1967).

[67] Gottlob Frege. *Grundgesetze der Arithmetik*, volume I. Hermann Pohle, Jena, 1893.

[68] Gottlob Frege. *Grundgesetze der Arithmetik*, volume II. Hermann Pohle, Jena, 1903.

[69] Gottlob Frege. *The Basic Laws of Arithmetic: Exposition of the System*. University of California Press, Berkeley, 1964. doi: 10.2307/jj.8501490. Translated and edited, with an introduction, by Montgomery Furth.

[70] Dov M. Gabbay. *Labelled Deductive Systems*. Oxford University Press, Oxford, 1996.

[71] Dov M. Gabbay and Ruy J.G.B. de Queiroz. Extending the Curry–Howard interpretation to linear, relevant and other resource logics. *Journal of Symbolic Logic*, 57(4):1319–1365, 1992. doi: 10.2307/2275370.

[72] Gerhard Gentzen. Untersuchungen über das logische schließen. *Mathematische Zeitschrift*, 39:176–210, 405–431, 1935. doi: 10.1007/BF01 201353. English translation: "Investigations into Logical Deduction" in Szabo (1969).

[73] Jean-Yves Girard. Linear logic. *Theoretical Computer Science*, 50:1–102, 1987.

[74] Jean-Yves Girard, Yves Lafont, and Paul Taylor. *Proofs and Types*, volume 7 of *Cambridge Tracts in Theoretical Computer Science*. Cambridge University Press, 1989.

[75] Kurt Gödel. Über eine bisher noch nicht benützte erweiterung des finiten standpunktes. *Dialectica*, 12:280–287, 1958. doi: 10.1111/j.1746-8361.1958.tb01464.x. English translation: "On a hitherto unexploited extension of the finitary standpoint" in Gödel's Collected Works, Vol. II.

[76] Marco Grandis and Robert Paré. Limits in double categories. *Cahiers de Topologie et Géométrie Différentielle Catégoriques*, 40(3):162–220, 1999.

[77] Alexander Grothendieck. Pursuing stacks (à la poursuite des champs). Manuscript, letter to Daniel Quillen, 1983.

[78] Yves Guiraud and Philippe Malbos. Higher-dimensional categories with finite derivation type. *Theory and Applications of Categories*, 22(18): 420–478, 2009.

[79] Allen Hatcher. *Algebraic Topology*. Cambridge University Press, Cambridge, 2002.

[80] Arend Heyting. Die formalen regeln der intuitionistischen logik. *Sitzungsberichte der Preussischen Akademie der Wissenschaften, Physikalisch-mathematische Klasse*, pages 42–56, 1930.

[81] Arend Heyting. *Intuitionism: An Introduction*. North-Holland, Amsterdam, 1956.

[82] J. Roger Hindley and Jonathan P. Seldin. *Introduction to Combinators and λ-Calculus*, volume 1 of *London Mathematical Society Student Texts*. Cambridge University Press, 1986.

[83] J. Roger Hindley and Jonathan P. Seldin. *Lambda-Calculus and Combinators: An Introduction*. Cambridge University Press, 2nd edition, 2008.

[84] Jaakko Hintikka. Language-games for quantifiers. In N. Rescher, editor, *Studies in Logical Theory*, pages 46–72. Blackwell, 1968.

[85] Jaakko Hintikka. Quantifiers vs. quantification theory. In Esa Saarinen, editor, *Game-Theoretical Semantics*, volume 5 of *Synthese Language Library*. Springer, Dordrecht, 1979.

[86] Martin Hofmann and Thomas Streicher. The groupoid interpretation of type theory. In Giovanni Sambin and Jan M. Smith, editors, *Twenty-Five Years of Constructive Type Theory*, pages 83–111. Oxford University Press, 1998.

[87] William A. Howard. The formulae-as-types notion of construction. In Jonathan P. Seldin and J. Roger Hindley, editors, *To H.B. Curry: Essays on Combinatory Logic, Lambda Calculus and Formalism*, pages 479–490. Academic Press, London, 1980. Original manuscript from 1969.

[88] Gérard Huet. Confluent reductions: Abstract properties and applications to term rewriting systems. *Journal of the ACM*, 27(4):797–821, 1980. doi: 10.1145/322217.322230.

[89] André Joyal. Quasi-categories and Kan complexes. *Journal of Pure and Applied Algebra*, 175:207–222, 2002. doi: 10.1016/S0022-4049(02)001 35-4.

[90] André Joyal. The theory of quasi-categories and its applications. In *Advanced Course on Simplicial Methods in Higher Categories*. Centre de Recerca Matemàtica, Bellaterra, 2008. Lecture notes.

[91] Mikhail M. Kapranov and Vladimir A. Voevodsky. ∞-groupoids and homotopy types. *Cahiers de Topologie et Géométrie Différentielle Catégoriques*, 32(1):29–46, 1991.

[92] G. Max Kelly. *Basic Concepts of Enriched Category Theory*, volume 64 of *London Mathematical Society Lecture Note Series*. Cambridge University Press, 1982. Republished in Reprints in Theory and Applications of Categories, No. 10, 2005.

[93] Donald E. Knuth and Peter B. Bendix. Simple word problems in universal algebras. In John Leech, editor, *Computational Problems in Abstract Algebra*, pages 263–297. Pergamon Press, 1970.

[94] Yuji Kobayashi. Finite homotopy bases of one-relator monoids. *Journal of Algebra*, 229(2):547–569, 2000. doi: 10.1006/jabr.1999.8251.

[95] Stephen Lack. A 2-categories companion. In John C. Baez and J. Peter May, editors, *Towards Higher Categories*, volume 152 of *IMA Volumes in Mathematics and its Applications*, pages 105–191. Springer, 2010.

[96] Joachim Lambek. The mathematics of sentence structure. *American Mathematical Monthly*, 65(3):154–170, 1958.

[97] Joachim Lambek and Philip J. Scott. *Introduction to Higher Order Categorical Logic*, volume 7 of *Cambridge Studies in Advanced Mathematics*. Cambridge University Press, Cambridge, 1986.

[98] Saunders Mac Lane. *Categories for the Working Mathematician*, volume 5 of *Graduate Texts in Mathematics*. Springer-Verlag, New York, 2nd edition, 1998.

[99] Tom Leinster. *Higher Operads, Higher Categories*, volume 298 of *London Mathematical Society Lecture Note Series*. Cambridge University Press, 2004.

[100] Daniel R. Licata. *Dependently Typed Programming with Domain-Specific Logics*. PhD thesis, Carnegie Mellon University, 2011. CMU-CS-11-151.

[101] Daniel R. Licata and Michael Shulman. Calculating the fundamental group of the circle in homotopy type theory. In *Proceedings of the 28th Annual ACM/IEEE Symposium on Logic in Computer Science (LICS)*, pages 223–232, 2013. doi: 10.1109/LICS.2013.28.

[102] Paul Lorenzen. Konstruktive begründung der mathematik. *Mathematische Zeitschrift*, 53(2):162–202, 1950.

[103] Paul Lorenzen. *Einführung in die operative Logik und Mathematik*, volume 78 of *Die Grundlehren der mathematischen Wissenschaften*. Springer-Verlag, Berlin-Göttingen-Heidelberg, 1955. VII + 298 pp.

[104] Paul Lorenzen. *Normative Logic and Ethics*. Number 236* in B.I.-Hochschultaschenbücher. Systematische Philosophie. Bibliographisches Institut, Mannheim/Zürich, 1969.

[105] Peter LeFanu Lumsdaine. Weak ω-categories from intensional type theory. *Logical Methods in Computer Science*, 6(3):1–19, 2010. doi: 10.2168/LMCS-6(3:24)2010.

[106] Jacob Lurie. *Higher Topos Theory*, volume 170 of *Annals of Mathematics Studies*. Princeton University Press, 2009. doi: 10.1515/9781400830558.

[107] Jacob Lurie. *Higher Algebra*. Self-published, 2017. Available at `https://www.math.ias.edu/~lurie/papers/HA.pdf`.

[108] Georges Maltsiniotis. Grothendieck ∞-groupoids, and still another definition of ∞-categories, 2010.

[109] Per Martin-Löf. An intuitionistic theory of types: Predicative part. In H.E. Rose and J.C. Shepherdson, editors, *Logic Colloquium '73*, volume 80 of *Studies in Logic and the Foundations of Mathematics*, pages 73–118. North-Holland, Amsterdam, 1975.

[110] Per Martin-Löf. Constructive mathematics and computer programming. In L. J. Cohen et al., editors, *Logic, Methodology and Philosophy of Science VI*, pages 153–175. North-Holland, 1982.

[111] Per Martin-Löf. *Intuitionistic Type Theory*. Bibliopolis, Naples, 1984. Notes by Giovanni Sambin of lectures given in Padua, June 1980.

[112] Per Martin-Löf. Logic and ethics. In Thomas Piecha and Peter Schroeder-Heister, editors, *Proof-Theoretic Semantics: Assessment and Future Perspectives*, pages 227–235. University of Tübingen, 2019. Proceedings of the Third Tübingen Conference on Proof-Theoretic Semantics, 27–30 March 2019, University of Tübingen. DOI: 10.15496/publikation-35319.

[113] Per Martin-Löf. Correctness of assertion and validity of inference. Lecture at the Rolf Schock Symposium, Stockholm, 2022. Slightly edited transcript of lecture delivered on 26 October 2022. `https://pml.flu.cas.cz/uploads/PML-Stockholm26Oct22.pdf`.

[114] J. Peter May. *A Concise Course in Algebraic Topology*. Chicago Lectures in Mathematics. University of Chicago Press, 1999.

[115] François Métayer. Cofibrant objects among higher-dimensional categories. *Homology, Homotopy and Applications*, 10(1):181–203, 2008. doi: 10.4310/HHA.2008.v10.n1.a7.

[116] M. H. A. Newman. On theories with a combinatorial definition of "equivalence". *Annals of Mathematics*, 43(2):223–243, 1942. doi: 10.2307/1968867.

[117] Bengt Nordström, Kent Petersson, and Jan M. Smith. *Programming in Martin-Löf's Type Theory: An Introduction*. Oxford University Press, 1990.

[118] Ulf Norell et al. Agda, 2024. URL `https://wiki.portal.chalmers.se/agda/pmwiki.php`.

[119] Christine Paulin-Mohring. Inductive definitions in the system Coq: Rules and properties. In M. Bezem and J.F. Groote, editors, *Typed Lambda Calculi and Applications (TLCA '93)*, volume 664 of *LNCS*, pages 328–345. Springer, 1993. doi: 10.1007/BFb0037116.

[120] Jacques Penon. Approche polygraphique des ∞-catégories non strictes. *Cahiers de Topologie et Géométrie Différentielle Catégoriques*, 40(1): 31–80, 1999.

[121] Dag Prawitz. *Natural Deduction: A Proof-Theoretical Study*. Number 3 in Stockholm Studies in Philosophy. Almqvist & Wiksell, Stockholm, 1965.

[122] Dag Prawitz. Ideas and results in proof theory. In J.E. Fenstad, editor, *Proceedings of the Second Scandinavian Logic Symposium*, pages 235–307. North-Holland, Amsterdam, 1971.

[123] Stephen J. Pride. Low-dimensional homotopy theory for monoids. *International Journal of Algebra and Computation*, 5(6):631–649, 1995. doi: 10.1142/S0218196795000252.

[124] Arthur F. Ramos. *Explicit Computational Paths in Type Theory*. PhD thesis, Universidade Federal de Pernambuco, 2019. Abstract published in Bulletin of Symbolic Logic, 25(2):213–214, 2019. doi:10.1017/bsl.2019.2.

[125] Arthur F. Ramos, Ruy J.G.B. de Queiroz, and Anjolina G. de Oliveira. Sequences of rewrites: A categorical interpretation. arXiv:1412.2105, 2014.

[126] Arthur F. Ramos, Ruy J.G.B. de Queiroz, and Anjolina G. de Oliveira. On the groupoid model of computational paths. arXiv:1506.02721, 2015.

[127] Arthur F. Ramos, Ruy J.G.B. de Queiroz, and Anjolina G. de Oliveira. On the identity type as the type of computational paths. *Logic Journal of the IGPL*, 25(4):562–584, 2017. doi: 10.1093/jigpal/jzx015.

[128] Arthur F. Ramos, Ruy J.G.B. de Queiroz, and Anjolina G. de Oliveira. Explicit computational paths. *South American Journal of Logic*, 4(2): 441–484, 2018. Preprint arXiv:1609.05079, 2016.

[129] Arthur F. Ramos, Ruy J.G.B. de Queiroz, Anjolina G. de Oliveira, and Tiago M.L. de Veras. On the use of computational paths in path spaces of homotopy type theory. arXiv:1803.01709, 2018.

[130] Arthur F. Ramos, Ruy J.G.B. de Queiroz, and Anjolina G. de Oliveira. Conversão de termos, homotopia, e estrutura de grupóide. In *Workshop Brasileiro de Lógica (WBL 2021)*, 2021. doi: 10.5753/wbl.2021.15776.

[131] Arthur F. Ramos, Ruy J.G.B. de Queiroz, and Anjolina G. de Oliveira. Computational paths and the fundamental groupoid of a type. In *Encontro de Teoria da Computação (ETC 2021)*, 2021. doi: 10.5753/etc.2021.163 71.

[132] Arthur F. Ramos, Anjolina G. de Oliveira, Ruy J.G.B. de Queiroz, and Tiago M.L. de Veras. Formalizing computational paths and fundamental groups in Lean. arXiv:2511.19142, 2025.

[133] Arthur F. Ramos, Anjolina G. de Oliveira, Ruy J.G.B. de Queiroz, and Tiago M.L. de Veras. A modular Lean 4 framework for confluence and strong normalization of lambda calculi with products and sums. arXiv:2512.09280, 2025.

[134] Arthur F. Ramos, Tiago M.L. de Veras, Ruy J.G.B. de Queiroz, and Anjolina G. de Oliveira. Computational paths form a weak ω-groupoid. arXiv:2512.00657, 2025.

[135] Arthur F. Ramos, Tiago M.L. de Veras, Ruy J.G.B. de Queiroz, and Anjolina G. de Oliveira. The Seifert–van Kampen theorem via computational paths: A formalized approach to computing fundamental groups. arXiv:2512.03175, 2025.

[136] Emily Riehl and Michael Shulman. A type theory for synthetic ∞-categories. *Higher Structures*, 1(1):147–224, 2017.

[137] Emily Riehl and Dominic Verity. *Elements of ∞-Category Theory*, volume 194 of *Cambridge Studies in Advanced Mathematics*. Cambridge University Press, 2022. doi: 10.1017/9781108936880.

[138] Herbert Seifert. Konstruktion dreidimensionaler geschlossener Räume. *Berichte der Sächsischen Akademie der Wissenschaften*, 83:26–66, 1931.

[139] Peter Selinger. A survey of graphical languages for monoidal categories. In Bob Coecke, editor, *New Structures for Physics*, volume 813 of *Lecture Notes in Physics*, pages 289–355. Springer, 2010.

[140] Morten Heine Sørensen and Paweł Urzyczyn. *Lectures on the Curry–Howard Isomorphism*, volume 149 of *Studies in Logic and the Foundations of Mathematics*. Elsevier, 2006.

[141] Craig C. Squier. Word problems and a homological finiteness condition for monoids. *Journal of Pure and Applied Algebra*, 49(1–2):201–217, 1987. doi: 10.1016/0022-4049(87)90129-0.

[142] James Dillon Stasheff. Homotopy associativity of H-spaces. I, II. *Transactions of the American Mathematical Society*, 108:275–312, 1963. doi: 10.2307/1993608.

[143] Richard Statman. Herbrand's theorem and Gentzen's notion of a direct proof. In Jon Barwise, editor, *Handbook of Mathematical Logic*, pages 897–912. North-Holland, 1977. doi: 10.1016/S0049-237X(08)71125-X.

[144] Richard Statman. Bounds for proof-search and speed-up in the predicate calculus. *Annals of Mathematical Logic*, 15:225–287, 1978.

[145] Ross Street. Limits indexed by category-valued 2-functors. *Journal of Pure and Applied Algebra*, 8(2):149–181, 1976. doi: 10.1016/0022-4049(76)90013-X.

[146] William W. Tait. Infinitely long terms of transfinite type. In J.N. Crossley and M.A.E. Dummett, editors, *Formal Systems and Recursive Functions*, pages 176–185, Amsterdam, 1965. North-Holland.

[147] William W. Tait. Intensional interpretations of functionals of finite type I. *Journal of Symbolic Logic*, 32(2):198–212, 1967. doi: 10.2307/2271658.

[148] Terese. *Term Rewriting Systems*, volume 55 of *Cambridge Tracts in Theoretical Computer Science*. Cambridge University Press, Cambridge, 2003.

[149] The Coq Development Team. The Coq proof assistant, 2024. URL `https://coq.inria.fr/`.

[150] The Univalent Foundations Program. *Homotopy Type Theory: Univalent Foundations of Mathematics*. Institute for Advanced Study, Princeton, NJ, 2013. URL `https://homotopytypetheory.org/book`.

[151] Anne Sjerp Troelstra and Dirk van Dalen. *Constructivism in Mathematics: An Introduction*, volume I–II of *Studies in Logic and the Foundations of Mathematics*. North-Holland, Amsterdam, 1988.

[152] Johan van Benthem. The lambek calculus. In Richard Oehrle, Emmon Bach, and Deirdre Wheeler, editors, *Categorial Grammars and Natural Language Structures*, pages 35–68. D. Reidel, 1988.

[153] Benno van den Berg and Richard Garner. Types are weak ω-groupoids. *Proceedings of the London Mathematical Society*, 102(2):370–394, 2011. doi: 10.1112/plms/pdq026.

[154] Egbert R. van Kampen. On the connection between the fundamental groups of some related spaces. *American Journal of Mathematics*, 55 (1–4):261–267, 1933.

[155] Andrea Vezzosi, Anders Mörtberg, and Andreas Abel. Cubical agda: A dependently typed programming language with univalence and higher inductive types. *Proceedings of the ACM on Programming Languages*, 3 (ICFP):87:1–87:29, 2019. doi: 10.1145/3341691.

[156] Vladimir Voevodsky. Univalent Foundations of Mathematics. In Lev D. Beklemishev and Ruy de Queiroz, editors, *Logic, Language, Information and Computation. WoLLIC 2011*, volume 6642 of *Lecture Notes in Computer Science*. Springer, Berlin, Heidelberg, 2011. doi: 10.1007/978-3-642-20920-8_4.

[157] Vladimir Voevodsky. A simple type system with two identity types. Unpublished notes, 2013. URL `https://www.math.ias.edu/vladimir/sites/math.ias.edu.vladimir/files/HTS.pdf`.

[158] Vladimir Voevodsky. Homotopy type theory and formalization of mathematics. Lecture slides, Institute for Advanced Study, 2014. Available at `https://www.math.ias.edu/vladimir/sites/math.ias.edu.vladimir/files/2014_IAS.pdf`.

[159] Matthew Z. Weaver. *A Constructive Model of Directed Univalence in Bicubical Sets*. PhD thesis, Carnegie Mellon University, 2021.

[160] Wittgenstein Archives at the University of Bergen. Wittgenstein, Ludwig: Interactive Dynamic Presentation (IDP) of Ludwig Wittgenstein's philosophical Nachlass [wittgensteinonline.no]. Edited under the direction of Alois Pichler, 2016. Bergen: Wittgenstein Archives at the University of Bergen. Accessed 22 May 2025. `http://wab.uib.no/`.

www.ingramcontent.com/pod-product-compliance
Ingram Content Group UK Ltd.
Pitfield, Milton Keynes, MK11 3LW, UK
UKHW020129290926
15471UKWH00003B/383

9 781848 905153